Enumerative Combinatorics of Young Tableaux

PURE AND APPLIED MATHEMATICS

A Program of Monographs, Textbooks, and Lecture Notes

MONOGRAPHS AND TEXTBOOKS IN PURE AND APPLIED MATHEMATICS

1. *K. Yano,* Integral Formulas in Riemannian Geometry (1970)*(out of print)*
2. *S. Kobayashi,* Hyperbolic Manifolds and Holomorphic Mappings (1970) *(out of print)*
3. *V. S. Vladimirov,* Equations of Mathematical Physics (A. Jeffrey, editor; A. Littlewood, translator) (1970) *(out of print)*
4. *B. N. Pshenichnyi,* Necessary Conditions for an Extremum (L. Neustadt, translation editor; K. Makowski, translator) (1971)
5. *L. Narici, E. Beckenstein, and G. Bachman,* Functional Analysis and Valuation Theory (1971)
6. *D. S. Passman,* Infinite Group Rings (1971)
7. *L. Dornhoff,* Group Representation Theory (in two parts). Part A: Ordinary Representation Theory. Part B: Modular Representation Theory (1971, 1972)
8. *W. Boothby and G. L. Weiss (eds.),* Symmetric Spaces: Short Courses Presented at Washington University (1972)
9. *Y. Matsushima,* Differentiable Manifolds (E. T. Kobayashi, translator) (1972)
10. *L. E. Ward, Jr.,* Topology: An Outline for a First Course (1972) *(out of print)*
11. *A. Babakhanian,* Cohomological Methods in Group Theory (1972)
12. *R. Gilmer,* Multiplicative Ideal Theory (1972)
13. *J. Yeh,* Stochastic Processes and the Wiener Integral (1973) *(out of print)*
14. *J. Barros-Neto,* Introduction to the Theory of Distributions (1973) *(out of print)*
15. *R. Larsen,* Functional Analysis: An Introduction (1973) *(out of print)*
16. *K. Yano and S. Ishihara,* Tangent and Cotangent Bundles: Differential Geometry (1973) *(out of print)*
17. *C. Procesi,* Rings with Polynomial Identities (1973)
18. *R. Hermann,* Geometry, Physics, and Systems (1973)
19. *N. R. Wallach,* Harmonic Analysis on Homogeneous Spaces (1973) *(out of print)*
20. *J. Dieudonné,* Introduction to the Theory of Formal Groups (1973)
21. *I. Vaisman,* Cohomology and Differential Forms (1973)
22. *B.-Y. Chen,* Geometry of Submanifolds (1973)
23. *M. Marcus,* Finite Dimensional Multilinear Algebra (in two parts) (1973, 1975)
24. *R. Larsen,* Banach Algebras: An Introduction (1973)
25. *R. O. Kujala and A. L. Vitter (eds.),* Value Distribution Theory: Part A; Part B: Deficit and Bezout Estimates by Wilhelm Stoll (1973)
26. *K. B. Stolarsky,* Algebraic Numbers and Diophantine Approximation (1974)
27. *A. R. Magid,* The Separable Galois Theory of Commutative Rings (1974)
28. *B. R. McDonald,* Finite Rings with Identity (1974)
29. *J. Satake,* Linear Algebra (S. Koh, T. A. Akiba, and S. Ihara, translators) (1975)

30. *J. S. Golan,* Localization of Noncommutative Rings (1975)
31. *G. Klambauer,* Mathematical Analysis (1975)
32. *M. K. Agoston,* Algebraic Topology: A First Course (1976)
33. *K. R. Goodearl,* Ring Theory: Nonsingular Rings and Modules (1976)
34. *L. E. Mansfield,* Linear Algebra with Geometric Applications: Selected Topics (1976)
35. *N. J. Pullman,* Matrix Theory and Its Applications (1976)
36. *B. R. McDonald,* Geometric Algebra Over Local Rings (1976)
37. *C. W. Groetsch,* Generalized Inverses of Linear Operators: Representation and Approximation (1977)
38. *J. E. Kuczkowski and J. L. Gersting,* Abstract Algebra: A First Look (1977)
39. *C. O. Christenson and W. L. Voxman,* Aspects of Topology (1977)
40. *M. Nagata,* Field Theory (1977)
41. *R. L. Long,* Algebraic Number Theory (1977)
42. *W. F. Pfeffer,* Integrals and Measures (1977)
43. *R. L. Wheeden and A. Zygmund,* Measure and Integral: An Introduction to Real Analysis (1977)
44. *J. H. Curtiss,* Introduction to Functions of a Complex Variable (1978)
45. *K. Hrbacek and T. Jech,* Introduction to Set Theory (1978)
46. *W. S. Massey,* Homology and Cohomology Theory (1978)
47. *M. Marcus,* Introduction to Modern Algebra (1978)
48. *E. C. Young,* Vector and Tensor Analysis (1978)
49. *S. B. Nadler, Jr.,* Hyperspaces of Sets (1978)
50. *S. K. Segal,* Topics in Group Rings (1978)
51. *A. C. M. van Rooij,* Non-Archimedean Functional Analysis (1978)
54. *L. Corwin and R. Szczarba,* Calculus in Vector Spaces (1979)
53. *C. Sadosky,* Interpolation of Operators and Singular Integrals: An Introduction to Harmonic Analysis (1979)
54. *J. Cronin,* Differential Equations: Introduction and Quantitative Theory (1980)
55. *C. W. Groetsch,* Elements of Applicable Functional Analysis (1980)
56. *I. Vaisman,* Foundations of Three-Dimensional Euclidean Geometry (1980)
57. *H. I. Freedman,* Deterministic Mathematical Models in Population Ecology (1980)
58. *S. B. Chae,* Lebesgue Integration (1980)
59. *C. S. Rees, S. M. Shah, and C. V. Stanojević,* Theory and Applications of Fourier Analysis (1981)
60. *L. Nachbin,* Introduction to Functional Analysis: Banach Spaces and Differential Calculus (R. M. Aron, translator) (1981)
61. *G. Orzech and M. Orzech,* Plane Algebraic Curves: An Introduction Via Valuations (1981)
62. *R. Johnsonbaugh and W. E. Pfaffenberger,* Foundations of Mathematical Analysis (1981)
63. *W. L. Voxman and R. H. Goetschel,* Advanced Calculus: An Introduction to Modern Analysis (1981)
64. *L. J. Corwin and R. H. Szcarba,* Multivariable Calculus (1982)
65. *V. I. Istrătescu,* Introduction to Linear Operator Theory (1981)
66. *R. D. Järvinen,* Finite and Infinite Dimensional Linear Spaces: A Comparative Study in Algebraic and Analytic Settings (1981)

67. *J. K. Beem and P. E. Ehrlich,* Global Lorentzian Geometry (1981)
68. *D. L. Armacost,* The Structure of Locally Compact Abelian Groups (1981)
69. *J. W. Brewer and M. K. Smith, eds.,* Emmy Noether: A Tribute to Her Life and Work (1981)
70. *K. H. Kim,* Boolean Matrix Theory and Applications (1982)
71. *T. W. Wieting,* The Mathematical Theory of Chromatic Plane Ornaments (1982)
72. *D. B. Gauld,* Differential Topology: An Introduction (1982)
73. *R. L. Faber,* Foundations of Euclidean and Non-Euclidean Geometry (1983)
74. *M. Carmeli,* Statistical Theory and Random Matrices (1983)
75. *J. H. Carruth, J. A. Hildebrant, and R. J. Koch,* The Theory of Topological Semigroups (1983)
76. *R. L. Faber,* Differential Geometry and Relativity Theory: An Introduction (1983)
77. *S. Barnett,* Polynomials and Linear Control Systems (1983)
78. *G. Karpilovsky,* Commutative Group Algebras (1983)
79. *F. Van Oystaeyen and A. Verschoren,* Relative Invariants of Rings: The Commutative Theory (1983)
80. *I. Vaisman,* A First Course in Differential Geometry (1984)
81. *G. W. Swan,* Applications of Optimal Control Theory in Biomedicine (1984)
82. *T. Petrie and J. D. Randall,* Transformation Groups on Manifolds (1984)
83. *K. Goebel and S. Reich,* Uniform Convexity, Hyperbolic Geometry, and Nonexpansive Mappings (1984)
84. *T. Albu and C. Năstăsescu,* Relative Finiteness in Module Theory (1984)
85. *K. Hrbacek and T. Jech,* Introduction to Set Theory, Second Edition, Revised and Expanded (1984)
86. *F. Van Oystaeyen and A. Verschoren,* Relative Invariants of Rings: The Noncommutative Theory (1984)
87. *B. R. McDonald,* Linear Algebra Over Commutative Rings (1984)
88. *M. Namba,* Geometry of Projective Algebraic Curves (1984)
89. *G. F. Webb,* Theory of Nonlinear Age-Dependent Population Dynamics (1985)
90. *M. R. Bremner, R. V. Moody, and J. Patera,* Tables of Dominant Weight Multiplicities for Representations of Simple Lie Algebras (1985)
91. *A. E. Fekete,* Real Linear Algebra (1985)
92. *S. B. Chae,* Holomorphy and Calculus in Normed Spaces (1985)
93. *A. J. Jerri,* Introduction to Integral Equations with Applications (1985)
94. *G. Karpilovsky,* Projective Representations of Finite Groups (1985)
95. *L. Narici and E. Beckenstein,* Topological Vector Spaces (1985)
96. *J. Weeks,* The Shape of Space: How to Visualize Surfaces and Three-Dimensional Manifolds (1985)
97. *P. R. Gribik and K. O. Kortanek,* Extremal Methods of Operations Research (1985)
98. *J.-A. Chao and W. A. Woyczynski, eds.,* Probability Theory and Harmonic Analysis (1986)
99. *G. D. Crown, M. H. Fenrick, and R. J. Valenza,* Abstract Algebra (1986)
100. *J. H. Carruth, J. A. Hildebrant, and R. J. Koch,* The Theory of Topological Semigroups, Volume 2 (1986)

101. *R. S. Doran and V. A. Belfi,* Characterizations of C*-Algebras: The Gelfand-Naimark Theorems (1986)
102. *M. W. Jeter,* Mathematical Programming: An Introduction to Optimization (1986)
103. *M. Altman,* A Unified Theory of Nonlinear Operator and Evolution Equations with Applications: A New Approach to Nonlinear Partial Differential Equations (1986)
104. *A. Verschoren,* Relative Invariants of Sheaves (1987)
105. *R. A. Usmani,* Applied Linear Algebra (1987)
106. *P. Blass and J. Lang,* Zariski Surfaces and Differential Equations in Characteristic $p > 0$ (1987)
107. *J. A. Reneke, R. E. Fennell, and R. B. Minton.* Structured Hereditary Systems (1987)
108. *H. Busemann and B. B. Phadke,* Spaces with Distinguished Geodesics (1987)
109. *R. Harte,* Invertibility and Singularity for Bounded Linear Operators (1988).
110. *G. S. Ladde, V. Lakshmikantham, and B. G. Zhang,* Oscillation Theory of Differential Equations with Deviating Arguments (1987)
111. *L. Dudkin, I. Rabinovich, and I. Vakhutinsky,* Iterative Aggregation Theory: Mathematical Methods of Coordinating Detailed and Aggregate Problems in Large Control Systems (1987)
112. *T. Okubo,* Differential Geometry (1987)
113. *D. L. Stancl and M. L. Stancl,* Real Analysis with Point-Set Topology (1987)
114. *T. C. Gard,* Introduction to Stochastic Differential Equations (1988)
115. *S. S. Abhyankar,* Enumerative Combinatorics of Young Tableaux (1988)
116. *H. Strade and R. Farnsteiner,* Modular Lie Algebras and Their Representations (1988)

Other Volumes in Preparation

Enumerative Combinatorics of Young Tableaux

SHREERAM S. ABHYANKAR
Purdue University
West Lafayette, Indiana

MARCEL DEKKER, INC. New York and Basel

Library of Congress Cataloging-in-Publication Data

Abhyankar, Shreeram Shankar.
Enumerative combinatorics of Young tableaux.

(Monographs and textbooks in pure and applied mathematics ; 115)
Includes index.
1. Young tableaux. 2. Combinatorial enumeration problems. I. Title. II. Series.
QA165.A24 1988 511'.6 87-27487
ISBN 0-8247-7816-2

MARCEL DEKKER, INC.
270 Madison Avenue, New York, New York 10016

Current printing (last digit):
10 9 8 7 6 5 4 3 2 1

PRINTED IN THE UNITED STATES OF AMERICA

To my wife YVONNE

Preface

A saturated increasing sequence of subspaces of a fixed vector space is called a flag. The geometric object formed by the set of all flags in a given vector space is called the flag manifold associated with that vector space. This flag manifold can be decomposed into certain natural configurations called Schubert varieties, which are labeled by permutations of letters as numerous as the dimension of the given vector space.

In the spring of 1982, V. Deodhar of Indiana University got me interested in the structure of singularities of Schubert varieties of flag manifolds. Reflecting on this structure, I was led to the investigation of certain determinantal ideals. Then in the summer of 1982, I had some discussions about these determinantal ideals with M. Giusti and M. Merle in Paris. These discussions directed me to a study of Young tableaux, which are certain tabular arrangements of integers introduced by Alfred Young at the end of the nineteenth century. I studied Young tableaux first in the writings of DeConcini-Eisenbud-Procesi and then in the writings of Gian-Carlo Rota and others. Eventually, I concluded that enumeration of Young tableaux and certain related objects would not only be fruitful for the investigation of determinantal ideals but would also be fascinating in itself.

This project was carried out during 1982 and 1983. The results were described first in a lecture in Paris in June 1983 under the auspices of the joint U.S.-France conference on singularities, and then in a lecture in Kyoto in July 1984 under the auspices of the joint U.S.-Japan conference on singularities. Some of the material was also included in the courses I gave during 1983 at the Universities of Poona and Bombay in India and the University of Nice in France. Moreover, during 1984 and 1985, all the material was covered in courses which I taught at Bhaskaracharya Pratishthana in Pune in India and at Purdue University.

In Chapter 1 we describe formulas for enumerating certain types of objects, and we prepare the groundwork for establishing these formulas. The objects may be certain tabular arrangements of integers called Young tableaux, or they may be certain types of monomials, or they may be something else. Although what Alfred Young introduced were mainly unitableaux, we enumerate bitableaux as well as multitableaux of higher width. In Chapter 1 we also give an exhaustive treatment of binomial coefficients.

Using the groundwork prepared in Chapter 1, the enumerative formulas are established in Chapter 2.

In Chapter 3 we present a universal identity satisfied by the minors of any matrix. This may be regarded as belonging to the combinatorics of Young tableaux.

In Chapter 4 we give several applications of the enumerative formulas and universal identity. These applications include enumerative proofs of the straightening law of Doubilet-Rota-Stein. They also include enumerative proofs of the second fundamental theorem of invariant theory and its generalizations. Finally, they include computations of Hilbert functions of polynomial ideals.

Being logically self-contained, with the relevant references cited within the general text, this monograph does not contain a separate bibliography. However, in *Young Tableaux in Combinatorics, Invariant Theory, and Algebra*, edited by Kung and published by Academic Press in 1982, the reader can find references to several articles about Young tableaux by various authors.

The material of this monograph is arranged such that each of the four chapters can be read independently of the others. Cross-references are provided within each chapter. Moreover, the author presumes no prior knowledge of Young tableaux or any other mathematical terms. Throughout the monograph, several problems are posed and certain Motivational Remarks and Informational Remarks are made; they are listed at the beginning of the relevant chapter and section. This monograph may be used as a text for a graduate or advanced undergraduate course.

My thanks are due to D. V. Kulkarni and Devadatta Kulkarni for stimulating conversations and help with the manuscript. I am also grateful to the National Science Foundation and the Office of Naval Research for supporting the work on this monograph.

Shreeram S. Abhyankar

Contents

General Introduction

Young tableaux occur in diverse aspects of mathematics, statistics, theoretical physics, and computer science. Basically, a (Young) tableau is a tabular arrangement of positive integers such as the following:

<table>
<tr><td>1</td><td>3</td><td>4</td><td>5</td><td>9</td></tr>
<tr><td>2</td><td>5</td><td>6</td><td>7</td><td>9</td></tr>
<tr><td>3</td><td>5</td><td>6</td><td>8</td></tr>
<tr><td>3</td><td>9</td></tr>
<tr><td>7</td></tr>
<tr><td>8</td></tr>
</table>

In a tableau it is usually required to have a strict increase along the rows. The tableau is said to be *standard* if, as in the example above, the row lengths are nonincreasing and there is no decrease along the columns.

The example above is a *unitableau*. A tableau may have two similar sides, i.e., two sides of the same shape but with possibly different entries, such as the diagram below, and then it is called a *bitableau*. Or it may have any number of similar sides and then we may call it a *multitableau*, or simply a tableau; the number of sides may be called the *width* of the tableau. Thus a unitableau is a tableau of width 1, a bitableau is a tableau of width 2, and so on.

7	6	5	3	2	1	3	4	5	9
8	7	5	4	2	2	5	6	7	9
	8	7	5	4	3	5	6	8	
			6	5	3	9			
				5	7				
				7	8				

The number of entries on each side may be called the *area* of the tableau. The largest row length may be called the *length* of the tableau, and the number of rows may be called the *depth* of the tableau. For instance, in both examples above, the area is 18, the length is 5, and the depth is 6; in the first example the width is 1, and in the second example the width is 2. The sides of a tableau of width q may be visualized as the q pages in a loose-leaf notebook, or as the q cards in a rotating circular card catalog. The tableau may be said to be *bounded* by a q-tuple $m = (m(1), m(2), \ldots, m(q))$ of positive integers if, for $j = 1, 2, \ldots, q$, all the entries on the jth side are $\le m(j)$. Thus, in the second example above, where $q = 2$ and where we regard the left-hand side as the first side and the right-hand side as the second side, the tableau is bounded by any pair of integers $(m(1), m(2))$ with $m(1) \ge 8$ and $m(2) \ge 9$.

Young tableaux (sometimes called *Young diagrams*, or *Young symmetrizers*, or *Young operators*) were introduced by Alfred Young at the end of the nineteenth century in a book on invariant theory which he coauthored with his friend J. H. Grace in 1903 and which has now been reprinted by Chelsea. Actually, they are implicitly contained in Young's 1901 paper in the *Journal of the London Mathematical Society*, to which reference is made in the invariant theory book. These original tableaux were unitableaux which were strongly standard in the sense that a strict increase was also required along the columns. For instance, we could be considering standard unitableaux of area n among whose entries each integer from 1 to n occurs exactly once. Clearly, these tableaux may be thought of as some kind of generalized permutations of the integers from 1 to n. Indeed, they were very effectively used by Young from 1901 to 1935 for describing the irreducible representations of S_n where, as usual, S_n is the group of all the n! permutations of the set of integers from 1 to n. This work of Young is contained in a series of papers which were called Quantitative Substitutional Analysis and which are now usually referred to as QSA. It was this work of Young which caught the attention of Hermann Weyl, who from 1930 to 1940 applied it to quantum mechanics with great success and who described it in his books *Group Theory and Quantum Mechanics* and *Classical Groups*. During the last 20 years there has been a great revival of the use of Young tableaux in various aspects of theoretical physics, such as quantum mechanics and the theory of elementary particles.

Initially, Young was led to the study of the irreducible representations of S_n while investigating the structure of various terms in Gordon series, which are certain polynomial identities occurring in the classical invariant theory of binary forms and which were discovered about 1865. Gordon series were generalized by A. Capelli about 1890 for invariant theory of forms in any number of variables. Young showed how the Capelli series can be derived from a formal identity involving strongly standard unitableaux. This discussion about Capelli series started in the book *Invariant Theory* and culminated in the last few parts of QSA. The ideas of Alfred Young were exposed and furthered by each of four of his students: D. E. Littlewood, G. Robinson, D. E. Rutherford, and H. W. Turnbull.

About 20 years ago Young tableaux were resurrected by Gian-Carlo Rota and his school. In this reincarnation they appeared mainly as bitableaux and they found more direct applications to invariant theory. In this category, in addition to Rota's immediate school consisting of P. Doubilet, J. P. S. Kung, J. Stein, etc., we may also count the relevant work of C. DeConcini, J. A. Eagon, D. Eisenbud, M. Hochster, J. I. Igusa, K. R. Mount, C. Procesi, R. P. Stanley, and others.

During the last 20 years Young tableaux have also been used in various aspects of statistics, such as in the so-called ballot problem and in problems dealing with lattice paths. These applications may be found in the recent books of S. G. Mohanty and T. V. Narayana entitled *Lattice Path Counting* and *Lattice Path Combinatorics*, respectively.

Again during the last 20 years, D. E. Knuth, C. Schenstead, and others have initiated the combinatorial use of Young tableaux in various aspects of computer science, such as sorting and searching. These applications of Young tableaux to computer science, including the so-called inserting and bumping algorithms of Schenstead, may be found in the third volume of Knuth's encyclopedic book, *The Art of Computer Programming*.

In algebraic geometry Young tableaux were used by W. V. D. Hodge about 1947 to study Schubert varieties, which are certain natural subvarieties of the Grassmann variety $G_{r,n}$. It may be noted that points of $G_{r,n}$ correspond to r-dimensional subspaces of a given n-dimensional vector space L; it may also be noted that both Grassmann and Schubert were schoolteachers in the nineteenth century. Now a sequence $L_1 \subset L_2 \subset \cdots \subset L_n$, where L_i is an i-dimensional subspace of L, is called a *flag* in L. The geometric object formed by the set of all flags in L is called the *flag manifold* associated with L. This flag manifold is a nonsingular subvariety of the cartesian product of the Grassmann varieties $G_{1,n}, G_{2,n}, \ldots, G_{n,n}$. The flag manifold can be decomposed into certain natural configurations which are indexed by the group S_n; that is, they are indexed by permutations of letters as numerous as the dimension of the given vector space. These natural configurations are generalizations of Schubert subvarieties of Grassmann varieties, so they are called *Schubert varieties in flag manifolds*. These Schubert varieties are closures of the so-called Bruhat cells. It may be noted that the idea of expressing the flag manifold as a disjoint union of Bruhat cells can be done in a more general setting of algebraic groups.

Let us note that a tableau of width q and depth 1 may be called a *multivector of depth* q; if q = 1, then we may call it a *univector*, and if q = 2, then we may call it a *bivector*, and so on. Now let a be a multivector of width q and length p such that a is bounded by the q-tuple m. A standard tableau of depth q may be said to be *predominated* by a if the tableau obtained by putting a on top of the given tableau is again standard. For example, the bitableau of the second example on page xii is predominated by the following bivector:

7	6	4	3	1	1	2	4	5	6

My enumerative work started by first constructing a formula $F^{(8)}(q, m,p,a,V)$ for counting the set stab(q,m,p,a,V) of all standard tableau of width q and area V which are bounded by m and predominated by a. This formula is an example of what may be thought of as *determinantal polynomials in binomial coefficients*. Then, by a series of transformations, I successively converted $F^{(8)}$ into several other determinantal polynomials in binomial coefficients which are labeled as $F^{(7*)}$, $F^{(7)}$, $F^{(6)}$, $F^{(5k)}$, $F^{(4*k)}$, $F^{(4k)}$, ..., $F^{(1k)}$ with k = 1, 2, ..., q. Out of these $F^{(4*k)}$, $F^{(4k)}$, ..., $F^{(1k)}$ are valid only for q = 2.

Before specializing to the case q = 2, which is significant in the study of determinantal ideals, let us note that most of the previous work on Young tableaux was restricted to unitableaux and bitableaux. I have initiated the theory of tableau of higher width with the hope that it may be useful in studying higher-dimensional determinants.

To continue with the case of q = 2, let X be an m(1) by m(2) matrix whose entries X_{ij} are indeterminates over a field K. Let K[X] be the ring of polynomials in the m(1)m(2) variables X_{ij} with coefficients in K. Now a bivector bounded by m is precisely the code for indicating a minor of X. For example, the bivector depicted above indicates the 5 by 5 minor of X which is formed by picking the rows and columns of X whose row numbers and column numbers are 1, 3, 4, 6, 7 and 1, 2, 4, 5, 6, respectively. Now a bitableau bounded by m is a sequence of bivectors bounded by m; upon taking the product of the minors which correspond to these bivectors we get a monomial in the minors of X corresponding to the given bitableau; this monomial in the minors of X is said to be *standard* if the given bitableau is standard. The straightening law of Doubilet-

Rota-Stein says that the standard monomials in the minors of X form a K-basis of K[X]; this law is a culmination of ideas going back to Clebsch, Gordon, and Young

Continuing the enumerative work, I considered a set of ordinary monomials of degree V in the variables X_{ij} satisfying certain constraints determined by the bivector a, and I denoted this set by mon(2,m,p,a,V). I proved that $F^{(1k)}(m,p,a,V)$ and mon(2,m,p,a,V) satisfy the same recurrence relations, and hence $F^{(1k)}(m,p,a,V)$ also counts mon(2,m,p,a,V), and therefore the two sets stab(2,m,p,a,V) and mon(2,m,p,a,V) are in one-to-one correspondence. Next, by using the combinatorics of Young tableaux and by generalizing the quadratic relations defining a Grassmann variety, I obtained a universal determinantal identity satisfied by different-size minors of X. Finally I made several applications of these enumerative formulas and the universal determinantal identity. These applications include enumerative proofs of the straightening law as well as the second fundamental theorem of invariant theory and its generalizations. They also include calculations of Hilbert functions of determinantal ideals in K[X].

To explain some of these applications in greater detail, for any positive integer p, let $\hat{G}[p + 1]$ be the set of all p + 1 by p + 1 minors of X and let I[p + 1] be the ideal in K[X] generated by $\hat{G}[p + 1]$. Moreover, with the bivector a of length p we associate a certain set $\hat{G}(p,a)$ of different-size minors of X and we consider the ideal I(p,a) in K[X] generated by $\hat{G}(p,a)$. Thinking for a moment of X as an m(1) by m(2) rectangle, we call certain subsets Y of X *saturated subsets*. For example, the subset Y enclosed within two ladders is a saturated subset:

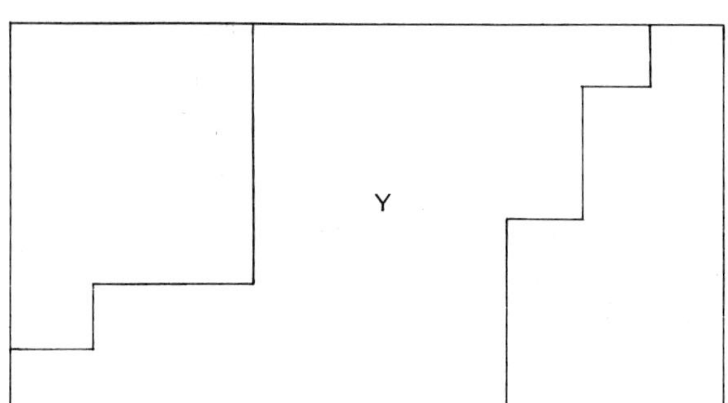

Let $I^*[p + 1, Y]$ and $I^*(p,a,Y)$ be the ideals in $K[Y]$ generated by those members of $\hat{G}[p + 1]$ and $\hat{G}(p,a)$ which lie entirely within the saturated subset Y. Now a part of the second fundamental theorem of invariant theory, which was originally proved by Eduardo Pascal in 1888, says that $I[p + 1]$ is a prime ideal in $K[X]$. In addition to giving several new proofs of the primeness of $I[p + 1]$, the applications include the new results according to which $I(p,a)$ is a prime ideal in $K[X]$, whereas $I^*[p + 1, Y]$ and $I^*(p,a,Y)$ are prime ideals in $K[Y]$; this is what I had conjectured while investigating the structure of the singularities of Schubert varieties in flag manifolds.

Finally, I proved that $F^{(1k)}(m,p,a,V)$ is the Hilbert function as well as the Hilbert polynomial of $I(p,a)$ in $K[X]$. Although I say "finally," this is what I really proved first and then the rest followed as a consequence. These calculations with $F^{(1k)}$ also give very satisfying verifications of several basic theorems of Hilbert in algebraic geometry.

In conclusion, a philosophical remark may be in order. Some time ago my experience with the problem of resolution of singularities in algebraic geometry taught me the value of high school algebra, by which I then meant formal manipulations with polynomials and power series. I have now enlarged this notion of high school algebra by thinking of the binomial coefficient $\binom{V}{A}$ as a polynomial in V of degree A; or better still, $\binom{V}{A}$ is to play the role of the basic monomial V^A. The resulting manipulations, often with a view that they are an end in themselves, constitute the backbone of this monograph.

Enumerative Combinatorics of Young Tableaux

1
Young Tableaux and Determinantal Polynomials in Binomial Coefficients

1. INTRODUCTION

One of the aims of this first chapter is to describe several formulas for counting certain types of tableaux which were introduced by Alfred Young at the turn of the nineteenth century and which were resurrected by Gian-Carlo Rota and others during the last decade. These formulas may very well be described as examples of what may be called determinantal polynomials in binomial coefficients. Some of these formulas also enumerate certain types of monomials.

The determinantal polynomials are labeled as

$$F^{(Lk)}(m,p,a,V) \qquad \text{with } L = 1, 1^*, 2, 2^*, 3, 3^*, 4, 4^* \text{ and } k = 1, 2$$

and

$$F^{(5k)}(q,m,p,a,V) \qquad \text{with } k = 1, 2, \ldots, q$$

and

$$F^{(L)}(q,m,p,a,V) \qquad \text{with } L = 6, 7, 7^*, 8$$

and are introduced in Subsections (2.7) and (2.10). These determinantal polynomials are defined in terms of certain apparently infinite summations because formal manipulations with summations are easier to handle when the summation indices range freely. In Subsection (2.6) we prepare the groundwork for showing that the relevant summations are essentially finite; i.e., all except a finite number of summands in these summations are zero. The fact that the summations are essentially finite is then established in Subsections (2.9) and (2.11). Of the determinantal polynomials above, $F^{(2k)}$ and $F^{(8)}$ may be called the Nice formulas because they were introduced in my 1983 Lectures at the University of Nice in France; the notes on those lectures by Andre Galligo have appeared in Volume 42 of *Rendiconti del Seminario Matematico di Universita et Polytecnico di Torino* for the year 1984.

In Subsections (2.2) and (2.3) we introduce the concept of a tableau. In brief let q be a positive integer and let m = (m(1), m(2), ..., m(q)) be a sequence of positive integers. By a multivector a of width q we essentially mean a multisequence

$$a(k,1) < a(k,2) < \cdots < a(k,p)$$

of positive integers, for k = 1, 2, ..., q, where the nonnegative integer p is called the length of a. We say that a is bounded by m to mean that

$$a(k,i) \le m(k) \qquad \text{for } k = 1, 2, \ldots, q \text{ and } i = 1, 2, \ldots, p.$$

Given any other multivector b of width q and length p', we define $a \le b$ to mean that $p \ge p'$ and

$$a(k,i) \le b(k,i) \qquad \text{for } k = 1, 2, \ldots, q \text{ and } i = 1, 2, \ldots, p'.$$

By a tableau T of width q we essentially mean a sequence

$$\begin{matrix} T[1] \\ T[2] \\ \vdots \\ T[d] \end{matrix}$$

of multivectors of width q where the nonnegative integer d is called the depth of T. We say that T is standard to mean that

$$T[1] \le T[2] \le \cdots \le T[d]$$

and the length of T[e] is nonzero for e = 1, 2, ..., d. We say that T is bounded by m to mean that T[e] is bounded by m for e = 1, 2, ..., d. We say that T is predominated by a to mean that $a \le T[e]$ for e = 1, 2, ..., d. The sum of the lengths

$$\text{length of } T[1] + \text{length of } T[2] + \cdots + \text{length of } T[d]$$

is called the area of T. By stab(q,m,p,a,V) we denote the set of all standard tableaux of width q which are bounded by m and whose area is V and which are predominated by the multivector a of width q and length p. By Stab(q,m,p,a,V) we denote the cardinality of stab(q,m,p,a,V). By mon(q,m,p,a,V) we denote the corresponding set of monomials satisfying certain constraints as cited above, and by Mon(q,m,p,a,V) we denote the cardinality of mon(q,m,p,a,V).

In Section 3 we prepare the groundwork for proving the conditional equivalence of the determinantal polynomials

$$F^{(5k)},\ F^{(6)},\ F^{(7*)},\ F^{(7)},\ F^{(8)}.$$

In Section 3 we also discuss some consequences of the multilinearity of the determinant as a function of its rows or columns. Certain pertinent remarks concerning the usual distributive law are made in (3.31) and (3.32).

In Section 4 we discuss some properties of binomial coefficients and thereby prepare the groundwork for proving the equivalence and recursiveness of the determinantal polynomials

$$F^{(1k)},\ F^{(1*k)},\ \ldots,\ F^{(4*k)}.$$

In Chapter 2 the material of Sections 3 and 4 will be put together to show that for every nonnegative integer V we have

$$\begin{aligned}\mathrm{Stab}(q,m,p,a,V) &= F^{(8)}(q,m,p,a,V)\\ &= F^{(7)}(q,m,p,a,V)\end{aligned}$$

and to show that if q is even, then for every nonnegative integer V we have

$$\begin{aligned}\mathrm{Stab}(q,m,p,a,V) &= F^{(7*)}(q,m,p,a,V)\\ &= F^{(6)}(q,m,p,a,V)\\ &= F^{(5k)}(q,m,p,a,V) \qquad \text{for } k = 1, 2, \ldots, q\end{aligned}$$

and to show that given any $L \in \{1,1*,2,2*,3,3*,4,4*\}$ and $k \in [1,2]$, for every nonnegative integer V, we have

$$\mathrm{Stab}(2,m,p,a,V) = F^{(Lk)}(m,p,a,V) = \mathrm{Mon}(2,m,p,a,V).$$

In Chapter 3 we prove a certain universal determinantal identity. In Chapter 4 we give several applications of the enumeration formulas above and the said universal determinantal identity. These applications include enumerative proofs of the Straightening Law of Doubilet-Rota-Stein. They include enumerative proofs of the Second Fundamental Theorem of Invariant Theory and its generalizations. Finally, they include computations of Hilbert functions of polynomial ideals.

As the second aim of Chapter 1, in Section 4 we give a systematic treatment of some aspects of binomial coefficients; this treatment is much more extensive than needed for establishing the enumerative formulas above. In (4.62) and (4.65) we pose some problems related to the material of Section 4, and in (4.63) and (4.64) we make certain relevant Motivational Remarks. The definition of binomial coefficients is reviewed in Subsection (2.5), and a pertinent Remark is made in (2.5.13). It may be noted that Section 4 is independent of Section 3.

2. NOTATION AND TERMINOLOGY

In this section we introduce some notation and terminology to be used in the remainder of this chapter.

(2.1) Generalities As usual we put

Q = the set of all rational numbers
Z = the set of all integers
N = the set of all nonnegative integers
N* = the set of all positive integers

For any set X, we put

card(X) = the cardinal number of X

and we define

card(X) < ∞ to mean that X is a finite set

and we define

$\operatorname{card}(X) = \infty$ to mean that X is an infinite set.

Recall that given any sets X and Y, a *map* $f: X \to Y$ (i.e., a map f from X to Y, or a map f of X into Y) is a rule which, to every $x \in X$, associates a unique $f(x) \in Y$; when we want to denote the associated value by a possibly different notation (e.g., f[x]), this will be mentioned explicitly and in that case f will be called a *mapping*.

For any two integers A and B, by [A,B] we denote the *closed integral segment* between A and B, i.e.,

$$[A,B] = \{D \in Z : A \leq D \leq B\}$$

and we note that

[A,B] is nonempty iff $A \leq B$.

For every $p \in N$ we put

$$Z(p) = \text{the set of all maps from } [1,p] \text{ to } Z$$

and we also put

$$N(p) = \{d \in Z(p) : d(i) \in N \quad \text{for all } i \in [1,p]\}$$

and

$$N^*(p) = \{d \in Z(p) : d(i) \in N^* \quad \text{for all } i \in [1,p]\}.$$

For every $p \in N^*$ and $D \in Z$ we put

$$Z(p,D) = \{d \in Z(p) : d(1) + d(2) + \cdots + d(p) = D\}$$

and

$$Z[p,D] = \{d \in Z(p) : d(1) + d(2) + \cdots + d(p) \leq D\}$$

and we also put

$$N(p,D) = Z(p,D) \cap N(p) \quad \text{and} \quad N[p,D] = Z[p,D] \cap N(p).$$

Given any $p \in N$, by W(p) we denote the set of all *permutations* of [1,p], i.e.,

$$W(p) = \text{the set of all bijective maps of } [1,p] \text{ onto itself}$$

and for every $w \in W(p)$ we put

$\mathrm{sgn}(w) = 1$ or -1 according as w is even or odd

and we also put

$W[p]$ = the set of all maps of $W(p)$ into itself.

Given any set v, first we put

$P(v)$ = the set of all subsets of v,

and second, for every $A \in Z$ we put

$$P(v,A) = \{a \in P(v) : \mathrm{card}(a) = A\}$$

and we note that then

for all $A \in Z \setminus N$ we have $P(v,A) = \emptyset$.

Given any $p \in N$, first we put

$$J(p) = P([1,p]),$$

and second, for every $A \in Z$ we put

$$J(p,A) = P([1,p],A),$$

and third, for every $t \in J(p)$ we put

$$J^*(p,t) = \{u \in J(p) : t \subset u\},$$

and fourth, for every $t \in J(p)$ and $A \in Z$ we put

$$J^*(p,t,A) = J^*(p,t) \cap J(p,A).$$

Finally, by a ring we shall always mean a commutative ring with identity.

(2.2) Multivectors For convenience of speaking, given any $q \in N$, a member of $Z(q)$ [resp., $N(q)$, $N^*(q)$] may be called a *premultinumber* (resp., *multinumber*, *positive multinumber*) *of width* q; similarly, a member of $Z(1)$ [resp., $N(1)$, $N^*(1)$] may be called a *preuninumber* (resp., *uninumber*, *positive uninumber*); likewise, a member of $Z(2)$ [resp., $N(2)$, $N^*(2)$] may be called a *prebinumber* (resp., *binumber, positive binumber*).

Given any $q \in N$ we consider the *infinite integral rectangle of width* q which we denote by $\mathrm{rec}(q)$ and which we define by putting

$$\text{rec}(q) = \text{the set of all pairs } (k,i) \text{ with } k \in [1,q] \text{ and } i \in Z.$$

Given any $q \in N$ and $p \in N$ we consider the *positive integral rectangle of width* q *and length* p which we denote by rec(q,p) and which we define by putting

$$\text{rec}(q,p) = \text{the set of all pairs } (k,i) \text{ with } k \in [1,q] \text{ and } i \in [1,p]$$

and we note that then

$$\text{rec}(q,p) = \{(k,i) \in \text{rec}(q) : i \in [1,p]\}.$$

Given any $q \in N^*$ and $p \in N$, by a *premultivector of width* q *and length* p we mean a mapping a of rec(q,p) into Z which, to every $k \in [1,q]$ and $i \in [1,p]$, associates $a(k,i) \in Z$. Given any $q \in N^*$, by a *premultivector of width* q we mean a premultivector a of width q and length p for some $p \in N$, and we then put

$$\text{len}(a) = p.$$

Given any $q \in N^*$, first we put

$$\text{pre}(q) = \text{the set of all premultivectors of width } q,$$

and second, for every $p \in N$ we put

$$\text{pre}[q,p] = \text{the set of premultivectors of width } q \text{ and length } p$$

and we note that

$$\text{pre}(q) = \coprod_{p \in N} \text{pre}[q,p]$$

where as usual

$\coprod$ denotes disjoint union.

Given any $q \in N^*$, for any $a \in \text{pre}(q)$ and $b \in \text{pre}(q)$ we define

$$a \le b \text{ to mean } \begin{cases} \text{len}(a) \ge \text{len}(b) \quad \text{and} \\ a(k,i) \le b(k,i) \text{ for all } k \in [1,q] \text{ and} \\ i \in [1,\text{len}(b)] \end{cases}$$

and we note that this is a partial order on pre(q). Given any $q \in N^*$, for any $a \in \text{pre}(q)$ and $m \in Z(q)$ we define

$$a \le m \text{ to mean } \begin{cases} a(k,i) \le m(k) \\ \text{for all } k \in [1,q] \text{ and } i \in [1,\text{len}(a)] \end{cases}$$

and we may express this by saying that a *is bounded by* m. Given any $q \in N^*$ and $a \in \text{pre}(q)$, we say that a is *positive* to mean that

$$a(k,i) > 0 \quad \text{for all } k \in [1,q] \text{ and } i \in [1,\text{len}(a)].$$

Given any $q \in N^*$, by a *multivector of width* q we mean a positive premultivector a of width q such that

$$a(k,i) < a(k, i+1) \quad \text{for all } k \in [1,q] \text{ and } i \in [1, \text{len}(a)-1].$$

Given any $q \in N^*$, first we define the sets

$$\text{vec}(q) \subset \text{popre}(q)$$

by putting

$$\text{popre}(q) = \text{the set of all positive premultivectors of width } q$$

and

$$\text{vec}(q) = \text{the set of all multivectors of width } q$$

and second, for every $p \in N$ we define

$$\text{popre}[q,p] = \text{pre}[q,p] \cap \text{popre}(q)$$

and

$$\text{vec}[q,p] = \text{pre}[q,p] \cap \text{vec}(q),$$

and third, for every $m \in Z(q)$ we define

$$\text{popre}(q,m) = \{a \in \text{popre}(q) : a \le m\}$$

and

$$\text{vec}(q,m) = \{a \in \text{vec}(q) : a \le m\},$$

and fourth, for every $m \in Z(q)$ and $p \in N$ we define

$$\text{popre}(q,m,p) = \text{popre}(q,m) \cap \text{popre}[q,p]$$

and

$$\text{vec}(q,m,p) = \text{vec}(q,m) \cap \text{vec}[q,p],$$

and fifth, for every $m \in Z(q)$ and $p \in N$ and $a \in \text{vec}[q,p]$ we define

$$\text{vec}(q,m,p,a) = \{b \in \text{vec}(q,m,p) : a \le b\}.$$

By a *preunivector* (resp., *univector*) we mean a premultivector (resp., multivector) of width 1. By a *prebivector* (resp., *bivector*) we mean a premultivector (resp., multivector) of width 2.

Let us note that given any $q \in N^*$ and $m \in N^*(q)$ and $p \in N^*$ and $a \in \mathrm{vec}(q,m,p)$ and $k \in [1,q]$, we have the sequence of integers

$$a(k,1) < a(k,2) < \cdots < a(k,p) \le m(k)$$

and hence for all $i \in [1,p]$ we have

$$m(k) - a(k,i) \ge p - i$$

and therefore for all $i \in [1,p]$ and $j \in [1,p]$ we have

$$m(k) - a(k,i) + i - j \ge p - i + (i - j) = p - j \ge 0$$

and for all $i \in [1,p]$ we have

$$m(k) - a(k,i) \ge p - i \ge 0;$$

thus

$$\left[\begin{array}{l}\text{given any } q \in N^* \text{ and } m \in N^*(q) \text{ and } p \in N^* \text{ and} \\ a \in \mathrm{vec}(q,m,p) \text{ and } k \in [1,q] \text{ and } i \in [1,p], \text{ we} \\ \text{have } m(k) - a(k,i) \ge p - i\end{array}\right. \tag{2.2.1}$$

and

$$\left[\begin{array}{l}\text{given any } q \in N^* \text{ and } m \in N^*(q) \text{ and } p \in N^* \text{ and} \\ a \in \mathrm{vec}(q,m,p) \text{ and } k \in [1,q] \text{ and } i \in [1,p] \text{ and} \\ j \in [1,p], \text{ we have } m(k) - a(k,i) + i - j \in N\end{array}\right. \tag{2.2.2}$$

and

$$\left[\begin{array}{l}\text{given any } q \in N^* \text{ and } m \in N^*(q) \text{ and } p \in N^* \text{ and} \\ a \in \mathrm{vec}(q,m,p) \text{ and } k \in [1,q], \text{ we have} \\ m(k) - a(k,i) \in N.\end{array}\right. \tag{2.2.3}$$

(2.3) Tableaux Given any $q \in N^*$ and $d \in N$, by a *pretableau of width* q *and depth* d we mean a mapping T of $[1,d]$ into $\mathrm{pre}(q)$ which, to every $j \in [1,d]$, associates $T[j] \in \mathrm{pre}(q)$. Given any $q \in N^*$, by a *pretableau of width* q we mean a pretableau T of width q and depth d for some $d \in N$, and we then put

$$\mathrm{dep}(T) = d.$$

Given any $q \in N^*$ and given any pretableau T of width q, first we introduce the *length of* T, which we denote by len(T) and which we define by putting

$$\text{len}(T) = \begin{cases} \max\{\text{len}(T[j]) : j \in [1,\text{dep}(T)]\} & \text{if dep}(T) \neq 0 \\ 0 & \text{if dep}(T) = 0, \end{cases}$$

and second, we introduce the *area of* T, which we denote by are(T) and which we define by putting

$$\text{are}(T) = \sum_{j\in[1,\text{dep}(T)]} \text{len}(T[j])$$

(where we recall that by convention the sum of an empty family is zero), and third, for every $i \in N$ we introduce the ith *depth of* T, which we denote by dep[i](T) and which we define by putting

$$\text{dep}[i](T) = \text{card}(\{j \in [1,\text{dep}(T)] : \text{len}(T[j]) = i\}),$$

and fourth, for any $m \in Z(q)$ we define

$$T \le m \text{ to mean } T[j] \le m \text{ for all } j \in [1,\text{dep}(T)]$$

and we may express this by saying that T *is bounded by* m, and fifth, for any $a \in \text{pre}(q)$ we define

$$a \le T \text{ to mean } a \le T[j] \text{ for all } j \in [1,\text{dep}(T)]$$

and we may express this by saying that T *is predominated by* a.

Given any $q \in N^*$ and given any pretableau T of width q, by a *shape* of T we mean a pair (p,v) with $p \in N$ and $v \in N(p)$, such that $\text{len}(T) \le p$ and $\text{dep}[i](T) = v(i)$ for all $i \in [1,p]$; if (p,v) is a shape of T, then we may say that T is a pretableau of shape (p,v) [we cannot call (p,v) *the* shape of T because, although only in a silly manner, T does not determine (p,v) uniquely]. Given any $q \in N^*$ and given any pretableau T of width q, we say that T is *pure* to mean that

$$\text{len}(T[j]) = \text{len}(T) \qquad \text{for all } j \in [1,\text{dep}(T)].$$

Given any $q \in N^*$ and given any pretableau T of width q we say that T is *positive* to mean that

$$T[j] \in \text{popre}(q) \qquad \text{for all } j \in [1,\text{dep}(T)].$$

Given any $q \in N^*$, by a *tableau of width* q we mean a pretableau T of width q such that

$$T[j] \in \text{vec}(q) \quad \text{for all } j \in [1,\text{dep}(T)].$$

Given any $q \in N^*$ and given any tableau T of width q, we say that T is *quasistandard* to mean that

$$T[j] \le T[j+1] \quad \text{for all } j \in [1, \text{dep}(T) - 1].$$

Given any $q \in N^*$ and given any tableau T of width q, we say that T is *standard* to mean that T is quasistandard and

$$\text{len}(T[j]) > 0 \quad \text{for all } j \in [1,\text{dep}(T)].$$

Given any $q \in N^*$, we define the sets

$$\text{stab}(q) \subset \text{qtab}(q) \subset \text{tab}(q) \subset \text{popab}(q) \subset \text{pab}(q)$$

by putting

$$\text{pab}(q) = \text{the set of all pretableaux of width } q$$

and

$$\text{popab}(q) = \text{the set of all positive pretableaux of width } q$$

and

$$\text{tab}(q) = \text{the set of all tableaux of width } q$$

and

$$\text{qtab}(q) = \text{the set of all quasistandard tableaux of width } q$$

and

$$\text{stab}(q) = \text{the set of all standard tableaux of width } q.$$

Given any $q \in N^*$, first for every $V \in Z$ we define

$$\text{pib}[q,V] = \{T \in \text{pab}(q) : \text{are}(T) = V\}$$

and we note that

$$\text{pab}(q) = \coprod_{V \in N} \text{pib}[q,V] \text{ whereas } \text{pib}[q,V] = \emptyset \text{ for all } V \in Z\backslash N$$

and second, for every $d \in Z$ we define

$$\text{pid}[q,d] = \{T \in \text{pab}(q) : \text{dep}(T) = d\}$$

and we note that

$$\mathrm{pab}(q) = \coprod_{d \in N} \mathrm{pid}[q,d] \text{ whereas } \mathrm{pid}[q,d] = \emptyset \text{ for all } d \in Z \setminus N,$$

and third, for every $p \in N$ we define

$$\mathrm{pip}[q,p] = \{T \in \mathrm{pab}(q) : T \text{ is pure and } \mathrm{len}(T) = p\},$$

and fourth, for every $p \in N$ and $v \in N(p)$ we define

$$\mathrm{pis}[q,p,v] = \{T \in \mathrm{pab}(q) : (p,v) \text{ is a shape of } T\}.$$

Given any $q \in N^*$ and given any $f \in \{\mathrm{pop},t,\mathrm{qt},\mathrm{st}\}$, upon letting

$$F = \begin{cases} \mathrm{Pop} & \text{if } f = \mathrm{pop} \\ T & \text{if } f = t \\ \mathrm{Qt} & \text{if } f = \mathrm{qt} \\ \mathrm{St} & \text{if } f = \mathrm{st} \end{cases}$$

first for every $m \in Z(q)$ we define

$$\mathrm{fab}(q,m) = \{T \in \mathrm{fab}(q) : T \le m\},$$

and second, for every $m \in Z(q)$ and $p \in N$ we define

$$\mathrm{fab}(q,m,p) = \{T \in \mathrm{fab}(q,m) : \mathrm{len}(T) \le p\},$$

and third, for every $m \in Z(q)$ and $p \in N$ and $a \in \mathrm{pre}[q,p]$ we define

$$\mathrm{fab}(q,m,p,a) = \{T \in \mathrm{fab}(q) : a \le T \le m\}$$

and we note that then

$$\mathrm{fab}(q,m,p,a) \subset \mathrm{fab}(q,m,p),$$

and fourth, for every $V \in Z$ we define

$$\mathrm{fib}[q,V] = \mathrm{fab}(q) \cap \mathrm{pib}[q,V],$$

and fifth, for every $m \in Z(q)$ and $V \in Z$ we define

$$\mathrm{fib}(q,m,V) = \mathrm{fab}(q,m) \cap \mathrm{fib}[q,V]$$

and

$$\mathrm{Fib}(q,m,V) = \mathrm{card}(\mathrm{fib}(q,m,V)),$$

and sixth, for every $m \in Z(q)$ and $p \in N$ and $V \in Z$ we define

$$fab[q,m,p,V] = fab(q,m,p) \cap fib[q,V]$$

and

$$Fab[q,m,p,V] = card(fab[q,m,p,V]),$$

and seventh, for every $m \in Z(q)$ and $p \in N$ and $a \in pre[q,p]$ and $V \in Z$ we define

$$fab(q,m,p,a,V) = fab(q,m,p,a) \cap fib[q,V]$$

and

$$Fab(q,m,p,a,V) = card(fab(q,m,p,a,V)),$$

and eighth, for every $d \in Z$ we define

$$fid[q,d] = fab(q) \cap pid[q,d],$$

and ninth, for every $m \in Z(q)$ and $d \in Z$ we define

$$fid(q,m,d) = fab(q,m) \cap fid[q,d]$$

and

$$Fid(q,m,d) = card(fid(q,m,d)),$$

and tenth, for every $m \in Z(q)$ and $p \in N$ and $d \in Z$ we define

$$fad[q,m,p,d] = fab(q,m,p) \cap fid[q,d]$$

and

$$Fad[q,m,p,d] = card(fad[q,m,p,d]),$$

and eleventh, for every $m \in Z(q)$ and $p \in N$ and $a \in pre[q,p]$ we define

$$fad(q,m,p,a,d) = fab(q,m,p,a) \cap fid[q,d]$$

and

$$Fad(q,m,p,a,d) = card(fad(q,m,p,a,d)),$$

and twelfth, for every $p \in N$ we define

$$fip[q,p] = fab(q) \cap pip[q,p],$$

and thirteenth, for every $m \in Z(q)$ and $p \in N$ we define

$$fap(q,m,p) = fab(q,m) \cap fip[q,p],$$

and fourteenth, for every $m \in Z(q)$ and $p \in N$ and $a \in pre[q,p]$ we define

$$fap(q,m,p,a) = fab(q,m,p,a) \cap fip[q,p],$$

and fifteenth, for every $m \in Z(q)$ and $p \in N$ and $d \in Z$ we define

$$fap[q,m,p,d] = fap(q,m,p) \cap fid[q,d]$$

and

$$Fap[q,m,p,d] = card(fap[q,m,p,d]),$$

and sixteenth, for every $m \in Z(q)$ and $p \in N$ and $a \in pre[q,p]$ and $d \in Z$ we define

$$fap(q,m,p,a,d) = fap(q,m,p,a) \cap fid[q,d]$$

and

$$Fap(q,m,p,a,d) = card(fap(q,m,p,a,d)),$$

and seventeenth, for every $p \in N$ and $v \in N(p)$ we define

$$fis[q,p,v] = fab(q) \cap pis[q,p,v],$$

and eighteenth, for every $m \in Z(q)$ and $p \in N$ and $v \in N(p)$ we define

$$fas[q,m,p,v] = fis[q,p,v] \cap fab(q,m)$$

and

$$Fas[q,m,p,v] = card(fas[q,m,p,v]),$$

and ninteenth, for every $m \in Z(q)$ and $p \in N$ and $a \in pre[q,p]$ and $v \in N(p)$ we define

$$fas(q,m,p,a,v) = fis[q,p,v] \cap fab(q,m,p,a)$$

and

$$Fas(q,m,p,a,v) = card(fas(q,m,p,a,v)).$$

For example, given any $q \in N^*$ and $m \in Z(q)$ and $p \in N$ and $a \in pre[q,p]$, first for every $V \in Z$ we have

$$stab(q,m,p,a,V) = \text{the set of all standard tableaux of width } q \text{ and area } V \text{ such that } a \le T \le m$$

and

$$\mathrm{Stab}(q,m,p,a,V) = \mathrm{card}(\mathrm{stab}(q,m,p,a,V)),$$

and second, for every $v \in N(p)$ we have

$\mathrm{stas}(q,m,p,a,v)$ = the set of all standard tableaux T of width q and shape (p,v) such that $a \le T \le m$

and

$$\mathrm{Stas}(q,m,p,a,v) = \mathrm{card}(\mathrm{stas}(q,m,p,a,v)).$$

By a *preunitableau* (resp., *unitableau*) we mean a pretableau (resp., tableau) of width 1. By a *prebitableau* (resp., *bitableau*) we mean a pretableau (resp., tableau) of width 2.

(2.4) Index and Protomonomials For every $q \in N^*$ and for every finite subset Y of $Z(q)$ we introduce the *index of* Y, which we denote ind(Y) and which we define by putting

ind(Y) = the largest nonnegative integer j for which there exist elements $y_1, y_2, \ldots, y_j$ in Y such that $y_i(k) < y_{i+1}(k)$ for all $i \in [1, j-1]$ and $k \in [1,q]$

and we note that then

$$\mathrm{ind}(Y) \in N$$

and we also note that

$$\mathrm{ind}(Y) = 0 \text{ iff } Y \text{ is empty.}$$

Given any $q \in N^*$ and $m \in Z(q)$, we consider the q-*dimensional positive integral cube bounded by* m, which we denote by cub(q,m) and which we define by putting

$$\mathrm{cub}(q,m) = \{y \in Z(q) : 1 \le y(k) \le m(k) \text{ for all } k \in [1,q]\}$$

and we note that

$$\mathrm{card}(\mathrm{cub}(q,m)) = \begin{cases} \prod_{k=1}^{q} m(k) & \text{if } m \in N(q) \\ 0 & \text{if } m \notin N(q). \end{cases}$$

Given any $q \in N^*$ and $m \in Z(q)$ and $p \in N$ and $a \in pre[q,p]$ and $k \in [1,q]$ and $i \in [1,p]$, first we define

$$tru[q,m,p,a,k,i]$$

to be the element of $Z(q)$ obtained by putting

$$tru[q,m,p,a,k,i](\hat{k}) = \begin{cases} m(\hat{k}) & \text{if } \hat{k} \in [1,q]\setminus\{k\} \\ a(k,i) - 1 & \text{if } \hat{k} = k \end{cases}$$

and second, we put

$$truc(q,m,p,a,k,i) = cub(q,tru[q,m,p,a,k,i]);$$

here

$$tru[q,m,p,a,k,i]$$

may be called the (a,k,i)th *truncation of* m, whereas

$$truc(q,m,p,a,k,i)$$

may be called the (a,k,i)th *truncation of* $cub(q,m)$ or it may also be called the truncated cube determined by (q,m,p,a,k,i).

Given any $q \in N^*$ and $m \in Z(q)$, by a *protomonomial on* $cub(q,m)$ we mean a map of $cub(q,m)$ into N. [Here the word "protomonomial" is meant to suggest the exponent system of a monomial; for example, if $(X_y)_{y \in cub(q,m)}$ is a family of elements X_y in a ring R and if t is a protomonomial on $cub(q,m)$, then

$$\prod_{y \in cub(q,m)} X_y^{t(y)}$$

is the corresponding monomial in the said family.]

Given any $q \in N^*$ and $m \in Z(q)$, first we define

$$mon(q,m) = \text{the set of all protomonomials on } cub(q,m),$$

and second, for every $t \in mon(q,m)$ we define

$$supp(t) = \{y \in cub(q,m) : t(y) \neq 0\}$$

and

$$abs(t) = \sum_{y \in cub(q,m)} t(y)$$

where supp and abs are meant to suggest *support* and *absolute value*, respectively, and third, for every $V \in Z$ we define

$$\mathrm{mon}[[q,m,V]] = \{t \in \mathrm{mon}(q,m) : \mathrm{abs}(t) = V\}$$

and

$$\mathrm{Mon}[[q,m,V]] = \mathrm{card}(\mathrm{mon}[[q,m,V]]),$$

and fourth, for every $p \in N$ we define

$$\mathrm{mon}(q,m,p) = \{t \in \mathrm{mon}(q,m) : \mathrm{ind}(\mathrm{supp}(t)) \le p\},$$

and fifth, for every $p \in N$ and $V \in Z$ we define

$$\mathrm{mon}[q,m,p,V] = \mathrm{mon}(q,m,p) \cap \mathrm{mon}[[q,m,V]]$$

and

$$\mathrm{Mon}[q,m,p,V] = \mathrm{card}(\mathrm{mon}[q,m,p,V]),$$

and sixth, for every $p \in N$ and $a \in \mathrm{pre}[q,p]$ we define

$$\begin{aligned}\mathrm{mon}(q,m,p,a) = \{t \in \mathrm{mon}(q,m,p) :\ & \mathrm{ind}(\mathrm{supp}(t) \cap \mathrm{truc}(q,m,p,a,k,i)) \le i - 1 \\ & \text{for all } k \in [1,q] \text{ and } i \in [1,p]\}\end{aligned}$$

and seventh, for every $p \in N$ and $a \in \mathrm{pre}[q,p]$ and $V \in Z$ we define

$$\mathrm{mon}(q,m,p,a,V) = \mathrm{mon}(q,m,p,a) \cap \mathrm{mon}[[q,m,V]]$$

and

$$\mathrm{Mon}(q,m,p,a,V) = \mathrm{card}(\mathrm{mon}(q,m,p,a,V)).$$

(2.5) Binomial Coefficients For any A in Z and for any V in any overring of Q, we define the *ordinary binomial coefficient* $\binom{V}{A}$ by putting

$$\binom{V}{A} = \begin{cases} \dfrac{V(V-1)\cdots(V-A+1)}{A!} & \text{if } A \ge 0 \\ 0 & \text{if } A < 0 \end{cases}$$

and we define the *twisted binomial coefficient* $\begin{bmatrix} V \\ A \end{bmatrix}$ by putting

$$\begin{bmatrix} V \\ A \end{bmatrix} = \begin{cases} \dfrac{(V+1)(V+2)\cdots(V+A)}{A!} & \text{if } A \ge 0 \\ 0 & \text{if } A < 0 \end{cases}$$

We note that by convention the product of an empty family is 1 and hence

$$0! = 1 \tag{2.5.1}$$

and

$$\text{for any } V \text{ in any overring of } Q \text{ we have } \binom{V}{0} = 1 \tag{2.5.1'}$$

and

$$\text{for any } V \text{ in any overring of } Q \text{ we have } \begin{bmatrix} V \\ 0 \end{bmatrix} = 1. \tag{2.5.1*}$$

For reference we also record that part of the definition says that

$$\left[\begin{array}{l}\text{for any } A \text{ in } Z \text{ with } A < 0, \text{ and for any } V \text{ in any} \\ \text{overring of } Q, \text{ we have } \binom{V}{A} = 0\end{array}\right. \tag{2.5.2'}$$

and

$$\left[\begin{array}{l}\text{for any } A \text{ in } Z \text{ with } A < 0, \text{ and for any } V \text{ in any} \\ \text{overring of } Q, \text{ we have } \begin{bmatrix} V \\ A \end{bmatrix} = 0.\end{array}\right. \tag{2.5.2*}$$

Given any A in Z and given any V in any overring of Q, by definition we see that

$$\text{if } A \geq 0,$$

then

$$\begin{aligned}
&[(V + A)][(V + A) - 1]\cdots[(V + A) - A + 1] \\
&= (V + A)(V + A - 1)\cdots(V + 1) \\
&= (V + 1)(V + 2)\cdots(V + A) \\
&= (-1)^A(-V - 1)(-V - 2)\cdots(-V - A) \\
&= (-1)^A[(-V - 1)][(-V - 1) - 1]\cdots[(-V - 1) - A + 1],
\end{aligned}$$

and upon dividing the first and the third and the fifth lines by A! we see that

$$\text{if } A \geq 0, \text{ then } \binom{V + A}{A} = \begin{bmatrix} V \\ A \end{bmatrix} = (-1)^A\binom{-V - 1}{A},$$

and by definition we also see that

$$\text{if } A < 0, \text{ then } \binom{V + A}{A} = 0 = \begin{bmatrix} V \\ A \end{bmatrix} = 0 = (-1)^A \binom{-V - 1}{A}$$

and thus in both the cases we have

$$\binom{V + A}{A} = \begin{bmatrix} V \\ A \end{bmatrix} = (-1)^A \binom{-V - 1}{A} \tag{1}$$

and upon substituting $V - A$ for V in the first two terms of (1) we get

$$\binom{V}{A} = \begin{bmatrix} V - A \\ A \end{bmatrix} \tag{2}$$

and upon substituting $-V - 1$ for V in the last two terms of (1) we get

$$\begin{bmatrix} -V - 1 \\ A \end{bmatrix} = (-1)^A \binom{V}{A}$$

and upon multiplying both sides of the equation above by $(-1)^A$ we get

$$(-1)^A \begin{bmatrix} -V - 1 \\ A \end{bmatrix} = \binom{V}{A}. \tag{3}$$

Now by (2) we see that

$$\left[\begin{array}{l} \text{for any } A \text{ in } Z \text{ and for any } V \text{ in any overring of } Q \\ \text{we have } \binom{V}{A} = \begin{bmatrix} V - A \\ A \end{bmatrix} \end{array}\right. \tag{2.5.3'}$$

and by (1) we see that

$$\left[\begin{array}{l} \text{for any } A \text{ in } Z \text{ and for any } V \text{ in any overring of } Q \\ \text{we have } \begin{bmatrix} V \\ A \end{bmatrix} = \binom{V + A}{A} \end{array}\right. \tag{2.5.3*}$$

and by (3) we see that

$$\left[\begin{array}{l} \text{for any } A \text{ in } Z \text{ and for any } V \text{ in any overring of } Q \\ \text{we have } \binom{V}{A} = (-1)^A \begin{bmatrix} -V - 1 \\ A \end{bmatrix} \end{array}\right. \tag{2.5.4'}$$

and by (1) we see that

$$\left[\begin{array}{l} \text{for any } A \text{ in } Z \text{ and for any } V \text{ in any overring of } Q \\ \text{we have } \begin{bmatrix} V \\ A \end{bmatrix} = (-1)^A \binom{-V - 1}{A}. \end{array}\right. \tag{2.5.4*}$$

Given any $A \in Z$ and $V \in Z$, we see that

if $V \geq A \geq 0$,

then as in the very classical proof

$$\begin{pmatrix} V \\ A \end{pmatrix} = \frac{V(V-1)\cdots(V-A+1)}{A!} \qquad \text{(by definition)}$$

$$= \frac{V(V-1)\cdots(V-A+1)}{A!} \text{ times } \frac{(V-A)(V-A-1)\cdots(1)}{(V-A)(V-A-1)\cdots(1)}$$

$$= \frac{V!}{A!(V-A)!};$$

thus

$$\left[\begin{array}{l}\text{if } A \in Z \text{ and } V \in Z \text{ are such that } V \geq A \geq 0, \\ \text{then } \begin{pmatrix} V \\ A \end{pmatrix} = \dfrac{V!}{A!(V-A)!}.\end{array}\right. \qquad (2.5.5')$$

Upon substituting $V + A$ for V in (2.5.5') and then using (2.5.3*) we see that

$$\left[\begin{array}{l}\text{if } A \in Z \text{ and } V \in Z \text{ are such that } V \geq 0 \text{ and } A \geq 0, \\ \text{then } \begin{bmatrix} V \\ A \end{bmatrix} = \dfrac{(V+A)!}{V!\ A!}.\end{array}\right. \qquad (2.5.5^*)$$

Now

if $A \in Z$ and $V \in Z$ are such that $A > V \geq 0$,

then

$$\begin{pmatrix} V \\ A \end{pmatrix} = \frac{V(V-1)\cdots(1)(0)(-1)\cdots(V-A+1)}{A!}$$

$$= 0$$

where the last equation follows by noting that

$$\left[\begin{array}{l}\text{(0) does occur in the numerator of RHS as indicated} \\ \text{because } V \geq 0 \text{ and } V - A + 1 \leq 0;\end{array}\right.$$

thus

$$\left[\begin{array}{l}\text{if } A \in Z \text{ and } V \in Z \text{ are such that } A > V \geq 0, \\ \text{then } \begin{pmatrix} V \\ A \end{pmatrix} = 0.\end{array}\right. \qquad (2.5.6')$$

Upon substituting V + A for V in (2.5.6') and then using (2.5.3*) we see that

$$\left[\begin{array}{l}\text{if } A \in Z \text{ and } V \in Z \text{ are such that } A > V + A \geq 0, \\ \text{then } \begin{bmatrix} V \\ A \end{bmatrix} = 0.\end{array}\right. \tag{2.5.6*}$$

In view of (2.5.5') we see that

$$\left[\begin{array}{l}\text{if } A \in Z \text{ and } V \in Z \text{ are such that } V \geq A \geq 0, \text{ then} \\ V \geq V - A \geq 0 \text{ and } \begin{pmatrix} V \\ A \end{pmatrix} = \dfrac{V!}{A!\ (V - A)!} = \begin{pmatrix} V \\ V - A \end{pmatrix}.\end{array}\right. \tag{2.5.7'}$$

By (2.5.5*) we see that

$$\left[\begin{array}{l}\text{if } A \in Z \text{ and } V \in Z \text{ are such that } V \geq 0 \text{ and } A \geq 0, \\ \text{then } \begin{bmatrix} V \\ A \end{bmatrix} = \dfrac{(V + A)!}{V!\ A!} = \begin{bmatrix} A \\ V \end{bmatrix}.\end{array}\right. \tag{2.5.7*}$$

Let us note that

$$\left[\begin{array}{l}\text{if } A \in Z \text{ and } V \in Z \text{ are such that } V \geq 0 \text{ and } A \notin [0,V], \\ \text{then by (2.5.2') and (2.5.6') we get } \begin{pmatrix} V \\ A \end{pmatrix} = 0\end{array}\right. \tag{1'}$$

and

$$\left[\begin{array}{l}\text{if } A \in Z \text{ and } V \in Z \text{ are such that } V \geq 0 \text{ and } A \notin [0,V], \\ \text{then clearly, } V - A \notin [0,V]\end{array}\right. \tag{2'}$$

and

$$\left[\begin{array}{l}\text{if } A \in Z \text{ and } V \in Z \text{ are such that } V + A \geq 0 \text{ and } A \notin \\ [0,\ V + A], \text{ then by (2.5.2*) and (2.5.6*) we get } \begin{bmatrix} V \\ A \end{bmatrix} = 0\end{array}\right. \tag{1*}$$

and

$$\left[\begin{array}{l}\text{if } A \in Z \text{ and } V \in Z \text{ are such that } V + A \geq 0 \text{ and} \\ A \notin [0,\ V + A], \text{ then clearly, } V \notin [0,\ V + A].\end{array}\right. \tag{2*}$$

Now by (1') and (2') we see that

$$\left[\begin{array}{l}\text{if } A \in Z \text{ and } V \in Z \text{ are such that } V \geq 0 \text{ and } A \notin [0,V], \\ \text{then } V - A \notin [0,V] \text{ and } \begin{pmatrix} V \\ A \end{pmatrix} = 0 = \begin{pmatrix} V \\ V - A \end{pmatrix}\end{array}\right. \tag{2.5.8'}$$

and by (1*) and (2*) we see that

$$\left[\begin{array}{l}\text{if } A \in Z \text{ and } V \in Z \text{ are such that } V + A \geq 0 \text{ and} \\ A \notin [0, V + A], \text{ then } V \notin [0, V + A] \text{ and } \begin{bmatrix} V \\ A \end{bmatrix} = 0 = \begin{bmatrix} A \\ V \end{bmatrix}.\end{array}\right. \tag{2.5.8*}$$

Given any $A \in Z$ and $V \in Z$, by definition we see that if $V \geq A \geq 0$, then

$$\begin{aligned}\binom{V}{A} &= \frac{V(V-1)\cdots(V-A+1)}{A!} \\ &\neq 0 \qquad \text{(because each term in the numerator is positive)}\end{aligned}$$

whereas if $A \geq 0 > V$, then

$$\begin{aligned}\binom{V}{A} &= \frac{V(V-1)\cdots(V-A+1)}{A!} \\ &\neq 0 \qquad \text{(because each term in the numerator is negative)}\end{aligned}$$

and hence in view of (2.5.2') and (2.5.6') we conclude that

$$\left[\begin{array}{l}\text{given any } A \in Z \text{ and } V \in Z \text{ we have that:} \\ \binom{V}{A} \neq 0 \text{ iff either } V \geq A \geq 0 \text{ or } A \geq 0 > V.\end{array}\right. \tag{2.5.9'}$$

Upon substituting $V + A$ for V in (2.5.9') and then using (2.5.3*) we see that

$$\left[\begin{array}{l}\text{given any } A \in Z \text{ and } V \in Z \text{ we have that:} \\ \begin{bmatrix} V \\ A \end{bmatrix} \neq 0 \text{ iff either } A \geq 0 \leq V \text{ or } A \geq 0 > V + A.\end{array}\right. \tag{2.5.9*}$$

Now by (2.5.7') and (2.5.8') we see that

$$\left[\begin{array}{l}\text{if } A \in Z \text{ and } V \in Z \text{ are such that } V \geq 0, \\ \text{then } \binom{V}{A} = \binom{V}{V - A}\end{array}\right. \tag{2.5.10'}$$

and by (2.5.7*) and (2.5.8*) we see that

$$\left[\begin{array}{l}\text{if } A \in Z \text{ and } V \in Z \text{ are such that } V + A \geq 0, \\ \text{then } \begin{bmatrix} V \\ A \end{bmatrix} = \begin{bmatrix} A \\ V \end{bmatrix}\end{array}\right. \tag{2.5.10*}$$

and by (2.5.2') we see that

$$\left[\begin{array}{l}\text{if } A \in Z \text{ and } V \in Z \text{ are such that } A < 0 \text{ and} \\ V - A < 0, \text{ then } \binom{V}{A} = 0 = \binom{V}{V - A}\end{array}\right. \tag{2.5.11'}$$

and by (2.5.2*) we see that

$$\left[\begin{array}{l}\text{if } A \in Z \text{ and } V \in Z \text{ are such that } A < 0 \text{ and } V < 0, \\ \text{then } \begin{bmatrix}V\\A\end{bmatrix} = 0 = \begin{bmatrix}A\\V\end{bmatrix}.\end{array}\right. \tag{2.5.11*}$$

Finally,

$$\left[\begin{array}{l}\text{to round off the assertions of (2.5.10') and} \\ \text{(2.5.11') we note that given any } A \in Z \text{ and } V \in Z, \\ \text{by (2.5.2') and (2.5.9') we have that if } V < 0 \\ \text{and } A \geq 0, \text{ then } \binom{V}{A} \neq 0 = \binom{V}{V - A}, \text{ whereas if} \\ V < 0 \text{ and } V - A \geq 0, \text{ then } \binom{V}{A} = 0 \neq \binom{V}{V - A}\end{array}\right. \tag{2.5.12'}$$

and

$$\left[\begin{array}{l}\text{to round off the assertions of (2.5.10*) and} \\ \text{(2.5.11*) we note that given any } A \in Z \text{ and } V \in Z, \\ \text{by (2.5.2*) and (2.5.9*) we have that if} \\ V + A < 0 \text{ and } A \geq 0, \text{ then } \begin{bmatrix}V\\A\end{bmatrix} \neq 0 = \begin{bmatrix}A\\V\end{bmatrix}, \text{ whereas} \\ \text{if } V + A < 0 \text{ and } V \geq 0, \text{ then } \begin{bmatrix}V\\A\end{bmatrix} = 0 \neq \begin{bmatrix}A\\V\end{bmatrix}.\end{array}\right. \tag{2.5.12*}$$

REMARK (2.5.13) The classical characterizations of the binomial coefficients $\binom{V}{A}$ and $\begin{bmatrix}V\\A\end{bmatrix}$ as given in (2.5.5') and (2.5.5*) use V!, and hence they are valid only for V in N. The characterizations of $\binom{V}{A}$ and $\begin{bmatrix}V\\A\end{bmatrix}$ that we have employed to define these entities are preferable because they do not use V! and so are valid for any V in any overring of Q. *In fact, we can even regard* $\binom{V}{A}$ *and* $\begin{bmatrix}V\\A\end{bmatrix}$ *as polynomials in an indeterminate* V *with coefficients in* Q. When an expression or identity involving binomial coefficients like

$$\binom{V + B}{A} \text{ and/or } \begin{bmatrix} V + B \\ A \end{bmatrix} \qquad \text{with } A \in Z \text{ and } B \in Q$$

is to be regarded as an expression or identity in the polynomial ring in an indeterminate V with coefficients in Q, we shall indicate this by a phrase such as "in Q[V]." For example, referring to (2.5.3*) we may say that

$$\text{in } Q[V] \text{ we have } \begin{bmatrix} V \\ A \end{bmatrix} = \binom{V + A}{A}$$

to mean that the equation above is an identity of polynomials in an indeterminate V with coefficients in Q. Or, as another example, in the definition at the beginning of this subsection (2.5), we could have said that "For any $A \in Z$ we define $\binom{V}{A} \in Q[V]$ by putting ...."

(2.6) Elementary Properties of Binomial Coefficients In this subsection we record some consequences of (2.5.2'), (2.5.2*), (2.5.6'), (2.5.10'), and (2.5.10*); in Subsections (2.7) and (2.10) we deal with certain apparently infinite summations; the consequences are used to show that all except a finite number of summands in these summations are zero. *The reader may wish to skip this entire subsection and return to it if and when necessary for reference.*

First, by (2.5.2') we see that

(1) $\left[\begin{array}{l}\text{if D, E, and R are any integers such that } \binom{E}{D + E - R} \neq 0, \\ \text{then } D + E - R \geq 0.\end{array}\right.$

Now clearly,

$$\left[\begin{array}{l}\text{if D, E, and R are any integers such that } E \leq R \text{ and} \\ D + E - R \geq 0, \text{ then } D \geq 0\end{array}\right.$$

and hence by (1) we see that

$$\left[\begin{array}{l}\text{if D, E, and R are any integers such that } E \leq R \\ \text{and } \binom{E}{D + E - R} \neq 0, \text{ then } D \geq 0.\end{array}\right. \qquad (2.6.11)$$

Second, by (2.5.6') we see that

$$\left[\begin{array}{l}\text{if D, E, and R are any integers such that } E \geq 0 \\ \text{and } \binom{E}{D + E - R} \neq 0, \text{ then } D + E - R \leq E\end{array}\right.$$

and clearly,

$$\left[\begin{array}{l}\text{if D, E, and R are any integers such that } D + E - R \leq E, \\ \text{then } D \leq R\end{array}\right.$$

and hence

$$\left[\begin{array}{l}\text{if D, E, and R are any integers such that } E \geq 0 \\ \text{and } \binom{E}{D + E - R} \neq 0, \text{ then } D \leq R.\end{array}\right. \tag{2.6.12}$$

Third, by (2.6.11) and (2.6.12) we conclude that

$$\left[\begin{array}{l}\text{given any integer R we have} \\ \left\{D \in Z : \left\{E \in [0,R] : \binom{E}{D + E - R} \neq 0\right\} \neq \emptyset\right\} \subset [0,R].\end{array}\right. \tag{2.6.13}$$

Fourth, by (2.5.2') we see that

(2) $$\left[\begin{array}{l}\text{if D and E are any integers such that } \binom{E}{D} \neq 0, \\ \text{then } D \geq 0.\end{array}\right.$$

By (2.5.6') we see that

(3) $$\left[\begin{array}{l}\text{if D, E, and R are any integers such that } 0 \leq E \leq R \\ \text{and } \binom{E}{D} \neq 0, \text{ then } D \leq R.\end{array}\right.$$

By (2) and (3) we conclude that

$$\left[\begin{array}{l}\text{given any integer R we have} \\ \left\{D \in Z : \left\{E \in [0,R] : \binom{E}{D} \neq 0\right\} \neq \emptyset\right\} \subset [0,R].\end{array}\right. \tag{2.6.14}$$

Fifth, by (2.5.6') we see that

(4) $$\left[\begin{array}{l}\text{if A, A', B', and D are any integers such that} \\ A + A' - B' \geq 0 \text{ and } D > A + A' - B', \text{ then } \binom{A + A' - B'}{D} = 0.\end{array}\right.$$

Now

(5) if A, B, A', B', and D are any integers such that $A + A' - B' \geq 0$ and $D > A + A' - B'$, then
$$\begin{bmatrix} B - D \\ A \end{bmatrix}\binom{A + A' - B'}{D} = \begin{bmatrix} A \\ B - D \end{bmatrix}\binom{A + A' - B'}{D}$$

because by (4) we see that each side is equal to zero. By (2.5.10*) we see that

(6) if A, B, and D are integers such that $(B - D) + A \geq 0$, then $\begin{bmatrix} B - D \\ A \end{bmatrix} = \begin{bmatrix} A \\ B - D \end{bmatrix}$.

Now

(7) if A, B, A', B', and D are any integers such that $0 \leq B + B' - A'$ and $D \leq A + A' - B'$, then by adding the two inequalities we would get $D \leq B + A$ and then by subtracting D from both sides we would get $0 \leq (B - D) + A$ and hence we would get $(B - D) + A \geq 0$.

By (6) and (7) we see that

(8) if A, B, A', B', and D are any integers such that $0 \leq B + B' - A'$ and $D \leq A + A' - B'$, then
$$\begin{bmatrix} B - D \\ A \end{bmatrix} = \begin{bmatrix} A \\ B - D \end{bmatrix}.$$

Upon multiplying both sides of the equation in (8) by $\binom{A + A' - B'}{D}$ we see that

(9) if A, B, A', B', and D are any integers such that $0 \leq B + B' - A'$ and $D \leq A + A' - B'$, then
$$\begin{bmatrix} B - D \\ A \end{bmatrix}\binom{A + A' - B'}{D} = \begin{bmatrix} A \\ B - D \end{bmatrix}\binom{A + A' - B'}{D}.$$

By (5) and (9) we see that

(10) if A, B, A', B', and D are any integers such that $A + A' - B' \geq 0 \leq B + B' - A'$, then
$$\begin{bmatrix} B - D \\ A \end{bmatrix}\binom{A + A' - B'}{D} = \begin{bmatrix} A \\ B - D \end{bmatrix}\binom{A + A' - B'}{D}.$$

By (2.5.10') we see that

(11) $$\left[\begin{array}{l}\text{if } A, A;, B', \text{ and } D \text{ are any integers such that} \\ A + A' - B' \ge 0, \text{ then} \\ \binom{A + A' - B'}{D} = \binom{A + A' - B'}{A + A' - B' - D}.\end{array}\right.$$

Upon multiplying both sides of the equation in (11) by $\begin{bmatrix} A \\ B - D \end{bmatrix}$ we see that

(12) $$\left[\begin{array}{l}\text{if } A, B, A', B', \text{ and } D \text{ are any integers such that} \\ A + A' - B' \ge 0, \text{ then} \\ \begin{bmatrix} A \\ B - D \end{bmatrix}\binom{A + A' - B'}{D} = \begin{bmatrix} A \\ B - D \end{bmatrix}\binom{A + A' - B'}{A + A' - B' - D}.\end{array}\right.$$

Upon multiplying both sides of the equation in (11) by $\begin{bmatrix} B - D \\ A \end{bmatrix}$ we see that

(13) $$\left[\begin{array}{l}\text{if } A, B, A', B', \text{ and } D \text{ are any integers such that} \\ A + A' - B' \ge 0, \text{ then} \\ \begin{bmatrix} B - D \\ A \end{bmatrix}\binom{A + A' - B'}{D} = \begin{bmatrix} B - D \\ A \end{bmatrix}\binom{A + A' - B'}{A + A' - B' - D}.\end{array}\right.$$

By (10), (12), and (13) we conclude that

$$\left[\begin{array}{l}\text{if } A, B, A', B', \text{ and } D \text{ are any integers such that} \\ A + A' - B' \ge 0 \le B + B' - A', \text{ then} \\ \begin{bmatrix} B - D \\ A \end{bmatrix}\binom{A + A' - B'}{D} = \begin{bmatrix} A \\ B - D \end{bmatrix}\binom{A + A' - B'}{D} \\ \qquad = \begin{bmatrix} A \\ B - D \end{bmatrix}\binom{A + A' - B'}{A + A' - B' - D} \\ \qquad = \begin{bmatrix} B - D \\ A \end{bmatrix}\binom{A + A' - B'}{A + A' - B' - D}\end{array}\right. \tag{2.6.15}$$

Sixth, we claim that

$$\left[\begin{array}{l}\text{given any } p \in N^* \text{ and } r \in Z(p), \text{ and given any} \\ g'[i,j] \in Z \text{ and } h'[i,j] \in Z \text{ for all } i \in [1,p] \text{ and} \\ j \in [1,p], \text{ we have} \\ \quad \{e \in Z(p) : \det K^{(2)}(e) \ne 0\} \subset N[p, r(1) + r(2) + \cdots + r(p)] \\ \text{where for every } e \in Z(p) \text{ we have let } K^{(2)}(e) \text{ be the} \\ p \text{ by } p \text{ matrix whose } (i,j)\text{th entry is}\end{array}\right. \tag{2.6.16}$$

$$\left|\binom{g'[i,j]}{r(i)-e(i)}\binom{h'[i,j]}{e(i)}\right|.$$

To prove (2.6.16), we note that

(14) $\left[\begin{array}{l}\text{if } e \in Z(p) \text{ and } i \in [1,p] \text{ are such that } e(i) < 0, \text{ then} \\ \text{for every } j \in [1,p], \text{ by (2.5.2') we have } \binom{h'[i,j]}{e(i)} = 0 \\ \text{and hence we have } \binom{g'[i,j]}{r(i)-e(i)}\binom{h'[i,j]}{e(i)} = 0\end{array}\right.$

and we note that

(15) $\left[\begin{array}{l}\text{if } e \in Z(p) \text{ and } i \in [1,p] \text{ are such that } r(i)-e(i) < 0, \\ \text{then for every } j \in [1,p], \text{ by (2.5.2') we have } \binom{g'[i,j]}{r(i)-e(i)} = 0 \\ \text{and hence we have } \binom{g'[i,j]}{r(i)-e(i)}\binom{h'[i,j]}{e(i)} = 0.\end{array}\right.$

By (14) and (15) we see that

(16) $\left[\begin{array}{l}\text{if } e \in Z(p) \text{ and } i \in [1,p] \text{ are such that } e(i) \notin [0,r(i)], \\ \text{then for every } j \in [1,p] \text{ we have } \binom{g'[i,j]}{r(i)-e(i)}\binom{h'[i,j]}{e(i)} = 0.\end{array}\right.$

Now (16) is equivalent to saying that

$$\left[\begin{array}{l}\text{if } e \in Z(p) \text{ and } i \in [1,p] \text{ are such that } e(i) \notin [0,r(i)], \\ \text{then the ith row of } K^{(2)}(e) \text{ consists entirely of zeros}\end{array}\right.$$

and hence we see that

$$\left[\begin{array}{l}\text{if } e \in Z(p) \text{ is such that for some i we have} \\ e(i) \notin [0,r(i)], \text{ then } \det K^{(2)}(e) = 0\end{array}\right.$$

and thus we conclude that

$$\{e \in Z(p) : \det K^{(2)}(e) \neq 0\} \subset \{e \in Z(p) : e(i) \in [0,r(i)] \text{ for all } i \in [1,p]\}$$

and clearly, we have

$$\begin{aligned}&\{e \in Z(p) : e(i) \in [0,r(i)] \text{ for all } i \in [1,p]\} \\ &= \{e \in N(p) : e(i) \le r(i) \text{ for all } i \in [1,p] \\ &\subset \{e \in N(p) : e(1) + e(2) + \cdots + e(p) \le r(1) + r(2) + \cdots + r(p) \\ &= N[p,r(1) + r(2) + \cdots + r(p)]\end{aligned}$$

and therefore we see that

$$\{e \in Z(p) : \det K^{(2)}(e) \neq 0\} \subset N[p, r(1) + r(2) + \cdots + r(p)].$$

Seventh, we claim that

$$\begin{bmatrix}\text{given any } p \in N^* \text{ and } r \in Z(p) \text{ and given any} \\ g'[i,j] \in Z \text{ and } h'[i,j] \in Z \text{ for all } i \in [1,p] \text{ and} \\ j \in [1,p], \text{ we have } \{e \in Z(p) : \det K^{(2^*)}(e) \neq 0\} \subset \\ N[p, r(1) + r(2) + \cdots + r(p)], \text{ where for every} \\ e \in Z(p) \text{ we have } K^{(2^*)}(e) \text{ to be the p by p matrix} \\ \text{whose (i,j)th entry is } \binom{g'[i,j]}{r(i) - e(j)}\binom{h'[i,j]}{e(j)}.\end{matrix} \tag{2.6.17}$$

To prove (2.6.17), by (2.5.2') we see that

$$\begin{bmatrix}\text{if } e \in Z(p) \text{ and } j \in [1,p] \text{ are such that } e(j) < 0, \\ \text{then for every } i \in [1,p] \text{ we have } \binom{h'[i,j]}{e(j)} = 0\end{matrix}$$

and therefore we see that

$$\begin{bmatrix}\text{if } e \in Z(p) \text{ and } j \in [1,p] \text{ are such that } e(j) < 0, \text{ then} \\ \text{for every } i \in [1,p] \text{ we have } \binom{g'[i,j]}{r(i) - e(j)}\binom{h'[i,j]}{e(j)} = 0\end{matrix}$$

and this is equivalent to saying that

$$\begin{bmatrix}\text{if } e \in Z(p) \text{ and } j \in [1,p] \text{ are such that } e(j) < 0, \text{ then} \\ \text{the jth column of } K^{(2^*)}(e) \text{ consists entirely of zeros}\end{matrix}$$

and hence we see that

$$\begin{bmatrix}\text{if } e \in Z(p) \text{ is such that for some } j \in [1,p] \text{ we have} \\ e(j) < 0, \text{ then } \det K^{(2^*)}(e) = 0\end{matrix}$$

and thus we conclude that

$$(17) \quad \{e \in Z(p) : K^{(2^*)}(e) \neq 0\} \subset N(p)$$

and upon expanding $\det K^{(2^*)}(e)$ by columns we see that for every $e \in Z(p)$ we have

$$\det K^{(2*)}(e) = \sum_{w\in W(p)} \operatorname{sgn}(w) \prod_{j=1}^{p} \binom{g'[w(j),j]}{r(w(j)) - e(j)} \binom{h'[w(j),j]}{e(j)}$$

$$= \sum_{w\in W(p)} \left[\operatorname{sgn}(w) \prod_{j=1}^{p} \binom{h'[w(j),j]}{e(j)}\right] \prod_{j=1}^{p} \binom{g'[w(j),j]}{r(w(j)) - e(j)}$$

and hence

$$(18)\quad \begin{cases} \{e \in Z(p) : \det K^{(2*)}(e) \neq 0\} \\ \subset \bigcup_{w\in W(p)} \left\{e \in Z(p) : \prod_{j=1}^{p} \binom{g'[w(j),j]}{r(w(j)) - e(j)} \neq 0\right\} \end{cases}$$

and now, given any $w \in W(p)$, by (2.5.2') we have

$$\left\{e \in Z(p) : \prod_{j=1}^{p} \binom{g'[w(j),j]}{r(w(j)) - e(j)} \neq 0\right\}$$
$$\subset \{e \in Z(p) : r(w(j)) - e(j) \geq 0 \text{ for all } j \in [1,p]\}$$

and clearly, we have

$$\begin{aligned} &\{e \in Z(p) : r(w(j)) - e(j) \geq 0 \text{ for all } j \in [1,p]\} \\ &= \{e \in Z(p) : e(j) \leq r(w(j)) \text{ for all } j \in [1,p]\} \\ &\subset \{e \in Z(p) : e(1) + e(2) + \cdots + e(p) \\ &\qquad \leq r(w(1)) + r(w(2)) + \cdots + r(w(p))\} \\ &= Z[p, r(w(1)) + r(w(2)) + \cdots + r(w(p))] \end{aligned}$$

and also clearly, we have

$$r(w(1)) + r(w(2)) + \cdots + r(w(p)) = r(1) + r(2) + \cdots + r(p)$$

and therefore we see that

$$\left\{e \in Z(p) : \prod_{j=1}^{p} \binom{g'[w(j),j]}{r(w(j)) - e(j)} \neq 0\right\}$$
$$\subset Z[p, r(1) + r(2) + \cdots + r(p)]$$

and this being so for every $w \in W(p)$, by (18) we get

$$\{e \in Z(p) : \det K^{(2*)}(e) \neq 0\} \subset Z[p, r(1) + \cdots + r(p)]$$

and hence by (17) we get

$$\{e \in Z(p) : \det K^{(2*)}(e) \neq 0\} \subset N[p, r(1) + r(2) + \cdots + r(p)].$$

Eighth, we claim that

given any $p \in N^*$ and $t \in Z(p)$, and given any $u(i,j,e^*) \in Q$ for all $i \in [1,p]$ and $j \in [1,p]$ and $e^* \in Z$, we have

$$\{e \in Z(p) : \det K^{(1')}(e) \neq 0\} \subset N(p) \tag{2.6.18}$$

where for every $e \in Z(p)$ we have let $K^{(1')}(e)$ be the p by p matrix whose (i,j)th entry is $\begin{bmatrix} t(i) \\ e(j) \end{bmatrix} u(i,j,e(j))$.

To prove (2.6.18), by (2.5.2*) we see that

if $e \in Z(p)$ and $j \in [1,p]$ are such that $e(j) < 0$, then for every $i \in [1,p]$ we have $\begin{bmatrix} t(i) \\ e(j) \end{bmatrix} = 0$

and therefore we see that

if $e \in Z(p)$ and $j \in [1,p]$ are such that $e(j) < 0$, then for every $i \in [1,p]$ we have $\begin{bmatrix} t(i) \\ e(j) \end{bmatrix} u(i,j,e(j)) = 0$

and this is equivalent to saying that

if $e \in Z(p)$ and $j \in [1,p]$ are such that $e(j) < 0$, then the ith column of $K^{(1')}(e)$ consists entirely of zeros

and hence we see that

if $e \in Z(p)$ is such that for some $j \in [1,p]$ we have $e(j) < 0$, then $\det K^{(1')}(e) = 0$

and thus we conclude that

$$\{e \in Z(p) : \det K^{(1')}(e) \neq 0\} \subset N(p).$$

Ninth, we claim that

given any $p \in N^*$ and $r \in Z(p)$ and $s \in Z(p)$, and given any $g[i,j] \in Z$ for all $i \in [1,p]$ and $j \in [1,p]$, we have

$$\left[\begin{array}{l} \{e \in Z(p) : \det K^{(1*)}(e) \neq 0\} \\ \subset N[p, r(1) + r(2) + \cdots + r(p)] \\ \text{where for every } e \in Z(p) \text{ we have let } K^{(1*)}(e) \\ \text{be the } p \text{ by } p \text{ matrix whose } (i,j)\text{th entry is} \\ \begin{bmatrix} s(i) \\ e(j) \end{bmatrix} \begin{pmatrix} g[i,j] \\ r(j) - e(j) \end{pmatrix}. \end{array}\right. \qquad (2.6.19)$$

To prove (2.6.19), by (2.5.2') we see that

$$\left[\begin{array}{l} \text{if } e \in Z(p) \text{ and } j \in [1,p] \text{ are such that } r(j) - e(j) < 0, \\ \text{then for every } i \in [1,p] \text{ we have } \begin{pmatrix} g[i,j] \\ r(j) - e(j) \end{pmatrix} = 0 \end{array}\right.$$

and therefore we see that

$$\left[\begin{array}{l} \text{if } e \in Z(p) \text{ and } j \in [1,p] \text{ are such that } r(j) - e(j) < 0, \\ \text{then for every } i \in [1,p] \text{ we have } \begin{bmatrix} s(i) \\ e(j) \end{bmatrix} \begin{pmatrix} g[i,j] \\ r(j) - e(j) \end{pmatrix} = 0 \end{array}\right.$$

and this is equivalent to saying that

$$\left[\begin{array}{l} \text{if } e \in Z(p) \text{ and } j \in [1,p] \text{ are such that } r(j) - e(j) < 0, \\ \text{then the } j\text{th column of } K^{(1*)}(e) \text{ consists entirely of zeros} \end{array}\right.$$

and hence we see that

$$\left[\begin{array}{l} \text{if } e \in Z(p) \text{ is such that for some } j \in [1,p] \text{ we have} \\ r(j) - e(j) < 0, \text{ then } \det K^{(1*)}(e) = 0 \end{array}\right.$$

and thus we conclude that

$$\{e \in Z(p) : \det K^{(1*)}(e) \neq 0\} \subset \{e \in Z(p) : r(j) - e(j) \geq 0 \text{ for all } j \in [1,p]\}$$

and clearly, we have

$$\begin{aligned} &\{e \in Z(p) : r(j) - e(j) \geq 0 \text{ for all } j \in [1,p]\} \\ &= \{e \in Z(p) : e(j) \leq r(j) \text{ for all } j \in [1,p]\} \\ &\subset \{e \in Z(p) : e(1) + e(2) + \cdots + e(p) \\ &\qquad\qquad \leq r(1) + r(2) + \cdots + r(p)\} \\ &= Z[p, r(1) + r(2) + \cdots + r(p)] \end{aligned}$$

and therefore we see that

$$\{e \in Z(p) : \det K^{(1*)}(e) \neq 0\} \subset Z[p, r(1) + r(2) + \cdots + r(p)]$$

and upon taking t = s and $u(i,j,e^*) = \binom{g[i,j]}{r(j) - e^*}$ in (2.6.18) we get

$$\{e \in Z(p) : \det K^{(1*)}(e) \neq 0\} \subset N(p)$$

and now by the two inclusions above we deduce that

$$\{e \in Z(p) : \det K^{(1*)}(e) \neq 0\} \subset N[p, r(1) + r(2) + \cdots + r(p)].$$

Tenth, we claim that

$$\left[\begin{array}{l} \text{given any } p \in N^* \text{ and } r \in Z(p) \text{ and given any} \\ g[i,j] \in Z \text{ for all } i \in [1,p] \text{ and } j \in [1,p]\text{, we have} \\ \quad \{e \in Z(p) : \det K^{(1)}(e) \neq 0\} \subset Z[p, r(1) + r(2) \\ \qquad\qquad\qquad\qquad + \cdots + r(p)] \\ \text{where for every } e \in Z(p) \text{ we have let } K^{(1)}(e) \text{ be the} \\ p \text{ by } p \text{ matrix whose } (i,j)\text{th entry is } \binom{g[i,j]}{r(j) - e(i)}. \end{array}\right. \qquad (2.6.20)$$

To prove (2.6.20), upon expanding det $K^{(1)}(e)$ by rows we see that for every $e \in Z(p)$ we have

$$\det K^{(1)}(e) = \sum_{w \in W(p)} \operatorname{sgn}(w) \prod_{i=1}^{p} \binom{g[i,w(i)]}{r(w(i)) - e(i)}$$

and hence

$$(19) \quad \left[\begin{array}{l} \{e \in Z(p) : \det K^{(1)}(e) \neq 0\} \\ \subset \bigcup_{w \in W(p)} \left\{e \in Z(p) : \prod_{i=1}^{p} \binom{g[i,w(i)]}{r(w(i)) - e(i)} \neq 0\right\} \end{array}\right.$$

and now, given any $w \in W(p)$, by (2.5.2') we have

$$\left\{e \in Z(p) : \prod_{i=1}^{p} \binom{g[i,w(i)]}{r(w(i)) - e(i)} \neq 0\right\}$$
$$\subset \{e \in Z(p) : r(w(i)) - e(i) \geq 0 \quad \text{for all } i \in [1,p]\}$$

and clearly, we have

$$\begin{aligned}&\{e \in Z(p) : r(w(i)) - e(i) \geq 0 \text{ for all } i \in [1,p]\}\\ &= \{e \in Z(p) : e(i) \leq r(w(i)) \text{ for all } i \in [1,p]\}\\ &\subset \{e \in Z(p) : e(1) + e(2) + \cdots + e(p)\\ &\qquad\qquad \leq r(w(1)) + r(w(2)) + \cdots + r(w(p))\}\\ &= Z[p, r(w(1)) + r(w(2)) + \cdots + r(w(p))]\end{aligned}$$

and also clearly, we have

$$r(w(1)) + r(w(2)) + \cdots + r(w(p)) = r(1) + r(2) + \cdots + r(p)$$

and therefore we see that

$$\begin{aligned}&\left\{e \in Z(p) : \prod_{i=1}^{p} \binom{g[i,w(i)]}{r(w(i)) - e(i)} \neq 0\right\}\\ &\subset Z[p, r(1) + r(2) + \cdots + r(p)]\end{aligned}$$

and, this being so for every $w \in W(p)$, by (19) we get

$$\{e \in Z(p) : \det K^{(1)}(e) \neq 0\} \subset Z[p, r(1) + r(2) + \cdots + r(p)].$$

To record some consequences (2.6.15) to (2.6.20) in a form which is convenient for reference, for a moment

(20) $\left[\begin{array}{l}\text{let there be given any } p \in N^*\text{, and let there be given}\\ \text{any } r \in Z(p) \text{ and } s \in Z(p) \text{ and } r' \in Z(p) \text{ and } s' \in Z(p)\text{,}\\ \text{and let us put } R = r(1) + r(2) + \cdots + r(p) \text{ and } R' =\\ r'(1) + r'(2) + \cdots + r'(p) \text{ and } S' = s'(1) + s'(2) +\\ \cdots + s'(p)\end{array}\right.$

and

(21) $\left[\begin{array}{l}\text{in the situation of (20), for every } e \in Z(p) \text{ let}\\ \hat{H}^{(1)}(e) = \left(\prod_{i=1}^{p} \begin{bmatrix} s(i) \\ e(i) \end{bmatrix}\right) \det \hat{G}^{(1)}(e) \text{ and } \hat{H}^{(1^*)}(e) = \det \hat{G}^{(1^*)}(e)\\ \text{where } \hat{G}^{(1)}(e) \text{ and } \hat{G}^{(1^*)}(e) \text{ are the p by p matrices whose}\\ \text{(i,j)th entries, respectively, are}\\ \binom{r(j) + r'(j) - s'(i)}{r(j) - e(i)} \quad \text{and} \quad \begin{bmatrix} s(i) \\ e(j) \end{bmatrix} \binom{r(j) + r'(j) - s'(i)}{r(j) - e(j)}\end{array}\right.$

and

(22) in the situation of (20), for every $e \in Z(p)$ let

$$\hat{H}^{(2)}(e) = \det \hat{G}^{(2)}(e) \text{ and } \hat{H}^{(2*)}(e) = \det \hat{G}^{(2*)}(e)$$

where $\hat{G}^{(2)}(e)$ and $\hat{G}^{(2*)}(e)$ are the p by p matrices whose (i,j)th entries, respectively, are

$$\binom{r(i) + r'(i) - s'(j)}{r(i) - e(i)}\binom{s(j) + s'(j) - r'(i)}{e(i)}$$

and

$$\binom{r(i) + r'(i) - s'(j)}{r(i) - e(j)}\binom{s(j) + s'(j) - r'(i)}{e(j)}$$

and

(23*) in the situation of (20), for every $i \in [1,p]$ and $j \in [1,p]$ and $d^* \in Z$ let

$$\hat{K}_{ij}(d^*) = \begin{bmatrix} s(i) - d^* \\ r(j) \end{bmatrix}\binom{r(j) + r'(j) - s'(i)}{d^*}$$

and

(23**) in the situation of (20) and (23*), for every $d \in Z(p)$ and $z \in W[p]$ and $w \in W(p)$ let

$$\hat{G}_w^{(3z)}(d) = \prod_{i=1}^{p} \hat{K}_{iw(i)}(d(z(w)(i))),$$

and for every $d \in Z(p)$ and $z \in W[p]$ let

$$\hat{H}^{(3z)}(d) = \sum_{w \in W(p)} \operatorname{sgn}(w)\hat{G}_w^{(3z)}(d)$$

and

(23) in the situation of (20) and (23*), for every $d \in Z(p)$ let

$$\hat{H}^{(3)}(d) = \det \hat{G}^{(3)}(d) \quad \text{and} \quad \hat{H}^{(3*)}(d) = \det \hat{G}^{(3*)}(d)$$

where $\hat{G}^{(3)}(d)$ and $\hat{G}^{(3*)}(d)$ are the p by p matrices whose (i,j)th entries, respectively, are $\hat{K}_{ij}(d(i))$ and $\hat{K}_{ij}(d(j))$

and

(23') given any $p \in N^*$, let $z'[3^*] \in W[p]$ be defined by putting

$$z'[3^*](w) = w \qquad \text{for every } w \in W(p),$$

and let $z'[3] \in W[p]$ be defined by putting, for every $w \in W(p)$,

$z'[3](w)$ = the identity map of $[1,p]$ onto itself, i.e.,

$z'[3](w)(i) = i$ for all $i \in [1,p]$

and

(23'') given any $p \in N^*$, for every $i \in [1,p]$ and $j \in [1,p]$ and $d^* \in Z$ let us take any $K_{ij}(d^*) \in Q$, and for every $d \in Z(p)$ and $z \in W[p]$ and $w \in W(p)$ let us put

$$K_w^{(z)}(d) = \prod_{i=1}^{p} K_{iw(i)}(d(z(w)(i))),$$

and for every $d \in Z(p)$ let $K(d)$ and $K^*(d)$ be the p by p matrices whose (i,j)th entries, respectively, are $K_{ij}(d(i))$ and $K_{ij}(d(j))$

and

(24) in the situation of (20), for every $v \in Z(p)$ let

$$\hat{H}^{(4)}(v) = \left(\prod_{i=1}^{p}\begin{bmatrix} r(i) \\ v(i)\end{bmatrix}\right)\det \hat{G}^{(4)}(v) \quad \text{and}$$

$$\hat{H}^{(4*)}(v) = \det \hat{G}^{(4*)}(v)$$

where $\hat{G}^{(4)}(v)$ and $\hat{G}^{(4*)}(v)$ are the p by p matrices whose (i,j)th entries, respectively, are

$$\begin{bmatrix} s(j) \\ v(i) + s'(j) - r'(i)\end{bmatrix} \quad \text{and} \quad \begin{bmatrix} r(i) \\ v(j)\end{bmatrix}\begin{bmatrix} s(j) \\ v(j) + s'(j) - r'(i)\end{bmatrix}$$

and

(24') in the situation of (20), for every $v \in Z(p)$ let $\tilde{G}^{(4)}(v)$ be the p by p matrix whose (i,j)th entry is

$$\begin{bmatrix} r(j) \\ v(i) + r'(j) - s'(i)\end{bmatrix}$$

and

in the situation of (20), for every $v \in Z(p)$ let

$$\hat{H}^{(6)}(v) = [\det \tilde{G}^{(4)}(v)][\det \hat{G}^{(4)}(v)]$$

(24'') where $\tilde{G}^{(4)}(v)$ and $\hat{G}^{(4)}(v)$ are the p by p matrices whose (i,j)th entries, respectively, are

$$\begin{bmatrix} r(j) \\ v(i) + r'(j) - s'(i) \end{bmatrix} \text{ and } \begin{bmatrix} s(j) \\ v(i) + s'(j) - r'(i) \end{bmatrix}.$$

Now first, we claim that

$$\left[\begin{array}{l}\text{in the situation of (20) and (21), for every} \\ L \in \{1,1^*\} \text{ we have} \\ \quad \{e \in Z(p) : \hat{H}^{(L)}(e) \neq 0\} \subset N[p,R].\end{array}\right. \tag{2.6.21}$$

To prove (2.6.21), by taking $g[i,j] = r(j) + r'(j) - s'(i)$ in (2.6.19) we get

$$\{e \in Z(p) : \hat{H}^{(1^*)}(e) \neq 0\} \subset N[p,R];$$

moreover, by taking $g[i,j] = r(j) + r'(j) - s'(i)$ in (2.6.20) we get

$$\{e \in Z(p) : \det \hat{G}^{(1)}(e) \neq 0\} \subset Z[p,R]$$

and by (2.5.2*) we get

$$\left\{e \in Z(p) : \prod_{i=1}^{p} \begin{bmatrix} s(i) \\ e(i) \end{bmatrix} \neq 0\right\} \subset N(p)$$

and by the two inclusions above we conclude that

$$\{e \in Z(p) : \hat{H}^{(1)}(e) \neq 0\} \subset N[p,R].$$

Second, we claim that

$$\left[\begin{array}{l}\text{in the situation of (20) and (22). for every } L \in \{2,2^*\} \\ \text{we have } \{e \in Z(p) : \hat{H}^{(L)}(e) \neq 0\} \subset N[p,R].\end{array}\right. \tag{2.6.22}$$

The proof of (2.6.22) follows by taking $g'[i,j] = r(i) + r'(i) - s'(j)$ and $h'[i,j] = s(j) + s'(j) - r'(i)$ in (2.6.16) and (2.6.17).

Third, we claim that

in the situation of (20) and (23*), *if*

$$(**) \quad \left[\begin{array}{l} s(i^*) + s'(i^*) - r'(j^*) \geq 0 \leq r(j^*) \\ + r'(j^*) - s'(i^*) \text{ for all } i^* \in [1,p] \\ \text{and } j^* \in [1,p], \end{array}\right.$$

then for every $i \in [1,p]$ and $j \in [1,p]$ and $d^* \in Z$ we have

$$\left[\begin{aligned}\hat{K}_{ij}(d^*) &= \begin{bmatrix} r(j) \\ s(i) - d^* \end{bmatrix} \binom{r(j) + r'(j) - s'(i)}{d^*} \\ &= \begin{bmatrix} r(j) \\ s(i) - d^* \end{bmatrix} \binom{r(j) + r'(j) - s'(i)}{r(j) + r'(j) - s'(i) - d^*} \\ &= \begin{bmatrix} s(i) - d^* \\ r(j) \end{bmatrix} \binom{r(j) + r'(j) - s'(i)}{r(j) + r'(j) - s'(i) - d^*}\end{aligned}\right. \tag{2.6.23*}$$

The proof of (2.6.23*) follows by taking A = r(j) and B = s(i) and A' = r'(j) and B' = s'(i) and D = d* in (2.6.15).

Fourth, we claim that

$$\left[\begin{array}{l}\text{in the situation of (20), (23*) and (23**),} \\ \text{given any } z \in W[p] \text{, we have} \\ \quad \{d \in Z(p) : \hat{H}^{(3z)}(d) \neq 0\} \subset N(p), \\ \text{and } \textit{if } (^{**}) \text{, } \textit{then} \text{ we have} \\ \quad \{d \in Z(p) : \hat{H}^{(3z)}(d) \neq 0\} \subset N[p, R + R' - S'].\end{array}\right. \tag{2.6.23**}$$

To prove (2.6.23**), by the definition of $\hat{H}^{(3z)}(d)$ we see that for every $d \in Z(p)$ we have

$$\begin{aligned}\hat{H}^{(3z)}(d) = \sum_{w \in W(p)} &\left(\operatorname{sgn}(w) \prod_{i=1}^{p} \begin{bmatrix} s(i) - d(z(w)(i)) \\ r(w(i)) \end{bmatrix}\right) \\ &\times \prod_{i=1}^{p} \binom{r(w(i)) + r'(w(i)) - s'(i)}{d(z(w)(i)}\end{aligned}$$

and hence

$$(23_1) \quad \left[\begin{array}{l}\{d \in Z(p) : \hat{H}^{(3z)}(d) \neq 0\} \subset \bigcup\limits_{w \in W(p)} \Big\{d \in Z(p) : \\ \prod\limits_{i=1}^{p} \dbinom{r(w(i) + r'(w(i)) - s'(i)}{d(z(w)(i))} \neq 0\Big\}\end{array}\right.$$

and now, given any $w \in W(p)$, by (2.5.2') we have

$$\begin{aligned}&\left\{d \in Z(p) : \prod_{i=1}^{p} \binom{r(w(i) + r'(w(i)) - s'(i)}{d(z(w)(i))} \neq 0\right\} \\ &\subset \{d \in Z(p) : d(z(w)(i)) \geq 0 \text{ for all } i \in [1,p]\}\end{aligned}$$

and clearly, we have

$$\{d \in Z(p) : d(z(w)(i)) \geq 0 \text{ for all } i \in [1,p]\} = N(p)$$

and hence we see that

$$\left\{d \in Z(p) : \prod_{i=1}^{p} \binom{r(w(i)) + r'(w(i)) - s'(i)}{d(z(w)(i))} \neq 0\right\} \subset N(p)$$

and this being so for every $w \in W(p)$, by (23_1) we get

$$(23_2) \quad \{d \in Z(p) : \hat{H}^{(3z)}(d) \neq 0\} \subset N(p)$$

and again, given any $w \in W(p)$, by (2.5.6') we have

$$\left[\begin{array}{l} \text{if } (**), \text{ then} \\ \left\{d \in Z(p) : \prod_{i=1}^{p} \binom{r(w(i)) + r'(w(i)) - s'(i)}{d(z(w)(i))} \neq 0\right\} \\ \subset \{d \in Z(p) : d(z(w)(i)) \leq r(w(i)) + r'(w(i)) - s'(i) \\ \qquad\qquad\qquad \text{for all } i \in [1,p]\} \end{array}\right.$$

and clearly, we have

$$\{d \in Z(p) : d(z(w)(i)) \leq r(w(i)) + r'(w(i)) - s'(i) \text{ for all } i \in [1,p]\} \subset \left\{d \in Z(p) : \sum_{i=1}^{p} d(z(w)(i)) \leq \sum_{i=1}^{p} [r(w(i)) + r'(w(i)) - s'(i)]\right\}$$

and also clearly, we have

$$\sum_{i=1}^{p} d(z(w)(i)) = \sum_{i=1}^{p} d(i)$$

and

$$\sum_{i=1}^{p} [r(w(i)) + r'(w(i)) - s'(i)] = R + R' - S'$$

and therefore we see that

$$\left[\text{if } (**), \text{ then } \left\{d \in Z(p) : \prod_{i=1}^{p} \binom{r(w(i)) + r'(w(i)) - s'(i)}{d(z(w)(i))} \neq 0\right\} \subset Z[p, R + R' - S']\right]$$

and this being so for every $w \in W(p)$, by (23_1) we get that

$$\begin{bmatrix} \textit{if } (**), \textit{ then} \\ \{d \in Z(p) : \hat{H}^{(3z)}(d) \neq 0\} \subset Z[p, R + R' - S'] \end{bmatrix}$$

and hence by (23_2) we conclude that

$$\begin{bmatrix} \textit{if } (**), \textit{ then} \\ \{d \in Z(p) : \hat{H}^{(3z)}(d) \neq 0\} \subset N[p, R + R' - S']. \end{bmatrix}$$

Fifth, we claim that

$$\begin{bmatrix} \text{in the situation of (23') and (23''), given any} \\ d \in Z(p)\text{, we have} \\ \det K(d) = \sum_{w \in W(p)} \operatorname{sgn}(w) K_w^{(z'[3])}(d) \\ \text{and} \\ \det K^*(d) = \sum_{w \in W(p)} \operatorname{sgn}(w) K_w^{(z'[3^*])}(d). \end{bmatrix} \qquad (2.6.23'')$$

To prove (2.6.23''), upon expanding det K(d) by rows we have

$$\det K(d) = \sum_{w \in W(p)} \operatorname{sgn}(w) \prod_{i=1}^{p} K_{iw(i)}(d(i))$$

and by the definition of $K_w^{(z)}(d)$ and $z'[3]$ we see that for every $w \in W(p)$ we have

$$K_w^{(z'[3])}(d) = \prod_{i=1}^{p} K_{iw(i)}(d(i))$$

and therefore we get

$$\det K(d) = \sum_{w \in W(p)} \operatorname{sgn}(w) K_w^{(z'[3])}(d)$$

and now upon expanding $K^*(d)$ by rows we have

$$\det K^*(d) = \sum_{w \in W(p)} \operatorname{sgn}(w) \prod_{i=1}^{p} K_{iw(i)}(d(w(i)))$$

and by the definition of $K^{(z)}(d)$ and $z'[3^*]$ we see that for every $w \in W(p)$ we have

$$K_w^{(z'[3^*])}(d) = \prod_{i=1}^{p} K_{iw(i)}(d(w(i)))$$

and therefore we get

$$\det K^*(d) = \sum_{w \in W(p)} \operatorname{sgn}(w) K_w^{(z'[3^*])}(d).$$

Sixth, we claim that

$$\left[\begin{array}{l}\text{in the situation of (20), (23*), (23**), (23),}\\ \text{and (23'), for every } L \in \{3,3^*\} \text{ and } d \in Z(p)\\ \text{we have } \hat{H}^{(L)}(d) = \hat{H}^{(3z'[L])}(d).\end{array}\right. \tag{2.6.23'}$$

The proof of (2.6.23') follows by taking $K_{ij}(d^*) = \hat{K}_{ij}(d^*)$ in (2.6.23").

Seventh, we claim that

$$\left[\begin{array}{l}\text{in the situation of (20), (23*), (23**), and (23),}\\ \text{given any } L \in \{3,3^*\} \cup \{3z : z \in W[p]\}, \text{ we have}\\ \quad \{d \in Z(p) : \hat{H}^{(L)}(d) \neq 0\} \subset N(p)\\ \text{and } \textit{if } (^{**}), \textit{ then} \text{ we have}\\ \quad \{d \in Z(p) : \hat{H}^{(L)}(d) \neq 0\} \subset N[p, R + R' - S'].\end{array}\right. \tag{2.6.23}$$

The proof of (2.6.23) follows from (2.6.23**) and (2.6.23').

Eighth, we claim that

$$\left[\begin{array}{l}\text{in the situation of (20) and (24), for every}\\ L \in \{4,4^*\} \text{ we have } \{v \in Z(p) : \hat{H}^{(L)}(v) \neq 0\} \subset N(p).\end{array}\right. \tag{2.6.24}$$

To prove (2.6.24), by taking $t = r$ and

$$u(i,j,e^*) = \begin{bmatrix} s(j) \\ e^* + s'(j) - r'(i) \end{bmatrix}$$

in (2.6.18) we get

$$\{v \in Z(p) : \hat{H}^{(4^*)}(v) \neq 0\} \subset N(p);$$

moreover, by (2.5.2') we get

$$\left\{v \in Z(p) : \prod_{i=1}^{p} \begin{bmatrix} r(i) \\ v(i) \end{bmatrix} \neq 0\right\} \subset N(p)$$

and hence we get

$$\{v \in Z(p) : \hat{H}^{(4)}(v) \neq 0\} \subset N(p).$$

Ninth, we claim that

$$\left[\begin{array}{l} \text{in the situation of (20) and (24') we have} \\ \{v \in Z(p) : \det \tilde{G}^{(4)}(v) \neq 0\} \\ \subset \{v \in Z(p) : v(i) \geq s'(i) - \hat{R} \text{ for all } i \in [1,p]\} \\ \text{where } \hat{R} = \max\{r'(j) : j \in [1,p]\}. \end{array}\right. \tag{2.6.24'}$$

To prove (2.6.24'), by (2.5.2*) we see that

$$\left[\begin{array}{l} \text{if } v \in Z(p) \text{ and } i \in [1,p] \text{ and } j \in [1,p] \text{ are such that} \\ v(i) + r'(j) - s'(i) < 0, \text{ then } \begin{bmatrix} r(j) \\ v(i) + r'(j) - s'(i) \end{bmatrix} = 0 \end{array}\right.$$

and hence we see that

$$\left[\begin{array}{l} \text{if } v \in Z(p) \text{ and } i \in [1,p] \text{ are such that } v(i) + r'(j) - \\ s'(i) < 0 \text{ for every } j \in [1,p], \text{ then the } i\text{th row of} \\ \tilde{G}^{(4)}(v) \text{ consists entirely of zeros} \end{array}\right.$$

and therefore we see that

$$\left[\begin{array}{l} \text{if } v \in Z(p) \text{ is such that for some } i \in [1,p] \text{ we have} \\ v(i) + r'(j) - s'(i) < 0 \text{ for every } j \in [1,p], \text{ then} \\ \det \tilde{G}^{(4)}(v) = 0 \end{array}\right.$$

and thus we get that

$$\begin{aligned} &\{v \in Z(p) : \det \tilde{G}^{(4)}(v) \neq 0\} \\ &\subset \{v \in Z(p) : \text{for each } i \in [1,p], \text{ there exists } j(i) \in [1,p] \\ &\qquad\qquad \text{such that } v(i) + r'(j(i)) - s'(i) \geq 0\} \end{aligned}$$

and clearly, we have

$$\begin{aligned}
&\{v \in Z(p) : \text{for each } i \in [1,p], \text{ there exists } j(i) \in [1,p] \\
&\qquad\qquad \text{such that } v(i) + r'(j(i)) - s'(i) \geq 0\} \\
&\subset \{v \in Z(p) : \text{for every } i \in [1,p] \text{ we have } v(i) + \hat{R} - s'(i) \geq 0\} \\
&= \{v \in Z(p) : v(i) \geq s'(i) - \hat{R} \text{ for all } i \in [1,p]\}
\end{aligned}$$

and so we conclude that

$$\begin{aligned}
&\{v \in Z(p) : \det \tilde{G}^{(4)}(v) \neq 0\} \\
&\subset \{v \in Z(p) : v(i) \geq s'(i) - R \text{ for all } i \in [1,p]\}.
\end{aligned}$$

Tenth, we claim that

$$\left[\begin{aligned}
&\text{in the situation of (20) and (24'') we have} \\
&\{v \in Z(p) : \hat{H}^{(6)}(v) \neq 0\} \\
&\subset \{v \in Z(p) : v(i) \geq \max(s'(i) - \hat{R},\ r'(i) - \hat{S}) \\
&\qquad\qquad\qquad \text{for all } i \in [1,p]\} \\
&\text{where } \hat{R} = \max\{r'(j) : j \in [1,p]\} \\
&\text{and } \hat{S} = \max\{s'(j) : j \in [1,p]\}.
\end{aligned}\right. \tag{2.6.24''}$$

To prove (2.6.24''), clearly we have

$$\begin{aligned}
&\{v \in Z(p) : \hat{H}^{(6)}(v) \neq 0\} \\
&\subset \{v \in Z(p) : \det \tilde{G}^{(4)}(v) \neq 0\} \cap \{v \in Z(p) : \det \hat{G}^{(4)}(v) \neq 0\}
\end{aligned}$$

and by (2.6.24') we have

$$\begin{aligned}
&\{v \in Z(p) : \det \tilde{G}^{(4)}(v) \neq 0\} \\
&\subset \{v \in Z(p) : v(i) \geq s'(i) - \hat{R} \text{ for all } i \in [1,p]\}
\end{aligned}$$

and upon interchanging $(r(j),r'(j),s'(i))$ and $(s(j),s'(j),r'(i))$, by (2.6.24') we get

$$\begin{aligned}
&\{v \in Z(p) : \det \tilde{G}^{(4)}(v) \neq 0\} \\
&\subset \{v \in Z(p) : v(i) \geq r'(i) - \hat{S} \text{ for all } i \in [1,p]\}
\end{aligned}$$

and clearly, we have

$$\begin{aligned}
&\{v \in Z(p) : v(i) \geq s'(i) - \hat{R} \text{ for all } i \in [1,p]\} \cap \{v \in Z(p) : \\
&\qquad\qquad\qquad v(i) \geq r'(i) - \hat{S} \text{ for all } i \in [1,p]\} \\
&\subset \{v \in Z(p) : v(i) \geq \max(s'(i) - \hat{R},\ r'(i) - \hat{S}) \\
&\qquad\qquad\qquad \text{for all } i \in [1,p]\}\}
\end{aligned}$$

and thus we conclude that

$$\{v \in Z(p) : \hat{H}^{(6)}(v) \neq 0\}$$
$$\subset \{v \in Z(p) : v(i) \geq \max(s'(i) - \hat{R},\ r'(i) - \hat{S}) \text{ for all } i \in [1,p]$$

(2.7) Determinantal Polynomials of Any Width In this subsection and in Subsection (2.10) we define the functions

$$F^{(1k)},\ F^{(1*k)},\ \ldots,\ F^{(4*k)},\ F^{(5k)},\ F^{(6)},\ F^{(7)},\ F^{(7*)},\ F^{(8)}$$

which may be described as examples of *determinantal polynomials in binomial coefficients*. Eventually, it will be shown that they count the number of tableaux with certain assigned conditions. Out of these, $F^{(5k)}$, $F^{(6)}$, ..., $F^{(8)}$ make sense for tableaux of any width, and they are introduced in this subsection. The remaining ones, i.e., $F^{(1k)}$, $F^{(1*k)}$, ..., $F^{(4*k)}$, make sense only for bitableaux, and they will be introduced in Subsection (2.10).

For every $p \in N^*$ and $V \in Z$ we put

$$z(p,V,8) = \left\{v \in N(p) : \sum_{i=1}^{p} iv(i) = V\right\}$$

and

$$Z(p,V,7^*) = Z(p,\ V + [p(p - 1)/2])$$

and

$$Z(p,V,7) = \{v \in Z(p,V,7^*) : v(1) > v(2) > \cdots > v(p) \geq 0\}.$$

Given any $q \in N^*$ and $m \in N^*(q)$ and $p \in N^*$ and $a \in \mathrm{vec}(q,m,p)$, first, for every $k \in [1,q]$ and $i \in [1,p]$ and $j \in [1,p]$ and $v \in Z(p)$ we define

$$G_{ij}^{(8k)}(q,m,p,a,v) = \left[\begin{matrix} m(k) - a(k,j) \\ v(i) + v(i + 1) + \cdots + v(p) + j - i \end{matrix}\right]$$

and

$$G_{ij}^{(7k)}(q,m,p,a,v) = \left[\begin{matrix} m(k) - a(k,j) \\ v(i) + j - p \end{matrix}\right]$$

and

$$G_{ij}^{(6k)}(q,m,p,a,v) = \begin{bmatrix} m(k) - a(k,j) \\ v(i) + j - i \end{bmatrix},$$

and second, for every $L \in [6,8]$ and $k \in [1,q]$ and $v \in Z(p)$ we define

$G^{(Lk)}(q,m,p,a,v)$ = the p by p matrix whose (i,j)th entry is $G_{ij}^{(Lk)}(q,m,p,a,v)$,

and third, for every $L \in [6,8]$ and $v \in Z(p)$ we define

$$H^{(L)}(q,m,p,a,v) = \prod_{k=1}^{q} \det G^{(Lk)}(q,m,p,a,v),$$

and fourth, for every $k \in [1,q]$ and $v \in Z(p)$ we define

$$G^{[5k]}(q,m,p,a,v) = \prod_{i=1}^{p} \begin{bmatrix} m(k) - a(k,i) \\ v(i) \end{bmatrix}$$

and

$$H^{(5k)}(q,m,p,a,v)$$
$$= G^{[5k]}(q,m,p,a,v) \prod_{n\in[1,q]\setminus\{k\}} \det G^{(6n)}(q,m,p,a,v),$$

and fifth, for every $V \in Z$ we define

$$F^{(8)}(q,m,p,a,V) = \sum_{v\in Z(p,V,8)} H^{(8)}(q,m,p,a,v)$$

and

$$F^{(7)}(q,m,p,a,V) = \sum_{v\in Z(p,V,7)} H^{(7)}(q,m,p,a,v)$$

and

$$F^{(7*)}(a,m,p,a,V) = (1/p!) \sum_{v\in Z(p,V,7*)} H^{(7)}(q,m,p,a,v)$$

and

$$F^{(6)}(q,m,p,a,V) = (1/p!) \sum_{v\in Z(p,V)} H^{(6)}(q,m,p,a,v),$$

and sixth, for every $k \in [1,q]$ and $V \in Z$ we define

$$F^{(5k)}(q,m,p,a,V) = \sum_{v \in Z(p,V)} H^{(5k)}(q,m,p,a,v).$$

We now proceed to make some observations concerning the summations involved in the definitions above.

Given any $q \in N^*$ and $m \in N^*(q)$ and $p \in N^*$ and $a \in \text{vec}(q,m,p)$ and $k \in [1,q]$, first, by (2.5.2*) we see that

$$\{v \in Z(p) : G^{[5k]}(q,m,p,a,v) \neq 0\} \subset N(p), \tag{2.7.5}$$

and second, by taking

$$\begin{cases} r(j) = m(k) - a(k,j) \text{ and } s(j) = 0 \text{ and } r'(j) = j \\ \text{and } s'(j) = j \text{ for all } j \in [1,p] \end{cases}$$

in (2.6.24') we see that

$$\begin{cases} \{v \in Z(p) : \det G^{(6k)}(q,m,p,a,v) \neq 0\} \\ \subset \{v \in Z(p) : v(i) \geq i - p \text{ for all } i \in [1,p]\}, \end{cases} \tag{2.7.6}$$

and third, by taking

$$\begin{cases} r(j) = m(k) - a(k,j) \text{ and } s(j) = 0 \text{ and } r'(j) = j \\ \text{and } s'(j) = p \text{ for all } j \in [1,p] \end{cases}$$

in (2.6.24') we see that

$$\{v \in Z(p) : \det G^{(7k)}(q,m,p,a,v) \neq 0\} \subset N(p). \tag{2.7.7*}$$

Now the summation in the definition of $F^{(5k)}$ above is over the infinite set $Z(p,V)$; however, by (2.7.5) we see that the set of all v in $Z(p,V)$ for which the summand $H^{(5k)}(q,m,p,a,v)$ is nonzero is contained in the finite set $N(p,V)$ and hence the summation makes sense because it is *essentially finite* and we have

$$F^{(5k)}(q,m,p,a,V) = \sum_{v \in N(p,V)} H^{(5k)}(q,m,p,a,v).$$

Before making similar observations about the summations involved in the definitions of $F^{(6)}$ and $F^{(7*)}$, let us introduce the following general

(2.8) Summation Conventions If f: $\hat{X} \to \hat{Q}$ is any map (or mapping) of any set $\hat{X}$ into any additive abelian group $\hat{Q}$, and if X is any subset of $\hat{X}$, and if we are considering an expression like

$$\sum_{x \in X} f(x)$$

then we may use a phrase such as

the summation above is essentially finite

to mean that

$$\text{card}(\{x \in X : f(x) \neq 0\}) < \infty;$$

in this connection we observe that

$$\left[\begin{array}{l}\text{if card}(\{x \in X : f(x) \neq 0\}) < \infty \text{ and if } Y \subset X^* \text{ is such that} \\ \{x \in X : f(x) \neq 0\} \subset Y, \text{ then} \\ \qquad \sum_{x \in X} f(x) = \sum_{x \in X \cap Y} f(x)\end{array}\right.$$

and we note that the observation above may be used tacitly.

Moreover, a statement like

this thing equals that thing

in which that thing involves a summation will mean that

[the summation in that thing is essentially finite
and this thing is equal to that thing.

And so on.

(2.9) Essential Finiteness of the Summations in the Definitions of Certain Determinantal Polynomials of any Width Now we can make the following observations concerning the summations involved in the definitions given in (2.7).

OBSERVATION (2.9.5) Given any $q \in N^*$ and $m \in N^*(q)$ and $p \in N^*$ and $a \in \text{vec}(q,m,p)$ and $k \in [1,q]$, by (2.7.5) we see that

$$\{v \in Z(p) : H^{(5k)}(q,m,p,a,v) \neq 0\} \subset N(p)$$

and for every $V \in Z$ we clearly have

$$\mathrm{card}(Z(p,V) \cap N(p)) < \infty$$

and hence the summation in the definition of $F^{(5k)}(q,m,p,a,V)$ is essentially finite.

OBSERVATION (2.9.6) Given any $q \in N^*$ and $m \in N^*(q)$ and $p \in N^*$ and $a \in \mathrm{vec}(q,m,p)$, by (2.7.6) we see that

$$\{v \in Z(p) : H^{(6)}(q,m,p,a,v) \neq 0\}$$
$$\subset \{v \in Z(p) : v(i) \geq i - p \text{ for all } i \in [1,p]\}$$

and for every $V \in Z$ we clearly have

$$\mathrm{card}(Z(p,V) \cap \{v \in Z(p) : v(i) \geq i - p \text{ for all } i \in [1,p]\}) < \infty$$

and hence the summation in the definition of $F^{(6)}(q,m,p,a,V)$ is essentially finite.

OBSERVATION (2.9.7)* Given any $q \in N^*$ and $m \in N^*(q)$ and $p \in N^*$ and $a \in \mathrm{vec}(q,m,p)$, by (2.7.7*) we see that

$$\{v \in Z(p) : H^{(7)}(q,m,p,a,v) \neq 0\} \subset N(p)$$

and hence the summation in the definition of $F^{(7*)}(q,m,p,a,V)$ is essentially finite.

(2.10) Determinantal Polynomials of Width 2 In this subsection we introduce the functions

$$F^{(1k)}, F^{(1*k)}, \ldots, F^{(4*k)}$$

which may be described as examples of *width 2 determinantal polynomials in binomial coefficients*. Eventually, it will be shown that they count the number of bitableaux with certain assigned restrictions.

Given any $m \in N^*(2)$ and $p \in N^*$ and $a \in \mathrm{vec}(2,m,p)$, first, for every $k \in [1,2]$ we define

$$C[m,p,a,k] = \sum_{i=1}^{p} [m(k) - a(k,i)]$$

and second, we define

$$C(m,p,a) = C[m,p,a,1] + C[m,p,a,2] + p - 1.$$

Given any $m \in N^*(2)$ and $p \in N^*$ and $a \in vec(2,m,p)$ and $k \in [1,2]$, put

$$k' = \begin{cases} 2 & \text{if } k = 1 \\ 1 & \text{if } k = 2, \end{cases}$$

First, for every $i \in [1,p]$ and $j \in [1,p]$ and $e \in Z(p)$ we define

$$G_{ij}^{(1k)}(m,p,a,e) = \binom{m(k) - a(k,j) + j - i}{m(k) - a(k,j) - e(i)},$$

and second, for every $e \in Z(p)$ we define

$$G^{(1k)}(m,p,a,e) = \text{the } p \text{ by } p \text{ matrix whose } (i,j)\text{th entry is } G_{ij}^{(1k)}(m,p,a,e)$$

and

$$G^{[1k]}(m,p,a,e) = \prod_{i=1}^{p} \left[\begin{matrix} m(k') - a(k',i) \\ e(i) \end{matrix}\right]$$

and

$$H^{(1k)}(m,p,a,e) = G^{[1k]}(m,p,a,e) \det G^{(1k)}(m,p,a,e),$$

and third, for every $i \in [1,p]$ and $j \in [1,p]$ and $e \in Z(p)$ we define

$$G_{ij}^{(1^*k)}(m,p,a,e) = \left[\begin{matrix} m(k') - a(k',i) \\ e(j) \end{matrix}\right] \binom{m(k) - a(k,j) + j - i}{m(k) - a(k,j) - e(j)},$$

and fourth, for every $e \in Z(p)$ we define

$$G^{(1^*k)}(m,p,a,e) = \text{the } p \text{ by } p \text{ matrix whose } (i,j)\text{th entry is } G_{ij}^{(1^*k)}(m,p,a,e)$$

and

$$H^{(1^*k)}(m,p,a,e) = \det G^{(1^*k)}(m,p,a,e),$$

and fifth, for every $L \in \{1,1^*\}$ and $E \in Z$ we define

$$H_E^{(Lk)}(m,p,a) = \sum_{e \in Z(p,E)} H^{(Lk)}(m,p,a,e),$$

and sixth, for every $L \in \{1,1^*\}$ and $D \in Z$ and $E \in Z$ we define

$$H_E^{[Lk]}(m,p,a,D) = (-1)^{C[m,p,a,k]-E}\binom{E}{D + E - C[m,p,a,k]}$$

and seventh, for every $L \in \{1,1^*\}$ and $D \in Z$ we define

$$F_D^{(Lk)}(m,p,a) = \sum_{E \in Z} H_E^{[Lk]}(m,p,a,D)H_E^{(Lk)}(m,p,a),$$

and eighth, for every $i \in [1,p]$ and $j \in [1,p]$ and $e \in Z(p)$ we define

$$G_{ij}^{(2k)}(m,p,a,e) = \binom{m(k) - a(k,i) + i - j}{m(k) - a(k,i) - e(i)}\binom{m(k') - a(k',j) + j - i}{e(i)}$$

and

$$G_{ij}^{(2^*k)}(m,p,a,e) = \binom{m(k) - a(k,i) + i - j}{m(k) - a(k,i) - e(j)}\binom{m(k') - a(k',j) + j - i}{e(j)},$$

and ninth, for every $L \in \{2,2^*\}$ and $e \in Z(p)$ we define

$G^{(Lk)}(m,p,a,e)$ = the p by p matrix whose (i,j)th entry is $G_{ij}^{(Lk)}(m,p,a,e)$

and

$$H^{(Lk)}(m,p,a,e) = \det G^{(Lk)}(m,p,a,e),$$

and tenth, for every $L \in \{2,2^*\}$ and $E \in Z$ we define

$$H_E^{(Lk)}(m,p,a) = \sum_{e \in Z(p,E)} H^{(Lk)}(m,p,a,e),$$

and eleventh, for every $L \in \{2,2^*\}$ and $D \in Z$ we define

$$F_D^{(Lk)}(m,p,a) = \sum_{E \in Z} \binom{E}{D} H_E^{(Lk)}(m,p,a),$$

and twelfth, for every $i \in [1,p]$ and $j \in [1,p]$ and $d \in Z(p)$ we define

$$G_{ij}^{(3k)}(m,p,a,d) = \begin{bmatrix} m(k') - a(k',i) - d(i) \\ m(k) - a(k,j) \end{bmatrix} \binom{m(k) - a(k,j) + j - i}{d(i)}$$

and

$$G_{ij}^{(3^*k)}(m,p,a,d) = \begin{bmatrix} m(k') - a(k',i) - d(j) \\ m(k) - a(k,j) \end{bmatrix} \binom{m(k) - a(k,j) + j - i}{d(j)},$$

and thirteenth, for every $L \in \{3,3^*\}$ and $d \in Z(p)$ we define

$$G^{(Lk)}(m,p,a,d) = \text{the } p \text{ by } p \text{ matrix whose } (i,j)\text{th entry is } G_{ij}^{(Lk)}(m,p,a,d)$$

and

$$H^{(Lk)}(m,p,a,d) = \det G^{(Lk)}(m,p,a,d),$$

and fourteenth, for every $z \in W[p]$ and $w \in W(p)$ and $d \in Z(p)$ we define

$$G_w^{(3zk)}(m,p,a,d) = \prod_{i=1}^{p} \begin{bmatrix} m(k') - a(k',i) - d(z(w)(i)) \\ m(k) - a(k,w(i)) \end{bmatrix} \times \binom{m(k) - a(k,w(i)) + w(i) - i}{d(z(w)(i))}$$

and fifteenth, for every $z \in W[p]$ and $d \in Z(p)$ we define

$$H^{(3zk)}(m,p,a,d) = \sum_{w \in W(p)} \operatorname{sgn}(w) G_w^{(3zk)}(m,p,a,d),$$

and sixteenth, for every $L \in \{3,3^*\} \cup \{3z : z \in W[p]\}$ and $D \in Z$ we define

$$F_D^{(Lk)}(m,p,a) = \sum_{d \in Z(p,D)} H^{(Lk)}(m,p,a,d),$$

and seventeenth, for every $i \in [1,p]$ and $j \in [1,p]$ and $v \in Z(p)$ we define

$$G_{ij}^{(4k)}(m,p,a,v) = \begin{bmatrix} m(k') - a(k',j) \\ v(i) + j - i \end{bmatrix},$$

and eighteenth, for every $v \in Z(p)$ we define

$$G^{(4k)}(m,p,a,v) = \text{the } p \text{ by } p \text{ matrix whose } (i,j)\text{th entry is } G_{ij}^{(4k)}(m,p,a,v)$$

and

$$G^{[4k]}(m,p,a,v) = \prod_{i=1}^{p} \begin{bmatrix} m(k) - a(k,i) \\ v(i) \end{bmatrix}$$

and

$$H^{(4k)}(m,p,a,v) = G^{[4k]}(m,p,a,v) \det G^{(4k)}(m,p,a,v),$$

and ninteenth, for every $i \in [1,p]$ and $j \in [1,p]$ and $v \in Z(p)$ we define

$$G_{ij}^{(4*k)}(m,p,a,v) = \begin{bmatrix} m(k) - a(k,i) \\ v(j) \end{bmatrix} \begin{bmatrix} m(k') - a(k',j) \\ v(j) + j - i \end{bmatrix},$$

and twentieth, for every $v \in Z(p)$ we define

$$G^{(4*k)}(m,p,a,v) = \text{the } p \text{ by } p \text{ matrix whose } (i,j)\text{th entry is } G_{ij}^{(4*k)}(m,p,a,v)$$

and

$$H^{(4*k)}(m,p,a,v) = \det G^{(4*k)}(m,p,a,v)$$

and twenty-first, for every $L \in \{4,4^*\}$ and $V \in Z$ we define

$$F^{(Lk)}(m,p,a,V) = \sum_{v \in Z(p,V)} H^{(Lk)}(m,p,a,v).$$

Let us note that given any $m \in N^*(2)$ and $p \in N^*$ and $a \in \text{vec}(2,m,p)$, by the definitions introduced in (2.7), for every $v \in Z(p)$ we have

$$H^{(6)}(2,m,p,a,v) = \prod_{k=1}^{2} \det G^{(4k)}(m,p,a,v)$$

and for every $V \in N$ we have

$$F^{(6)}(2,m,p,a,V) = (1/p!) \sum_{v \in Z(p,V)} H^{(6)}(2,m,p,a,v).$$

Now given any $m \in N^*(2)$ and $p \in N^*$ and $a \in vec(2,m,p)$ and given any

$$L \in \{1,1^*,2,2^*,3,3^*\} \cup \{3z : z \in W[p]\}$$

and given any $k \in [1,2]$, upon putting

$$k' = \begin{cases} 2 & \text{if } k = 1 \\ 1 & \text{if } k = 2, \end{cases}$$

first, for every $C' \in Z$ and $V \in Z$ we define

$$F^{((Lk))}(m,p,a,C',V) = \sum_{D \in Z} (-1)^D F_D^{(Lk)}(m,p,a)\begin{bmatrix} C' - D \\ V \end{bmatrix},$$

and second, for every $V \in Z$ we define

$$F^{(Lk)}(m,p,a,V) = \sum_{D \in Z} (-1)^D F_D^{(Lk)}(m,p,a)\begin{bmatrix} C(m,p,a) - D \\ V \end{bmatrix},$$

and third, for every $C' \in Z$, in $Q[V]$ we define

$$F^{[[Lk]]}(m,p,a,C',V) = \sum_{D \in Z} (-1)^D F_D^{(Lk)}(m,p,a)\begin{bmatrix} V \\ C' - D \end{bmatrix},$$

and finally, in $Q[V]$ we define

$$F^{[Lk]}(m,p,a,V) = \sum_{D \in Z} (-1)^D F_D^{(Lk)}(m,p,a)\begin{bmatrix} V \\ C(m,p,a) - D \end{bmatrix}$$

where the meaning of the phrase

in $Q[V]$

is as explained in (2.5.13).

In the next subsection we discuss the essential finiteness of the summations involved in the definitions above. *To prepare the groundwork for that discussion, in the rest of this subsection let there be fixed* any $m \in N^*(2)$ and $p \in N^*$ and $a \in vec(2,m,p)$ and $k \in [1,2]$ and let us put

$$\hat{C} = C[m,p,a,k] \qquad \text{and} \qquad k' = \begin{cases} 2 & \text{if } k = 1 \\ 1 & \text{if } k = 2 \end{cases}$$

and let r, s, r', and s' be the members of $Z(p)$ obtained by putting

$$\left[\begin{array}{l} r(i) = m(k) - a(k,i) \text{ and } s(i) = m(k') - a(k',i) \text{ for all} \\ i \in [1,p], \text{ and } r'(i) = i \text{ and } s'(i) = i \text{ for all } i \in [1,p]. \end{array}\right.$$

Now by (2.6.21) we see that

$$\left[\begin{array}{l} \text{for every } L \in \{1,1^*\} \text{ we have} \\ \{e \in Z(p) : H^{(Lk)}(m,p,a,e) \neq 0\} \subset N[p,\hat{C}] \end{array}\right. \tag{2.10.1}$$

and by (2.6.22) we see that

$$\left[\begin{array}{l} \text{for every } L \in \{2,2^*\} \text{ we have} \\ \{e \in Z(p) : H^{(Lk)}(m,p,a,e) \neq 0\} \subset N[p,\hat{C}] \end{array}\right. \tag{2.10.2}$$

and by (2.2.2) and (2.6.23) we see that

$$\left[\begin{array}{l} \text{for every } L \in \{3,3^*\} \cup \{3z : z \in W[p]\} \text{ we have} \\ \{d \in Z(p) : H^{(Lk)}(m,p,a,d) \neq 0\} \subset N[p,\hat{C}] \end{array}\right. \tag{2.10.3}$$

and by (2.6.24) we see that

$$\left[\begin{array}{l} \text{for every } L \in \{4,4^*\} \text{ we have} \\ \{v \in Z(p) : H^{(Lk)}(m,p,a,v) \neq 0\} \subset N(p) \end{array}\right. \tag{2.10.4}$$

and by (2.6.24') we see that

$$\left[\begin{array}{l} \{v \in Z(p) : \det G^{(4k)}(m,p,a,v) \neq 0\} \\ \subset \{v \in Z(p) : v(i) \geq i - p \text{ for all } i \in [1,p]\}. \end{array}\right. \tag{2.10.4'}$$

Upon taking $\hat{C}$ for R, by (2.6.13) we see that

$$\left[\begin{array}{l} \text{for every } L \in \{1,1^*\} \text{ we have} \\ \{D \in Z : \{E \in [0,\hat{C}] : H_E^{[Lk]}(m,p,a,D) \neq 0\} \neq \emptyset\} \subset [0,\hat{C}] \end{array}\right. \tag{2.10.5}$$

and by (2.6.14) we see that

$$\left\{D \in Z : \left\{E \in [0,\hat{C}] : \binom{E}{D} \neq 0\right\} \neq \emptyset\right\} \subset [0,\hat{C}]. \tag{2.10.6}$$

By (2.2.2) and (2.6.23*) we see that

$$\left[\begin{array}{l} \text{for every } d^* \in Z \text{ we have} \\ \begin{bmatrix} m(k') - a(k',i) - d^* \\ m(k) - a(k,j) \end{bmatrix} \binom{m(k) - a(k,j) + j - i}{d^*} \end{array}\right.$$

$$
\begin{aligned}
&= \begin{bmatrix} m(k) - a(k,j) \\ m(k') - a(k',i) - d^* \end{bmatrix} \begin{pmatrix} m(k) - a(k,j) + j - i \\ d^* \end{pmatrix} \\
&= \begin{bmatrix} m(k) - a(k,j) \\ m(k') - a(k',i) - d^* \end{bmatrix} \begin{pmatrix} m(k) - a(k,j) + j - i \\ m(k) - a(k,j) + j - i - d^* \end{pmatrix} \\
&= \begin{bmatrix} m(k') - a(k',i) - d^* \\ m(k) - a(k,j) \end{bmatrix} \begin{pmatrix} m(k) - a(k,j) + j - i \\ m(k) - a(k,j) + j - i - d^* \end{pmatrix}
\end{aligned}
\qquad (2.10.7)
$$

Finally, by (2.6.23') we see that

for every $L \in \{3,3^*\}$ and $d \in Z(p)$ we have

$$H^{(Lk)}(m,p,a,d) = H^{(3z'[L]k)}(m,p,a,d)$$

where $z'[3^*] \in W[p]$ is defined by putting

$$z'[3^*](w) = w \qquad \text{for all } w \in W(p)$$

and where $z'[3] \in W[p]$ is defined by putting, (2.10.8)

for every $w \in W(p)$,

$$z'[3](w) = \text{the identity map of } [1,p] \text{ onto itself,}$$

i.e.,

$$z'[3](w)(i) = i \qquad \text{for all } i \in [1,p].$$

(2.11) Essential Finiteness of the Summations in the Definitions of Certain Determinantal Polynomials of Width 2 Now we are ready to make the following observations [i.e., (2.11.11) to (2.11.16)] concerning the summations involved in the definitions given in (2.10). Before doing so, in (2.11.0) we record some properties of the constants C which were defined in (2.10), and in (2.11.1) to (2.11.3) we record several alternative characterizations of some of the functions that were introduced in (2.10) as G and H with different subscripts and superscripts.

OBSERVATION (2.11.0) Let there be given any $m \in N^*(2)$ and $p \in N^*$ and $a \in \text{vec}(2,m,p)$. Now first, by (2.2.1) and the definition of $C[m,p,a,k]$ we see that

(1) for every $k \in [1,2]$ we have $C[m,p,a,k] \geq p(p - 1)/2$,

and second, by (1) we see that

(2) for every $k \in [1,2]$ we have $C[m,p,a,k] \in N$,

and third, by (1) and the definition of C(m,p,a) we see that

$$(3)\quad \begin{bmatrix}\text{for every } k \in [1,2] \text{ we have} \\ C(m,p,a) \geq C[m,p,a,k] + [p(p-1)/2] + (p-1)\end{bmatrix}$$

and fourth, by (3) we see that

(4) for every $k \in [1,2]$ we have $C(m,p,a) \geq C[m,p,a,k]$,

and finally, by (2) and the definition of C(m,p,a) we see that

(5) $C(m,p,a) \in N$.

REMARK (2.11.0') In Observations (2.11.1), (2.11.2), and (2.11.3) we use the well-known fact that the determinant of a square matrix equals the determinant of its transpose.

OBSERVATION (2.11.1) Let there be given any $m \in N^*(2)$ and $p \in N^*$ and $a \in \text{vec}(2,m,p)$ and $k \in [1,2]$ and $e \in Z(p)$, and let us put

$$k' = \begin{cases} 2 & \text{if } k = 1 \\ 1 & \text{if } k = 2. \end{cases}$$

Now by (2.2.2) and (2.5.10') we see that for every $i \in [1,p]$ and $j \in [1,p]$ we have

$$G_{ij}^{(1k)}(m,p,a,e) = \binom{m(k) - a(k,j) + j - i}{e(i) + j - i}$$

and in view of (2.11.0') we also see that

$$H^{(1k)}(m,p,a,e)$$
$$= \left(\prod_{j=1}^{p} \begin{bmatrix} m(k') - a(k',j) \\ e(j) \end{bmatrix}\right) \text{ times } \begin{pmatrix}\text{det of the transpose} \\ \text{of } G^{(1k)}(m,p,a,e)\end{pmatrix}$$

and for every $i \in [1,p]$ and $j \in [1,p]$, the (i,j)th entry of the transpose has two different expressions which are obtained by interchanging i and j in the foregoing two expressions of $G_{ij}^{(1k)}$.

By (2.2.2) and (2.5.10') we see that for every $i \in [1,p]$ and $j \in [1,p]$ we have

$$G_{ij}^{(1*k)}(m,p,a,e) = \begin{bmatrix} m(k') - a(k',i) \\ e(j) \end{bmatrix}\binom{m(k) - a(k,j) + j - i}{e(j) + j - i}$$

and in view of (2.11.0') we also see that

$$H^{(1*k)}(m,p,a,e) = \text{det of the transpose of } G^{(1*k)}(m,p,a,e)$$

and for every $i \in [1,p]$ and $j \in [1,p]$, the (i,j)th entry of the transpose has two different expressions which are obtained by interchanging i and j in the two expressions of $G_{ij}^{(1*k)}$ above.

OBSERVATION (2.11.2) Let there be given any $m \in N^*(2)$ and $p \in N^*$ and $a \in \text{vec}(2,m,p)$ and $k \in [1,2]$ and $e \in Z(p)$, and let us put

$$k' = \begin{cases} 2 & \text{if } k = 1 \\ 1 & \text{if } k = 2. \end{cases}$$

Now by (2.2.2) and (2.5.10') we see that for every $i \in [1,p]$ and $j \in [1,p]$ we have

$$\begin{aligned} &G_{ij}^{(2k)}(m,p,a,e) \\ &= \binom{m(k) - a(k,i) + i - j}{e(i) + i - j}\binom{m(k') - a(k',j) + j - i}{e(i)} \\ &= \binom{m(k) - a(k,i) + i - j}{e(i) + i - j}\binom{m(k') - a(k',j) + j - i}{m(k') - a(k',j) + j - i - e(i)} \\ &= \binom{m(k) - a(k,i) + i - j}{m(k) - a(k,i) - e(i)}\binom{m(k') - a(k',j) + j - i}{m(k') - a(k',j) + j - i - e(i)} \end{aligned}$$

and in view of (2.11.0') we also see that

$$H^{(2k)}(m,p,a,e) = \text{det of the transpose of } G^{(2k)}(m,p,a,e)$$

and for every $i \in [1,p]$ and $j \in [1,p]$, the (i,j)th entry of the transpose has four different expressions which are obtained by interchanging i and j in the four expressions of $G_{ij}^{(2k)}$ above.

By (2.2.2) and (2.5.10') we see that for every $i \in [1,p]$ and $j \in [1,p]$ we have

$$\begin{aligned} &G_{ij}^{(2*k)}(m,p,a,e) \\ &= \binom{m(k) - a(k,i) + i - j}{e(j) + i - j}\binom{m(k') - a(k',j) + j - i}{e(j)} \\ &= \binom{m(k) - a(k,i) + i - j}{e(j) + i - j}\binom{m(k') - a(k',j) + j - i}{m(k') - a(k',j) + j - i - e(j)} \end{aligned}$$

$$= \begin{pmatrix} m(k) - a(k,i) + i - j \\ m(k) - a(k,i) - e(j) \end{pmatrix} \begin{pmatrix} m(k') - a(k',j) + j - i \\ m(k') - a(k',j) + j - i - e(j) \end{pmatrix}$$

and in view of (2.11.0') we also see that

$$H^{(2*k)}(m,p,a,e) = \text{det of the transpose of } G^{(2*k)}(m,p,a,e)$$

and for every $i \in [1,p]$ and $j \in [1,p]$, the (i,j)th entry of the transpose has four different expressions which are obtained by interchanging i and j in the four expressions of $G_{ij}^{(2*k)}$ above.

OBSERVATION (2.11.3') Let there be given any $m \in N^*(2)$ and $p \in N^*$ and $a \in \text{vec}(2,m,p)$ and $k \in [1,2]$. Now by (2.10.8) we see that for every $L \in \{3,3^*\}$ and $d \in Z(p)$ we have

$$H^{(Lk)}(m,p,a,d) = H^{(3z'[L]k)}(m,p,a,d),$$

where $z'[3^*] \in W[p]$ is defined by putting

$$z'[3^*](w) = w \qquad \text{for all } w \in W(p),$$

and where $z'[3] \in W[p]$ is defined by putting, for every $w \in W(p)$,

$$z'[3](w) = \text{the identity map of } [1,p] \text{ onto itself,}$$

i.e.,

$$z'[3](w)(i) = i \qquad \text{for all } i \in [1,p].$$

OBSERVATION (2.11.3) Let there be given any $m \in N^*(2)$ and $p \in N^*$ and $a \in \text{vec}(2,m,p)$ and $k \in [1,2]$ and $d \in Z(p)$, and let us put

$$k' = \begin{cases} 2 & \text{if } k = 1 \\ 1 & \text{if } k = 2. \end{cases}$$

Now, upon taking $d^* = d(i)$ in (2.10.7), we see that for every $i \in [1,p]$ and $j \in [1,p]$ we have

$$G_{ij}^{(3k)}(m,p,a,d)$$

$$= \begin{bmatrix} m(k) - a(k,j) \\ m(k') - a(k',i) - d(i) \end{bmatrix} \begin{pmatrix} m(k) - a(k,j) + j - i \\ d(i) \end{pmatrix}$$

$$= \begin{bmatrix} m(k) - a(k,j) \\ m(k') - a(k',i) - d(i) \end{bmatrix} \begin{pmatrix} m(k) - a(k,j) + j - i \\ m(k) - a(k,j) + j - i - d(i) \end{pmatrix}$$

$$= \begin{bmatrix} m(k') - a(k',i) - d(i) \\ m(k) - a(k,j) \end{bmatrix} \begin{pmatrix} m(k) - a(k,j) + j - i \\ m(k) - a(k,j) + j - i - d(i) \end{pmatrix}$$

and in view of (2.11.0') we also see that

$$H^{(3k)}(m,p,a,d) = \text{det of the transpose of } G^{(3k)}(m,p,a,d)$$

and for every $i \in [1,p]$ and $j \in [1,p]$, the (i,j)th entry of the transpose has four different expressions which are obtained by interchanging i and j in the four expressions of $G_{ij}^{(3k)}$ above.

Upon taking $d^* = d(j)$ in (2.10.7), we see that for every $i \in [1,p]$ and $j \in [1,p]$ we have

$$G_{ij}^{(3^*k)}(m,p,a,d)$$

$$= \begin{bmatrix} m(k) - a(k,j) \\ m(k') - a(k',i) - d(j) \end{bmatrix} \begin{pmatrix} m(k) - a(k,j) + j - i \\ d(j) \end{pmatrix}$$

$$\begin{bmatrix} m(k) - a(k,j) \\ m(k') - a(k',i) - d(j) \end{bmatrix} \begin{pmatrix} m(k) - a(k,j) + j - i \\ m(k) - a(k,j) + j - i - d(j) \end{pmatrix}$$

$$\begin{bmatrix} m(k') - a(k',i) - d(j) \\ m(k) - a(k,j) \end{bmatrix} \begin{pmatrix} m(k) - a(k,j) + j - i \\ m(k) - a(k,j) + j - i - d(j) \end{pmatrix}$$

and in view of (2.11.0') we also see that

$$H^{(3^*k)}(m,p,a,d) = \text{det of the transpose of } G^{(3^*k)}(m,p,a,d)$$

and for every $i \in [1,p]$ and $j \in [1,p]$, the (i,j)th entry of the transpose has four different expressions which are obtained by interchanging i and j in the four expressions of $G_{ij}^{(3^*k)}$ above.

Given any $z \in W[p]$ and $w \in W(p)$, upon taking $j = w(i)$ and $d^* = d(z(w)(i))$ in (2.10.7) and then producting as i varies over $[1,p]$, we see that

$$G_w^{(3zk)}(m,p,a,d)$$

$$= \prod_{i=1}^{p} \begin{bmatrix} m(k) - a(k,w(i)) \\ m(k') - a(k',i) - d(z(w)(i)) \end{bmatrix} \begin{pmatrix} m(k) - a(k,w(i)) + w(i) - i \\ d(z(w)(i)) \end{pmatrix}$$

$$= \prod_{i=1}^{p} \begin{bmatrix} m(k) - a(k,w(i)) \\ m(k') - a(k',i) - d(z(w)(i)) \end{bmatrix} \begin{pmatrix} m(k) - a(k,w(i)) + w(i) - i \\ m(k) - a(k,w(i)) + w(i) - i - d(z(w)(i)) \end{pmatrix}$$

$$= \prod_{i=1}^{p} \begin{bmatrix} m(k') - a(k',i) - d(z(w)(i)) \\ m(k) - a(k,w(i)) \end{bmatrix} \begin{pmatrix} m(k) - a(k,w(i)) + w(i) - i \\ m(k) - a(k,w(i)) + w(i) - i - d(z(w)(i)) \end{pmatrix}$$

OBSERVATION (2.11.11) Let there be given any $m \in N^*(2)$ and $p \in N^*$ and $a \in \mathrm{vec}(2,m,p)$ and $k \in [1,2]$ and $L \in \{1,1^*\}$, and let us put

$$\hat{C} = C[m,p,a,k].$$

Now by (2.10.1) we see that

$$(1)\quad \{e \in Z(p) : H^{(Lk)}(m,p,a,e) \neq 0\} \subset N[p,\hat{C}]$$

and by (2.10.5) we see that

$$(2)\quad \{D \in Z : \{E \in [0,\hat{C}] : H_E^{[Lk]}(m,p,a,D) \neq 0\} \neq \emptyset\} \subset [0,\hat{C}]$$

and by (4) of (2.11.0) we see that

(3) for every $D \in [0,\hat{C}]$ we have $C(m,p,a) - D \geq 0$.

By (1), first we see that for every $E \in Z$, the summation in the definition of $H_E^{(Lk)}(m,p,a)$ is essentially finite, and second, we see that

$$(4)\quad \{E \in Z : H_E^{(Lk)}(m,p,a) \neq 0\} \subset [0,\hat{C}].$$

By (4), first we see that for every $D \in Z$, the summation in the definition of $F_D^{(Lk)}(m,p,a)$ is essentially finite, and second, in view of (2), we see that

$$(5)\quad \{D \in Z : F_D^{(Lk)}(m,p,a) \neq 0\} \subset [0,\hat{C}].$$

By (5), first we see that

(6) [the summations in the definitions of $F^{((Lk))}(m,p,a,C',V)$, $F^{(Lk)}(m,p,a,V)$, $F^{[[Lk]]}(m,p,a,C',V)$ and $F^{[Lk]}(m,p,a,V)$ are essentially finite,

and second, in view of (3) and (2.5.7*), we see that

(7) for every $V \in N$ we have $F^{(Lk)}(m,p,a,V) = F^{[Lk]}(m,p,a,V)$.

OBSERVATION (2.11.12) Let there be given any $m \in N^*(2)$ and $p \in N^*$ and $a \in \mathrm{vec}(2,m,p)$ and $k \in [1,2]$ and $L \in \{2,2^*\}$, and let us put

$\hat{C} = C[m,p,a,k]$.

Now by (2.10.2) we see that

(1) $\{e \in Z(p) : H^{(Lk)}(m,p,a,e) \neq 0\} \subset N[p,\hat{C}]$

and by (2.10.6) we see that

(2) $\left\{D \in Z : \left\{E \in [0,\hat{C}] : \binom{E}{D} \neq 0\right\} \neq \emptyset\right\} \subset [0,\hat{C}]$

and by (4) of (2.11.0) we see that

(3) for every $D \in [0,\hat{C}]$ we have $C(m,p,a) - D \geq 0$.

By (1), first we see that for every $E \in Z$, the summation in the definition of $H_E^{(Lk)}(m,p,a)$ is essentially finite, and second, we see that

(4) $\{E \in Z : H_E^{(Lk)}(m,p,a) \neq 0\} \subset [0,\hat{C}]$.

By (4), first we see that for every $D \in Z$, the summation in the definition of $F_D^{(Lk)}(m,p,a)$ is essentially finite, and second, in view of (2), we see that

(5) $\{D \in Z : F_D^{(Lk)}(m,p,a) \neq 0\} \subset [0,\hat{C}]$.

By (5), first we see that

(6) [the summations in the definitions of $F^{((Lk))}(m,p,a,C',V)$, $F^{(Lk)}(m,p,a,V)$, $F^{[[Lk]]}(m,p,a,C',V)$, and $F^{[Lk]}(m,p,a,V)$ are essentially finite,

and second, in view of (3) and (2.5.7*), we see that

(7) for every $V \in N$ we have $F^{(Lk)}(m,p,a,V) = F^{[Lk]}(m,p,a,V)$.

OBSERVATION (2.11.13) Let there be given any $m \in N^*(2)$ and $p \in N^*$ and $a \in \text{vec}(2,m,p)$ and $k \in [1,2]$ and $L \in \{3,3^*\} \cup \{3z : z \in W[p]\}$, and let us put

$\hat{C} = C[m,p,a,k]$.

Now by (2.10.3) we see that

(1) $\{d \in Z(p) : H^{(Lk)}(m,p,a,d) \neq 0\} \subset N[p,\hat{C}]$

and by (4) of (2.11.0) we see that

(2) for every $D \in [0,\hat{C}]$ we have $C(m,p,a) - D \geq 0$.

By (1), first we see that for every $D \in Z$, the summation in the definition of $F_D^{(Lk)}(m,p,a)$ is essentially finite, and second, we see that

(3) $\{D \in Z : F_D^{(Lk)}(m,p,a) \neq 0\} \subset [0,\hat{C}]$.

By (3), first we see that

(4) $\left[\right.$the summations in the definitions of $F^{((Lk))}(m,p,a,C',V)$, $F^{(Lk)}(m,p,a,V)$, $F^{[[Lk]]}(m,p,a,C',V)$, and $F^{[Lk]}(m,p,a,V)$ are essentially finite,

and second, in view of (2) and (2.5.7*), we see that

(5) for every $V \in N$ we have $F^{(Lk)}(m,p,a,V) = F^{[Lk]}(m,p,a,V)$.

OBSERVATION (2.11.14) Let there be given any $m \in N^*(2)$ and $p \in N^*$ and $a \in \text{vec}(2,m,p)$ and $k \in [1,2]$ and $L \in \{4,4^*\}$. Now by (2.10.4) we see that

$$\{v \in Z(p) : H^{(Lk)}(m,p,a,v) \neq 0\} \subset N(p)$$

and hence the summation in the definition of $F^{(Lk)}(m,p,a,V)$ is essentially finite.

OBSERVATION (2.11.16) Let there be given any $m \in N^*(2)$ and $p \in N^*$ and $a \in \text{vec}(2,m,p)$. Now by (2.10.4') we see that

$$\{v \in Z(p) : H^{(6)}(2,m,p,a,v) \neq 0\} \subset \{v \in Z(p) : v(i) \geq i - p \text{ for all } i \in [1,p]\}$$

and hence the summation in the definition of $F^{(6)}(2,m,p,a,V)$ is essentially finite.

3. SUMMATION LEMMAS

In (3.1) to (3.3) we shall state some obvious, but basic principles of summation; using these principles, in (3.11) to (3.28) we prepare the groundwork for proving the conditional equivalence of the determinantal polynomials

$$F^{(8)},\ F^{(7)},\ F^{(7*)},\ F^{(6)},\ F^{(5k)}$$

which were introduced in (2.7).

As a direct consequence of the multilinearity of the determinant as a function of its rows or columns, in (3.33) to (3.38) we show that if every entry of a p by p determinant is given as a sum of several elements, all these elements can be collated in numerous ways, and each such collating method gives rise to an expression of the said determinant as a sum of many other p by p determinants. Now the idea of multilinearity arises essentially out of the usual distributive law, so the reader's attention is especially drawn to Remarks (3.31) and (3.32), where we talk about this law. The material (3.31) to (3.38) is totally independent of the material (3.1) to (3.30).

LEMMA (3.1) Let there be given any sets X' and X^*, and any map $f: X^* \to Q$, and any injective map $A: X' \to X^*$. Also, let there be given any $X \subset X'$. Now if $\operatorname{card}(X) < \infty$, then

$$\sum_{v\in A(X)} f(v) = \sum_{v\in X} f(A(v)).$$

Proof: Obvious.

LEMMA (3.2) Let there be given any sets X' and X^*, and any map $f: X^* \to Q$ and any injective map $A: X' \to X^*$. Also, let there be given any $X \subset X'$. Now if

$$\operatorname{card}(\{v \in A(X) : f(v) \neq 0\}) < \infty,$$

then

$$\sum_{v\in A(X)} f(v) = \sum_{v\in X} f(A(v))$$

where both the summations are essentially finite.

Proof: Obvious.

LEMMA (3.3) Let there be given any set X^*, and any map $F: X^* \to Q$. Also let there be given any $Y \subset X^*$ and any $Y' \subset X^*$. Finally, let there be given any finite set W^* and for every $w \in W^*$, let there be given any $Y'_w \subset Y'$ such that

$$Y' = \coprod_{w \in W^*} Y'_w .$$

Now if

$$\{v \in Y : f(v) \neq 0\} \subset Y' \subset Y$$

and

$$\text{card}(\{v \in Y'_w : f(v) \neq 0\}) < \infty \qquad \text{for all } w \in W^*,$$

then

$$\sum_{v \in Y} f(v) = \sum_{w \in W^*} \sum_{v \in Y'_w} f(v)$$

where all the summations are essentially finite.

Proof: Obvious.

LEMMA (3.11) Let there be given any $p \in N^*$. For every $e \in Z(p)$ and $t \in Z(p)$, let $u[e,t] \in Z(p)$ be defined by putting

$$u[e,t](i) = e(i) + t(i) \qquad \text{for all } i \in [1,p]. \tag{3.11.1'}$$

Now if $f: Z(p) \to Q$ is any map such that

$$\text{card}(\{e \in Z(p) : f(e) \neq 0\}) < \infty,$$

then for every $t \in Z(p)$ we have

$$\sum_{e \in Z(p)} f(e) = \sum_{e \in Z(p)} f(u[e,t])$$

where both the summations are essentially finite.

Proof: In (3.2) take $X^* = X' = X = Z(p)$ and let $A: X' \to X^*$ be the map obtained by putting $A(e) = u[e,t]$ for all $e \in X'$.

LEMMA (3.12) Let there be given any $p \in N^*$. For every $d \in Z(p)$ and $g \in W(p)$, let $h[d,g] \in Z(p)$ be defined by putting

$$h[d,g](i) = d(g(i)) \quad \text{for all } i \in [1,p]. \tag{3.12.1'}$$

Now for every map $f: Z(p) \to Q$ and every $g \in W(p)$ and every $D \in Z$ we have

$$\sum_{d \in N(p,D)} f(d) = \sum_{d \in N(p,D)} f(h[d,g]).$$

Proof: In (3.1) take $X^* = X' = Z(p)$ and $X = N(p,D)$ and let $A: X' \to X^*$ be the map obtained by putting $A(d) = h[d,g]$ for all $d \in X'$.

LEMMA (3.13) Let p and u be as in (3.11). Let $f: Z(p) \to Q$ be any map such that for some $D \in Z$ we have

$$\{v \in Z(p) : f(v) \neq 0\} \subset \{v \in Z(p) : v(i) \geq D \text{ for all } i \in [1,p]\}.$$

Now for every $T \in Z$ and $t \in Z(p,T)$ and $V \in Z$ we have

$$\sum_{v \in Z(p,V+T)} f(v) = \sum_{v \in Z(p,V)} f(u[v,t])$$

where both the summations are essentially finite.

Proof: In (3.2) take $X^* = X' = Z(p)$ and $X = Z(p,V)$ and let $A: X' \to X^*$ be the map obtained by putting $A(v) = u[v,t]$ for all $v \in X'$.

LEMMA (3.14) Let p and h be as in (3.12). Let $f: Z(p) \to Q$ be any map such that for some $D \in Z$ we have

$$\{v \in Z(p) : f(v) \neq 0\} \subset \{v \in Z(p) : v(i) \geq D \text{ for all } i \in [1,p]\}.$$

Now for every $g \in W(p)$ and $V \in Z$ we have

$$\sum_{v \in Z(p,V)} f(v) = \sum_{v \in Z(p,V)} f(h[v,g])$$

where both the summations are essentially finite.

Proof: In (3.2) take $X^* = X' = Z(p)$ and $X = Z(p,V)$ and let $A: X' \to X^*$ be the map obtained by putting $A(v) = h[v,g]$ for all $v \in X'$.

LEMMA (3.15) Let there be given any $p \in N^*$. For every $v \in Z(p)$ and $w \in W(p)$ and $r \in Z(p)$, let $h'[v,w,r] \in Z(p)$ be defined by putting

$$h'[v,w,r](i) = v(w(i)) - r(i) + r(w(i)) \quad \text{for all } i \in [1,p] \tag{3.15.1'}$$

and let $h^*[v,w,r] \in Z(p)$ be defined by putting

$$h^*[v,w,r](i) = v(w(i)) + r(w(i)) \quad \text{for all } i \in [1,p]. \tag{3.15.2'}$$

Let $f: Z(p) \to Q$ be any map, and let u and h be as in (3.11) and (3.12). Now, given any $g \in W(p)$ and $r \in Z(p)$, upon letting $f_g: Z(p) \to Q$ be the map obtained by putting

$$f_g(v) = f(h[v,g]) \quad \text{for all } v \in Z(p) \tag{3.15.3'}$$

we have

$$f_g(u[v,r]) = f(h^*[v,g,r]) \quad \text{for all } v \in Z(p) \tag{3.15.4}$$

and upon letting t_{gr} be the member of $Z(p)$ obtained by putting

$$t_{gr}(i) = r(i) - r(g^{-1}(i)) \quad \text{for all } i \in [1,p] \tag{3.15.5'}$$

we have

$$f_g(u[v,t_{gr}]) = f(h'[v,g,r]) \quad \text{for all } v \in Z(p). \tag{3.15.6}$$

Proof: Upon substituting $u[v,r]$ for v in (3.15.3') we see that

(1) $f_g(u[v,r]) = f(h[u[v,r],g])$ for all $v \in Z(p)$.

Now given any $v \in Z(p)$, for every $j \in [1,p]$ we get

$h[u[v,r],g](j)$

$= u[v.r](g(j))$ [by taking $d = u[v,r]$ and $j = i$ in (3.12.1')]

$= v(g(j)) + r(g(j))$ [by taking $e = v$ and $t = r$ and $i = g(j)$ in (3.11.1')]

$= h^*[v,g,r](j)$ [by taking $w = g$ and $i = j$ in (3.15.2')]

and hence we have

$$h[u[v,r],g] = h^*[v,g,r]$$

and therefore in view of (1) we get (3.15.4).

Upon substituting $u[v,t_{gr}]$ for v in (3.15.3') we see that

(2) $f_g(u[v,t_{gr}]) = f(h[u[v,t_{gr}],g])$ for all $v \in Z(p)$.

Now given any $v \in Z(p)$, for every $j \in [1,p]$ we get

$h[u[v,t_{gr}],g](j)$

$= u[v,t_{gr}](g(j))$ [by taking $d = u[v,t_{gr}]$ and $j = i$ in (3.12.1')]

$= v(g(j)) + t_{gr}(g(j))$ [by taking $e = v$ and $t = t_{gr}$ and $i = g(j)$ in (3.11.1')]

$= v(g(j)) + [r(g(j)) - r(g^{-1}(g(j)))]$ [by taking $i = g(j)$ in (3.15.5')]

$= v(g(j)) - r(j) + r(g(j))$

$= h'[v,g,r](j)$ [by taking $w = g$ and $i = j$ in (3.15.1')]

and hence we have

$$h[u[v,t_{gr}],g] = h'[v,g,r]$$

and therefore in view of (2) we get (3.15.6).

LEMMA (3.16) Let p and h' be as in (3.15). Let $f: Z(p) \to Q$ be any map such that for some $D \in Z$ we have

$$\{v \in Z(p) : f(v) \neq 0\} \subset \{v \in Z(p) : v(i) \geq D \text{ for all } i \in [1,p]\}.$$

Now for every $g \in W(p)$ and $r \in Z(p)$ and $V \in Z$ we have

$$\sum_{v \in Z(p,V)} f(v) = \sum_{v \in Z(p,V)} f(h'[v,g,r])$$

where both the summations are essentially finite.

Proof: Let the notation be as in (3.15). Now by (3.14) we see that

(1) $\left[\sum_{v \in Z(p,V)} f(v) = \sum_{v \in Z(p,V)} f_g(v)\right.$ where both the summations are essentially finite.

Also, clearly,

$$\{v \in Z(p) : f_g(v) \neq 0\} \subset \{v \in Z(p) : v(i) \geq D \text{ for all } i \in [1,p]\}$$

and

$$t_{gr} \in Z(p,0)$$

and therefore by (3.13) we see that

(2) $\left[\begin{array}{l}\displaystyle\sum_{v\in Z(p,V)} f_g(v) = \sum_{v\in Z(p,V)} f_g(u[v,t_{gr}]) \\ \text{where both the summations are essentially finite.}\end{array}\right.$

Now in view of (3.15.6), our assertion follows from (1) and (2).

LEMMA (3.17) Let p and h' be as in (3.15). Let $f'\colon Z(p) \to Q$ be any map. Now given any $w \in W(p)$ and $r \in Z(p)$, upon letting $g = w^{-1}$ and upon letting $\hat{f}'\colon Z(p) \to Q$ to be the map obtained by putting

$$\hat{f}'(v) = f'(h'[v,w,r]) \quad \text{for all } v \in Z(p) \tag{3.17.1'}$$

we have

$$\hat{f}'(h'[v,g,r]) = f'(v) \quad \text{for all } v \in Z(p) \tag{3.17.2}$$

Proof: Upon substituting $h'[v,g,r]$ for v in (3.17.1') we see that

(1) $\hat{f}'(h'[v,g,r]) = f'(h'[h'[v,g,r],w,r]) \quad$ for all $v \in Z(p)$.

Now given any $v \in Z(p)$, for every $j \in [1,p]$ we get

$$\begin{aligned}
&h'[h'[v,g,r],w,r](j)\\
&= h'[v,g,r](w(j)) - r(j) + r(w(j))\\
&\quad [\text{by taking } v = h'[v,g,r] \text{ and } i = j \text{ in } (3.15.1')]\\
&= [v(g(w(j))) - r(w(j)) + r(g(w(j)))] - r(j) + r(w(j))\\
&\quad [\text{by taking } w = g \text{ and } i = w(j) \text{ in } (3.15.1')]\\
&= v(j) \quad (\text{because } g = w^{-1})
\end{aligned}$$

and hence we have

$$h'[h'[v,g,r],w,r] = v$$

and therefore in view of (1) we get (3.17.2).

LEMMA (3.18) Let p, u, h', h* be as in (3.11) and (3.15). Let $f^*\colon Z(p) \to Q$ be any map. Now given any $w \in W(p)$ and $r \in Z(p)$, upon letting $g = w^{-1}$ and upon letting $\hat{f}^*\colon Z(p) \to Q$ to be the map obtained by putting

$$\hat{f}^*(v) = f^*(h^*[v,w,r]) \quad \text{for all } v \in Z(p) \tag{3.18.1'}$$

we have

$$\hat{f}^*(h'[v,g,r]) = f^*(u[v,r]) \quad \text{for all } v \in Z(p). \tag{3.18.2}$$

Proof: Upon substituting $h'[v,g,r]$ for v in (3.18.1') we see that

(1) $\hat{f}^*(h'[v,g,r]) = f^*(h^*[h'[v,g,r],w,r])$ for all $v \in Z(p)$.

Now given any $v \in Z(p)$, for every $j \in [1,p]$ we get

$h^*[h'[v,g,r],w,r](j)$

$= h'[v,g,r](w(j)) + r(w(j))$

[by taking $v = h'[v,g,r]$ and $i = j$ in (3.15.2')]

$= [v(g(w(j))) - r(w(j)) + r(g(w(j)))] + r(w(j))$

[by taking $w = g$ and $i = w(j)$ in (3.15.1')]

$= v(j) + r(j)$ (because $g = w^{-1}$)

$= u[v,r](j)$ [by taking $e = v$ and $t = r$ and $i = j$ in (3.11.1')]

and hence we have

$$h^*[h'[v,g,r],w,r] = u[v,r]$$

and therefore in view of (1) we get (3.18.2).

LEMMA (3.19) Let p, u, h', h^* be as in (3.11) and (3.15). Let $f'\colon Z(p) \to Q$ and $f^*\colon Z(p) \to Q$ be any maps. Now, given any $w \in W(p)$ and $r \in Z(p)$, upon letting $g = w^{-1}$ and upon letting $\hat{f}\colon Z(p) \to Q$ be the map obtained by putting

$$\hat{f}(v) = f'(h'[v,w,r])f^*(h^*[v,w,r]) \quad \text{for all } v \in Z(p) \tag{3.19.1'}$$

we have

$$\hat{f}(h'[v,g,r]) = f'(v)f^*(u[v,r]) \quad \text{for all } v \in Z(p). \tag{3.19.2}$$

Proof: Follows from (3.17) and (3.18).

LEMMA (3.20) Let p, u, h', h^* be as in (3.11) and (3.15). Let $f'\colon Z(p) \to Q$ be any map such that for some $E \in Z$ we have

$$\{v \in Z(p) : f'(v) \neq 0\} \subset \{v \in Z(p) : v(i) \geq E \text{ for all } i \in [1,p]\}.$$

Let $f^*: Z(p) \to Q$ be any map. Now for every $w \in W(p)$ and $r \in Z(p)$ and $V \in Z$ we have

$$\sum_{v \in Z(p,V)} f'(h'[v,w,r])f^*(h^*[v,w,r]) = \sum_{v \in Z(p,V)} f'(v)f^*(u[v,r])$$

where both the summations are essentially finite.

Proof: Given any $w \in W(p)$ and $r \in Z(p)$, let g and $\hat{f}$ be as in (3.19). Now, upon letting

$$D = E - \max\{r(w(i)) - r(i) : i \in [1,p]\},$$

by (3.15.1') we see that

$$\{v \in Z(p) : f'(h'[v,w,r]) \neq 0\} \subset \{v \in Z(p) : v(i) \geq D \text{ for all } i \in [1,p]\}$$

and hence by (3.19.1') we see that

$$\{v \in Z(p) : \hat{f}(v) \neq 0\} \subset \{v \in Z(p) : v(i) \geq D \text{ for all } i \in [1,p]\}$$

and therefore, in view of (3.19.1') and (3.19.2), our assertion follows by taking $f = \hat{f}$ in (3.16).

LEMMA (3.21) Let p, u, h, h^* be as in (3.11), (3.12), and (3.15), and let there be given any $q \in N^*$. Let $f: Z(p) \to Q$ be any map such that for every $v \in Z(p)$ and $g \in W(p)$ we have

$$f(h[v,g]) = [\mathrm{sgn}(g)]^{q-1}f(v). \tag{3.21.1'}$$

Now for every $v \in Z(p)$ and $g \in W(p)$ and $r \in Z(p)$ we have

$$f(u[v,r]) = [\mathrm{sgn}(g)]^{q-1}f(h^*[v,g,r]). \tag{3.21.2}$$

Proof: Given any $v \in Z(p)$ and $g \in W(p)$ and $r \in Z(p)$, upon taking $u[v,r]$ for v in (3.21.1'), we see that

(1) $f(h[u[v,r],g]) = [\mathrm{sgn}(g)]^{q-1}f(u[v,r])$

and upon using the notation of (3.15) and upon taking $u[v,r]$ for v in (3.15.3'), we see that

(2) $f_g(u[v,r]) = f(h[u[v,r],g])$

and clearly we have

(3) $[\mathrm{sgn}(g)]^{q-1} = 1/[\mathrm{sgn}(g)]^{q-1}$

and now, in view of (3.15.4), by (1), (2), and (3) we get (3.21.2).

LEMMA (3.22) Let p, u, h, h' be as in (3.11), (3.12), and (3.15), and let there be given any $q \in N^*$, and let there be given any map $f': Z(p) \to Q$ such that for some $E \in Z$ we have

$$\{v \in Z(p) : f'(v) \neq 0\} \subset \{v \in Z(p) : v(i) \geq E \text{ for all } i \in [1,p]\},$$

and let there be given any map $f^*: Z(p) \to Q$ such that for every $v \in Z(p)$ and $w \in W(p)$ we have

$$f^*(h[v,w]) = [\mathrm{sgn}(w)]^{q-1} f^*(v),$$

and for every $v \in Z(p)$ and $w \in W(p)$ and $r \in Z(p)$ let us put

$$f''_w(v,r) = \mathrm{sgn}(w) f'(h'[v,w,r]),$$

and for every $v \in Z(p)$ and $r \in Z(p)$ let us put

$$f''(v,r) = \sum_{w \in W(p)} f''_w(v,r).$$

Then we have the following.

(3.22.1) For every $w \in W(p)$ and $r \in Z(p)$ and $V \in Z$ we have

$$\sum_{v \in Z(p,V)} f''_w(v,r) f^*(u[v,r]) = [\mathrm{sgn}(w)]^q \sum_{v \in Z(p,V)} f'(v) f^*(u[v,r])$$

where both the summations are essentially finite.

(3.22.2) For every $r \in Z(p)$ and $V \in Z$, upon letting

$$\hat{F}(V,r) = (1/p!) \sum_{v \in Z(p,V)} f''(v,r) f^*(u[v,r])$$

and

$$\tilde{F}(V,r) = \sum_{v \in Z(p,V)} f'(v) f^*(u[v,r])$$

we have that the two summations above are essentially finite and

$$\hat{F}(V,r) = \begin{cases} \tilde{F}(V,r) & \text{if either } q \text{ is even or } p = 1 \\ 0 & \text{if } q \text{ is odd and } p \geq 2. \end{cases}$$

Proof: Let h^* be as in (3.15). Now given $v \in Z(p)$ and $w \in W(p)$ and $r \in Z(p)$, upon taking $g = w$ and $f = f^*$ in (3.21), we see that

$$f^*(u[v,r]) = [\mathrm{sgn}(w)]^{q-1} f^*(h^*[v,w,r])$$

and therefore (3.22.1) follows from (3.20). Clearly, (3.22.2) follows from (3.22.1).

LEMMA (3.23) Let p, u, h, h' be as in (3.11), (3.12), and (3.15), and let there be given any $q \in N^*$ and $k \in [1,q]$, and let there be given any map $f'\colon Z(p) \to Q$ such that for some $E \in Z$ we have

$$\{v \in Z(p) : f'(v) \neq 0\} \subset \{v \in Z(p) : v(i) \geq E \text{ for all } i \in [1,p]\},$$

and for every $n \in [1,q]\backslash\{k\}$ let there be given any map $f^{*(n)}\colon Z(p) \to Q$ such that for every $v \in Z(p)$ and $w \in W(p)$ we have

$$f^{*(n)}(h[v,w]) = \mathrm{sgn}(w) f^{*(n)}(v),$$

and let $f^*\colon Z(p) \to Q$ be the map obtained by putting

$$f^*(v) = \prod_{n \in [1,q]\backslash\{k\}} f^{*(n)}(v) \qquad \text{for all } v \in Z(p),$$

and for every $v \in Z(p)$ and $w \in W(p)$ and $r \in Z(p)$ let us put

$$f''_w(v,r) = \mathrm{sgn}(w) f'(h'[v,w,r]),$$

and for every $v \in Z(p)$ and $r \in Z(p)$ let us put

$$f''(v,r) = \sum_{w \in W(p)} f''_w(v,r).$$

Then we have (3.22.1) and (3.22.2).

Proof: Follows from (3.22).

LEMMA (3.24) Let there be given any $q \in N^*$ and $p \in N^*$ and $t' \in \mathrm{pre}[q,p]$ and $k \in [1,q]$. For every $n \in [1,q]$ and $j \in [1,p]$ let there be given any map $B_j^{(n)}\colon Z \to Q$. Assume that there exists $E \in Z$ such that

$$\{v^* \in Z : B_j^{(k)}(v^*) \neq 0\} \subset \{v^* \in Z : v^* \geq E\} \qquad \text{for all } j \in [1,p].$$

Now first, for every $n \in [1,q]$ and $i \in [1,p]$ and $j \in [1,p]$ and $v \in Z(p)$ let us put

$$\hat{G}_{ij}^{(5n)}(v) = B_j^{(n)}(v(i) + t'(n,j) - t'(k,i)),$$

and second, for every $n \in [1,q]$ and $v \in Z(p)$ let us put

$$\hat{G}^{(5n)}(v) = \text{the } p \text{ by } p \text{ matrix whose } (i,j)\text{th entry is } \hat{G}_{ij}^{(5n)}(v),$$

and third, for every $w \in W(p)$ and $v \in Z(p)$ let us put

$$\hat{H}_w^{(5k)}(v) = \text{sgn}(w)\left[\prod_{i=1}^{p} B_i^{(k)}(v(w(i)) + t'(k,i) - t'(k,w(i)))\right] \prod_{n\in[1,q]\setminus\{k\}} \det \hat{G}^{(5n)}(v),$$

and fourth, for every $v \in Z(p)$ let us put

$$\hat{H}^{(5k)}(v) = \left[\prod_{i=1}^{p} B_i^{(k)}(v(i))\right] \prod_{n\in[1,q]\setminus\{k\}} \det \hat{G}^{(5n)}(v)$$

and

$$\hat{H}^{(5)}(v) = \prod_{n=1}^{q} \det \hat{G}^{(5n)}(v).$$

Then we have the following.

(3.24.1) For every $w \in W(p)$ and $V \in Z$ we have

$$\sum_{v\in Z(p,V)} \hat{H}_w^{(5k)}(v) = [\text{sgn}(w)]^q \sum_{v\in Z(p,V)} \hat{H}^{(5k)}(v)$$

where both the summations are essentially finite.

(3.24.2) For every $V \in Z$, upon letting

$$\hat{F}^{(5)}(V) = (1/p!) \sum_{v\in Z(p,V)} \hat{H}^{(5)}(v)$$

and

$$\hat{F}^{(5k)}(V) = \sum_{v \in Z(p,V)} \hat{H}^{(5k)}(v)$$

we have that the two summations above are essentially finite and

$$\hat{F}^{(5)}(V) = \begin{cases} \hat{F}^{(5k)}(V) & \text{if either } q \text{ is even or } p = 1 \\ 0 & \text{if } q \text{ is odd and } p \geq 2. \end{cases}$$

Proof: Let u, h, h' be as in (3.11), (3.12), and (3.15). Let f': $Z(p) \to Q$ be the map obtained by putting

$$(1') \quad f'(v) = \prod_{j=1}^{p} B_j^{(k)}(v(j)) \qquad \text{for all } v \in Z(p).$$

Now clearly

$$(2') \quad \{v \in Z(p) : f'(v) \neq 0\} \subset \{v \in Z(p) : v(i) \geq E \text{ for all } i \in [1,p]\}.$$

Let r be the member of Z(p) obtained by putting

$$(3') \quad r(i) = -t'(k,i) \qquad \text{for all } i \in [1,p].$$

In view of (1') and (2') and the definition of h' we see that for every $v \in Z(p)$ and $w \in W(p)$ we have

$$(4) \quad f'(h'[v,w,r]) = \prod_{j=1}^{p} B_j^{(k)}(v(w(j)) + t'(k,j) - t'(k,w(j))).$$

Upon expanding det $\hat{G}^{(5k)}(v)$ by columns, in view of (4) we see that for every $v \in Z(p)$ we have

$$(5) \quad \det \hat{G}^{(5k)}(v) = \sum_{w \in W(p)} \mathrm{sgn}(w) f'(h'[v,w,r]).$$

By general properties of determinants we know that

(6) [if we apply any permutation w to the rows of any p by p matrix, then its det gets multiplied by sgn(w).

Given any $n \in [1,q]\backslash\{k\}$, by (6) we see that for every $v \in Z(p)$ and $w \in W(p)$ we have

(7) $\left[\begin{array}{l}\text{[det of the p by p matrix whose (i,j)th entry is} \\ B_j^{(n)}(v(w(i)) + t'(n,j))] = \text{sgn}(w) \text{ times [det of the p by p} \\ \text{matrix whose (i,j)th entry is } B_j^{(n)}(v(i) + t'(n,j))].\end{array}\right.$

For every $n \in [1,q]\backslash\{k\}$ let $f^{*(n)}: Z(p) \to Q$ be the map obtained by putting, for all $v \in Z(p)$,

(8') $\left[\begin{array}{l}f^{*(n)}(v) = \text{det of the p by p matrix whose (i,j)th entry is} \\ B_j^{(n)}(v(i) + t'(n,j)).\end{array}\right.$

By (7) and (8') and the definition of h we see that for every $v \in Z(p)$ and $w \in W(p)$ we have

(9) $\quad f^{*(n)}(h[v,w]) = \text{sgn}(w)f^{*(n)}(v).$

Let $f^*: Z(p) \to Q$ be the map obtained by putting

(10') $\quad f^*(v) = \prod_{n\in[1,q]\backslash\{k\}} f^{*(n)}(v) \quad$ for all $v \in Z(p)$.

By (3'), (8'), (10'), and the definition of u we see that for every $v \in Z(p)$ we have

(11) $\quad f^*(u[v,r]) = \prod_{n\in[1,q]\backslash\{k\}} \det \hat{G}^{(5n)}(v).$

By (4) and (11) we see that for every $v \in Z(p)$ and $w \in W(p)$ we have

(12) $\quad \hat{H}_w^{(5k)}(v) = \text{sgn}(w)f'(h'[v,w,r])f^*(u[v,r]).$

By (1') and (11) we see that for every $v \in Z(p)$ we have

(13) $\quad \hat{H}^{(5k)}(v) = f'(v)f^*(u[v,r]).$

By (5) and (11) we see that for every $v \in Z(p)$ we have

(14) $\quad \hat{H}^{(5)}(v) = \left[\sum_{w\in W(p)} \text{sgn}(w)f'(h'[v,w,r])\right]f^*(u[v,r]).$

Now in view of (12), (13), and (14), our assertions follow from (3.23).

PROPOSITION (3.25) Let there be given any $q \in N^*$ and $p \in N^*$ and $t \in \text{pre}[q,p]$ and $t' \in \text{pre}[q,p]$ and $k \in [1,q]$. Now first, for every $n \in [1,q]$ and $i \in [1,p]$ and $j \in [1,p]$ and $v \in Z(p)$ let us put

$$\hat{G}_{ij}^{(5n)}(v) = \begin{bmatrix} t(n,j) \\ v(i) + t'(n,j) - t'(k,i) \end{bmatrix},$$

and second, for every $n \in [1,q]$ and $v \in Z(p)$ let us put

$$\hat{G}^{(5n)}(v) = \text{the } p \text{ by } p \text{ matrix whose } (i,j)\text{th entry is } \hat{G}_{ij}^{(5n)}(v),$$

and third, for every $w \in W(p)$ and $v \in Z(p)$ let us put

$$\hat{H}_{w}^{(5k)}(v) = \text{sgn}(w)\left(\prod_{i=1}^{p}\begin{bmatrix} t(k,i) \\ v(w(i)) + t'(k,i) - t'(k,w(i)) \end{bmatrix}\right) \times \prod_{n\in[1,q]\setminus\{k\}} \det \hat{G}^{(5n)}(v),$$

and fourth, for every $v \in Z(p)$ let us put

$$\hat{H}^{(5k)}(v) = \left(\prod_{i=1}^{p}\begin{bmatrix} t(k,i) \\ v(i) \end{bmatrix}\right) \prod_{n\in[1,q]\setminus\{k\}} \det \hat{G}^{(5n)}(v)$$

and

$$\hat{H}^{(5)}(v) = \prod_{n=1}^{q} \det \hat{G}^{(5n)}(v).$$

Then we have (3.24.1) and (3.24.2).

Proof: In view of (2.5.2*), upon taking $E = 0$ and upon taking $B_j^{(n)}(v^*) = \begin{bmatrix} t(n,j) \\ v^* \end{bmatrix}$ for all $n \in [1,q]$ and $j \in [1,p]$ and $v^* \in Z$ our assertions follow from (3.24).

LEMMA (3.26) Let p and h be as in (3.12), and let there be given any $q \in N^*$, and let there be given any map $f\colon Z(p) \to Q$ such that

$$\{v \in Z(p) : f(v) \neq 0\} \subset \{v \in N(p) : v(i) \neq v(j) \text{ for all } i \neq j \text{ in } [1,p]\}$$

and such that

$$f(h[v,w]) = [\text{sgn}(w)]^q f(v) \quad \text{for all } v \in Z(p) \text{ and } w \in W(p).$$

Now, given any $V \in Z$, upon referring to (2.7) for the definitiona of $Z(p,V,7^*)$ and $Z(p,V,7)$ and upon letting

$$\hat{F}^{(7^*)}(V) = (1/p!) \sum_{v \in Z(p,V,7^*)} f(v)$$

and

$$\hat{F}^{(7)}(v) = \sum_{v \in Z(p,V,7)} f(v)$$

we have that the two summations above are essentially finite and

$$\hat{F}^{(7^*)}(V) = \begin{cases} \hat{F}^{(7)}(V) & \text{if either } q \text{ is even or } p = 1 \\ 0 & \text{if } q \text{ is odd and } p \geq 2. \end{cases}$$

Proof: For every $w \in W(p)$, we clearly get a bijective map A_w: $Z(p) \to Z(p)$ by putting

$$A_w(v) = h[v,w] \qquad \text{for all } v \in Z(p).$$

Note that now

(1) $f(A_w(v)) = [\mathrm{sgn}(w)]^q f(v)$ for all $v \in Z(p)$ and $w \in W(p)$.

Let there be fixed any $D \in Z$. Now upon letting

$$X_D = \{v \in Z(p,D) : v(1) > v(2) > \cdots > v(p) \geq 0\}$$

and

$$Y_D = \{v \in Z(p,D) : v(1) > v(2) > \cdots > v(p)\}$$

we clearly have

(2) $\{v \in Y_D : f(v) \neq 0\} \subset X_D \subset Y_D$ and $\mathrm{card}(X_D) < \infty$.

By (2) we see that

(3) $\mathrm{card}(\{v \in Y_D : f(v) \neq 0\}) < \infty$

and

(4) $\begin{cases} \sum_{v \in Y_D} f(v) = \sum_{v \in X_D} f(v) \\ \text{where both the summations are essentially finite.} \end{cases}$

Given any $w \in W(p)$, in view of (3), upon taking

$$X^* = X' = Z(p) \quad \text{and} \quad X = Y_D \quad \text{and} \quad A = A_w$$

in (3.2) we see that

$$\sum_{v \in A_w(Y_D)} f(v) = \sum_{v \in Y_D} f(A_w(v))$$

where both the summations are essentially finite and hence in view of (1) we see that

$$\sum_{v \in A_w(Y_D)} f(v) = [\operatorname{sgn}(w)]^q \sum_{v \in Y_D} f(v)$$

where both the summations are essentially finite and therefore in view of (4) we conclude that

(5) $\begin{cases} \text{for every } w \in W(p) \text{ we have} \\ \sum_{v \in A_w(Y_D)} f(v) = [\operatorname{sgn}(w)]^q \sum_{v \in X_D} f(v) \\ \text{where both the summations are essentially finite.} \end{cases}$

By (5) we have

(6) $\operatorname{card}(\{v \in A_w(Y_D) : f(v) \neq 0\}) < \infty \quad$ for all $w \in W(p)$.

Now upon letting

$$\hat{Y}_D = \{v \in Z(p,D) : v(i) \neq v(j) \quad \text{for all } i \neq j \text{ in } [1,p]\}$$

we clearly have

$$\{v \in Z(p,D) : f(v) \neq 0\} \subset \hat{Y}_D \subset Z(p,D)$$

and

$$\hat{Y}_D = \coprod_{w \in W(p)} A_w(Y_D)$$

and hence, in view of (6), upon taking

$$X^* = Z(p) \text{ and } Y = Z(p,D) \text{ and } Y' = Y_D \text{ and } W^* = W(p) \text{ and } Y'_w = A_w(Y_D)$$

in (3.3) we see that

(7) $$\begin{cases} \displaystyle\sum_{v\in Z(p,D)} f(v) = \sum_{w\in W(p)} \sum_{v\in A_w(Y_D)} f(v) \\ \text{where all the summations are essentially finite.} \end{cases}$$

By the definitions of $Z(p,V,7^*)$ and $Z(p,V,7)$ we see that

(8) $$\begin{cases} \text{if } D = V + [p(p-1)/2], \text{ then} \\ Z(p,D) = Z(p,V,7^*) \text{ and } X_D = Z(p,V,7). \end{cases}$$

Upon letting

$$\hat{F}^*(D) = (1/p!) \sum_{v\in Z(p,D)} f(v) \quad \text{and} \quad \hat{F}(D) = \sum_{v\in X_D} f(v)$$

by (5) and (7) we see that the two summations above are essentially finite and

$$\hat{F}^*(D) = \begin{cases} \hat{F}(D) & \text{if either } q \text{ is even or } p = 1 \\ 0 & \text{if } q \text{ is odd and } p \geq 2 \end{cases}$$

and therefore, in view of (8), our assertion follows by taking $D = V + [p(p-1)/2]$.

LEMMA (3.27) Let there be given any $q \in N^*$ and $p \in N^*$ and $t \in \text{pre}[q,p]$. For every $n \in [1,q]$ let there be given any map $B_j^{(n)} : Z \to Q$. Assume that there exists $k \in [1,q]$ such that

$$\{v^* \in Z : B_j^{(k)}(v^*) \neq 0\} \subset N \qquad \text{for all } j \in [1,p].$$

Now first, for every $n \in [1,q]$ and $i \in [1,p]$ and $j \in [1,p]$ and $v \in Z(p)$ let us put

$$\hat{G}_{ij}^{(6n)}(v) = B_j^{(n)}(v(i) + j - i)$$

and

$$\hat{G}_{ij}^{(7n)}(v) = B_j^{(n)}(v(i) + j - p)$$

and

$$\hat{G}_{ij}^{(8n)}(v) = B_j^{(n)}(v(i) + v(i+1) + \cdots + v(p) + j - i),$$

and second, for every $L \in [6,8]$ and $n \in [1,q]$ and $v \in Z(p)$ let us put

$$\hat{G}^{(Ln)}(v) = \text{the } p \text{ by } p \text{ matrix whose } (i,j)\text{th entry is } \hat{G}_{ij}^{(Ln)}(v),$$

and third, for every $L \in [6,8]$ let $\hat{H}^{(L)}: Z(p) \to Q$ be the map obtained by putting

$$\hat{H}^{(L)}(v) = \prod_{n=1}^{q} \det \hat{G}^{(Ln)}(v) \qquad \text{for all } v \in Z(p).$$

Then we have the following.

(3.27.1) For every $V \in Z$, upon referring to (2.7) for the definition of $Z(p,V,7^*)$ and upon letting

$$\hat{F}^{(7^*)}(V) = (1/p!) \sum_{v \in Z(p,V,7^*)} \hat{H}^{(7)}(v)$$

and

$$\hat{F}^{(6)}(V) = (1/p!) \sum_{v \in Z(p,V)} \hat{H}^{(6)}(v)$$

we have that the two summations above are essentially finite and

$$\hat{F}^{(7^*)}(V) = \hat{F}^{(6)}(V).$$

(3.27.2) For every $V \in Z$, upon referring to (2.7) for the definitions of $Z(p,V,7)$ and $Z(p,V,8)$ and upon letting

$$\hat{F}^{(7)}(V) = \sum_{v \in Z(p,V,7)} \hat{H}^{(7)}(v)$$

and

$$\hat{F}^{(8)}(V) = \sum_{v \in Z(p,V,8)} \hat{H}^{(8)}(v)$$

we have

$$\hat{F}^{(7)}(V) = \hat{F}^{(8)}(V).$$

(3.27.3) For every $V \in Z$, with the notation of (3.27.1) and (3.27.2), we have

$$\hat{F}^{(7^*)}(V) = \begin{cases} \hat{F}^{(7)}(V) & \text{if either } q \text{ is even or } p = 1 \\ 0 & \text{if } q \text{ is odd and } p \geq 2. \end{cases}$$

Proof: Now

$$\left[\begin{array}{l}\text{if } v \in Z(p) \text{ and } i \in [1,p] \text{ are such that } v(i) < 0, \\ \text{then } v(i) + j - p < 0 \text{ for all } j \in [1,p]\end{array}\right.$$

and hence

$$\left[\begin{array}{l}\text{if } v \in Z(p) \text{ and } i \in [1,p] \text{ are such that } v(i) < 0, \\ \text{then } \hat{G}_{ij}^{(7k)}(v) = 0 \text{ for all } j \in [1,p]\end{array}\right.$$

and therefore

$$\{v \in Z(p) : \det \hat{G}^{(7k)}(v) \neq 0\} \subset N(p)$$

and hence

(1) $\{v \in Z(p) : \hat{H}^{(7)}(v) \neq 0\} \subset N(p).$

Upon letting $T = p(p - 1)/2$ we clearly have

$$Z(p,V,7^*) = Z(p, V + T) \quad \text{for all } V \in Z$$

and upon letting u be as in (3.11) and upon letting t be the member of $Z(p)$ obtained by putting $t(i) = p - i$ for all $i \in [1,p]$ we clearly have

$$t \in Z(p,T)$$

and

$$\hat{H}^{(6)}(v) = \hat{H}^{(7)}(u[v,t]) \quad \text{for all } v \in Z(p)$$

and therefore, in view of (1), upon taking $D = 0$ and $f = \hat{H}^{(7)}$ in (3.13) we get (3.27.1).

Upon letting $A: Z(p) \to Z(p)$ be the map obtained by putting $A(v)(i) = v(i) + v(i+1) + \cdots + v(p) + p - i$ for all $v \in Z(p)$ and $i \in [1,p]$ we clearly have

$$A(Z(p,V,8)) \subset Z(p,V,7) \qquad \text{for all } V \in Z$$

and

$$\hat{H}^{(8)}(v) = \hat{H}^{(7)}(A(v)) \qquad \text{for all } v \in Z(p)$$

and upon letting $A': Z(p) \to Z(p)$ be the map obtained by putting

$$A'(v)(i) = \begin{cases} v(p) & \text{for all } v \in Z(p) \text{ and } i = p \\ v(i) - v(i+1) - 1 & \text{for all } v \in Z(p) \text{ and } i \in [1, p-1] \end{cases}$$

we clearly have

$$A'(Z(p,V,7)) \subset Z(p,V,8) \qquad \text{for all } V \in Z$$

and

$$A(A'(v)) = v = A'(A(v)) \qquad \text{for all } v \in Z(p)$$

and hence A is bijective and

$$A(Z(p,V,8)) = Z(p,V,7) \qquad \text{for all } V \in Z$$

and therefore (3.27.2) follows from (3.1) by taking $X' = X^* = Z(p)$ and $f = \hat{H}^{(7)}$ and $X = Z(p,V,8)$.

Now

$$\left[\begin{array}{l}\text{if } v \in Z(p) \text{ and } i \in [1,p] \text{ and } i' \in [1,p] \text{ are such that } v(i) = \\ v(i'), \text{ then the ith row of } \hat{G}^{(7k)}(v) \text{ coincides with its i'th row}\end{array}\right.$$

and hence

$$\left[\begin{array}{l}\text{if } v \in Z(p) \text{ is such that } v(i) = v(i') \text{ for some } i \neq i' \text{ in} \\ [1,p], \text{ then } \det \hat{G}^{(7k)}(v) = 0\end{array}\right.$$

and therefore

$$\begin{aligned} &\{v \in Z(p) : \hat{H}^{(7)}(v) \neq 0\} \\ &\subset \{v \in Z(p) : v(i) \neq v(j) \text{ for all } i \neq j \text{ in } [1,p]\} \end{aligned}$$

and hence in view of (1) we conclude that

(2) $\begin{cases}\{v \in Z(p) : \hat{H}^{(7)}(v) \neq 0\} \\ \subset \{v \in N(p) : v(i) \neq v(j) \text{ for all } i \neq j \text{ in } [1,p]\}.\end{cases}$

By general properties of determinants we know that

[if we apply any permutation w to the rows of any p by p matrix, then its det gets multiplied by sgn(w)

and hence, upon letting h be as in (3.12), we see that for every $n \in [1,q]$ and $v \in Z(p)$ and $w \in W(p)$ we have

$$\det \hat{G}^{(7n)}(h[v,w]) = \text{sgn}(w)\det \hat{G}^{(7n)}(v)$$

and therefore we get

(3) $\hat{H}^{(7)}(h[v,w]) = [\text{sgn}(w)]^q\hat{H}^{(7)}(v)$ for all $v \in Z(p)$ and $w \in W(p)$.

In view of (2) and (3), upon taking $f = \hat{H}^{(7)}$ in (3.26) we get (3.27.3).

PROPOSITION (3.28) Let there be given any $q \in N^*$ and $p \in N^*$ and $t \in \text{pre}[q,p]$. Now first, for every $n \in [1,q]$ and $i \in [1,p]$ and $j \in [1,p]$ and $v \in Z(p)$ let

$$\hat{G}_{ij}^{(6n)}(v) = \begin{bmatrix} t(n,j) \\ v(i) + j - i \end{bmatrix} \quad \text{and} \quad \hat{G}_{ij}^{(7n)}(v) = \begin{bmatrix} t(n,j) \\ v(i) + j - p \end{bmatrix}$$

and

$$\hat{G}_{ij}^{(8n)}(v) = \begin{bmatrix} t(n,j) \\ v(i) + v(i+1) + \cdots + v(p) + j - i \end{bmatrix},$$

and second, for every $L \in [6,8]$ and $n \in [1,q]$ and $v \in Z(p)$ let

$$\hat{G}^{(Ln)}(v) = \text{the p by p matrix whose (i,j)th entry is } \hat{G}_{ij}^{(Ln)}(v),$$

and third, for every $L \in [6,8]$ and $v \in Z(p)$ let

$$\hat{H}^{(L)}(v) = \prod_{n=1}^{q} \det \hat{G}^{(Ln)}(v).$$

Then we have (3.27.1) and (3.27.2) and (3.27.3).

Proof: In view of (2.5.2*), our assertion follows from (3.27) by taking

$$B_j^{(n)}(v^*) = \begin{bmatrix} t(n,j) \\ v^* \end{bmatrix} \quad \text{for all } n \in [1,q] \text{ and } j \in [1,p] \text{ and } v^* \in Z.$$

REMARK (3.31) The usual *distributive law*

$$(A + B)C = AC + BC$$

gives rise to the more general distributive laws (or, schoolbook exercises on opening parenthesis), such as

$$(A + B)(A' + B') = AA' + AB' + BA' + BB'$$

and

$$(A + B)(A' + B')(A^* + B^* + C^*) = AA'A^* + AA'B^* + \cdots + BB'C^*$$

and so on. To put this into proper perspective, we may proceed thus.

Let there be given any $p \in N^*$. Let $Y_1, Y_2, \ldots, Y_p$ be any sets and let

$$T = \begin{bmatrix} \text{the set of all maps } e: [1,p] \to \cup_{k=1}^{p} Y_k \\ \text{such that } e(k) \in Y_k \text{ for all } k \in [1,p] \end{bmatrix}$$

(in other words, let T be the cartesian product of $Y_1, Y_2, \ldots, Y_p$). For every $k \in [1,p]$ and $e^* \in Y_k$, let there be given any $f^{(k)}(e^*) \in Q$.

Now *the generalized distributive law* can be stated by saying that if $S, X_1, X_2, \ldots, X_p$ are such that

$$\begin{bmatrix} X_k \subset Y_k \text{ and } \operatorname{card}(X_k) < \infty \text{ for all } k \in [1,p], \\ \text{and } S = \{e \in T : e(k) \in X_k \text{ for all } k \in [1,p]\}, \end{bmatrix}$$

then we have

$$\prod_{k=1}^{p} \sum_{e^* \in X_k} f^{(k)}(e^*) = \sum_{e \in S} \prod_{k=1}^{p} f^{(k)}(e(k)). \qquad (3.31.1)$$

REMARK (3.32) By using a different notation for the cartesian product, (3.31) can be rephrased thus. Let there be given any $q \in N^*$, and let there be given any sets $T, Y_1, Y_2, \ldots, Y_q$, and for every $e \in T$ and $k \in [1,q]$, let there be given any $z[e,k] \in Y_k$;

assume that given any $k \in [1,q]$ and $r \in Y_k$, there exists $e \in T$ such that $z[e,k] = r$; also assume that $e \neq e'$ in T, there exists $k \in [1,q]$ such that $z[e,k] \neq z[e',k]$ (in other words, let T be the cartesian product of $Y_1, Y_2, \ldots, Y_q$). For every $k \in [1,q]$ and $r \in Y_k$ let there be given any $f^{(k)}(r) \in Q$.

Now *the generalized distributive law* can be stated by saying that if $S, X_1, X_2, \ldots, X_q$ are such that

$$\begin{cases} X_k \subset Y_k \text{ and } \operatorname{card}(X_k) < \infty \text{ for all } k \in [1,q], \\ \text{and } S = \{e \in T : z[e,k] \in X_k \text{ for all } k \in [1,q]\}, \end{cases} \tag{3.32.1}$$

then we have

$$\prod_{k=1}^{q} \sum_{r \in X_k} f^{(k)}(r) = \sum_{e \in S} \prod_{k=1}^{p} f^{(k)}(z[e,k]). \tag{3.32.1}$$

LEMMA (3.33) Let there be given any $p \in N^*$. For every $i \in [1,p]$ let there be given any set Y_i, and let

$$T = \begin{cases} \text{the set of all maps } e\colon [1,p] \to \cup_{i=1}^{p} Y_i \\ \text{such that } e(i) \in Y_i \text{ for all } i \in [1,p]. \end{cases}$$

For every $i \in [1,p]$ and $j \in [1,p]$ let there be given any map $L_{ij}\colon Y_i \to Q$ such that

$$\operatorname{card}(\{e^* \in Y_i : L_{ij}(e^*) \neq 0\}) < \infty.$$

Let $\hat{K}$ be the p by p matrix whose (i,j)th entry is

$$\sum_{e^* \in Y_i} L_{ij}(e^*).$$

For every $e \in T$ let $K'(e)$ be the p by p matrix whose (i,j)th entry is $L_{ij}(e(i))$. Then we have

$$\operatorname{card}(\{e \in T : \det K'(e) \neq 0\}) < \infty$$

and

$$\det \hat{K} = \sum_{e \in T} \det K'(e).$$

REMARK-PROOF Briefly speaking, the equation above follows by expanding det $\hat{K}$ and applying the distributive law (3.31.1) to each one of the resulting p! terms. Alternatively, our assertion follows from the multiadditivity of the determinant as a function of its rows, where multiadditivity means multilinearity relative to Z; for a more detailed explanation of this second approach, we may proceed as follows.

Let Q(p) be the set of all maps of [1,p] into Q. Now Q(p) becomes an additive abelian group by putting

$$(x + y)(k) = x(k) + y(k) \quad \text{for all } x \in Q(p) \text{ and } y \in Q(p) \text{ and } k \in [1,p].$$

Upon letting 0^* be the member of Q(p) obtained by putting

$$0^*(k) = 0 \quad \text{for all } k \in [1,p]$$

we see that 0^* is the zero element of Q(p).

Let f be a multiadditive Q-valued function of p variables, each of which varies over Q(p); in other words, let there be given

$$f(x_1,x_2,\dots,x_p) \in Q \quad \text{for all } x_1 \in Q(p),\ x_2 \in Q(p),\ \dots,\ x_p \in Q(p),$$

such that

(1) $\left[\begin{array}{l}\text{for every } y_1 \in Q(p),\ y_2 \in Q(p),\ \dots,\ y_p \in Q(p), \text{ and}\\ \text{for every } k \in [1,p] \text{ and } z_k \in Q(p), \text{ we have}\\ f(y_1,\dots,y_{k-1},y_k + z_k,y_{k+1},\dots,y_p) = f(y_1,y_2,\dots,y_p) +\\ f(y_1,\dots,y_{k-1},z_k,y_{k+1},\dots,y_p).\end{array}\right.$

Note that then

(2) $\left[\begin{array}{l}\text{for every } k \in [1,p], \text{ and for every}\\ y_1 \in Q(p),\ \dots,\ y_{k-1} \in Q(p),\ y_{k+1} \in Q(p),\ \dots,\ y_p \in Q(p),\\ \text{we must have } f(y_1,\dots,y_{k-1},\ 0^*,\ y_{k+1},\dots,y_p) = 0.\end{array}\right.$

For every $i \in [1,p]$ let $A_i\colon Y_i \to Q(p)$ be the map obtained by putting

(3') $A_i(e^*)(j) = L_{ij}(e^*) \quad$ for all $e^* \in Y_i$ and $j \in [1,p]$.

In view of our assumption, there exists

(4) $X_i \subset Y_i$ for all $i \in [1,p]$

such that

(5) $\mathrm{card}(X_i) < \infty$ for all $i \in [1,p]$

and

(6) $\{e^* \in Y_i : A_i(e^*) \neq 0^*\} \subset X_i$ for all $i \in [1,p]$.

Let

(7) $S = \{e \in T : e(i) \in X_i \text{ for all } i \in [1,p]\}$.

By (1), (4), (5), and (7) we see that

(8) $$\begin{cases} f\left(\sum_{e^*\in X_1} A_1(e^*), \ldots, \sum_{e^*\in X_i} A_i(e^*), \ldots, \sum_{e^*\in X_p} A_p(e^*)\right) \\ = \sum_{e\in S} f(A_1(e(1)), \ldots, A_i(e(i)), \ldots, A_p(e(p))). \end{cases}$$

By (4), (5), and (6) we see that

(9) $$\begin{cases} \{e^* \in Y_i : A_i(e^*) \neq 0^*\} \subset X_i \subset Y_i \text{ and } \mathrm{card}(X_i) < \infty \\ \text{for all } i \in [1,p] \end{cases}$$

and hence

(10) $$\begin{cases} f\left(\sum_{e^*\in Y_1} A_1(e^*), \ldots, \sum_{e^*\in Y_i} A_i(e^*), \ldots, \sum_{e^*\in Y_p} A_p(e^*)\right) \\ \quad = f\left(\sum_{e^*\in X_1} A_1(e^*), \ldots, \sum_{e^*\in X_i} A_i(e^*), \ldots, \sum_{e^*\in X_p} A_p(e^*)\right). \end{cases}$$

By (2), (6), and (7) we see that

$$\{e \in T : f(A_1(e(1)), \ldots, A_i(e(i)), \ldots, A_p(e(p))) \neq 0\} \subset S$$

and therefore in view of (4) and (5) we get

(11) $$\begin{cases} \{e \in T : f(A_1(e(1)), \ldots, A_i(e(i)), \ldots, A_p(e(p))) \neq 0\} \\ \subset S \subset T \text{ and } \mathrm{card}(S) < \infty \end{cases}$$

and hence

$$(12)\quad \begin{cases} \sum_{e\in T} f(A_1(e(1)),\ldots,A_i(e(i)),\ldots,A_p(e(p))) \\ = \sum_{e\in S} f(A_1(e(1)),\ldots,A_i(e(i)),\ldots,A_p(e(p))). \end{cases}$$

By (8) to (12) we see that

$$(13)\quad \begin{cases} \operatorname{card}(\{e^* \in Y_i : A_i(e^*) \neq 0^*\}) < \infty \text{ for all } i \in [1,p] \text{ and} \\ \operatorname{card}(\{e \in T : f(A_1(e(1))),\ldots,A_i(e(i)),\ldots,A_p(e(p))) \neq 0\}) < \infty \end{cases}$$

and

$$(14)\quad \begin{cases} f\left(\sum_{e^*\in Y_1} A_1(e^*), \ldots, \sum_{e^*\in Y_i} A_i(e^*), \ldots, \sum_{e^*\in Y_p} A_p(e^*)\right) \\ = \sum_{e\in T} f(A_1(e(1)), \ldots, A_i(e(i)), \ldots, A_p(e(p))). \end{cases}$$

For all $x_1 \in Q(p), \ldots, x_i \in Q(p), \ldots, x_p \in Q(p)$, let

$$(15')\quad \begin{cases} \hat{f}(x_1,\ldots,x_i,\ldots,x_p) = \text{the det of the p by p matrix} \\ \text{whose (i,j)th entry is } x_i(j). \end{cases}$$

Now det as a function of its rows is multilinear and so in particular multiadditive; in other words, $\hat{f}$ is a multiadditive Q-valued function of p variables, each of which varies over Q(p); therefore, our assertion follows by taking $f = \hat{f}$ in (13) and (14).

LEMMA (3.34) Let there be given any $p \in N^*$. For every $j \in [1,p]$ let there be given any set Y_j, and let

$$T = \begin{cases} \text{the set of all maps } e\colon [1,p] \to \cup_{j=1}^{p} Y_j \text{ such that} \\ e(j) \in Y_j \text{ for all } j \in [1,p]. \end{cases}$$

For every $i \in [1,p]$ and $j \in [1,p]$ let there be given any map L_{ij}: $Y_j \to Q$ such that

$$\operatorname{card}(\{e^* \in Y_j : L_{ij}(e^*) \neq 0\}) < \infty.$$

Let $\hat{K}$ be the p by p matrix whose (i,j)th entry is

$$\sum_{e^*\in Y_j} L_{ij}(e^*).$$

For every $e \in T$ let $K^*(e)$ be the p by p matrix whose (i,j)th entry is $L_{ij}(e(j))$. Then we have

$$\mathrm{card}(\{e \in T : \det K^*(e) \neq 0\}) < \infty$$

and

$$\det \hat{K} = \sum_{e \in T} \det K^*(e).$$

REMARK-PROOF: In the Remark-Proof of (3.33), everywhere change row to column, and everywhere, except in (3') and (15'), change i to j, and change (3') to

(3*) $A_j(e^*)(i) = L_{ij}(e^*)$ for all $e^* \in Y_j$ and $i \in [1,p]$

and change (15') to

(15*) $\begin{cases} f(x_1,\ldots,x_j,\ldots,x_p) = \text{the det of the p by p matrix whose} \\ \text{(i,j)th entry is } x_j(i). \end{cases}$

LEMMA (3.35) Let there be given any $p \in N^*$. For every $i \in [1,p]$ and $j \in [1,p]$ let there be given any set X_{ij}, and any map $K_{ij}: X_{ij} \to Q$ such that

$$\mathrm{card}(\{e^* \in X_{ij} : K_{ij}(e^*) \neq 0\}) < \infty.$$

Let $\hat{K}$ be the p by p matrix whose (i,j)th entry is

$$\sum_{e^* \in X_{ij}} K_{ij}(e^*).$$

For every $i \in [1,p]$ let there be given any set Y_i, and let

$$T = \begin{cases} \text{the set of all maps } e\colon [1,p] \to \cup_{i=1}^{p} Y_i \text{ such that} \\ e(i) \in Y_i \text{ for all } i \in [1,p]. \end{cases}$$

For every $i \in [1,p]$ and $j \in [1,p]$ let there be given any bijective map $A_{ij}: Y_{ij} \to X_{ij}$ with $Y_{ij} \subset Y_i$, and let $L_{ij}: Y_i \to Q$ be the map obtained by putting

$$L_{ij}(e^*) = \begin{cases} K_{ij}(A_{ij}(e^*)) & \text{if } e^* \in Y_{ij} \\ 0 & \text{if } e^* \in Y_i \setminus Y_{ij}. \end{cases}$$

For every $e \in T$ let $K'(e)$ be the p by p matrix whose (i,j)th entry is $L_{ij}(e(i))$. Then we have

$$\text{card}(\{e \in T : \det K'(e) \neq 0\}) < \infty$$

and

$$\det \hat{K} = \sum_{e \in T} \det K'(e).$$

Proof: For every $i \in [1,p]$ and $j \in [1,p]$ we clearly have

$$\text{card}(\{e^* \in Y_i : L_{ij}(e^*) \neq 0\}) < \infty$$

and

$$\sum_{e^* \in Y_i} L_{ij}(e^*) = \sum_{e^* \in X_{ij}} K_{ij}(e^*)$$

and therefore our assertion follows from (3.33).

LEMMA (3.36) Let there be given any $p \in N^*$. For every $i \in [1,p]$ and $j \in [1,p]$ let there be given any set X_{ij}, and any map $K_{ij}: X_{ij} \to Q$ such that

$$\text{card}(\{e^* \in X_{ij} : K_{ij}(e^*) \neq 0\}) < \infty.$$

Let $\hat{K}$ be the p by p matrix whose (i,j)th entry is

$$\sum_{e^* \in X_{ij}} K_{ij}(e^*).$$

For every $j \in [1,p]$ let there be given any set Y_j, and let

$$T = \begin{bmatrix} \text{the set of all maps } e: [1,p] \to \cup_{j=1}^{p} Y_j \text{ such that} \\ e(j) \in Y_j \text{ for all } j \in [1,p]. \end{bmatrix}$$

For every $i \in [1,p]$ and $j \in [1,p]$ let there be given any bijective map $A_{ij}: Y_{ij} \to X_{ij}$ with $Y_{ij} \subset Y_j$, and let $L_{ij}: Y_j \to Q$ be the map obtained by putting

$$L_{ij}(e^*) = \begin{cases} K_{ij}(A_{ij}(e^*)) & \text{if } e^* \in Y_{ij} \\ 0 & \text{if } e^* \in Y_j \backslash Y_{ij}. \end{cases}$$

For every $e \in T$ let $K^*(e)$ be the p by p matrix whose (i,j)th entry is $L_{ij}(e(j))$. Then we have

$$\text{card}(\{e \in T : \det K^*(e) \neq 0\}) < \infty$$

and

$$\det \hat{K} = \sum_{e \in T} \det K^*(e).$$

Proof: For every $i \in [1,p]$ and $j \in [1,p]$ we clearly have

$$\text{card}(\{e^* \in Y_j : L_{ij}(e^*) \neq 0\}) < \infty$$

and

$$\sum_{e^* \in Y_j} L_{ij}(e^*) = \sum_{e^* \in X_{ij}} K_{ij}(e^*)$$

and therefore our assertion follows from (3.34).

LEMMA (3.37) Let there be given any $p \in N^*$. For every $i \in [1,p]$ and $j \in [1,p]$ let there be given any set X_{ij}, and any map K_{ij}: $X_{ij} \to Q$ such that

$$\text{card}(\{e^* \in X_{ij} : K_{ij}(e^*) \neq 0\}) < \infty.$$

Let $\hat{K}$ be the p by p matrix whose (i,j)th entry is

$$\sum_{e^* \in X_{ij}} K_{ij}(e^*).$$

Let there be given any set Y, and let T be the set of all maps of [1,p] into Y. For every $i \in [1,p]$ and $j \in [1,p]$ let there be given any bijective map A_{ij}: $Y_{ij} \to X_{ij}$ with $Y_{ij} \subset Y$, and let L_{ij}: $Y \to Q$ be the map obtained by putting

$$L_{ij}(e^*) = \begin{cases} K_{ij}(A_{ij}(e^*)) & \text{if } e^* \in Y_{ij} \\ 0 & \text{if } e^* \in Y \backslash Y_{ij}. \end{cases}$$

For every $e \in T$ let $K'(e)$ and $K^*(e)$ be the p by p matrices whose (i,j)th entries are $L_{ij}(e(i))$ and $L_{ij}(e(j))$, respectively. Then we have

$$\operatorname{card}(\{e \in T : \det K'(e) \neq 0\}) < \infty$$

and

$$\operatorname{card}(\{e \in T : \det K^*(e) \neq 0\}) < \infty$$

and

$$\sum_{e \in T} \det K'(e) = \det \hat{K} = \sum_{e \in T} \det K^*(e).$$

Proof: Upon taking $Y_i = Y$ for all $i \in [1,p]$ in (3.35) we see that

$$\operatorname{card}(\{e \in T : \det K'(e) \neq 0\}) < \infty$$

and

$$\det \hat{K} = \sum_{e \in T} \det K'(e).$$

Upon taking $Y_j = Y$ for all $j \in [1,p]$ in (3.36) we see that

$$\operatorname{card}(\{e \in T : \det K^*(e) \neq 0\}) < \infty$$

and

$$\det \hat{K} = \sum_{e \in T} \det K^*(e).$$

LEMMA (3.38) Let there be given any $p \in N^*$. Let there be given any set X, and let T be the set of all maps of [1,p] into X. For every $i \in [1,p]$ and $j \in [1,p]$ let there be given any map $K_{ij}: X \to Q$ such that

$$\operatorname{card}(\{e^* \in X : K_{ij}(e^*) \neq 0\}) < \infty.$$

Let $\hat{K}$ be the p by p matrix whose (i,j)th entry is

$$\sum_{e^* \in X} K_{ij}(e^*).$$

For every $e \in T$ let $K'(e)$ and $K^*(e)$ be the p by p matrices whose (i,j)th entries are $K_{ij}(e(i))$ and $K_{ij}(e(j))$, respectively. Then we have

$$\text{card}(\{e \in T : \det K'(e) \neq 0\}) < \infty$$

and

$$\text{card}(\{e \in T : \det K^*(e) \neq 0\}) < \infty$$

and

$$\sum_{e \in T} \det K'(e) = \det \hat{K} = \sum_{e \in T} \det K^*(e).$$

Proof: Follows from (3.37) by taking $Y = X$ and by taking $Y_{ij} = X_{ij} = X$ and A_{ij} = the identity map of X onto X for all $i \in [1,p]$ and $j \in [1,p]$.

4. BINOMIAL LEMMAS

In this section we give a systematic treatment of some aspects of binomial coefficients, and in (4.64) we pose a related problem. Before reading this section, the reader may find it useful to browse in Motivational Remark (4.63); similarly, the reader should not omit Informational Remarks (4.62) and (4.65).

SPECIAL BIPRODUCT LEMMA (4.1') Given any $A \in Z$ and $B \in Z$, in $Q[V]$ we have

$$\binom{V}{A}\binom{V-A}{B} = \binom{V}{A+B}\begin{bmatrix}B\\A\end{bmatrix} = \binom{V}{A+B}\begin{bmatrix}A\\B\end{bmatrix} = \binom{V}{B}\binom{V-B}{A}.$$

Proof: By (2.5.2') and (2.5.2*) we see that

(1) $\left[\begin{array}{l}\text{if } A < 0, \text{ then the first and second and fourth expressions}\\ \text{are equal to zero,}\end{array}\right.$

whereas

(2) $\left[\begin{array}{l}\text{if } B < 0, \text{ then the first and third and fourth expressions}\\ \text{are equal to zero.}\end{array}\right.$

By (2.5.6*) we see that

(3) $\left[\begin{array}{l}\text{if } A < 0 \le A + B, \text{ then } \begin{bmatrix}A\\B\end{bmatrix} = 0\\ \text{and hence the third expression is zero,}\end{array}\right.$

whereas

(4) $\left[\text{if } B < 0 \le A + B, \text{ then } \begin{bmatrix} B \\ A \end{bmatrix} = 0\right.$ and hence the second expression is zero.

By (2.5.2') we see that

(5) [if $A + B < 0$, then the second and third expressions are equal to zero.

By (1) to (5) we see that

(6) [if either $A < 0$ or $B < 0$, then all the four expressions are equal to zero.

Finally, if $A \ge 0 \le B$ then by (2.5.5*) and the definition of binomial coefficients, we see that

$$
\begin{aligned}
\binom{V}{A}\binom{V-A}{B} &= \left[\frac{V(V-1)\cdots(V-A+1)}{A!}\right] \\
&\quad \times \left[\frac{(V-A)(V-A-1)\cdots(V-A-B+1)}{B!}\right] \\
&= \left[\frac{V(V-1)\cdots(V-A-B+1)}{(A+B)!}\right]\left[\frac{(A+B)!}{A!\,B!}\right] = \binom{V}{A+B}\begin{bmatrix} B \\ A \end{bmatrix} \\
&= \left[\frac{V(V-1)\cdots(V-A-B+1)}{(A+B)!}\right]\left[\frac{(A+B)!}{A!\,B!}\right] = \binom{V}{A+B}\begin{bmatrix} A \\ B \end{bmatrix} \\
&= \left[\frac{V(V-1)\cdots(V-B+1)}{B!}\right] \\
&\quad \times \left[\frac{(V-B)(V-B-1)\cdots(V-B-A+1)}{A!}\right] \\
&= \binom{V}{B}\binom{V-B}{A}.
\end{aligned}
$$

SPECIAL TWISTED BIPRODUCT LEMMA (4.1*) Given any $A \in Z$ and $B \in Z$, in $Q[V]$ we have

$$
\begin{bmatrix} V \\ A \end{bmatrix}\begin{bmatrix} V+A \\ B \end{bmatrix} = \begin{bmatrix} V \\ A+B \end{bmatrix}\begin{bmatrix} B \\ A \end{bmatrix} = \begin{bmatrix} V \\ A+B \end{bmatrix}\begin{bmatrix} A \\ B \end{bmatrix} = \begin{bmatrix} V \\ B \end{bmatrix}\begin{bmatrix} V+B \\ A \end{bmatrix}.
$$

NOTE In view of (2.5.3*), the assertion follows by substituting $V + A + B$ for V in (4.1'). As an alternative, we shall now give a direct proof of (4.1*), and then we could invoke (2.5.3') to see that (4.1') would follow from (4.1*) by substituting $V - A - B$ for V.

Proof: By (2.5.2') and (2.5.2*) we see that

(1) $\left[\begin{array}{l}\text{if } A < 0\text{, then the first and second and fourth}\\ \text{expressions are equal to zero,}\end{array}\right.$

whereas

(2) $\left[\begin{array}{l}\text{if } B < 0\text{, then the first and third and fourth}\\ \text{expressions are equal to zero.}\end{array}\right.$

By (2.5.6*) we see that

(3) $\left[\begin{array}{l}\text{if } A < 0 \le A + B\text{, then } \begin{bmatrix}A\\B\end{bmatrix} = 0\\ \text{and hence the third expression is zero,}\end{array}\right.$

whereas

(4) $\left[\begin{array}{l}\text{if } B < 0 \le A + B\text{, then } \begin{bmatrix}B\\A\end{bmatrix} = 0\\ \text{and hence the second expression is zero.}\end{array}\right.$

By (2.5.2*) we see that

(5) $\left[\begin{array}{l}\text{if } A + B < 0\text{, then the second and third}\\ \text{expressions are equal to zero.}\end{array}\right.$

By (1) to (5) we see that

(6) $\left[\begin{array}{l}\text{if either } A < 0 \text{ or } B < 0\text{, then all the four}\\ \text{expressions are equal to zero.}\end{array}\right.$

Finally, if $A \ge 0 \le B$, then by (2.5.3*) and the definition of twisted binomial coefficients we see that

$$\begin{aligned}
\begin{bmatrix}V\\A\end{bmatrix}\begin{bmatrix}V + A\\B\end{bmatrix} &= \left\{\frac{(V + 1)(V + 2)\cdots(V + A)}{A!}\right\}\\
&\quad\times\left\{\frac{(V + A + 1)(V + A + 2)\cdots(V + A + B)}{B!}\right\}\\
&= \left\{\frac{(V + 1)(V + 2)\cdots(V + A + B)}{(A + B)!}\right\}\left\{\frac{(A + B)!}{A!\ B!}\right\} = \begin{bmatrix}V\\A + B\end{bmatrix}\begin{bmatrix}B\\A\end{bmatrix}\\
&= \left\{\frac{(V + 1)(V + 2)\cdots(V + A + B)}{(A + B)!}\right\}\left\{\frac{(A + B)!}{A!\ B!}\right\} = \begin{bmatrix}V\\A + B\end{bmatrix}\begin{bmatrix}A\\B\end{bmatrix}\\
&= \left\{\frac{(V + 1)(V + 2)\cdots(V + B)}{B!}\right\}\\
&\quad\times\left\{\frac{(V + B + 1)(V + B + 2)\cdots(V + B + A)}{A!}\right\}\\
&= \begin{bmatrix}V\\B\end{bmatrix}\begin{bmatrix}V + B\\A\end{bmatrix}.
\end{aligned}$$

SWITCHING LEMMA (4.2') Given any $U \in Z$ and $T \in Z$, in $Q[V]$ we have

$$\binom{V}{U}\binom{U}{T} = \binom{V}{T}\binom{V-T}{U-T} = \binom{V}{U-T}\binom{V-U+T}{T} = \binom{V}{U}\binom{U}{U-T}.$$

Proof: In view of (2.5.3*), our assertion follows by taking $A = T$ and $B = U - T$ in (4.1').

TWISTED SWITCHING LEMMA (4.2*) Given any $U \in Z$ and $T \in Z$, in $Q[V]$ we have

$$\left[{V \atop U}\right]\binom{U}{T} = \left[{V \atop T}\right]\left[{V+T \atop U-T}\right] = \left[{V \atop U-T}\right]\left[{V+U-T \atop T}\right] = \left[{V \atop U}\right]\binom{U}{U-T}.$$

Proof: In view of (2.5.3*), our assertion follows by taking $A = T$ and $B = U - T$ in (4.1*).

SPECIAL SWITCHING LEMMA (4.3) Given any $p \in N$ and $T \in [0,p]$, for every $U \in Z$ we have

$$\binom{p}{U}\binom{U}{T} = \binom{p}{T}\binom{p-T}{U-T}.,$$

NOTE This is only a special case of the first part of (4.2') obtained by assuming that $V = p \in N$ and $T \in [0,p]$.

REMARK (4.4) Given any $T \in Z$ and $U \in Z$ and $V \in Z$, in view of (2.5.10'), we are tempted to investigate whether the first part of (4.2') remains valid after changing

$$\binom{V-T}{U-T} \text{ to } \binom{V-T}{V-U}.$$

Note that here we are forced to assume V to be in Z because, in the second binomial coefficient exhibited above, V occurs in the denominator.

Now first, by (2.5.8') we see that

(1) $\binom{V}{T} = 0$ iff either $T < 0$ or $0 \le V < T$,

and second, by (2.5.10') and (2.5.11') and (2.5.12') we see that

(2) $\binom{V-T}{U-T} = \binom{V-T}{V-U}$ iff either $T \le V$ or $V < U < T$,

and third, it is clear that

$$(3) \quad \binom{V}{T}\binom{V-T}{U-T} = \binom{V}{T}\binom{V-T}{V-U} \text{ iff either } \binom{V}{T} = 0 \text{ or } \binom{V-T}{U-T} = \binom{V-T}{V-U}$$

and fourth, it is clear that

$$(4) \quad \left[\begin{array}{l}\text{if either } T < 0 \text{ or } 0 \le V < T \text{ or } T \le V \text{ or } V < U < T,\\ \text{then either } T < 0 \text{ or } 0 \le V \text{ or } V < U < T,\end{array}\right.$$

and fifth, it is clear that

$$(5) \quad \left[\begin{array}{l}\text{if either } T < 0 \text{ or } 0 \le V \text{ or } V < U < T, \text{ then either}\\ T < 0 \text{ or } 0 \le V < T \text{ or } T \le V \text{ or } V < U < T.\end{array}\right.$$

By (1) to (5) we see that

$$(6) \quad \left[\begin{array}{l}\binom{V}{T}\binom{V-T}{U-T} = \binom{V}{T}\binom{V-T}{V-U}\\ \text{iff either } T < 0 \text{ or } 0 \le V \text{ or } V < U < T.\end{array}\right.$$

In view of (6), by the first part of (4.2') we see that

$$\left[\begin{array}{l}\binom{V}{U}\binom{U}{T} = \binom{V}{T}\binom{V-T}{V-U}\\ \text{iff either } T < 0 \text{ or } 0 \le V \text{ or } V < U < T\end{array}\right. \tag{4.4.1}$$

and therefore we see that

$$\left[\begin{array}{l}\binom{V}{U}\binom{U}{T} \ne \binom{V}{T}\binom{V-T}{V-U}\\ \text{if either } U \le V < 0 \le T \text{ or } V < 0 \le T \le U.\end{array}\right. \tag{4.4.2}$$

BASIC BINOMIAL IDENTITY (4.11') In $Q[V]$ we have

$$\binom{V}{A} = \binom{V-1}{A} + \binom{V-1}{A-1} \qquad \text{for all } A \in Z.$$

Proof: If $A \in Z\backslash N$, then by (2.5.2') we see that both sides are equal to zero. If $A = 0$, then by (2.5.1') and (2.5.2') we see that both sides are equal to 1. If $A \in N^*$, then

$$\text{RHS} = \frac{(V-1)(V-2)\cdots(V-A)}{A!} + \frac{(V-1)(V-2)\cdots(V-A+1)}{(A-1)!}$$

(by definition)

$$= \left[\frac{(V-1)(V-2)\cdots(V-A+1)}{(A-1)!}\right]\left[\frac{V-A}{A}+1\right]$$

$$= \left[\frac{(V-1)(V-2)\quad(V-A+1)}{(A-1)!}\right]\left[\frac{V}{A}\right]$$

$$= \frac{V(V-1)\cdots(V-A+1)}{A!}$$

$$= \text{LHS} \qquad \text{(by definition)}.$$

TWISTED BASIC BINOMIAL IDENTITY (4.11*) In Q[V] we have

$$\begin{bmatrix} V \\ A \end{bmatrix} = \begin{bmatrix} V-1 \\ A \end{bmatrix} + \begin{bmatrix} V \\ A-1 \end{bmatrix} \qquad \text{for all } A \in Z.$$

Proof: In view of (2.5.3*), our assertion follows by substituting V + A for V in (4.11').

GENERAL PROGRESSION LEMMA (4.12') Given any integers e, x, y, with $x \le y$, we have

$$(1) \qquad \sum_{s\in[x,y-1]} \binom{s}{e-1} = \binom{y}{e} - \binom{x}{e}.$$

Proof: Upon taking A = e and V = y in (4.11') we get

$$(11) \qquad \binom{y-1}{e-1} = \binom{y}{e} - \binom{y-1}{e}.$$

If $x < y$, then clearly

$$[\text{LHS of (1)}] - [\text{LHS of (1) with y replaced by y - 1}] = \binom{y-1}{e-1}$$

and

$$[\text{RHS of (1)}] - [\text{RHS of (1) with y replaced by y - 1}]$$
$$= \binom{y}{e} - \binom{y-1}{e}$$

and therefore by (11) we see that

(12) if $x < y$, then [LHS of (1)] - [LHS of (1) with y replaced by y - 1] = [RHS of (1)] - [RHS of (1) with y replaced by y - 1].

Now if $x = y$, then obviously the LHS of (1) and the RHS of (1) are both equal to zero, and therefore (1) follows from (12) by induction on $y - x$.

GENERAL TWISTED PROGRESSION LEMMA (4.12*) Given any integers e, x, y with $x \le y$, we have

$$(1) \qquad \sum_{s\in[x+1,y]} \begin{bmatrix} s \\ e-1 \end{bmatrix} = \begin{bmatrix} y \\ e \end{bmatrix} - \begin{bmatrix} x \\ e \end{bmatrix}.$$

Proof: Upon taking A = e and V = y in (4.11*) we get

$$(11) \qquad \begin{bmatrix} y \\ e-1 \end{bmatrix} = \begin{bmatrix} y \\ e \end{bmatrix} - \begin{bmatrix} y-1 \\ e \end{bmatrix}.$$

If $x < y$, then clearly

[LHS of (1)] - [LHS of (1) with y replaced by y - 1]

$$= \begin{bmatrix} y \\ e-1 \end{bmatrix}$$

and

[RHS of (1)] - [RHS of (1) with y replaced by y - 1]

$$= \begin{bmatrix} y \\ e \end{bmatrix} - \begin{bmatrix} y-1 \\ e \end{bmatrix}$$

and therefore by (11) we see that

(12) [if $x < y$, then [LHS of (1)] - [LHS of (1) with y replaced by y - 1] = [RHS of (1)] - [RHS of (1) with y replaced by y - 1].

Now if x = y, then obviously the LHS of (1) and the RHS of (1) are both equal to zero, and therefore (1) follows from (12) by induction on y - x.

PROGRESSION LEMMA (4.13') We have

$$\sum_{V\in[0,U-1]} \binom{V}{A} = \binom{U}{A+1} \qquad \text{for every } A \in N \text{ and } U \in N.$$

Proof: By (2.5.6') we see that

$$\binom{0}{A+1} = 0 \qquad \text{for all } A \in N$$

and therefore our assertion follows by taking e = A + 1 and x = 0 and y = U in (4.12').

TWISTED PROGRESSION LEMMA (4.13*) We have

$$\sum_{V\in[0,U]} \begin{bmatrix} V \\ A \end{bmatrix} = \begin{bmatrix} U \\ A+1 \end{bmatrix} \qquad \text{for every } A \in N \text{ and } U \in N.$$

Proof: By (2.5.6*) we see that

$$\begin{bmatrix} -1 \\ A+1 \end{bmatrix} = 0 \qquad \text{for all } A \in N$$

and therefore our assertion follows by taking $e = A + 1$ and $x = -1$ and $y = U$ in (4.12*).

CHOICE LEMMA (4.21') Given any $V \in N$, we have

$$\mathrm{card}(J(V,A)) = \binom{V}{A} \qquad \text{for all } A \in Z.$$

NOTE Traditionally, the LHS is called the number of ways of choosing A out of V things.

Proof: For $A < 0$, the assertion follows from (2.5.2'). For $V < A$, the assertion follows from (2.5.6'). If $0 < A \le V$, then clearly

$$J(V,A) = J(V-1, A) \coprod \{R \in J(V,A) : V \in R\}$$

and

$$\left[\begin{array}{l} R \mapsto R\backslash\{V\} \text{ gives a bijection of} \\ \{R \in J(V,A) : V \in R\} \text{ onto } J(V-1, A-1); \end{array}\right.$$

consequently,

$$(1) \quad \left[\begin{array}{l} \mathrm{card}(J(V,A)) = \mathrm{card}(J(V-1, A)) + \mathrm{card}(J(V-1, A-1)) \\ \text{for all } A \in [1,V]. \end{array}\right.$$

For $A = 0$ our assertion follows from (2.5.1'); therefore, for $A \in [0,V]$, our assertion follows from (1) and (4.11') by induction on V.

REPETITIVE CHOICE LEMMA (4.21*) Given any $V \in N$, we have

$$\mathrm{card}(N(V+1, A)) = \begin{bmatrix} V \\ A \end{bmatrix} \qquad \text{for all } A \in Z.$$

NOTE Traditionally, the LHS is called the number of ways of choosing A out of V + 1 things with repetitions allowed.

Proof: For $A < 0$ the assertion follows from (2.5.2*). If $A > 0$, then clearly

$$\left[\begin{array}{l} N(V+1, A) = \{R \in N(V+1, A) : R(V+1) = 0\} \\ \qquad\qquad\qquad \coprod \{R \in N(V+1, A) : R(V+1) \neq 0\} \end{array}\right.$$

and

$$\left[\begin{array}{l} R \mapsto \text{restriction of } R \text{ to } [1,V] \text{ gives a bijection of} \\ \{R \in N(V+1, A) : R(V+1) = 0\} \text{ onto } N(V,A) \end{array}\right.$$

and

$$\left[\begin{array}{l} R \mapsto R^* \text{ gives a bijection of } \{R \in N(V+1, A) : R(V+1) \neq 0\} \\ \text{onto } N(V+1, A-1) \text{ where } R^*: [1, V+1] \to N \text{ is the map} \\ \text{obtained by putting} \\ R^*(I) = \begin{cases} R(I) & \text{if } I \in [1,V] \\ R(I) - 1 & \text{if } I = V+1; \end{cases} \end{array}\right.$$

consequently,

$$(1)\quad \left[\begin{array}{l} \text{card}(N(V+1, A)) = \text{card}(N(V,A)) + \text{card}(N(V+1, A-1)) \\ \text{for all } A \in N^*. \end{array}\right.$$

For $A = 0$, the assertion follows from (2.5.1*); therefore, for $A \in N$, the assertion follows from (1) and (4.11*) by double induction.

REMARK (4.22') For an alternative formulation of (4.21'), given any finite set v, upon letting $V = \text{card}(v)$, we can find a bijection $w: v \to [1,V]$, and now for every $A \in Z$ we see that $R \mapsto w(R)$ gives a bijection of $P(v,A)$ onto $J(V,A)$. Therefore, by (4.21') we get

$$\text{card}(P(v,A)) = \begin{pmatrix} V \\ A \end{pmatrix} \qquad \text{for all } A \in Z.$$

REMARK (4.22)* For an alternative formulation of (4.21*), given any $V \in N$, let $Y_1, Y_2, \ldots, Y_{V+1}$ be indeterminates. Now for every $A \in Z$,

$$\left[\begin{array}{l} R \mapsto Y_1^{R(1)} Y_2^{R(2)} \cdots Y_{V+1}^{R(V+1)} \text{ gives a bijection of } N(V+1, A) \\ \text{onto the set of all monomials of degree } A \text{ in } Y_1, Y_2, \ldots, Y_{V+1} \end{array}\right.$$

and upon letting

$$K[Y_1,Y_2,\dots,Y_{V+1}]_A = \text{the set consisting of the zero polynomial and all homogeneous polynomials of degree } A \text{ in } Y_1,\ Y_2,\ \dots,\ Y_{V+1} \text{ with coefficients in a field } K$$

we see that

$$\left[\begin{array}{l}(Y_1^{R(1)}Y_2^{R(2)}\cdots Y_{V+1}^{R(V+1)})_{R\in N(V+1,A)} \text{ is a free K-basis of} \\ \text{the K-vector-space } K[Y_1,Y_2,\dots,Y_{V+1}]_A\end{array}\right.$$

and therefore by (4.21*) we conclude that

$$\begin{bmatrix}V\\A\end{bmatrix} = \text{the number of monomials of degree A in } Y_1,\ Y_2,\ \dots,\ Y_{V+1}$$
$$= \text{the K-vector-space dimension of } K[Y_1,Y_2,\dots,Y_{V+1}]_A.$$

REMARK (4.23') For yet another formulation of (4.21'), given any $V \in N$, let $V^* \in N(1)$ be defined by putting $V^*(1) = V$. Now for every $A \in N$ we see that $R \mapsto \{R(i) : i \in [1,A]\}$ gives a bijection of vec(1,V*,A) onto J(V,A) and therefore by (4.21') we get

$$\text{card}(\text{vec}(1,V^*,A)) = \begin{pmatrix}V\\A\end{pmatrix}.$$

REMARK (4.23)* For yet another formulation of (4.21*), given any $V \in N$, let $V' \in N(1)$ be defined by putting $V'(1) = V + 1$, and let $\hat{V} \in \text{vec}(1,V',1)$ be defined by putting $\hat{V}(1,1) = 1$, and for every $t \in \text{tab}(1)$ let $t^* \in N(V + 1)$ be defined by putting $t^*(i) = \text{card}(\{j \in [1,\text{dep}(t)] : t[j](1,1) = i\})$. Now for every $A \in Z$, $t \mapsto t^*$ gives a bijection of $\text{stab}(1,V',1,\hat{V},A)$ onto $N(V + 1, A)$ and therefore by (4.21*) we get

$$\text{card}(\text{stab}(1,V',1,\hat{V},A)) = \begin{bmatrix}V\\A\end{bmatrix}.$$

LEMMA (4.24) Given any $p \in N$ and $t \in J(p)$, upon letting $T = \text{card}(t)$, for every $U \in Z$ we have

$$\text{card}(J^*(p,t,U)) = \begin{pmatrix}p - T\\U - T\end{pmatrix}.$$

Proof: If $U < T$, then our assertion follows from (2.5.2'). If $U \geq T$, then our assertion follows by taking $v = [1,p]\backslash t$ and $A = U - T$ in (4.22') and by noting that $u \mapsto u\backslash t$ gives a bijection of $J^*(p,t,U)$ onto $P(v,A)$.

LEMMA (4.25) Let there be given any $p \in N$, and let there be given any map $\hat{X}: J(p) \to Q$, and let there be given any $U \in [0,p]$. Then for every $T \in [0,p]$ we have

$$\sum_{t \in J(p,T)} \hat{X}(t)\,\mathrm{card}(J^*(p,t,U)) = \sum_{u \in J(p,U)} \sum_{t \in P(u,T)} \hat{X}(t). \tag{4.25.1}$$

Also, we have

$$\begin{cases} \displaystyle\sum_{T \in [0,p]} \sum_{t \in J(p,T)} \hat{X}(t)\,\mathrm{card}(J^*(p,t,U)) \\ \displaystyle= \sum_{u \in J(p,U)} \sum_{t \in P(u)} \hat{X}(t). \end{cases} \tag{4.25.2}$$

Proof: Let

$J[p,U]$ = the set of all pairs (t,u) such that $t \in J(p)$ and $u \in J(p,U)$ and $t \subset u$

and for every $T \in [0,p]$ let

$J[p,T,U]$ = the set of all pairs (t,u) such that $t \in J(p,T)$ and $u \in J(p,U)$ and $t \subset u$

and note that then, clearly,

(1) $$J[p,U] = \coprod_{T \in [0,p]} J[p,T,U].$$

Now we get a mapping

$$f: J[p,U] \to J(p)$$

by putting

$$f(t,u) = t \quad \text{for all } (t,u) \in J[p,U].$$

By (1) we see that

(2) $$\sum_{(t,u) \in J[p,U]} \hat{X}(f(t,u)) = \sum_{T \in [0,p]} \sum_{(t,u) \in J[p,T,U]} \hat{X}(f(t,u)).$$

For every $T \in [0,p]$ we clearly have

$$J[p,T,U] = \coprod_{t \in J(p,T)} f^{-1}(t)$$

and hence we see that

$$(1')\quad \begin{cases} \text{for every } T \in [0,p] \text{ we have} \\ \sum_{(t,u) \in J[p,T,U]} \hat{X}(f(t,u)) = \sum_{t \in J(p,T)} \sum_{(t',u) \in f^{-1}(t)} \hat{X}(f(t',u)). \end{cases}$$

Now

$$\begin{cases} \text{given any } t \in J(p), \text{ for every } (t',u) \in f^{-1}(t) \text{ we have} \\ \hat{X}(f(t',u)) = \hat{X}(t), \text{ and moreover } (t',u) \mapsto u \text{ gives a} \\ \text{bijection of } f^{-1}(t) \text{ onto } J^*(p,t,U) \end{cases}$$

and therefore

$$(2')\quad \begin{cases} \text{given any } t \in J(p) \text{ we have} \\ \sum_{(t',u) \in f^{-1}(t)} \hat{X}(f(t',u)) = \hat{X}(t)\,\mathrm{card}(J^*(p,t,U)). \end{cases}$$

By (1') and (2') we see that

$$(3')\quad \begin{cases} \text{for every } T \in [0,p] \text{ we have} \\ \sum_{(t,u) \in J[p,T,U]} \hat{X}(f(t,u)) = \sum_{t \in J(p,T)} \hat{X}(t)\,\mathrm{card}(J^*(p,t,U)). \end{cases}$$

By (2) and (3') we see that

$$(3)\quad \begin{cases} \sum_{(t,u) \in J[p,U]} \hat{X}(f(t,u)) \\ = \sum_{T \in [0,p]} \sum_{t \in J(p,T)} \hat{X}(t)\,\mathrm{card}(J^*(p,t,U)). \end{cases}$$

We get a mapping

$$g: J[p,U] \to J(p,U)$$

by putting

$$g(t,u) = u \quad \text{for all } (t,u) \in J[p,U].$$

Because g is a mapping, we get

$$J[p,U] = \coprod_{u \in J(p,U)} g^{-1}(u)$$

and hence

$$(4)\quad \left[\begin{array}{l} \displaystyle\sum_{(t,u)\in J[p,U]} \hat{X}(f(t,u)) \\ \displaystyle = \sum_{u\in J(p,U)} \sum_{(t,u')\in g^{-1}(u)} \hat{X}(f(t,u')). \end{array}\right.$$

Now

$$\left[\begin{array}{l} \text{given any } u \in J(p,U)\text{, for every } (t,u') \in g^{-1}(u) \\ \text{we have } \hat{X}(f(t,u')) = \hat{X}(t)\text{, and moreover } (t,u') \mapsto t \\ \text{gives a bijection of } g^{-1}(u) \text{ onto } P(u) \end{array}\right.$$

and therefore

$$(5)\quad \left[\begin{array}{l} \text{given any } u \in J(p,U) \text{ we have} \\ \displaystyle\sum_{(t,u')\in g^{-1}(u)} \hat{X}(f(t,u')) = \sum_{t\in P(u)} \hat{X}(t). \end{array}\right.$$

By (4) and (5) we see that

$$(6)\quad \sum_{(t,u)\in J[p,U]} \hat{X}(f(t,u)) = \sum_{u\in J(p,U)} \sum_{t\in P(u)} \hat{X}(t).$$

By (3) and (6) we get (4.25.2).

Now let there be fixed any $T \in [0,p]$. We get a mapping

$$g_T\colon J[p,T,U] \to J(p,U)$$

by putting

$$g_T(t,u) = u \quad \text{for all } (t,u) \in J[p,T,U].$$

Because g_T is a mapping, we set

$$J[p,T,U] = \coprod_{u\in J(p,U)} g_T^{-1}(u)$$

and hence

$$(4')\quad \left[\begin{array}{l} \displaystyle\sum_{(t,u)\in J[p,T,U]} \hat{X}(f(t,u)) \\ \displaystyle = \sum_{u\in J(p,U)} \sum_{(t,u')\in g_T^{-1}(u)} \hat{X}(f(t,u')). \end{array}\right.$$

$$\left[\begin{array}{l}\text{given any } u \in J(p,U)\text{, for every } (t,u') \in g_T^{-1}(u) \\ \text{we have } \hat{X}(f(t,u')) = \hat{X}(t)\text{, and moreover} \\ (t,u') \mapsto t \text{ gives a bijection of } g_T^{-1}(u) \text{ onto } P(u,T)\end{array}\right.$$

and therefore

$$(5') \quad \left[\begin{array}{l}\text{given any } u \in J(p,U) \text{ we have} \\ \displaystyle\sum_{(t,u') \in g_T^{-1}(u)} \hat{X}(f(t,u')) = \sum_{t \in P(u,T)} \hat{X}(t).\end{array}\right.$$

By (4') and (5') we see that

$$(6') \quad \sum_{(t,u) \in J[p,T,U]} \hat{X}(f(t,u)) = \sum_{u \in J(p,U)} \sum_{t \in P(u,T)} \hat{X}(t).$$

By (3') and (6') we get (4.25.1).

LEMMA (4.26) Let there be given any $p \in N$, and let there be given any map $\hat{X}: J(p) \to Q$. Then for every $U \in [0,p]$ we have

$$\sum_{T \in [0,p]} \binom{p-T}{U-T} \sum_{t \in J(p,T)} \hat{X}(t) = \sum_{u \in J(p,U)} \sum_{t \in P(u)} \hat{X}(t).$$

Proof: Follows from (4.24) and (4.25.2).

REMARK (4.27) Lemmas (4.25) and (4.26) together with their proofs remain verbatim valid if we replace Q by any additive abelian group.

LEMMA (4.28) Given any $p \in N$ and $T \in [0,p]$, for every $U \in Z$ we have

$$\sum_{t \in J(p,T)} \operatorname{card}(J^*(p,t,U)) = \binom{p}{U}\binom{U}{T}.$$

Proof: By (4.21') we see that

$$\operatorname{card}(J(p,U)) = \binom{p}{U}$$

and by (4.22') we see that

$$\operatorname{card}(P(u,T)) = \binom{U}{T} \quad \text{for all } u \in J(p,U)$$

and therefore we see that

$$(1)\quad \left[\begin{array}{l}\text{if } \hat{X}(t) = 1 \text{ for all } t \in J(p), \text{ then} \\ \displaystyle\sum_{u\in J(p,U)} \sum_{t\in P(u)} \hat{X}(t) = \binom{p}{U}\binom{U}{T}.\end{array}\right.$$

In view of (1), upon taking $\hat{X}(t) = 1$ for all $t \in J(p)$, by (4.25.1) we get

$$\sum_{t\in J(p,T)} \operatorname{card}(J^*(p,t,U)) = \binom{p}{U}\binom{U}{T}.$$

REMARK (4.29) We shall now obtain alternative proofs of (4.3) and (4.24) by showing that these two lemmas can be deduced from each other. Note that the already given proof of (4.24) depends, directly or indirectly, only on (4.11'), (4.21'), and (4.22'). Also note that the already given proof of (4.28) depends, directly or indirectly, only on (4.11'), (4.21'), (4.22'), and (4.25).

Now clearly,

$$(1)\quad \left[\begin{array}{l}\text{for every } p \in N \text{ and } T \in [0,p] \text{ we have} \\ \operatorname{card}(J(p,T)) = \binom{p}{T} \in N^*.\end{array}\right.$$

Next we claim that

$$(2)\quad \left[\begin{array}{l}\text{given any } p \in N \text{ and } T \in [0,p] \text{ and } U \in Z, \\ \text{and given any } t \text{ and } t' \text{ in } J(p,T), \text{ we have} \\ \operatorname{card}(J^*(p,t,U)) = \operatorname{card}(J^*(p,t',U)).\end{array}\right.$$

To give a direct proof of (2), first we note that by the definition of cardinality, there exists a bijection g_1 of t onto t'. Also, clearly,

$$\operatorname{card}([1,p]\backslash t) = p - T = \operatorname{card}([1,p]\backslash t')$$

and hence there exists a bijection g_2 of $[1,p]\backslash t$ onto $[1,p]\backslash t'$. Now upon letting

$$g(y) = \begin{cases} g_1(y) & \text{if } y \in t \\ g_2(y) & \text{if } y \in [1,p]\backslash t \end{cases}$$

we get a bijection g of $[1,p]$ onto itself such that $g(t) = t'$.

Clearly, $s \mapsto g(s)$ gives a bijection of $J^*(p,t,U)$ onto $J^*(p,t',U)$ and therefore

$$\operatorname{card}(J^*(p,t,U)) = \operatorname{card}(J^*(p,t',U)).$$

Now in view of (1) and (4.28), we see that (4.24) implies (4.3). Conversely, in view of (1), (2), and (4.28), we see that (4.3) implies (4.24).

BINOMIAL THEOREM (4.31') Let Y be an indeterminate. Then in the formal power series ring $Q[[Y]]$ we have

$$(1+Y)^V = \sum_{A\in Z} \binom{V}{A} Y^A \qquad \text{for all } V \in Z.$$

NOTE This was proved by Newton in the seventeenth century; for the sake of completeness we shall give a proof. Recall that $Q[[Y]]$ consists of all expressions of the type $\Sigma_{A\in Z}\, r_A Y^A$ with $r_A \in Q$ for all $A \in Z$, such that $r_A = 0$ for all $A \in Z\setminus N$. Also recall that $(1+Y)^U$ is a unit in $Q[[Y]]$ for all $U \in N$, and hence $(1+Y)^V$ makes sense in $Q[[Y]]$ for all $V \in Z$.

Proof: For every $V \in Z$ we can write

$$(1) \quad (1+Y)^V = \sum_{A\in Z} B(A,V) Y^A \qquad \text{with } B(A,V) \in Q$$

and we want to show that then

$$(1^*) \quad B(A,V) = \binom{V}{A} \qquad \text{for every } V \in Z \text{ and } A \in Z.$$

By (1) we see that

$$(2) \quad B(A,0) = 0 \qquad \text{for every } A \in Z\setminus\{0\}$$

and

$$(3) \quad B(0,V) = 1 \qquad \text{for every } V \in Z$$

and

$$(4) \quad B(A,V) = 0 \qquad \text{for every } V \in Z \text{ and } A \in Z\setminus N.$$

By (2), (3), (2.5.1'), (2.5.2'), and (2.5.6') we see that

(5) $B(A,0) = \binom{0}{A}$ for every $A \in Z$.

By (4) and (2.5.2') we see that

$$B(A,V) = 0 = \binom{V}{A} \quad \text{for every } V \in Z \text{ and } A \in Z\backslash N$$

and hence, in view of (1*), it only remains to show that

(6') $B(A,V) = \binom{V}{A}$ for every $V \in N$ and $A \in Z$

and

(7') $B(A,V) = \binom{V}{A}$ for every $V \in Z\backslash N$ and $A \in N$.

Ncw

$$(1 + Y)^V = (1 + Y)(1 + Y)^{V-1} \quad \text{for every } V \in Z$$

and hence by (1) we get

(8) $B(A,V) = B(A, V - 1) + B(A - 1, V - 1)$
for every $V \in Z$ and $A \in Z$.

In view of (5), (8), and (4.11') we get (6') by induction on V. In view of (3), (5), (8), and (4.11') we also get (7') by decreasing induction on V and increasing induction on A.

TWISTED BINOMIAL THEOREM (4.31*) Let Y be an indeterminate. Then in the formal power series ring Q[[Y]] we have

$$1/(1 - Y)^{V+1} = \sum_{A \in Z} \begin{bmatrix} V \\ A \end{bmatrix} Y^A \quad \text{for all } V \in Z.$$

Proof: In view of (2.5.4*) our assertion follows by substituting -Y for Y, and -V - 1 for V, in (4.31').

PRODUCT LEMMA (4.32') Given any $p \in N^*$ and $D^* \in Z$ and $E' \in Z$ and $e' \in Z(p,E')$, we have

$$\sum_{d \in N(p,D^*)} \prod_{k=1}^{p} \binom{e'(k)}{d(k)} = \binom{E'}{D^*}.$$

Proof: Upon letting Y be an indeterminate, we have

$$\prod_{k=1}^{p} (1 + Y)^{e'(k)} = (1 + Y)^{E'}$$

and now, in view of (4.31'), our assertion follows by equating the coefficients of Y^{D^*} on the two sides of the equation above.

TWISTED PRODUCT PRELEMMA (4.32**) Given any $p \in N^*$ and $D^* \in Z$ and $E \in Z$ and $e \in Z(p,E)$, we have

$$\sum_{d \in N(p,D^*)} \prod_{k=1}^{p} \begin{bmatrix} e(k) \\ d(k) \end{bmatrix} = \begin{bmatrix} E + p - 1 \\ D^* \end{bmatrix}.$$

Proof: Upon letting Y be an indeterminate, we have

$$\prod_{k=1}^{p} [1/(1 - Y)^{e(k)+1}] = 1/(1 - Y)^{E+p}$$

and now, in view of (4.31*), our assertion follows by equating the coefficients of Y^{D^*} on the two sides of the equation above.

TWISTED PRODUCT LEMMA (4.32*) Let there be given any $p \in N^*$ and $D^* \in Z$ and $e' \in Z(p)$ and $t \in J(p)$, and let us put

$$T = \operatorname{card}(t) \qquad \text{and} \qquad E' = \left\{ \sum_{k \in [1,p] \setminus t} e'(k) \right\} - \left\{ \sum_{k \in t} e'(k) \right\}$$

and let us put

$$S(d) = \sum_{k \in t} d(k) \qquad \text{for all } d \in Z(p).$$

Then we have

$$\sum_{d \in N(p,D^*)} (-1)^{S(d)} \left\{ \prod_{k \in t} \begin{pmatrix} e'(k) \\ d(k) \end{pmatrix} \right\} \left\{ \prod_{k \in [1,p] \setminus t} \begin{bmatrix} e'(k) \\ d(k) \end{bmatrix} \right\}$$

$$= \begin{bmatrix} E' + p - 1 - T \\ D^* \end{bmatrix}.$$

NOTE We shall give two proofs, the first using Twisted Product Prelemma (4.32**), and the second using both Binomial Theorems (4.31') and (4.31*).

First Proof: Our assertion follows by taking

$$e(k) = \begin{cases} -e'(k) - 1 & \text{if } k \in t \\ e'(k) & \text{if } k \in [1,p]\backslash t \end{cases}$$

in (4.32**) and by noting that, in view of (2.5.4'), for every $d \in Z(p)$ and $k \in t$ we have

$$\left[\begin{matrix} -e'(k) - 1 \\ d(k) \end{matrix}\right] = (-1)^{d(k)} \left[\begin{matrix} e'(k) \\ d(k) \end{matrix}\right].$$

Second Proof: Upon letting Y be an indeterminate, we have

$$\left\{\prod_{k\in t} (1 - Y)^{e'(k)}\right\}\left\{\prod_{k\in[1,p]\backslash t} [1/(1 - Y)^{e'(k) + 1}]\right\}$$
$$= 1/(1 - Y)^{E'+p-T}$$

and now in view of (4.31') and (4.31*), our assertion follows by equating the coefficients of Y^{D^*} on the two sides of the equation above.

IDENTITY THEOREM (4.33) Let $p \in N^*$, and let $Q_1, Q_2, \ldots, Q_p$ be infinite subsets of Q, and let $F(V_1,V_2,\ldots,V_p)$ be a polynomial in indeterminates $V_1, V_2, \ldots, V_p$ with coefficients in Q. Assume that $F(e_1,e_2,\ldots,e_p) = 0$ for all $e_1 \in Q_1$, $e_2 \in Q_2$, ..., $e_p \in Q_p$. Then $F(V_1,V_2, \quad ,V_p) = 0$, and hence $F(e_1',e_2',\ldots,e_p') = 0$ for any elements $e_1', e_2', \ldots, e_p'$ in any overring of Q.

Proof: We shall make induction on p. For p = 1, the assertion is trivial. So let p > 1 and assume that the assertion is true for all values of p smaller than the given one. Now for some $I \in N$, we can write

$$(1) \quad F(V_1,V_2,\ldots,V_p) = \sum_{i\in[0,I]} F_i(V_1,V_2,\ldots,V_{p-1})V_p^i$$

where $F_i(V_1,V_2,\ldots,V_{p-1})$ is a polynomial in $V_1, V_2, \ldots, V_{p-1}$ with coefficients in Q. Given any $e_1 \in Q_1$, $e_2 \in Q_2$, ..., $e_{p-1} \in Q_{p-1}$, upon considering the polynomial $F^*(V_p)$ in one indeterminate V_p with coefficients in Q obtained by putting

$$F^*(V_p) = \sum_{i\in[0,I]} F_i(e_1,e_2,\ldots,e_{p-1})V_p^i,$$

by assumption we know that

(2) $F^*(e_p) = 0$ for all $e_p \in Q_p$

and hence by the p = 1 case we see that

$$F^*(V_p) = 0$$

and therefore by (2) we get

(3) $F_i(e_1,e_2,\ldots,e_{p-1}) = 0$ for all $i \in [0,I]$.

Now given any $i \in [0,I]$, by (3) we have

$$F_i(e_1,e_2,\ldots,e_{p-1}) = 0 \quad \text{for all } e_1 \in Q_1,\ e_2 \in Q_2,\ \ldots,\ e_{p-1} \in Q_{p-1}$$

and hence by the induction hypothesis we get

(4) $F_i(V_1,V_2,\ldots,V_{p-1}) = 0$.

Since (4) is true for all $i \in [0,I]$, by (1) we get

$$F(V_1,V_2,\ldots,V_p) = 0.$$

REMARK (4.34) For the purposes of the next two theorems (4.35') and (4.35*), given any $p \in N^*$ and $D \in Z$ and $r \in Z(p)$, we let

$$Z'(p,D,r) = \{d \in Z(p,D) : d(k) \geq r(k) \quad \text{for all } k \in [1,p]\}$$

and

$$Z^*[p-1, D, r] = \{d \in Z[p-1, D-r(p)] : d(k) \geq r(k) \quad \text{for all } k \in [1, p-1]\}$$

and we note that then clearly

$$Z'(p,D,r) \subset Z(p,D) \quad \text{and} \quad \operatorname{card}(Z'(p,D,r)) < \infty \tag{4.34.1}$$

and

$$Z^*[p-1, D, r] \subset Z(p-1) \text{ and } \operatorname{card}(Z^*[p-1, D, r]) < \infty. \tag{4.34.2}$$

PRODUCT THEOREM (4.35') Let Z' and Z^* be as in (4.34). Let there be given any $p \in N^*$ and $D \in Z$ and $D' \in Z$ and $d' \in Z(p,D')$. Let there be given any $r \in Z(p)$ such that $r(k) \leq -d'(k)$ for all $k \in [1,p]$. Let there be given any elements $e'(1), e'(2), \ldots, e'(p)$ in any overring of Q, and let us put

$$E' = \sum_{k \in [1,p]} e'(k).$$

Also, let us put

$$S'(d) = \sum_{k \in [1,p-1]} d(k) \qquad \text{for all } d \in Z(p-1).$$

Now first, we have

$$\left\{d \in Z(p,D) : \prod_{k=1}^{p} \binom{e'(k)}{d(k) + d'(k)} \neq 0\right\} \subset Z'(p,D,r) \tag{4.35'.1}$$

and second, we have

$$\sum_{d \in Z'(p,D,r)} \prod_{k=1}^{p} \binom{e'(k)}{d(k) + d'(k)} = \binom{E'}{D + D'}, \tag{4.35'.2}$$

and third, we have

$$\sum_{d \in Z(p,D)} \prod_{k=1}^{p} \binom{e'(k)}{d(k) + d'(k)} = \binom{E'}{D + D'} \tag{4.35'.3}$$

where, in view of (4.34.1) and (4.35'.1), the summation is essentially finite,

and fourth, we have

$$\left\{d \in Z(p-1) : \binom{e'(p)}{D - S'(d) + d'(p)} \prod_{k=1}^{p-1} \binom{e'(k)}{d(k) + d'(k)} \neq 0\right\} \subset Z^*[p-1, D, r], \tag{4.35'.4}$$

and fifth, we have

$$\sum_{d \in Z^*[p-1,D,r]} \binom{e'(p)}{D - S'(d) + d'(p)} \prod_{k=1}^{p-1} \binom{e'(k)}{d(k) + d'(k)} = \binom{E'}{D + D'}, \tag{4.35'.5}$$

and sixth, we have

$$\left[\begin{array}{l}\displaystyle\sum_{d\in Z(p-1)} \binom{e'(p)}{D - S'(d) + d'(p)} \prod_{k=1}^{p-1} \binom{e'(k)}{d(k) + d'(k)} = \binom{E;}{D + D'} \\ \text{where, in view of (4.34.2) and (4.35'.4), the} \\ \text{summation is essentially finite.}\end{array}\right. \qquad (4.35'.6)$$

Proof: For every $d \in Z(p)$ and $\hat{d} \in Z(p)$, let $u[d,\hat{d}] \in Z(p)$ be defined by putting

$$u[d,\hat{d}](k) = d(k) + \hat{d}(k) \qquad \text{for all } k \in [1,p].$$

For every $d \in Z(p - 1)$ and $\hat{D} \in Z$ and $\hat{d} \in Z(p)$, let $g[d,\hat{D},\hat{d}] \in Z(p)$ be defined by putting

$$g[d,\hat{D},\hat{d}](k) = \begin{cases} d(k) + \hat{d}(k) & \text{if } k \in [1, p - 1] \\ \hat{D} - S'(d) + \hat{d}(p) & \text{if } k = p. \end{cases}$$

For a moment let us assume that $e'(k) \in Z$ *for all* $k \in [1,p]$, and let us put $X^* = Z(p)$ and let us define $f: X^* \to Q$ to be the map obtained by putting

$$f(d) = \prod_{k=1}^{p} \binom{e'(k)}{d(k)} \qquad \text{for all } d \in X^*$$

and let us note that by (2.5.2') we have

$$\{d \in Z'(p, D + D', u[r,d']) : f(d) \neq 0\} \subset N(p, D + D')$$

and clearly, we have

$$N(p, D + D') \subset Z'(p, D + D', u[r,d'])$$

and therefore upon taking $D^* = D + D'$ in (4.32') we get

$$(1) \qquad \sum_{d\in Z'(p,D+D',u[r,d'])} f(d) = \binom{E'}{D + D'};$$

now, on the one hand, upon letting

$$X' = Z(p) \qquad \text{and} \qquad X = Z'(p,D,r)$$

and upon letting $A: X' \to X^*$ to be the bijective map obtained by putting

$$A(d) = u[d,d'] \qquad \text{for all } d \in X'$$

we clearly have

$$A(X) = Z'(p, D + D', u[r,d'])$$

and we clearly have

$$f(A(d)) = \prod_{k=1}^{p} \binom{e'(k)}{d(k) + d'(k)} \qquad \text{for all } d \in X'$$

and therefore by (1) and (3.1) we see that (4.35'.2) holds; on the other hand, upon letting

$$X' = Z(p - 1) \qquad \text{and} \qquad X = Z^*[p - 1, D, r]$$

and upon letting $A: X' \to X^*$ be the injective map obtained by putting

$$A(d) = g[d,D,d'] \qquad \text{for all } d \in X'$$

we clearly have

$$A(X) = Z'(p, D + D', u[r,d'])$$

and we clearly have

$$f(A(d)) = \binom{e'(p)}{D - S'(d) + d'(p)} \prod_{k=1}^{p-1} \binom{e'(k)}{d(k) + d'(k)}$$

$$\text{for all } d \in X'$$

and therefore by (1) and (3.1) we see that (4.35'.5) holds.

Thus we have shown that (4.35'.2) and (4.35'.5) are true under the assumption that $e'(k) \in Z$ for all $k \in [1,p]$; therefore, by (4.33) we conclude that (4.35'.2) and (4.35'.5) are true without any special assumption. By (2.5.2') we get (4.35'.1) and (4.35'.4). Now by (4.34.1) and (4.35'.1) and (4.35'.2) we get (4.35'.3), whereas by (4.34.2) and (4.35'.4) and (4.35'.5) we get (4.35'.6).

TWISTED PRODUCT THEOREM (4.35*) Let Z' and Z^* be as in (4.34). Let there be given any $p \in N^*$ and $D \in Z$ and $D' \in Z$ and $d' \in Z(p,D')$. Let there be given $r \in Z(p)$ such that $r(k) \le -d'(k)$ for all $k \in [1,p]$. Let there be given any elements $e'(1), e'(2), \ldots, e'(p)$ in any overring of Q, and let there be given any $t \in J(p)$, and let us put

$$T = \text{card}(t) \qquad \text{and} \qquad E' = \left\{\sum_{k\in[1,p]\setminus t} e'(k)\right\} - \left\{\sum_{k\in t} e'(k)\right\}$$

and let us put

$$S(d) = \sum_{k\in t} d(k) \qquad \text{for all } d \in Z(p),$$

and let us put

$$h(d) = (-1)^{S(d)}\left\{\prod_{k\in t}\binom{e'(k)}{d(k)+d'(k)}\right\}\left\{\prod_{k\in[1,p]\backslash t}\left[\begin{matrix}e'(k)\\ d(k)+d'(k)\end{matrix}\right]\right\}$$

for all $d \in Z(p)$.

Also, let us put

$$S'(d) = \sum_{k\in[1,p-1]} d(k) \qquad \text{for all } d \in Z(p-1).$$

Finally, on the one hand,

if $p \notin t$, then for all $d \in Z(p-1)$, let us put

$$S^*(d) = \sum_{k\in t}[d(k) + d'(k)]$$

and

$$h^*(d) = (-1)^{S^*(d)}\left\{\prod_{k\in t}\binom{e'(k)}{d(k)+d'(k)}\right\} \times\left\{\prod_{k\in[1,p-1]\backslash t}\left[\begin{matrix}e'(k)\\ d(k)+d'(k)\end{matrix}\right]\right\}\left[\begin{matrix}e'(p)\\ D-S'(d)+d'(p)\end{matrix}\right],$$

whereas on the other hand,

if $p \in t$, then for all $d \in Z(p-1)$, let us put

$$S^{**}(d) = D - S'(d) + d'(p) + \sum_{k\in t\backslash\{p\}}[d(k)+d'(k)]$$

and

$$h^{**}(d) = (-1)^{S^{**}(d)}\left\{\prod_{k\in[1,p]\backslash t}\left[\begin{matrix}e'(k)\\ d(k)+d'(k)\end{matrix}\right]\right\} \times\left\{\prod_{k\in t\backslash\{p\}}\binom{e'(k)}{d(k)+d'(k)}\right\}\binom{e'(p)}{D-S'(d)+d'(p)}.$$

Now we first have

$$\{d \in Z(p,D) : h(d) \neq 0\} \subset Z'(p,D,r), \tag{4.35*.1}$$

and second, we have

$$\sum_{d\in Z'(p,D,r)} h(d) = \left[\begin{matrix}E' + p - 1 - T\\ D + D'\end{matrix}\right], \tag{4.35*.2}$$

and third, we have

$$\left[\begin{array}{l}\displaystyle\sum_{d\in Z(p,D)} h(d) = \begin{bmatrix} E' + p - 1 - T \\ D + D' \end{bmatrix} \\ \text{where in view of (4.34.1) and (4.35*.1), the} \\ \text{summation is essentially finite}\end{array}\right. \qquad (4.35^*.3)$$

and fourth,

$$\left[\begin{array}{l}\text{if } p \notin t\text{, then we have} \\ \{d \in Z(p) : h^*(d) \neq 0\} \subset Z^*[p - 1, D, r],\end{array}\right. \qquad (4.35^*.4)$$

and fifth,

$$\left[\begin{array}{l}\text{if } p \notin t\text{, then we have} \\ \displaystyle\sum_{d\in Z^*[p-1,D,r]} h^*(d) = \begin{bmatrix} E' + p - 1 - T \\ D + D' \end{bmatrix},\end{array}\right. \qquad (4.35^*.5)$$

and sixth,

$$\left[\begin{array}{l}\text{if } p \notin t\text{, then we have} \\ \displaystyle\sum_{d\in Z(p-1)} h^*(d) = \begin{bmatrix} E' + p - 1 - T \\ D + D' \end{bmatrix} \\ \text{where in view of (4.34.2) and (4.35*.4), the} \\ \text{summation is essentially finite,}\end{array}\right. \qquad (4.35^*.6)$$

and seventh,

$$\left[\begin{array}{l}\text{if } p \in t\text{, then we have} \\ \{d \in Z(p - 1) : h^{**}(d) \neq 0\} \subset Z^*[p - 1, D, r],\end{array}\right. \qquad (4.35^*.7)$$

and eighth,

$$\left[\begin{array}{l}\text{if } p \in t\text{, then we have} \\ \displaystyle\sum_{d\in Z^*[p-1,D,r]} h^{**}(d) = \begin{bmatrix} E' + p - 1 - T \\ D + D' \end{bmatrix},\end{array}\right. \qquad (4.35^*.8)$$

and ninth,

$$\left[\begin{array}{l}\text{if } p \in t\text{, then we have} \\ \displaystyle\sum_{d\in Z(p-1)} h^{**}(d) = \begin{bmatrix} E' + p - 1 - T \\ D + D' \end{bmatrix} \\ \text{where in view of (4.34.2) and (4.35*.7), the} \\ \text{summation is essentially finite.}\end{array}\right. \qquad (4.35^*.9)$$

Proof: For every $d \in Z(p)$ and $\hat{d} \in Z(p)$, let $u[d,\hat{d}] \in Z(p)$ be defined by putting

$$u[d,\hat{d}](k) = d(k) + \hat{d}(k) \qquad \text{for all } k \in [1,p].$$

For every $d \in Z(p-1)$ and $\hat{D} \in Z$ and $\hat{d} \in Z(p)$, let $g[d,\hat{D},\hat{d}] \in Z(p)$ be defined by putting

$$g[d,\hat{D},\hat{d}](k) = \begin{cases} d(k) + \hat{d}(k) & \text{if } k \in [1, p-1] \\ \hat{D} - S'(d) + \hat{d}(p) & \text{if } k = p. \end{cases}$$

For a moment let us assume that $e'(k) \in Z$ *for all* $k \in [1,p]$, and let us put $X^* = Z(p)$ and let us define $f: X^* \to Q$ to be the map obtained by putting

$$f(d) = (-1)^{S(d)}\left\{\prod_{k\in t}\binom{e'(k)}{d(k)}\right\}\left\{\prod_{k\in[1,p]\setminus t}\begin{bmatrix}e'(k)\\ d(k)\end{bmatrix}\right\}$$
$$\text{for all } d \in X^*$$

and let us note that by (2.5.2') and (2.5.2*) we have

$$\{d \in Z'(p, D + D', u[r,d']) : f(d) \neq 0\} \subset N(p, D + D')$$

and clearly we have

$$N(p, D + D') \subset Z'(p, D + D', u[r,d'])$$

and therefore upon taking $D^* = D + D'$ in (4.32*) we set

$$(1) \qquad \sum_{d\in Z'(p,D+D',u[r,d'])} f(d) = \begin{bmatrix}E' + p - 1 - T\\ D + D'\end{bmatrix};$$

now, on the one hand, upon letting

$$X' = Z(p) \qquad \text{and} \qquad X = Z'(p,D,r)$$

and upon letting $A: X' \to X^*$ be the bijective map obtained by putting

$$A(d) = u[d,d'] \qquad \text{for all } d \in X'$$

we clearly have

$$A(X) = Z'(p, D + D', u[r,d'])$$

and we clearly have

$$f(A(d)) = h(d) \qquad \text{for all } d \in X'$$

and therefore by (1) and (3.1) we see that (4.35*.2) holds; on the other hand, upon letting

$$X' = Z(p - 1) \quad \text{and} \quad X = \check{Z}^*[p - 1, D, r]$$

and upon letting $A: X' \to X^*$ be the injective map obtained by putting

$$A(d) = g[d,D,d'] \quad \text{for all } d \in X'$$

we clearly have

$$A(X) = Z'(p, D + D', u[r,d'])$$

and for all $d \in X'$, we clearly have

$$f(A(d)) = \begin{cases} h^*(d) & \text{if } p \notin t \\ h^{**}(d) & \text{if } p \in t \end{cases}$$

and therefore by (1) and (3.1) we see that (4.35*.5) and (4.35*.8) hold.

Thus we have shown that (4.35*.2) and (4.35*.5) and (4.35*.8) are true under the assumption that $e'(k) \in Z$ for all $k \in [1,p]$; therefore, by (4.33) we conclude that (4.35*.2) and (4.35*.5) and (4.35*.8) are true without any special assumption. By (2.5.2') and (2.5.2*) we get (4.35*.1) and (4.35*.4) and (4.35*.7). Now by (4.34.1) and (4.35*.1) and (4.35*.2) we get (4.35*.3), whereas by (4.34.2) and (4.35*.4) and (4.35*.5) we get (4.35*.6), and finally, by (4.34.2) and (4.35*.7) and (4.35*.8) we get (4.35*.9).

COROLLARY OF TWISTED PRODUCT THEOREM (4.36) Given any $V \in Z$ and $U \in N$, and given any elements A and B in any overring of Q, we have

$$\sum_{v \in N(2,V)} \begin{bmatrix} A \\ v(1) \end{bmatrix} \begin{bmatrix} B \\ v(2) - U \end{bmatrix} = \begin{bmatrix} A + B + 1 \\ V - U \end{bmatrix}.$$

Proof: Follows from (4.35*.2) by taking

$$\begin{bmatrix} p = 2 \text{ and } t = \emptyset \text{ and } D = V \text{ and } d = v \text{ and } d'(1) = 0 \text{ and} \\ r(1) = 0 \text{ and } e'(1) = A \text{ and } d'(2) = -U \text{ and } r(2) = 0 \text{ and} \\ e'(2) = B. \end{bmatrix}$$

(TRANSLATION LEMMA =) TL (4.41') Given any $B \in Z$ and $R \in Z$, and given any elements B', R', V in any overring of Q, we have

$$\begin{cases} \binom{V + B'}{B} = \sum_{E \in Z} \binom{B' - R'}{R + E}\binom{V + R'}{B - R - E} \\ \text{where the summation is essentially finite because} \\ \left\{E \in Z : \binom{B' - R'}{R + E}\binom{V + R'}{B - R - E} \neq 0\right\} \subset [-R, B - R] \end{cases} \tag{4.41'.1}$$

and we have

$$\begin{cases} \binom{V + B'}{B} = \sum_{E \in Z} \binom{B' - R'}{R - E}\binom{V + R'}{B - R + E} \\ \text{where the summation is essentially finite because} \\ \left\{E \in Z : \binom{B' - R'}{R - E}\binom{V + R'}{B - R + E} \neq 0\right\} \subset [R - B, R]. \end{cases} \tag{4.41'.2}$$

NOTE We shall give two proofs, the first using Product Theorem (4.35'), and the second by induction.

First Proof: (4.41'.1) follows from (4.35'.6) by taking

$$\begin{cases} p = 2 \text{ and } D = 0 \text{ and } d(1) = E \text{ and } d'(1) = R \\ \text{and } r(1) = -d'(1) \text{ and } e'(1) = B' - R' \text{ and} \\ d'(2) = B - R \text{ and } r(2) = -d'(2) \text{ and } e'(2) = V + R'. \end{cases}$$

(4.41'.2) follows from (4.35'.6) by taking

$$\begin{cases} p = 2 \text{ and } D = 0 \text{ and } d(1) = E \text{ and } d'(1) = B - R \\ \text{and } r(1) = -d'(1) \text{ and } e'(1) = V + R' \text{ and} \\ d'(2) = R \text{ and } r(2) = -d'(2) \text{ and } e'(2) = B' - R'. \end{cases}$$

Second Proof: Let there be fixed any $U \in N$. By induction on D we shall show that for all $I \in Z$ and $D \in N$ we have

$$(1') \quad \binom{U + D}{I} = \sum_{E \in [0,D]} \binom{D}{E}\binom{U}{I - E}.$$

For $D = 0$, the LHS $= \binom{U}{I}$ and by (2.5.1') also the RHS $= \binom{U}{I}$. So let $D > 0$ and assume that (1') is true for $D - 1$ (and for all $I \in Z$). Now by the induction hypothesis we have

$$(1) \quad \binom{U + D - 1}{I} = \sum_{E \in [0,D-1]} \binom{D - 1}{E}\binom{U}{I - E}$$

and

$$(2)\quad \binom{U + D - 1}{I - 1} = \sum_{E\in[0,D-1]} \binom{D - 1}{E}\binom{U}{I - 1 - E}$$

By (2.5.1') we have $\binom{D - 1}{0} = 1 = \binom{D}{0}$ and hence we see that

$$[\text{zeroth term in RHS of (1)}] = \binom{D - 1}{0}\binom{U}{I} = \binom{D}{0}\binom{U}{I}$$

and therefore by (1) we get

$$(3)\quad \binom{U + D - 1}{I} = \binom{D}{0}\binom{U}{I} + \sum_{E\in[1,D-1]} \binom{D - 1}{E}\binom{U}{I - E}.$$

By (2.5.1') and (2.5.10') we have $\binom{D - 1}{D - 1} = 1 = \binom{D}{D}$ and hence see that

$$[\text{last term of RHS of (2)}] = \binom{D - 1}{D - 1}\binom{U}{I - D} = \binom{D}{D}\binom{U}{I - D}$$

and therefore by (2) we get

$$(4)\quad \binom{U + D - 1}{I - 1} = \binom{D}{D}\binom{U}{I - D} + \sum_{E\in[0,D-2]} \binom{D - 1}{E}\binom{U}{I - 1 - E}.$$

By translation of summation index we get

$$\sum_{E\in[0,D-2]} \binom{D - 1}{E}\binom{U}{I - 1 - E}$$

$$= \sum_{E'\in[1,D-1]} \binom{D - 1}{E' - 1}\binom{U}{I - E'} \qquad \text{(by putting } E = E' - 1\text{)}$$

$$= \sum_{E\in[1,D-1]} \binom{D - 1}{E - 1}\binom{U}{I - E} \qquad \text{(by putting } E' = E\text{)}$$

and hence by (4) we get

$$(5)\quad \binom{U + D - 1}{I - 1} = \binom{D}{D}\binom{U}{I - D} + \sum_{E\in[1,D-1]} \binom{D - 1}{E - 1}\binom{U}{I - E}$$

Now we have

[RHS of (3)] + [RHS of (5)]

$$= \binom{D}{0}\binom{U}{I} + \binom{D}{D}\binom{U}{I-D} + \sum_{E\in[1,D-1]} \left[\binom{D-1}{E} + \binom{D-1}{E-1}\right]\binom{U}{I-E}$$

$$= \binom{D}{0}\binom{U}{I} + \binom{D}{D}\binom{U}{I-D} + \sum_{E\in[1,D-1]} \binom{D}{E}\binom{U}{I-E} \qquad \text{[by (4.11')]}$$

$$= \sum_{E\in[0,D]} \binom{D}{E}\binom{U}{I-D}$$

and by (4.11') we also have

$$\binom{U+D}{I} = \text{[LHS of (3)] + [LHS of (5)]}$$

and therefore in view of (3) and (5) we conclude that

$$\binom{U+D}{I} = \sum_{E\in[0,D]} \binom{D}{E}\binom{U}{I-E}.$$

Thus the induction is completed and hence (1') is proved.

By (2.5.2') and (2.5.6') we see that

$$\left\{E \in Z : \binom{D}{E}\binom{U}{I-E} \neq 0\right\} \subset [0,D] \cap [0,I]$$

and hence by (1') we see that given any $I \in Z$, for all $U \in N$ and $D \in N$ we have

$$\binom{U+D}{I} = \sum_{E\in[0,I]} \binom{D}{E}\binom{U}{I-E}$$

and therefore by (4.33) we conclude that

(10') $\left[\begin{array}{l}\text{given any } I' \in Z \text{ and given any elements}\\ U' \text{ and } D' \text{ in any overring of } Q\text{, we have}\\ \dbinom{U'+D'}{I'} = \displaystyle\sum_{E\in[0,I']} \binom{D'}{E}\binom{U'}{I'-E}.\end{array}\right.$

Henceforth let there be given any $B \in Z$ and $R \in Z$, and let there be given any elements B', R', V in any overring of Q. Now upon taking $I' = B$ and $U' = V + R'$ and $D' = B' - R'$ in (10') we get

(10) $$\binom{V+B'}{B} = \sum_{E\in[0,B]} \binom{B'-R'}{E}\binom{V+R'}{B-E}.$$

By translation of the summation index we see that

RHS of (10)

$$= \sum_{E' \in [-R, B-R]} \binom{B' - R'}{R + E'} \binom{V + R'}{B - R - E'} \qquad \text{(by putting } E = R + E')$$

$$= \sum_{E \in [-R, B-R]} \binom{B' - R'}{R + E} \binom{V + R'}{B - R - E} \qquad \text{(by putting } E' = E)$$

and hence by (10) we get

$$(11) \quad \binom{V + B'}{B} = \sum_{E \in [-R, B-R]} \binom{B' - R'}{R + E} \binom{V + R'}{B - R - E}.$$

By (2.5.2') we see that

$$\left\{E \in Z : \binom{B' - R'}{R + E} \binom{V + R'}{B - R - E} \neq 0\right\} \subset [-R, B - R]$$

and hence by (11) we get (4.41'.1). By transformation of the summation index we see that

RHS of (10)

$$= \sum_{E' \in [R-B, R]} \binom{B' - R'}{R - E'} \binom{V + R'}{B - R + E'} \qquad \text{(by putting } E = R - E')$$

$$= \sum_{E \in [R-B, R]} \binom{B' - R'}{R - E} \binom{V + R'}{B - R + E} \qquad \text{(by putting } E' = E)$$

and hence by (10) we get

$$(12) \quad \binom{V + B'}{B} = \sum_{E \in [R-B, R]} \binom{B' - R'}{R - E} \binom{V + R'}{B - R + E}.$$

By (2.5.2') we see that

$$\left\{E \in Z : \binom{B' - R'}{R - E} \binom{V + R'}{B - R + E} \neq 0\right\} \subset [R - B, R]$$

and hence by (12) we get (4.41'.2).

(TWISTED TRANSLATION LEMMA =) TTL (4.41*) Given any $B \in Z$ and $R \in Z$, and given any elements B', R', V in any overring of Q, we have

$$\left|\begin{bmatrix} V - B' \\ B \end{bmatrix} = \sum_{E \in Z} (-1)^{R+E} \binom{B' - R'}{R + E} \begin{bmatrix} V - R' \\ B - R - E \end{bmatrix}\right.$$

where the summation is essentially finite because (4.41*.1)

$$\left\{E \in Z : \binom{B' - R'}{R + E}\begin{bmatrix} V - R' \\ B - R - E\end{bmatrix} \neq 0\right\} \subset [-R, B - R]$$

and we have

$$\begin{bmatrix}\begin{bmatrix} V - B' \\ B \end{bmatrix} = \sum_{E \in Z} (-1)^{R-E}\binom{B' - R'}{R - E}\begin{bmatrix} V - R' \\ B - R + E\end{bmatrix}\end{bmatrix}$$

where the summation is essentially finite because (4.41*.2)

$$\left\{E \in Z : \binom{B' - R'}{R - E}\begin{bmatrix} V - R' \\ B - R + E\end{bmatrix} \neq 0\right\} \subset [R - B, R].$$

NOTE We shall give two proofs, the first using Twisted Product Theorem (4.35*), and the second by induction.

First Proof: (4.41*.1) follows from (4.35*.6) by taking

p = 2 and t = {1} and D = 0 and d(1) = E and
d'(1) = R and r(1) = -d'(1) and e'(1) = B' - R'
and d'(2) = B - R and r'(2) = -d'(2) and e'(2) = V - R'.

(4.41*.2) follows from (4.35*.9) by taking

p = 2 and t = {2} and D = 0 and d(1) = E and
d'(1) = B - R and r'(1) = -d'(1) and e'(1) = V - R'
and d'(2) = R and r'(2) = -d'(2) and e'(2) = B' - R'.

Second Proof: Let there be fixed any U ∈ N. By induction on D we shall show that for all I ∈ Z and D ∈ N we have

$$(1^*) \quad \begin{bmatrix} U - D \\ I \end{bmatrix} = \sum_{E \in [0,D]} (-1)^E \binom{D}{E}\begin{bmatrix} U \\ I - E\end{bmatrix}.$$

For D = 0, the LHS = $\begin{bmatrix} U \\ I \end{bmatrix}$ and by (2.5.1') also the RHS = $\begin{bmatrix} U \\ I \end{bmatrix}$. So let D > 0 and assume that (1*) is true for D - 1 (and for all I ∈ Z). Now by the induction hypothesis we have

$$(1) \quad \begin{bmatrix} U - (D - 1) \\ I \end{bmatrix} = \sum_{E \in [0,D-1]} (-1)^E \binom{D - 1}{E}\begin{bmatrix} U \\ I - E\end{bmatrix}$$

and

(2) $\begin{bmatrix} U - (D - 1) \\ I - 1 \end{bmatrix} = \sum_{E \in [0,D-1]} (-1)^E \binom{D - 1}{E} \begin{bmatrix} U \\ I - 1 - E \end{bmatrix}.$

By (2.5.1') we have $\binom{D - 1}{0} = 1 = \binom{D}{0}$ and hence we see that

$$[\text{zeroth term in RHS of } (1)] = \binom{D - 1}{0} \begin{bmatrix} U \\ I \end{bmatrix} = \binom{D}{0} \begin{bmatrix} U \\ I \end{bmatrix}$$

and therefore by (1) we get

(3) $\begin{bmatrix} U - D + 1 \\ I \end{bmatrix} = \binom{D}{0} \begin{bmatrix} U \\ I \end{bmatrix} + \sum_{E \in [1,D-1]} (-1)^E \binom{D - 1}{E} \begin{bmatrix} U \\ I - E \end{bmatrix}.$

By (2.5.1') and (2.5.10') we have $\binom{D - 1}{D - 1} = 1 = \binom{D}{D}$ and hence we see that

$$[\text{last term in RHS of } (2)]$$

$$= (-1)^{D-1} \binom{D - 1}{D - 1} \begin{bmatrix} U \\ I - D \end{bmatrix} = -(-1)^D \binom{D}{D} \begin{bmatrix} U \\ I - D \end{bmatrix}$$

and therefore by (2) we get

(4) $\begin{bmatrix} U - D + 1 \\ I - 1 \end{bmatrix} = -(-1)^D \binom{D}{D} \begin{bmatrix} U \\ I - D \end{bmatrix} + \sum_{E \in [0,D-2]} (-1)^E \binom{D - 1}{E} \begin{bmatrix} U \\ I - 1 - E \end{bmatrix}.$

By translation of summation index we get

$$\sum_{E \in [0,D-2]} (-1)^E \binom{D - 1}{E} \begin{bmatrix} U \\ I - 1 - E \end{bmatrix}$$

$$= \sum_{E' \in [1,D-1]} (-1)^{E'-1} \binom{D - 1}{E' - 1} \begin{bmatrix} U \\ I - E' \end{bmatrix} \quad \text{(by putting } E = E' - 1)$$

$$= \sum_{E \in [1,D-1]} (-1)^{E-1} \binom{D - 1}{E - 1} \begin{bmatrix} U \\ I - E \end{bmatrix} \quad \text{(by putting } E' = E)$$

and hence by (4) we get

(5) $\begin{bmatrix} U - D + 1 \\ I - 1 \end{bmatrix} = -(-1)^D \binom{D}{D} \begin{bmatrix} U \\ I - D \end{bmatrix} + \sum_{E \in [1,D-1]} (-1)^{E-1} \binom{D - 1}{E - 1} \begin{bmatrix} U \\ I - E \end{bmatrix}.$

Now we have

$$
\begin{aligned}
&[\text{RHS of }(3)] - [\text{RHS of }(5)]\\
&= \binom{D}{0}\begin{bmatrix}U\\I\end{bmatrix} + (-1)^D\binom{D}{D}\begin{bmatrix}U\\I-D\end{bmatrix} + \sum_{E\in[1,D-1]}(-1)^E\left\{\binom{D-1}{E}\right.\\
&\quad + \left.\binom{D-1}{E-1}\right\}\begin{bmatrix}U\\I-E\end{bmatrix}\\
&= \binom{D}{0}\begin{bmatrix}U\\I\end{bmatrix} + (-1)^D\binom{D}{D}\begin{bmatrix}U\\I-D\end{bmatrix} + \sum_{E\in[1,D-1]}(-1)^E\binom{D}{E}\begin{bmatrix}U\\I-E\end{bmatrix}\\
&\qquad\qquad\qquad\qquad\qquad\qquad\qquad\qquad [\text{by }(4.11')]\\
&= \sum_{E\in[0,D]}(-1)^E\binom{D}{E}\begin{bmatrix}U\\I-E\end{bmatrix}
\end{aligned}
$$

and by (4.11*) we also have

$$\begin{bmatrix}U-D\\I\end{bmatrix} = [\text{LHS of }(3)] - [\text{LHS of }(5)]$$

and therefore in view of (3) and (5) we conclude that

$$\begin{bmatrix}U-D\\I\end{bmatrix} = \sum_{E\in[0,D]}(-1)^E\binom{D}{E}\begin{bmatrix}U\\I-E\end{bmatrix}.$$

Thus the induction is completed and hence (1*) is proved.

By (2.5.2'), (2.5.2*), and (2.5.6') we see that

$$\left\{E\in Z : \binom{D}{E}\begin{bmatrix}U\\I-E\end{bmatrix}\neq 0\right\}\subset[0,D]\cap[0,I]$$

and hence by (1*) we see that given any $I\in Z$, for all $U\in N$ and $D\in N$ we have

$$\begin{bmatrix}U-D\\I\end{bmatrix} = \sum_{E\in[0,I]}(-1)^E\binom{D}{E}\begin{bmatrix}U\\I-E\end{bmatrix}$$

and therefore by (4.33) we conclude that

(10*) given any $I^*\in Z$ and given any elements U^* and D^* in any overring of Q, we have
$$\begin{bmatrix}U^*-D^*\\I^*\end{bmatrix} = \sum_{E\in[0,I^*]}(-1)^E\binom{D^*}{E}\begin{bmatrix}U^*\\I^*-E\end{bmatrix}.$$

Henceforth let there be given any $B\in Z$ and $R\in Z$, and let there be given any elements B', R', V in any overring of Q. Now upon taking $I^*=B$ and $U^*=V-R'$ and $D^*=B'-R'$ in (10*) we get

$$(10)\quad \begin{bmatrix} V-B' \\ B \end{bmatrix} = \sum_{E\in[0,B]} (-1)^E \binom{B'-R'}{E} \begin{bmatrix} V-R' \\ B-E \end{bmatrix}.$$

By translation of the summation index we see that

RHS of (10)

$$= \sum_{E'\in[-R,B-R]} (-1)^{R+E'} \binom{B'-R'}{R+E'} \begin{bmatrix} V-R' \\ B-R-E' \end{bmatrix} \quad \text{(by putting } E = R+E'\text{)}$$

$$= \sum_{E\in[-R,B-R]} (-1)^{R+E} \binom{B'-R'}{R+E} \begin{bmatrix} V-R' \\ B-R-E \end{bmatrix} \quad \text{(by putting } E' = E\text{)}$$

and hence by (10) we get

$$(11)\quad \begin{bmatrix} V-B' \\ B \end{bmatrix} = \sum_{E\in[-R,B-R]} (-1)^{R+E} \binom{B'-R'}{R+E} \begin{bmatrix} V-R' \\ B-R-E \end{bmatrix}.$$

By (2.5.2') and (2.5.2*) we see that

$$\left\{E \in Z : \binom{B'-R'}{R+E} \begin{bmatrix} V-R' \\ B-R-E \end{bmatrix} \neq 0\right\} \subset [-R,\ B-R]$$

and hence by (11) we get (4.41*.1). By transformation of the summation index we see that

RHS of (10)

$$= \sum_{E'\in[R-B,R]} (-1)^{R-E'} \binom{B'-R'}{R-E'} \begin{bmatrix} V-R' \\ B-R+E' \end{bmatrix} \quad \text{(by putting } E = R-E'\text{)}$$

$$= \sum_{E\in[R-B,R]} (-1)^{R-E} \binom{B'-R'}{R-E} \begin{bmatrix} V-R' \\ B-R+E \end{bmatrix} \quad \text{(by putting } E' = E\text{)}$$

and hence by (10) we get

$$(12)\quad \begin{bmatrix} V-B' \\ B \end{bmatrix} = \sum_{E\in[R-B,R]} (-1)^{R-E} \binom{B'-R'}{R-E} \begin{bmatrix} V-R' \\ B-R+E \end{bmatrix}.$$

By (2.5.2') and (2.5.2*) we see that

$$\left\{E \in Z : \binom{B'-R'}{R-E} \begin{bmatrix} V-R' \\ B-R+E \end{bmatrix} \neq 0\right\} \subset [R-B,\ R]$$

and hence by (12) we get (4.41*.2).

VARIATION OF TL (4.42) Given any $p \in N$ and $U \in [0,p]$ and $B \in Z$, and given any element V in any overring of Q, we have

$$\binom{V+U}{B} = \sum_{I\in[0,p]} \binom{U}{p-I}\binom{V}{B-U+p-I}.$$

Proof: Upon taking

$$B' = U \text{ and } R' = 0 \text{ and } R = U - p \text{ and } E = I$$

in (4.41'.1) we see that

$$(1)\quad \left[\begin{array}{l}\displaystyle\binom{V+U}{B} = \sum_{I\in Z} \binom{U}{U-p+I}\binom{V}{B-U+p-I}\\ \text{where the summation is essentially finite.}\end{array}\right.$$

By (2.5.10') we see that

$$(2)\quad \binom{U}{U-p+I} = \binom{U}{p-I} \qquad \text{for all } I \in Z.$$

By (2.5.2') we see that

(3) if $I \in Z$ is such that $I > p$, then $\binom{U}{p-I} = 0$.

By (2.5.6') we see that

(4) if $I \in Z$ is such that $I < 0$, then $\binom{U}{p-I} = 0$.

Now our assertion follows from (1) to (4).

COROLLARY OF VARIATION OF TL (4.43) Given any $p \in N$ and $U \in [0,p]$, and given any D, E, R, R', S' in Z, we have

$$\sum_{I\in[0,p]} \binom{U}{p-I}\binom{E+R'-S'}{D-I+E-R} = \binom{E+U+R'-S'}{D+E+U-p-R}.$$

Proof: Follows from (4.42) by taking $B = D + E + U - p - R$ and $V = E + R' - S'$.

COROLLARY OF TTL (4.44) Given any $p \in N$ and $U \in [0,p]$ and $C' \in Z$ and $D \in Z$, in $Q[V]$ we have

$$\begin{bmatrix}V-U\\C'-D\end{bmatrix} = \sum_{I\in[0,p]} (-1)^I \binom{U}{I}\begin{bmatrix}V\\C'-D-I\end{bmatrix}.$$

Proof: Upon taking

$$B = C' - D \text{ and } R = 0 \text{ and } B' = U \text{ and } R' = 0 \text{ and } E = I$$

in (4.41*.1) we see that

$$\begin{bmatrix} V - U \\ C' - D \end{bmatrix} = \sum_{I \in Z} (-1)^I \binom{U}{I} \begin{bmatrix} V \\ C' - D - I \end{bmatrix},$$

where the summation is essentially finite, and now our assertion follows by noting that by (2.5.2') and (2.5.6') we have

$$\left\{ I \in Z : \binom{U}{I} \neq 0 \right\} \subset [0,p].$$

VARIATION OF TTL (4.45) Given any $I \in Z$ and given any element V in any overring of Q, for every $D \in N$ we have

$$\binom{V - D}{I} = \sum_{E \in [0,D]} (-1)^E \binom{D}{E} \binom{V - E}{I - E}.$$

NOTE We shall give two proofs, the first using TTL (4.41*) and the second by induction.

First Proof: Upon taking

$B' = D + I$ and $R' = I$ and $B = I$ and $R = 0$

in (4.41*.1) we see that

(1) $\left[\begin{bmatrix} V - D - I \\ I \end{bmatrix} = \sum_{E \in Z} (-1)^E \binom{D}{E} \begin{bmatrix} V - I \\ I - E \end{bmatrix} \right.$ where the summation is essentially finite.

By (2.5.3*) we see that

(2) $\begin{bmatrix} V - D - I \\ I \end{bmatrix} = \binom{V - D}{I}$

and we see that

(3) $\begin{bmatrix} V - I \\ I - E \end{bmatrix} = \binom{V - E}{I - E}$ for all $E \in Z$.

By (2.5.2') we see that

(4) if $E \in Z$ is such that $E < 0$, then $\binom{D}{E} = 0$.

By (2.5.6') we see that

(5) if $E \in Z$ is such that $E > D$, then $\begin{pmatrix} D \\ E \end{pmatrix} = 0$.

Now our assertion follows from (1) to (5).

Second Proof: We shall make induction on D. For D = 0, the LHS = $\begin{pmatrix} V \\ I \end{pmatrix}$ and by (2.5.1') also the RHS = $\begin{pmatrix} V \\ I \end{pmatrix}$. So let D > 0, and assume that the assertion is true for D - 1 (and for all $I \in Z$ and for all V in the said overring of Q). Now by the induction hypothesis we have

$$(1) \quad \begin{pmatrix} V - (D - 1) \\ I \end{pmatrix} = \sum_{E \in [0,D-1]} (-1)^E \begin{pmatrix} D - 1 \\ E \end{pmatrix} \begin{pmatrix} V - E \\ I - E \end{pmatrix}$$

and

$$(2) \quad \begin{pmatrix} (V - 1) - (D - 1) \\ I - 1 \end{pmatrix} = \sum_{E \in [0,D-1]} (-1)^E \begin{pmatrix} D - 1 \\ E \end{pmatrix} \begin{pmatrix} (V - 1) - E \\ (I - 1) - E \end{pmatrix}.$$

By (2.5.1') we have $\begin{pmatrix} D - 1 \\ 0 \end{pmatrix} = 1 = \begin{pmatrix} D \\ 0 \end{pmatrix}$ and hence we see that

$$[\text{zeroth term in RHS of (1)}] = \begin{pmatrix} D - 1 \\ 0 \end{pmatrix} \begin{pmatrix} V \\ I \end{pmatrix} = \begin{pmatrix} D \\ 0 \end{pmatrix} \begin{pmatrix} V \\ I \end{pmatrix}$$

and therefore by (1) we get

$$(3) \quad \begin{pmatrix} V - D + 1 \\ I \end{pmatrix} = \begin{pmatrix} D \\ 0 \end{pmatrix} \begin{pmatrix} V \\ I \end{pmatrix} + \sum_{E \in [1,D-1]} (-1)^E \begin{pmatrix} D - 1 \\ E \end{pmatrix} \begin{pmatrix} V - E \\ I - E \end{pmatrix}.$$

By (2.5.1') and (2.5.10') we have $\begin{pmatrix} D - 1 \\ D - 1 \end{pmatrix} = 1 = \begin{pmatrix} D \\ D \end{pmatrix}$ and hence we see that

$$[\text{last term in RHS of (2)}]$$
$$= (-1)^{D-1} \begin{pmatrix} D - 1 \\ D - 1 \end{pmatrix} \begin{pmatrix} V - D \\ I - D \end{pmatrix} = -(-1)^D \begin{pmatrix} D \\ D \end{pmatrix} \begin{pmatrix} V - D \\ I - D \end{pmatrix}$$

and therefore by (2) we get

$$(4) \quad \begin{pmatrix} V - D \\ I - D \end{pmatrix} = -(-1)^D \begin{pmatrix} D \\ D \end{pmatrix} \begin{pmatrix} V - D \\ I - D \end{pmatrix} + \sum_{E \in [0,D-2]} (-1)^E \begin{pmatrix} D - 1 \\ E \end{pmatrix} \begin{pmatrix} (V - 1) - E \\ (I - 1) - E \end{pmatrix}.$$

By translation of the summation index we get

$$\sum_{E\in[0,D-2]} (-1)^E \binom{D-1}{E} \binom{(V-1)-E}{(I-1)-E}$$

$$= \sum_{E'\in[1,D-1]} (-1)^{E'-1} \binom{D-1}{E'-1} \binom{V-E'}{I-E'} \quad \text{(by putting } E = E' - 1\text{)}$$

$$= \sum_{E\in[1,D-1]} (-1)^{E-1} \binom{D-1}{E-1} \binom{V-E}{I-E} \quad \text{(by putting } E' = E\text{)}$$

and hence by (4) we get

$$(5) \quad \binom{V-D}{I-D} = -(-1)^D \binom{D}{D} \binom{V-D}{I-D} + \sum_{E\in[1,D-1]} (-1)^{E-1} \binom{D-1}{E-1} \binom{V-E}{I-E}.$$

Now we have

$$[\text{RHS of } (3)] - [\text{RHS of } (5)]$$

$$= \binom{D}{0}\binom{V}{I} + (-1)^D \binom{D}{D}\binom{V-D}{I-D}$$

$$+ \sum_{E\in[1,D-1]} (-1)^E \left[\binom{D-1}{E} + \binom{D-1}{E-1}\right] \binom{V-E}{I-E}$$

$$= \binom{D}{0}\binom{V}{I} + (-1)^D \binom{D}{D}\binom{V-D}{I-D} + \sum_{E\in[1,D-1]} (-1)^E \binom{D}{E}\binom{V-E}{I-E}$$

[by (4.11')]

$$= \sum_{E\in[0,D]} (-1)^E \binom{D}{E}\binom{V-E}{I-E}$$

and by (4.11') we also have

$$\binom{V-D}{I} = [\text{LHS of } (3)] - [\text{LHS of } (5)]$$

and therefore in view of (3) and (5) we conclude that

$$\binom{V-D}{I} = \sum_{E\in[0,D]} (-1)^E \binom{D}{E}\binom{V-E}{I-E}.$$

This completes the induction.

COROLLARY OF VARIATION OF TTL (4.46) Given any $p \in N$ and $T \in [0,p]$ and $I \in Z$ we have

$$\binom{T}{I} = \sum_{U\in[0,p]} (-1)^{p-U} \binom{U}{p-I} \binom{p-T}{U-T}.$$

Proof: Upon taking $V = p$ and $D = p - T$ in (4.45) we see that

$$(1)\quad \binom{T}{I} = \sum_{E\in[0,p-T]} (-1)^E \binom{p-T}{E}\binom{p-E}{I-E}.$$

By (2.5.6') we see that

$$(2)\quad \binom{p-T}{E} = 0 \qquad \text{for all } E \in [p-T+1, p].$$

By (1) and (2) we conclude that

$$(3)\quad \binom{T}{I} = \sum_{E\in[0,p]} (-1)^E \binom{p-T}{E}\binom{p-E}{I-E}.$$

Now $E \mapsto p - E$ gives a bijection of $[0,p]$ onto itself and hence, upon writing U for $p - E$, by (3) we get

$$(4)\quad \binom{T}{I} = \sum_{U\in[0,p]} (-1)^{p-U} \binom{p-T}{p-U}\binom{U}{I-p+U}.$$

By (2.5.10') we see that

$$(5)\quad \binom{p-T}{p-U} = \binom{p-T}{U-T} \qquad \text{for all } U \in Z$$

and

$$(6)\quad \binom{U}{I-p+U} = \binom{U}{p-I} \qquad \text{for all } U \in N.$$

Our assertion follows from (4) to (6).

CONSEQUENCE OF TTL (4.47) Given any $D \in Z$ and $U \in Z$ and $T \in Z$, in $Q[V]$ we have

$$\begin{bmatrix} V-D \\ U \end{bmatrix}\binom{U+T}{D} = \sum_{E\in Z} (-1)^{U-E} \begin{bmatrix} V \\ E \end{bmatrix}\binom{U+T}{U-E}\binom{E+T}{D+E-U}$$

where the summation is essentially finite.

Proof: Upon taking $B' = D$ and $R' = 0$ and $B = R = U$ in (4.41*.2) we see that

$$\begin{bmatrix} V-D \\ U \end{bmatrix} = \sum_{E\in Z} (-1)^{U-E} \binom{D}{U-E}\begin{bmatrix} V \\ E \end{bmatrix}$$

where the summation is essentially finite, and now our assertion follows upon multiplying both sides of the equation above by $\binom{U+T}{D}$ and upon noting that by (4.2') we have

$$\binom{U+T}{D}\binom{D}{U-E} = \binom{U+T}{U-E}\binom{E+T}{D+E-U}.$$

COROLLARY OF CONSEQUENCE OF TTL (4.48) Given any $p \in N^*$ and $r \in Z(p)$ and $s \in Z(p)$ and $r' \in Z(p)$ and $s' \in Z(p)$, for every $i \in [1,p]$ and $j \in [1,p]$ let

$$A_{ij}(e^*) = \begin{bmatrix} s(i) \\ e^* \end{bmatrix}\binom{r(j) + r'(j) - s'(i)}{r(j) - e^*} \qquad \text{for all } e^* \in Z$$

and

$$B_{ij}(d^*,e^*) = (-1)^{r(j)-e^*}\binom{e^* + r'(j) - s'(i)}{d^* + e^* - r(j)}$$
$$\text{for all } d^* \in Z \text{ and } e^* \in Z$$

and

$$K_{ij}(d^*) = \begin{bmatrix} s(i) - d^* \\ r(j) \end{bmatrix}\binom{r(j) + r'(j) - s'(i)}{d^*} \qquad \text{for all } d^* \in Z.$$

Then for every $i \in [1,p]$ and $j \in [1,p]$ and $d^* \in Z$ we have

$$K_{ij}(d^*) = \sum_{e^* \in Z} A_{ij}(e^*)B_{ij}(d^*,e^*)$$

where the summation is essentially finite.

Proof: Follows from (4.47) by taking

$U = r(j)$ and $V = s(i)$ and $T = r'(j) - s'(i)$ and $D = d^*$ and $E = e^*$.

GENERAL (BIPRODUCT LEMMA =) BL (4.51') Given any $A \in Z$ and $B \in Z$ and $R \in Z$, and given any elements B' and V in any overring of Q, we have

$$\left[\begin{array}{l}\binom{V}{A}\binom{V+B'}{B} = \sum_{E \in Z}\binom{A+B'}{R+E}\begin{bmatrix} B - R - E \\ A \end{bmatrix}\binom{V}{A+B-R-E} \\ \text{where the summation is essentially finite because} \qquad (4.51'.1) \\ \left\{E \in Z : \binom{A+B'}{R+E}\begin{bmatrix} B - R - E \\ A \end{bmatrix}\binom{V}{A+B-R-E} \neq 0\right\} \subset [-R, B-R]\end{array}\right.$$

and we have

$$\binom{V}{A}\binom{V+B'}{B} = \sum_{E \in Z} \binom{A+B'}{R+E}\begin{bmatrix} A \\ B-R-E \end{bmatrix}\binom{V}{A+B-R-E}$$

where the summation is essentially finite because (4.51'.2)

$$\left\{E \in Z : \binom{A+B'}{R+E}\begin{bmatrix} B-R-E \\ A \end{bmatrix}\binom{V}{A+B-R-E} \neq 0\right\} \subset [-R, B-R]$$

and we have

$$\binom{V}{A}\binom{V+B'}{B} = \sum_{E \in Z} \binom{A+B'}{R-E}\begin{bmatrix} B-R+E \\ A \end{bmatrix}\binom{V}{A+B-R+E}$$

where the summation is essentially finite because (4.51'.3)

$$\left\{E \in Z : \binom{A+B'}{R-E}\begin{bmatrix} B-R+E \\ A \end{bmatrix}\binom{V}{A+B-R+E} \neq 0\right\} \subset [R-B, R]$$

and we have

$$\binom{V}{A}\binom{V+B'}{B} = \sum_{E \in Z} \binom{A+B'}{R-E}\begin{bmatrix} A \\ B-R+E \end{bmatrix}\binom{V}{A+B-R+E}$$

where the summation is essentially finite because (4.51'.4)

$$\left\{E \in Z : \binom{A+B'}{R-E}\begin{bmatrix} A \\ B-R+E \end{bmatrix}\binom{V}{A+B-R+E} \neq 0\right\} \subset [R-B, R].$$

Proof: (4.51'.1) and (4.51'.2) follow by taking $R' = -A$ in (4.41'.1) and then multiplying both sides of (4.41'.1) by $\binom{V}{A}$ and then noting that in view of (4.1'), for all $E \in Z$ we have

$$\binom{V}{A}\binom{V-A}{B-R-E} = \begin{bmatrix} B-R-E \\ A \end{bmatrix}\binom{V}{A+B-R-E}$$

$$= \begin{bmatrix} A \\ B-R-E \end{bmatrix}\binom{V}{A+B-R-E}.$$

Similarly, (4.51'.3) and (4.51'.4) follow by taking $R' = -A$ in (4.41'.2) and then multiplying both sides of (4.41'.2) by $\binom{V}{A}$ and then noting that in view of (4.1'), for all $E \in Z$ we have have

$$\binom{V}{A}\binom{V-A}{B-R+E} = \begin{bmatrix} B-R+E \\ A \end{bmatrix}\binom{V}{A+B-R+E}$$

$$= \begin{bmatrix} A \\ B-R+E \end{bmatrix}\binom{V}{A+B-R+E}.$$

GENERAL (TWISTED BIPRODUCT LEMMA =) TBL (4.51*) Given any $A \in Z$ and $B \in Z$ and $R \in Z$, and given any elements B' and V in any overring of Q, we have

$$\begin{bmatrix} V \\ A \end{bmatrix}\begin{bmatrix} V - B' \\ B \end{bmatrix} = \sum_{E \in Z} (-1)^{R+E}\binom{A + B'}{R + E}\begin{bmatrix} B - R - E \\ A \end{bmatrix}\begin{bmatrix} V \\ A + B - R - E \end{bmatrix}$$

where the summation is essentially finite because (4.51*.1)

$$\left\{E \in Z : \binom{A + B'}{R + E}\begin{bmatrix} B - R - E \\ A \end{bmatrix}\begin{bmatrix} V \\ A + B - R - E \end{bmatrix} \neq 0\right\} \subset [-R, B - R]$$

and we have

$$\begin{bmatrix} V \\ A \end{bmatrix}\begin{bmatrix} V - B' \\ B \end{bmatrix} = \sum_{E \in Z} (-1)^{R+E}\binom{A + B'}{R + E}\begin{bmatrix} A \\ B - R - E \end{bmatrix}\begin{bmatrix} V \\ A + B - R - E \end{bmatrix}$$

where the summation is essentially finite because (4.51*.2)

$$\left\{E \in Z : \binom{A + B'}{R + E}\begin{bmatrix} A \\ B - R - E \end{bmatrix}\begin{bmatrix} V \\ A + B - R - E \end{bmatrix} \neq 0\right\} \subset [-R, B - R]$$

and we have

$$\begin{bmatrix} V \\ A \end{bmatrix}\begin{bmatrix} V - B' \\ B \end{bmatrix} = \sum_{E \in Z} (-1)^{R-E}\binom{A + B'}{R - E}\begin{bmatrix} B - R + E \\ A \end{bmatrix}\begin{bmatrix} V \\ A + B - R + E \end{bmatrix}$$

where the summation is essentially finite because (4.51*.3)

$$\left\{E \in Z : \binom{A + B'}{R - E}\begin{bmatrix} B - R + E \\ A \end{bmatrix}\begin{bmatrix} V \\ A + B - R + E \end{bmatrix} \neq 0\right\} \subset [R - B, R]$$

and we have

$$\begin{bmatrix} V \\ A \end{bmatrix}\begin{bmatrix} V - B' \\ B \end{bmatrix} = \sum_{E \in Z} (-1)^{R-E}\binom{A + B'}{R - E}\begin{bmatrix} A \\ B - R + E \end{bmatrix}\begin{bmatrix} V \\ A + B - R + E \end{bmatrix}$$

where the summation is essentially finite because (4.51*.4)

$$\left\{E \in Z : \binom{A + B'}{R - E}\begin{bmatrix} A \\ B - R + E \end{bmatrix}\begin{bmatrix} V \\ A + B - R + E \end{bmatrix} \neq 0\right\} \subset [R - B, R].$$

Proof: (4.51*.1) and (4.51*.2) follow by taking $R' = -A$ in (4.41*.1) and then multiplying both sides of (4.41*.1) by $\begin{bmatrix} V \\ A \end{bmatrix}$ and then noting that in view of (4.1*), for all $E \in Z$ we have

$$\begin{bmatrix} V \\ A \end{bmatrix}\begin{bmatrix} V + A \\ B - R - E \end{bmatrix} = \begin{bmatrix} B - R - E \\ A \end{bmatrix}\begin{bmatrix} V \\ A + B - R - E \end{bmatrix}$$

$$= \begin{bmatrix} A \\ B - R - E \end{bmatrix}\begin{bmatrix} V \\ A + B - R - E \end{bmatrix}.$$

Similarly, (4.51*.3) and (4.51*.4) follow by taking R' = -A in (4.41*.2) and then multiplying both sides of (4.41*.2) by $\begin{bmatrix}V\\A\end{bmatrix}$ and then noting that in view of (4.1*), for all $E \in Z$ we have

$$\begin{bmatrix}V\\A\end{bmatrix}\begin{bmatrix}V + A\\B - R + E\end{bmatrix} = \begin{bmatrix}B - R + E\\A\end{bmatrix}\begin{bmatrix}V\\A + B - R + E\end{bmatrix}$$

$$= \begin{bmatrix}A\\B - R + E\end{bmatrix}\begin{bmatrix}V\\A + B - R + E\end{bmatrix}.$$

CONSEQUENCE OF GENERAL BL (4.52) Given any $V \in Z$ and $D \in Z$, and given any elements U and T in any overring of Q, we have

$$\begin{bmatrix}U\\V - D\end{bmatrix}\binom{U + T}{D} = \sum_{E \in Z}\binom{V - T}{V - E}\binom{U + T}{E}\binom{E}{D}$$

where the summation is essentially finite because

$$\left\{E \in Z : \binom{V - T}{V - E}\binom{U + T}{E}\binom{E}{D} \neq 0\right\} \subset [D,V].$$

NOTE We shall give two proofs, the first directly using (4.41'), and the second using (4.51*).

First Proof: By (4.41'.2) we know that given any $B \in Z$ and $R \in Z$, and given any elements B', R', V' in any overring of Q, we have

$$(1) \quad \binom{V' + B'}{B} = \sum_{E \in [R-B,R]}\binom{B' - R'}{R - E}\binom{V' + R'}{B - R + E}$$

and

$$(2) \quad \left\{E \in Z : \binom{B' - R'}{R - E}\binom{V' + R'}{B - R + E} \neq 0\right\} \subset [R - B, R].$$

By (2.5.3') we have

$$\binom{V' + B}{B} = \begin{bmatrix}V'\\B\end{bmatrix}$$

and hence upon taking B' = B = V - D and R = V and R' = T - D and V' = U in (1) and (2) we see that

$$(3) \quad \begin{bmatrix}U\\V - D\end{bmatrix} = \sum_{E \in [D,V]}\binom{V - T}{V - E}\binom{U + T - D}{E - D}$$

and

$$(4)\quad \left\{E \in Z : \binom{V - T}{V - E}\binom{U + T - D}{E - D} \neq 0\right\} \subset [D,V].$$

By (2.5.3*) and (4.1') we see that

$$(5)\quad \binom{U + T}{D}\binom{U + T - D}{E - D} = \binom{U + T}{E}\binom{E}{D} \qquad \text{for all } E \in Z.$$

Now, in view of (4) and (5), our assertion follows by multiplying both sides of (3) by $\binom{U + T}{D}$.

Second Proof: By (4.51'.3) we know that given any $A \in Z$ and $B \in Z$ and $R \in Z$, and given any elements B' and V' in any overring of Q, we have

$$(1)\quad \binom{V'}{A}\binom{V' + B'}{B} = \sum_{E\in[R-B,R]} \binom{A + B'}{R - E}\begin{bmatrix}B - R + E\\ A\end{bmatrix}\binom{V'}{A + B - R + E}$$

and

$$(2)\quad \left\{E \in Z : \binom{A + B'}{R - E}\begin{bmatrix}B - R + E\\ A\end{bmatrix}\binom{V'}{A + B - R + E} \neq 0\right\} \subset [R - B, R].$$

Upon taking $A = D$ and $B = V - D$ and $R = V$ and $B' = V - D - T$ and $V' = U + T$ in (1) and (2) we see that

$$(3)\quad \binom{U + T}{D}\binom{U + V - D}{V - D} = \sum_{E\in[D,V]} \binom{V - T}{V - E}\begin{bmatrix}E - D\\ D\end{bmatrix}\binom{U + T}{E}$$

and

$$(4)\quad \left\{E \in Z : \binom{V - T}{V - E}\begin{bmatrix}E - D\\ D\end{bmatrix}\binom{U + T}{E} \neq 0\right\} \subset [D,V].$$

Now our assertion follows from (3) and (4) by noting that in view of (2.5.3'), we have

$$\binom{U + V - D}{V - D} = \begin{bmatrix}U\\ V - D\end{bmatrix}$$

and in view of (2.5.3*), we have

$$\begin{bmatrix}E - D\\ D\end{bmatrix} = \binom{E}{D} \qquad \text{for all } E \in Z.$$

COROLLARY OF CONSEQUENCE OF GENERAL BL (4.53) Given any $p \in N^*$ and $r \in Z(p)$ and $s \in Z(p)$ and $r' \in Z(p)$ and $s' \in Z(p)$, for every $i \in [1,p]$ and $j \in [1,p]$ let

$$A_{ij}(e^*) = \binom{s(i) + s'(i) - r'(j)}{s(i) - e^*}\binom{r(j) + r'(j) - s'(i)}{e^*}$$
$$\text{for all } e^* \in Z$$

and

$$B_{ij}(d^*,e^*) = \binom{e^*}{d^*} \quad \text{for all } d^* \in Z \text{ and } e^* \in Z$$

and

$$K_{ij}(d^*) = \left[\begin{matrix} r(j) \\ s(i) - d^* \end{matrix}\right]\binom{r(j) + r'(j) - s'(i)}{d^*} \quad \text{for all } d^* \in Z.$$

Then for every $i \in [1,p]$ and $j \in [1,p]$ and $d^* \in Z$ we have

$$K_{ij}(d^*) = \sum_{e^* \in Z} A_{ij}(e^*)B_{ij}(d^*,e^*)$$

where the summation is essentially finite.

Proof: Follows from (4.52) by taking $U = r(j)$ and $V = s(i)$ and $T = r'(j) - s'(i)$ and $D = d^*$ and $E = e^*$.

CONSEQUENCE OF TL (4.54) Given any $U \in Z$ and $T \in Z$ and $D \in Z$, and given any element V in any overring of Q, we have

$$\left[\begin{matrix} V - D \\ U \end{matrix}\right]\binom{U + T}{U + T - D} = \sum_{E' \in Z} \binom{V - T}{E'}\binom{U + T}{U - E'}\binom{T + E'}{T + E' - D}$$

where the summation is essentially finite because

$$\left\{E' \in Z : \binom{V - T}{E'}\binom{U + T}{U - E'}\binom{T + E'}{T + E' - D} \neq 0\right\} \subset [0,U].$$

NOTE We shall give two proofs, the first using (4.41') and the second by the Leibnitz Rule.

First Proof: By (2.5.3') we have

$$\binom{U + V - D}{U} = \left[\begin{matrix} V - D \\ U \end{matrix}\right]$$

and hence upon taking B = U and R = U and B' = U - D and R' = -T and E = E' in (4.41'.2) we see that

$$(1)\quad \begin{bmatrix} V - D \\ U \end{bmatrix} = \sum_{E' \in [0,U]} \begin{pmatrix} U + T - D \\ U - E' \end{pmatrix} \begin{pmatrix} V - T \\ E' \end{pmatrix}$$

and

$$(2)\quad \left\{ E' \in Z : \begin{pmatrix} U + T - D \\ U - E' \end{pmatrix} \begin{pmatrix} V - T \\ E' \end{pmatrix} \neq 0 \right\} \subset [0,U].$$

By (4.2') we see that

$$(3)\quad \begin{pmatrix} U + T \\ U + T - D \end{pmatrix} \begin{pmatrix} U + T - D \\ U - E' \end{pmatrix} = \begin{pmatrix} U + T \\ U - E' \end{pmatrix} \begin{pmatrix} T + E' \\ T + E' - D \end{pmatrix}.$$

Now, in view of (2) and (3), our assertion follows by multiplying both sides of (1) by $\begin{pmatrix} U + T \\ U + T - D \end{pmatrix}$.

Second Proof: In view of (4.33), without loss of generality we may assume that $V \in Z$. We shall calculate in the meromorphic series field $Q((Y))$ in an indeterminate Y with coefficients in Q, where we recall that $Q((Y))$ consists of all expressions of the type $\Sigma_{A \in Z} r_A Y^A$ with $r_A \in Q$ for all $A \in Z$, such that $\mathrm{card}(\{A \in Z \setminus N : r_A \neq 0\}) < \infty$. For every $f \in Q((Y))$ and $E \in N$, let $f^{(E)}$ denote the Eth Y-derivative of f. Let there be given any $U' \in N$. Let us recall that by the Leibnitz Rule, for all f and g in $Q((Y))$ we have

$$(1)\quad (fg)^{(U')} = \sum_{E' \in [0,U']} \begin{pmatrix} U' \\ E' \end{pmatrix} f^{(E')} g^{(U'-E')}.$$

Let us also note that clearly,

$$(2)\quad (Y^B)^{(E')} = E'! \begin{pmatrix} B \\ E' \end{pmatrix} Y^{B-E'} \qquad \text{for all } B \in Z \text{ and } E' \in N$$

and

$$(3)\quad \begin{bmatrix} ((1 + Y)^B)^{(U'-E')} = (U' - E')! \begin{pmatrix} B \\ U' - E' \end{pmatrix} (1 + Y)^{B-U'+E'} \\ \text{for all } B \in Z \text{ and } E' \in [0,U']. \end{bmatrix}$$

By (4.31') we have

$$\sum_{D' \in Z} \begin{pmatrix} U' + T \\ U' + T - D' \end{pmatrix} Y^{U'+T-D'} = (1 + Y)^{U'+T}$$

and upon multiplying both sides of the equation above by Y^{V-T} we get

$$\sum_{D'\in Z} \binom{U' + T}{U' + T - D'} Y^{U'+V-D'} = Y^{V-T}(1 + Y)^{U'+T}$$

and upon taking the Uth Y-derivatives of both sides of the equation above, in view of (1) to (3) we conclude that

$$(4) \quad \begin{cases} U'! \sum_{D\in Z} \binom{U' + T}{U' + T - D'} \binom{U' + V - D'}{U'} Y^{V-D'} \\ = \sum_{E'\in[0,U']} \binom{U'}{E'} \left[E'! \binom{V - T}{E'} Y^{V-T-E'}\right] \\ \quad \times \left[(U' - E')! \binom{U' + T}{U' - E'} (1 + Y)^{T+E'}\right]. \end{cases}$$

By (2.5.5') we have

$$\binom{U'}{E'} E'!(U' - E')! = U'! \qquad \text{for all } E' \in [0,U']$$

and hence upon dividing both sides of (4) by U'! we get

$$(5) \quad \begin{cases} \sum_{D'\in Z} \binom{U' + T}{U' + T - D'} \binom{U' + V - D'}{U'} Y^{V-D'} \\ = \sum_{E'\in[0,U']} \binom{V - T}{E'} \binom{U' + T}{U' - E'} Y^{V-T-E'} (1 + Y)^{T+E'}. \end{cases}$$

For every $E' \in [0,U']$, by (4.31') we have

$$(1 + Y)^{T+E'} = \sum_{D'\in Z} \binom{T + E'}{T + E' - D'} Y^{T+E'-D'}$$

and upon multiplying both sides of the equation above by $Y^{V-T-E'}$ we get

$$Y^{V-T-E'}(1 + Y)^{T+E'} = \sum_{D'\in Z} \binom{T + E'}{T + E' - D'} Y^{V-D'}$$

and therefore by (5) we get

$$(6) \quad \begin{cases} \sum_{D'\in Z} \binom{U' + T}{U' + T - D'} \binom{U' + V - D'}{U'} Y^{V-D'} \\ = \sum_{D'\in Z} \left[\sum_{E'\in[0,U']} \binom{V - T}{E'} \binom{U' + T}{U' - E'} \binom{T + E'}{T + E' - D'}\right] Y^{V-D'}. \end{cases}$$

By (2.5.3') we have

$$\begin{pmatrix} U' + V - D \\ U' \end{pmatrix} = \begin{bmatrix} V - D \\ U' \end{bmatrix}$$

and hence upon equating the coefficients of Y^{V-D} on the two sides of (6) we get

$$(7)\quad \begin{bmatrix} V - D \\ U' \end{bmatrix}\begin{pmatrix} U' + T \\ U' + T - D \end{pmatrix} = \sum_{E' \in [0,U']} \begin{pmatrix} V - T \\ E' \end{pmatrix}\begin{pmatrix} U' + T \\ U' - E' \end{pmatrix}\begin{pmatrix} T + E' \\ T + E' - D \end{pmatrix}.$$

By (2.5.2*) we see that

$$\text{if } U \notin N, \text{ then } \begin{bmatrix} V - D \\ U \end{bmatrix} = 0$$

and therefore we conclude that

$$(8)\quad \text{if } U \notin N, \text{ then } \begin{bmatrix} V - D \\ U \end{bmatrix}\begin{pmatrix} U + T \\ U + T - D \end{pmatrix} = 0.$$

Clearly,

$$\text{if } U \notin N, \text{ then } [0,U] = \emptyset$$

and hence

$$(9)\quad \text{if } U \notin N, \text{ then } \sum_{E' \in [0,U]} \begin{pmatrix} V - T \\ E' \end{pmatrix}\begin{pmatrix} U + T \\ U - E' \end{pmatrix}\begin{pmatrix} T + E' \\ T + E' - D \end{pmatrix} = 0.$$

Now if $U \notin N$, then by (8) and (9), whereas if $U \in N$, then upon taking $U' = U$ in (7), we conclude that

$$(10)\quad \begin{bmatrix} V - D \\ U \end{bmatrix}\begin{pmatrix} U + T \\ U + T - D \end{pmatrix} = \sum_{E' \in [0,U]} \begin{pmatrix} V - T \\ E' \end{pmatrix}\begin{pmatrix} U + T \\ U - E' \end{pmatrix}\begin{pmatrix} T + E' \\ T + E' - D \end{pmatrix}$$

By (2.5.2') we see that

$$\text{if } E' \in Z \text{ is such that } E' < 0, \text{ then } \begin{pmatrix} V - T \\ E' \end{pmatrix} = 0,$$

whereas

$$\text{if } E' \in Z \text{ is such that } E' > U, \text{ then } \begin{pmatrix} U + T \\ U - E' \end{pmatrix} = 0$$

and therefore we conclude that

$$(11)\quad \left\{E' \in Z : \begin{pmatrix} V - T \\ E' \end{pmatrix}\begin{pmatrix} U + T \\ U - E' \end{pmatrix}\begin{pmatrix} T + E' \\ T + E' - D \end{pmatrix} \neq 0\right\} \subset [0,U].$$

Now our assertion follows from (10) and (11).

IMPLICATION OF CONSEQUENCE OF TL (4.55) Let there be given any $U \in Z$ and $T \in Z$ and $D \in Z$ such that $U + T \geq 0$. Also let there be given any element V in any overring of Q. Then we have

$$\begin{bmatrix} V - D \\ U \end{bmatrix}\binom{U + T}{D} = \sum_{E' \in Z} \binom{V - T}{E'}\binom{U + T}{T + E'}\binom{T + E'}{D}$$

where the summation is essentially finite because

$$\left\{E' \in Z : \binom{V - T}{E'}\binom{U + T}{T + E'}\binom{T + E'}{D} \neq 0\right\} \subset [0,U].$$

Proof: Since $U + T \geq 0$, by (2.5.10') we see that

$$\binom{U + T}{U + T - D} = \binom{U + T}{D}$$

and

$$\binom{U + T}{U - E'} = \binom{U + T}{T + E'} \qquad \text{for all } E' \in Z$$

and hence by (4.54) we conclude that

$$(1) \quad \begin{bmatrix} V - D \\ U \end{bmatrix}\binom{U + T}{D} = \sum_{E' \in [0,U]} \binom{V - T}{E'}\binom{U + T}{T + E'}\binom{T + E'}{T + E' - D}$$

and

$$(2) \quad \left\{E' \in Z : \binom{V - T}{E'}\binom{U + T}{T + E'}\binom{T + E'}{T + E' - D} \neq 0\right\} \subset [0,U].$$

By (2.5.2') we see that

(3) if $E' \in Z$ is such that $T + E' < 0$, then $\binom{U + T}{T + E'} = 0$.

By (2.5.10') we see that

(4) if $E' \in Z$ is such that $T + E \geq 0$, then $\binom{T + E'}{T + E' - D} = \binom{T + E'}{D}$.

By (3) and (4) we see that

$$(5) \quad \binom{U + T}{T + E'}\binom{T + E'}{T + E' - D} = \binom{U + T}{T + E'}\binom{T + E'}{D} \qquad \text{for all } E' \in Z.$$

Now our assertion follows from (1), (2), and (5).

COROLLARY OF IMPLICATION OF CONSEQUENCE OF TL (4.56) Let there be given any $U \in Z$ and $V \in Z$ and $T \in Z$ and $D \in Z$ such that $V - T \geq 0$ and $U + T \geq 0$. Then we have

$$\begin{bmatrix} V - D \\ U \end{bmatrix} \binom{U + T}{D} = \sum_{E' \in Z} \binom{V - T}{E'} \binom{U + T}{T + E'} \binom{T + E'}{D}$$

where the summation is essentially finite because

$$\left\{ E' \in Z : \binom{V - T}{E'} \binom{U + T}{T + E'} \binom{T + E'}{D} \neq 0 \right\} \subset [D - T, V - T].$$

Proof: Since $V - T \geq 0$, by (2.5.6') we see that

(1) if $E' \in Z$ is such that $E' > V - T$, then $\binom{V - T}{E'} = 0$.

By (2.5.2') we see that

(2) if $E' \in Z$ is such that $T + E' < 0$, then $\binom{U + T}{T + E'} = 0$.

By (2.5.6') we see that

(3) if $E' \in Z$ is such that $D > T + E' \geq 0$, then $\binom{T + E'}{D} = 0$.

By (2) and (3) we see that

(4) if $E' \in Z$ is such that $D > T + E'$, then $\binom{U + T}{T + E'} \binom{T + E'}{D} = 0$.

By (1) and (4) we see that

(5) $\left\{ E' \in Z : \binom{V - T}{E'} \binom{U + T}{T + E'} \binom{T + E'}{D} \neq 0 \right\} \subset [D - T, V - T].$

In view of (5), our assertion follows from (4.55).

REMARK (4.57) Consequence (4.52), in the form of its Corollary (4.53), will be used in the proof of Proposition (7.29.1) of Chapter 2. In the original proof of that proposition as given in our Nice lectures, which have appeared as "Combinatoire des tableaux de Young, varietes determinantielles et calcul de fonctions de Hilbert" in Volume 42 of *Rendiconti del Seminario Matematico di Universita et Polytecnico di Torino* for the year 1984, instead of (4.52) we had used Consequence (4.54), in the form of Implication (4.55). The proof of (4.55) given in Lemma 6 of Section 3 of the Nice lectures corresponds to the second proof of (4.54) given above. Lemma 6 of Section 3 of the Nice lectures was used directly only in the proof of the proposition of Section 3 of the Nice lectures and in that usage it was assumed that $V \in Z$ and $V - T \geq 0$; in other words, we had used (4.55) only in the form of its Corollary (4.56). Although we shall not use (4.54) to (4.56) anywhere in the present book, we have included them so as to give the following comparison between them and (4.52). Let us observe that the first proofs of (4.52) of (4.54), given above, differ from each other only very slightly.

As they stand, (4.52) and (4.54) are certainly mutually nonequivalent because in (4.52) we assume that $V \in Z$, whereas in (4.54) we allow $V \notin Z$. For the same reason, (4.52) and (4.55) are obviously mutually nonequivalent. *So henceforth let us assume that all four quantities* U, V, T, D *are in* Z. Let us note that now (4.54) says that

$$\left[\begin{array}{l} \begin{bmatrix} V-D \\ U \end{bmatrix} \dbinom{U+T}{U+T-D} = \displaystyle\sum_{E' \in [0,U]} \dbinom{V-T}{E'} \dbinom{U+T}{U-E'} \dbinom{T+E'}{T+E'-D} \\ \text{and} \\ \left\{E' \in Z : \dbinom{V-T}{E'} \dbinom{U+T}{U-E'} \dbinom{T+E'}{T+E'-D} \neq 0\right\} \subset [0,U], \end{array}\right. \tag{4.54'}$$

whereas (4.55) says that

$$\left[\begin{array}{l} \text{if } U + T \geq 0, \text{ then} \\ \begin{bmatrix} V-D \\ U \end{bmatrix} \dbinom{U+T}{D} = \displaystyle\sum_{E' \in [0,U]} \dbinom{V-T}{E'} \dbinom{U+T}{T+E'} \dbinom{T+E'}{D} \\ \text{and} \\ \left\{E' \in Z : \dbinom{V-T}{E'} \dbinom{U+T}{T+E'} \dbinom{T+E'}{D} \neq 0\right\} \subset [0,U] \end{array}\right. \tag{4.55'}$$

and finally, (4.56) says that

$$\left[\begin{array}{l} \text{if } V - T \geq 0 \leq U + T, \text{ then} \\ \begin{bmatrix} V-D \\ U \end{bmatrix} \dbinom{U+T}{D} = \displaystyle\sum_{E' \in [D-T,V-T]} \dbinom{V-T}{E'} \dbinom{U+T}{T+E'} \dbinom{T+E'}{D} \end{array}\right.$$

and (4.56')

$$\left\{E' \in Z : \binom{V - T}{E'}\binom{U + T}{T + E'}\binom{T + E'}{D} \neq 0\right\} \subset [D - T, V - T].$$

To make the comparison clearer, let us reformulate (4.52) in its equivalent form which is obtained by putting E = T + E' and which says that

$$\begin{bmatrix}U\\V - D\end{bmatrix}\binom{U + T}{D} = \sum_{E' \in [D-T,V-T]} \binom{V - T}{V - T - E'}\binom{U + T}{T + E'}\binom{T + E'}{D}$$

and (4.52')

$$\left\{E' \in Z : \binom{V - T}{V - T - E'}\binom{U + T}{T + E'}\binom{T + E'}{D} \neq 0\right\} \subset [D - T, V - T].$$

By (2.5.1') and (2.5.1*) and (2.5.2') we see that

if $V = D = 0 > U = -T$, then

[LHS of (4.54')] = 0 = [LHS of (4.55')]

and

[LHS of (4.52')] = 1

and thus (4.52') and (4.54') are still nonequivalent, and likewise (4.52') and (4.55') are still nonequivalent.

However, by (2.5.10') we see that

if $V - T \geq 0$, then

$$\binom{V - T}{V - T - E'} = \binom{V - T}{E'} \quad \text{for all } E' \in Z$$

and by (2.6.15) we see that

if $V - T \geq 0 \leq U + T$, then

[LHS of (4.52')] = [LHS of (4.55')]

and therefore (4.55') is equivalent to the case of (4.52') when $V - T \geq 0 \leq U + T$.

CONSEQUENCE OF TBL (4.58) Let there be given any $U \in Z$ and $V \in Z$ and $T \in Z$ and $D \in Z$, such that $V - T \geq 0$ and $U + T \geq 0$. Then

$$\begin{bmatrix}U\\D\end{bmatrix}\begin{bmatrix}V\\D - T\end{bmatrix} = \sum_{E \in Z} (-1)^E \begin{bmatrix}V - E\\U\end{bmatrix}\binom{U + T}{E}\begin{bmatrix}U + V - E\\D\end{bmatrix}$$

where the summation is essentially finite because

$$\left\{E \in Z : \begin{bmatrix} V - E \\ U \end{bmatrix} \begin{pmatrix} U + T \\ E \end{pmatrix} \begin{bmatrix} U + V - E \\ D \end{bmatrix} \neq 0\right\} \subset [0,V].$$

Proof: By (4.51*.1) we know that given any $A \in Z$ and $B \in Z$ and $R \in Z$, and given any elements B' and V' in any subring of Q, we have

$$(1)\quad \begin{bmatrix} V' \\ A \end{bmatrix} \begin{bmatrix} V' - B' \\ B \end{bmatrix} = \sum_{E \in [-R,B-R]} (-1)^{R+E} \begin{pmatrix} A + B' \\ R + E \end{pmatrix} \begin{bmatrix} B - R - E \\ A \end{bmatrix} \times \begin{bmatrix} V' \\ A + B - R - E \end{bmatrix}.$$

Upon taking $A = U$ and $B = V$ and $R = 0$ and $B' = T$ and $V' = D$ in (1) we see that

$$(2)\quad \begin{bmatrix} D \\ U \end{bmatrix} \begin{bmatrix} D - T \\ V \end{bmatrix} = \sum_{E \in [0,V]} (-1)^{E} \begin{bmatrix} V - E \\ U \end{bmatrix} \begin{pmatrix} U + T \\ E \end{pmatrix} \begin{bmatrix} D \\ U + V - E \end{bmatrix}.$$

By (2.5.10*) we see that

$$(3)\quad \text{if } D \geq 0 \leq U, \text{ then } \begin{bmatrix} D \\ U \end{bmatrix} = \begin{bmatrix} U \\ D \end{bmatrix}.$$

By assumption $V - T \geq 0$ and hence by (2.5.10*) we see that

$$(4)\quad \text{if } D \geq 0, \text{ then } \begin{bmatrix} V \\ D - T \end{bmatrix} = \begin{bmatrix} D - T \\ V \end{bmatrix}.$$

By (2.5.10*) we also see that

$$(5)\quad \left[\begin{array}{l} \text{if } D \geq 0 \leq U, \text{ then for all } E \in [0,V] \text{ we have} \\ \begin{bmatrix} D \\ U + V - E \end{bmatrix} = \begin{bmatrix} U + V - E \\ D \end{bmatrix}. \end{array}\right.$$

By (2) to (5) we see that

$$(6)\quad \left[\begin{array}{l} \text{if } D \geq 0 \leq U, \text{ then} \\ \begin{bmatrix} U \\ D \end{bmatrix} \begin{bmatrix} V \\ D - T \end{bmatrix} = \displaystyle\sum_{E \in [0,V]} (-1)^{E} \begin{bmatrix} V - E \\ U \end{bmatrix} \begin{pmatrix} U + T \\ E \end{pmatrix} \begin{bmatrix} U + V - E \\ D \end{bmatrix}. \end{array}\right.$$

For any $E \in Z$ we have

$$\begin{aligned} U + V - E &= (V - T) + (U + T - E) \\ &\geq U + T - E \quad \text{(because by assumption } V - T \geq 0) \end{aligned}$$

and therefore we see that

(7) if $E \in Z$ is such that $U + V - E < 0$, then $U + T - E < 0$.

By assumption $U + T \geq 0$ and hence by (2.5.6') we see that

(8) if $E \in Z$ is such that $U + T - E < 0$, then $\begin{pmatrix} U + T \\ E \end{pmatrix} = 0$.

By (7) and (8) we see that

(9) if $E \in Z$ is such that $U + V - E < 0$, then $\begin{pmatrix} U + T \\ E \end{pmatrix} = 0$.

By (2.5.6*) we see that

(10) if $E \in Z$ is such that $E > V$ and $U + V - E \geq 0$, then $\begin{bmatrix} V - E \\ U \end{bmatrix} = 0$.

By (9) and (10) we see that

(11) if $E \in Z$ is such that $E > V$, then $\begin{bmatrix} V - E \\ U \end{bmatrix}\begin{pmatrix} U + T \\ E \end{pmatrix} = 0$.

By (2.5.2') we see that

(12) if $E \in Z$ is such that $E < 0$, then $\begin{pmatrix} U + T \\ E \end{pmatrix} = 0$.

By (11) and (12) we see that

(13) $\left\{E \in Z : \begin{bmatrix} V - E \\ U \end{bmatrix}\begin{pmatrix} U + T \\ E \end{pmatrix}\begin{bmatrix} U + V - E \\ D \end{bmatrix} \neq 0\right\} \subset [0,V]$.

By (2.5.2*) we see that

$$\text{if } D < 0, \text{ then } \begin{bmatrix} U \\ D \end{bmatrix} = 0 = \begin{bmatrix} U + V - E \\ D \end{bmatrix} \quad \text{for all } E \in Z$$

and hence

(14) if $D < 0$, then [LHS of (6)] = 0 = [RHS of (6)].

By (2.5.2*) we see that

(15) if $D < T$, then $\begin{bmatrix} V \\ D - T \end{bmatrix} = 0$.

By assumption $U + T \geq 0$ and hence

(16) if $D \geq T$, then $U + D \geq 0$.

By (2.5.6*) we see that

(17) if $U < 0$ and $U + D \geq 0$, then $\begin{bmatrix} U \\ D \end{bmatrix} = 0$.

By (16) and (17) we see that

(18) if $U < 0$ and $D \geq T$, then $\begin{bmatrix} U \\ D \end{bmatrix} = 0$.

By (15) and (18) we see that

(19) if $U < 0$, then [LHS of (6)] = 0.

By (2.5.2*) we see that

$$\text{if } U < 0, \text{ then } \begin{bmatrix} V - E \\ U \end{bmatrix} = 0 \qquad \text{for all } E \in Z$$

and therefore we conclude that

(20) if $U < 0$, then [RHS of (6)] = 0.

Now our assertion follows from (6), (13), (14), (19), and (20).

COROLLARY OF CONSEQUENCE OF TBL (4.59) Given any $p \in N^*$ and $r \in Z(p)$ and $s \in Z(p)$ and $r' \in Z(p)$ and $s' \in Z(p)$, for every $i \in [1,p]$ and $j \in [1,p]$ let

$$A_{ij}(e^*) = \begin{bmatrix} s(i) - e^* \\ r(j) \end{bmatrix} \begin{pmatrix} r(j) + r'(j) - s'(i) \\ e^* \end{pmatrix} \quad \text{for all } e^* \in Z$$

and

$$B_{ij}(d^*,e^*) = (-1)^{e^*} \begin{bmatrix} r(j) + s(i) - e^* \\ d^* \end{bmatrix} \quad \text{for all } d^* \in Z \text{ and } e^* \in Z$$

and

$$K_{ij}(d^*) = \begin{bmatrix} r(j) \\ d^* \end{bmatrix} \begin{bmatrix} s(i) \\ d^* + s'(i) - r'(j) \end{bmatrix} \quad \text{for all } d^* \in Z.$$

Assume that

$$\left[\begin{array}{l} \text{for every } i \in [1,p] \text{ and } j \in [1,p] \text{ we have} \\ s(i) + s'(i) - r'(j) \geq 0 \leq r(j) + r'(j) - s'(i). \end{array}\right.$$

Then for every $i \in [1,p]$ and $j \in [1,p]$ and $d^* \in Z$ we have

$$K_{ij}(d^*) = \sum_{e^* \in Z} A_{ij}(e^*) B_{ij}(d^*,e^*)$$

where the summation is essentially finite.

Proof: Follows from (4.59) by taking U = r(j) and V = s(i) and T = r'(j) - s'(i) and D = d* and E = e*.

BIPRODUCT LEMMA (4.61') For every $A \in Z$ and $B \in Z$, let

$$S_E^{'(1)}(A,B) = \begin{cases} \dfrac{(A+B-E)!}{(A-E)!(B-E)!E!} & \text{if } E \in [0,\min(A,B)] \\ 0 & \text{if } E \in Z\setminus[0,\min(A,B)] \end{cases}$$

and

$$S_E^{'(2)}(A,B) = \begin{pmatrix} A \\ E \end{pmatrix}\begin{bmatrix} B-E \\ A \end{bmatrix} \qquad \text{for all } E \in Z$$

and

$$S_E^{'(3)}(A,B) = \begin{pmatrix} A \\ E \end{pmatrix}\begin{bmatrix} A \\ B-E \end{bmatrix} \qquad \text{for all } E \in Z$$

and

$$S_E^{'(4)}(A,B) = \begin{pmatrix} B \\ E \end{pmatrix}\begin{bmatrix} A-E \\ B \end{bmatrix} \qquad \text{for all } E \in Z$$

and

$$S_E^{'(5)}(A,B) = \begin{pmatrix} B \\ E \end{pmatrix}\begin{bmatrix} B \\ A-E \end{bmatrix} \qquad \text{for all } E \in Z.$$

Then given any $A \in Z$ and $B \in Z$, we have

(1) $S_E^{'(1)}(A,B) = 0 \qquad \text{for all } E \in Z\setminus[0,\min(A,B)]$

and for every $L \in [2,5]$ we have

(L) $S_E^{'(L)}(A,B) = S_E^{'(1)}(A,B) \qquad \text{for all } E \in Z \text{ with } E \le A + B.$

Moreover, given any $A \in Z$ and $B \in Z$, in Q[V] we have

(6) $\begin{pmatrix} V \\ A+B-E \end{pmatrix} = 0 \qquad \text{for all } E \in Z \text{ with } E > A + B$

and

(11) $\begin{pmatrix} V \\ A \end{pmatrix}\begin{pmatrix} V \\ B \end{pmatrix} = \sum_{E\in Z} S_E^{'(1)}(A,B)\begin{pmatrix} V \\ A+B-E \end{pmatrix}$

where, in view of (1), the summation is essentially finite.

Finally, given any $A \in Z$ and $B \in Z$ and $L \in [2,5]$, in $Q[V]$ we have

(1L) $\begin{cases} \binom{V}{A}\binom{V}{B} = \sum_{E \in Z} S_E^{\prime(L)}(A,B)\binom{V}{A+B-E} \\ \text{where, in view of (1), (L), and (6), the summation} \\ \text{is essentially finite.} \end{cases}$

Proof: (1) is obvious. (2) to (6) follow from (2.5.2'), (2.5.3*), (2.5.5'), and (2.5.6'). In view of (2) and (6), we get (11) by taking B' = 0 = R in (4.51'.1). Given any $L \in [2,5]$, by (L), (6), and (11) we get (1L).

TWISTED BIPRODUCT LEMMA (4.61*) For every $A \in Z$ and $B \in Z$, let

$$S_E^{*(1)}(A,B) = \begin{cases} \dfrac{(A+B-E)!}{(A-E)!(B-E)!E!} & \text{if } E \in [0,\min(A,B)] \\ 0 & \text{if } E \in Z\setminus[0,\min(A,B)] \end{cases}$$

and

$$S_E^{*(2)}(A,B) = \binom{A}{E}\begin{bmatrix} B-E \\ A \end{bmatrix} \quad \text{for all } E \in Z$$

and

$$S_E^{*(3)}(A,B) = \binom{A}{E}\begin{bmatrix} A \\ B-E \end{bmatrix} \quad \text{for all } E \in Z$$

and

$$S_E^{*(4)}(A,B) = \binom{B}{E}\begin{bmatrix} A-E \\ B \end{bmatrix} \quad \text{for all } E \in Z$$

and

$$S_E^{*(5)}(A,B) = \binom{B}{E}\begin{bmatrix} B \\ A-E \end{bmatrix} \quad \text{for all } E \in Z.$$

Then given any $A \in Z$ and $B \in Z$, we have

(1) $S_E^{*(1)}(A,B) = 0$ for all $E \in Z\setminus[0,\min(A,B)]$

and for every $L \in [2,5]$ we have

(L) $S_E^{*(L)}(A,B) = S_E^{*(1)}(A,B)$ for all $E \in Z$ with $E \le A + B$.

Moreover, given any $A \in Z$ and $B \in Z$, in $Q[V]$ we have

$$(6)\qquad \begin{bmatrix} V \\ A+B-E \end{bmatrix} = 0 \qquad \text{for all } E \in Z \text{ with } E > A + B$$

and

$$(11)\qquad \begin{bmatrix} \begin{bmatrix} V \\ A \end{bmatrix} \begin{bmatrix} V \\ B \end{bmatrix} = \sum_{E \in Z} (-1)^E S_E^{*(1)}(A,B) \begin{bmatrix} V \\ A+B-E \end{bmatrix} \\ \text{where in view of (1), the summation is essentially finite.} \end{bmatrix}$$

Finally, given any $A \in Z$ and $B \in Z$ and $L \in [2,5]$, in $Q[V]$ we have

$$(1L)\qquad \begin{bmatrix} \begin{bmatrix} V \\ A \end{bmatrix} \begin{bmatrix} V \\ B \end{bmatrix} = \sum_{E \in Z} (-1)^E S_E^{*(L)}(A,B) \begin{bmatrix} V \\ A+B-E \end{bmatrix} \\ \text{where in view of (1), (L), and (6), the summation} \\ \text{is essentially finite.} \end{bmatrix}$$

Proof: (1) is obvious. (2) to (6) follow from (2.5.2'), (2.5.2*), (2.5.3*), (2.5.5'), and (2.5.6'). In view of (2) and (6), we get (11) by taking $B' = 0 = R$ in (4.51*.1). Given any $L \in [2,5]$, by (L), (6), and (11) we get (1L).

REMARK (4.62) We may regard (4.35'.2) as a polynomial identity concerning polynomials in indeterminates $e'(1), e'(2), \ldots, e'(p)$ with coefficients in Q, and now, for every $m \in N(p)$, we may equate the coefficients of

$$e'(1)^{m(1)} e'(2)^{m(2)} \cdots e'(p)^{m(p)}$$

on the two sides of (4.35'.2); thus, for every $m \in N(p)$, we get a new identity concerning rational number and hence, upon multiplying by a common denominator, a new identity about integers; conversely, by independently establishing the said identities about integers, we could get a new proof of (4.35'.2). Let us also recall that the general case of (4.35'.2) followed from the case of (4.35'.2) when $e'(1), e'(2), \ldots, e'(p)$ take only integer values or, alternatively, only positive integer values.

Similarly, (4.35*.2) gives a new integer identity for every $m \in N(p)$. Note that (4.35'.3), (4.35'.5), and (4.35'.6) do not give anything extra because they are only alternative forms of (4.35'.2); likewise, (4.35*.3), (4.35*.5), (4.35*.6), (4.35*.8), and (4.35*.9) do not give anything extra because they are only alternative forms of (4.35*.2).

More generally, we may take any nonempty subset u of [1,p], and we may fix any $e'(j) \in Q$ for each $j \in [1,p]\backslash u$, and we may regard (4.35'.2) as a polynomial identity concerning polynomials in indeterminates $e'(i)_{i \in u}$ with coefficients in Q, and now, for every map m: $u \to N$, we may equate the coefficients of

$$\prod_{i \in u} e'(i)^{m(i)}$$

on the two sides of (4.35'.2), and then upon multiplying by a common denominator we would get a new identity concerning integers. Similarly for (4.35*.2).

In a like manner we can treat (4.1'), (4.1*), (4.2'), (4.2*), (4.11'), (4.11*), (4.41'), (4.41*), (4.42), (4.45), (4.47), (4.51'), (4.51*), (4.52), (4.54), (4.55), (4.61'), and (4.61*). For example, upon fixing $B' \in Q$ and $R' \in Q$, we may regard (4.41'.1) as an identity in Q[V] and upon equating coefficients of V^I and multiplying by a common denominator we get an integer identity for each $I \in N$; note that, in the second proof of (4.41'), the general case of (4.41'.1) followed from the case of (4.41'.1) when B', R', V take only integer values.

These are examples of the *principles of identities*, which says that *an identity concerning positive integers may give rise to a polynomial identity which, in turn, could give rise to several identities about integers. Conversely, by independently establishing the said identities about integers, there may ensue a different proof of the said polynomial identity.*

REMARK (4.63) We shall now motivate the various results which have been given in this section.

Let V be an indeterminate over Q. Now, upon regarding the polynomial ring Q[V] as a vector space over Q, clearly

$$(1') \quad \binom{V}{A}_{A \in N}$$

and

$$(1^*) \quad \begin{bmatrix} V \\ A \end{bmatrix}_{A \in N}$$

are bases of $Q[V]$. The Progression Lemmas (4.13') and (4.13*) show that for summation problems, (1') and (1*) are better bases than the usual basis

$$(1) \quad (V^A)_{A \in N}$$

as can be realized by noting that

$$(2) \quad \sum_{n=0}^{m} n^A = f_{A+1}(m) \qquad \text{for all } m \in N$$

where $f_{A+1}(V)$ is a polynomial of degree $A + 1$ in V with coefficients in Q, and where the nature of f_{A+1} is very complicated for large A and hence in school we learn to sum the LHS of (2) only for small A.

To explain the name of Progression Lemma, recall that an

Arithmetico-Arithmetico-····-Arithmetico-Arithmetic

series of rational numbers, where Arithmetico occurs $A - 1$ times, is a series whose nth term equals $g(n)$ for some polynomial $g(V)$ of degree at most A in V with coefficients in Q. Using the basis (1*), for example, we can write

$$g(V) = \sum_{I=0}^{A} g_I \begin{bmatrix} V \\ I \end{bmatrix} \qquad \text{with } g_I \in Q$$

and now the said series can be summed because by the Twisted Progression Lemma (4.13*) we get

$$\sum_{n=0}^{m} g(n) = \sum_{I=0}^{A} g_I \begin{bmatrix} m \\ I + 1 \end{bmatrix} \qquad \text{for all } m \in N.$$

In particular, upon taking $g(V)$ to be V^A we have

$$V^A = \sum_{I=0}^{A} h(A,I)\begin{bmatrix} V \\ I \end{bmatrix} \qquad \text{with } h(A,I) \in Q$$

and hence

$$\sum_{n=0}^{m} n^A = \sum_{I=0}^{A} h(A,I)\begin{bmatrix} m \\ I+1 \end{bmatrix} \qquad \text{for all } m \in N,$$

which proves (2).

Having seen the usefulness of the basis (1*), we may proceed to relate any two of its translates. In other words, for any two fixed quantities B' and R', we may want to express any member of the basis

$$(3^*) \qquad \begin{bmatrix} V - B' \\ B \end{bmatrix}_{B \in N}$$

as a Q-linear combination of the basis

$$(4^*) \qquad \begin{bmatrix} V - R' \\ D \end{bmatrix}_{D \in N}.$$

This is what we do in the Twisted Translation Lemma (4.41*). In the Translation Lemma (4.41') we do the analogous thing for translates of the basis (1').

Next, in the Twisted Biproduct Lemma (4.61*) we give a formula for expressing the product of any two members of (1*) as a Q-linear combination of (1*). Actually, we deduce (4.61*) by putting B' = 0 in the General Twisted Biproduct Lemma (4.51*), which expresses the product of any member of (1*) with any member (3*) as a Q-linear combination of (1*). In fact, first we prove a very easy case of (4.51*), which we call the Special Twisted Biproduct Lemma (4.1*), and then we get (4.51*) as a consequence of (4.1*) and (4.41*).

Similarly, upon replacing the basis (1*) by the basis (1'), first we prove the Special Biproduct Lemma (4.1'), and then we get the General Biproduct Lemma (4.51') as a consequence of (4.1') and (4.41'), and finally we deduce the Biproduct Lemma (4.61') by putting B' = 0 in (4.51').

This time around, the basis (1) seems superior to the bases (1') and (1*) because the analog of the Biproduct Lemma for (1) is the obvious formula

$$V^A V^B = V^{A+B}.$$

An attempt to generalize (4.61') and (4.61*), to what may be called the Multiproduct Lemma and the Twisted Multiproduct Lemma, respectively, gives rise to the following fundamental

PROBLEM (4.64) Let V be an indeterminate over Q, and let p be a positive integer. Now for any nonnegative integers $X_1, X_2, \ldots, X_p$, upon letting

$$Y = X_1 + X_2 + \cdots + X_p$$

we clearly get the Multiproduct Lemma, which says that in Q[V] we have

$$\prod_{i=1}^{p} \begin{bmatrix} V \\ X_i \end{bmatrix} = \sum_{E=0}^{Y} S'_E(X_1,X_2,\ldots,X_p) \begin{bmatrix} V \\ Y - E \end{bmatrix} \tag{11'}$$

and we clearly get the Twisted Multiproduct Lemma, which says that in Q[V] we have

$$\prod_{i=1}^{p} \begin{bmatrix} V \\ X_i \end{bmatrix} = \sum_{E=0}^{Y} (-1)^E S^*_E(X_1,X_2,\ldots,X_p) \begin{bmatrix} V \\ Y - E \end{bmatrix} \tag{11*}$$

where

$$S'_E(X_1,X_2,\ldots,X_p) \in Q \quad \text{and} \quad S^*_E(X_1,X_2,\ldots,X_p) \in Q.$$

The problem is to find (= describe = study) these functions S'_E *and* S^*_E.

Note that for any permutation w of {1,2,...,p} we clearly have

$$X_{w(1)} + X_{w(2)} + \cdots + X_{w(p)} = Y$$

and

$$\prod_{i=1}^{p} \binom{V}{X_{w(i)}} = \prod_{i=1}^{p} \binom{V}{X_i}$$

and

$$\prod_{i=1}^{p} \begin{bmatrix} V \\ X_{w(i)} \end{bmatrix} = \prod_{i=1}^{p} \begin{bmatrix} V \\ X_i \end{bmatrix}$$

and therefore

$$S_E'(X_{w(1)}, X_{w(2)}, \ldots, X_{w(p)}) = S_E'(X_1, X_2, \ldots, X_p)$$

and

$$S_E^*(X_{w(1)}, X_{w(2)}, \ldots, X_{w(p)}) = S_E^*(X_1, X_2, \ldots, X_p)$$

In other words, S_E' and S_E^* are Q-valued symmetric functions of the p nonnegative integer variables X_1, X_2, ..., X_p. So perhaps S_E' may be called the Eth *basic* (= similar to elementary but numerical) *symmetric function*, and S_E^* may be called the Eth *twisted basic symmetric function*. Note that although S_E' and S_E^* are not polynomial functions of X_1, X_2, ..., X_p, it is likely that they become so after we multiply them by suitable factorials; so we might be able to express S_E' and S_E^* in terms of the customary elementary symmetric functions $S_n(X_1, X_2, \ldots, X_p)$ defined by the equation

$$\prod_{i=1}^{p} (V - X_i) = \sum_{n=0}^{p} (-1)^n S_n(X_1, X_2, \ldots, X_p) V^n.$$

At any rate, the real problem is to develop a systematic theory of Q-*valued symmetric functions of nonnegative integer variables* X_1, X_2, ..., X_p.

Note that in (4.61'), for p = 2, we have given the binomial expressions $S_E'^{(2)}$, $S_E'^{(3)}$, $S_E'^{(4)}$, $S_E'^{(5)}$ of S_E' and we have also given the trinomial expression $S_E'^{(1)}$ of S_E'; out of these, only $S_E'^{(1)}$ *looks* symmetric, while the other four do not. Similarly in (4.61*), for p = 2, we have given the binomial expressions $S_E^{*(2)}$, $S_E^{*(3)}$, $S_E^{*(4)}$, $S_E^{*(5)}$ of S_E and we have also given the trinomial expression $S_E^{*(1)}$ of S_E^*; out of these only $S_E^{*(1)}$ *looks* symmetric, while the other four do not. In view of (4.61') and (4.61*) we also see that if p = 2, then $S_E' = S_E^*$ for all E.

Finally, let us observe that the analog of (11') and (11*) for the usual basis

$$(V^A)_{A\in N}$$

of Q[V] is the obvious Usual Multiproduct Lemma, which says that

$$\prod_{i=1}^{p} V^{X_i} = V^Y. \tag{11}$$

REMARK (4.65) From the one-variable case considered in (4.63), we may go to the case of the polynomial ring Q[V] in the *vector variable* V = (V(1),V(2),...,V(p)); in other words, let Q[V] be the polynomial ring Q[V(1),V(2),...,V(p)] in indeterminates V(1), V(2), ..., V(p) over Q. Now for any A ∈ N(p), we have the usual *monomial*

$$\prod_{i=1}^{p} V(i)^{A(i)}$$

of degree (= multidegree) A in V, and we have the *twisted monomial*

$$\prod_{i=1}^{p} \begin{bmatrix} V(i) \\ A(i) \end{bmatrix}$$

of degree (= multidegree) A in V, and we have the *untwisted monomial*

$$\prod_{i=1}^{A} \begin{pmatrix} V(i) \\ A(i) \end{pmatrix}$$

of degree (= multidegree) A in V. Correspondingly, (1) of (4.63) generalizes to the usual *monomial basis*

$$\left\{ \prod_{i=1}^{p} V(i)^{A(i)} \right\}_{A\in N(p)} \tag{4.65.1}$$

of Q[V], and (1*) of (4.63) generalizes to the *twisted monomial basis*

$$\left\{ \prod_{i=1}^{p} \begin{bmatrix} V(i) \\ A(i) \end{bmatrix} \right\}_{A\in N(p)} \tag{4.65.1*}$$

of Q[V], and (1') of (4.63) generalizes to the *untwisted monomial basis*

$$\left\{ \prod_{i=1}^{p} \binom{V(i)}{A(i)} \right\}_{A \in N(p)} \qquad (4.65.1')$$

of Q[V]; alternatively, (4.65.1') may be called the *binomial basis* of Q[V], and (4.65.1*) may be called the *twisted binomial basis* of Q[V]; the complete description of (4.65.1') could be *the basis of* Q[V], consisting of monomials *in the binomial coefficients involving* V, and similarly for (4.65.1*).

Note that

[in the principle of identities discussed in (4.62), we used the basis (4.65.1); upon replacing (4.65.1) by (4.65.1*) or (4.65.1') we get certain obvious variations of the principle. (4.65.2)

2

Enumeration of Young Tableaux

5. PREPARATION FOR COUNTING TABLEAUX

(5.1) Introduction In my 1983 lectures at the University of Nice in France, of which the Notes by Andre Galligo have appeared in Volume 42 of *Rendiconti del Seminario Matematico di Universita et Polytecnico di Torino* for the year 1984, I gave some enumerative formulas for counting certain types of tableaux which were introduced by Alfred Young at the end of the nineteenth century and which were resurrected by Gian-Carlo Rota and others during the last decade. In the Nice lectures, I had also drawn some ideal-theoretic consequences of the formulas.

In this chapter I establish some more enumerative formulas for counting tableaux. For the convenience of the reader, I include a self-contained treatment of the Nice formulas. The Nice formulas may very well be described as examples of what may be called determinantal polynomials in binomial coefficients. A major portion of this chapter is concerned with transforming the Nice formulas into various other determinantal polynomials in binomial coefficients and in establishing certain recurrence relations for the resulting determinantal polynomials. The relevant determinantal polynomials will be labeled as

$$F^{(Lk)}(m,p,a,V) \qquad \text{with } L = 1, 1^*, 2, 2^*, 3, 3^*, 4, 4^* \text{ and } k = 1, 2$$

and

$$F^{(5k)}(q,m,p,a,V) \quad \text{with } k = 1, 2, \ldots, q$$

and

$$F^{(L)}(q,m,p,a,V) \quad \text{with } L = 6, 7, 7^*, 8.$$

Out of these, $F^{(2k)}$ and $F^{(8)}$, may be called the Nice formulas. At any rate, "The theory of transformations of determinantal polynomials in binomial coefficients" could very well have been an alternative title for this chapter. In addition to being of interest in themselves, some of these transformed formulas count certain other objects. In fact, these objects are monomials satisfying some constraints on what we define to be the index of the support of a monomial.

To recall briefly the notion of a tableau, let q be a positive integer and let m = (m(1),m(2),...,m(q)) be a sequence of positive integers. By a multivector a of width q we essentially mean a multisequence

$$a(k,1) < a(k,2) < \cdots < a(k,p)$$

of positive integers, for k = 1, 2, ..., q, where the nonnegative integer p is called the length of a. We say that a is bounded by m to mean that

$$a(k,i) \le m(k) \quad \text{for } k = 1, 2, \ldots, q \text{ and } i = 1, 2, \ldots, p.$$

Given any other multivector b of width q and length p', we define $a \le b$ to mean that $p \ge p'$ and

$$a(k,i) \le b(k,i) \quad \text{for } k = 1, 2, \ldots, q \text{ and } i = 1, 2, \ldots, p'.$$

By a tableau T of width q we essentially mean a sequence

$$\begin{matrix} T[1] \\ T[2] \\ \vdots \\ T[d] \end{matrix}$$

of multivectors of width q, where the nonnegative integer d is called the depth of T. We say that T is standard to mean that

$$T[1] \le T[2] \le \cdots \le T[d]$$

and the length of T[e] is nonzero for e = 1, 2, ..., d. We say that T is bounded by m to mean that T[e] is bounded by m for e = 1, 2, ..., d. We say that T is predominated by a to mean that $a \le T[e]$ for e = 1, 2, ..., d. The sum of the lengths

$$\text{length of } T[1] + \text{length of } T[2] + \cdots + \text{length of } T[d]$$

is called the area of T. By stab(q,m,p,a,V) we denote the set of all standard tableaux of width q which are bounded by m and whose area is V and which are predominated by the multivector a of width q and length p. By Stab(q,m,p,a,V) we denote the cardinality of stab(q,m,p,a,V). By mon(q,m,p,a;V) we denote the corresponding set of monomials satisfying certain constraints as cited above, and by Mon(q,m,p,a,V) we denote the cardinality of mon(q,m,p,a,V).

In Subsections (5.2), (5.3), and (5.4) we summarize some preliminary material which is required in establishing the said enumerative formulas and recurrence relations and which was developed, respectively, in Sections 2, 3, and 4 of Chapter 1. In Subsection (5.5) we give a preview of the main portion of Section 5, which begins with Subsection (5.6).

In the main portion of Section 5 we build the strategy for proving that

$$\mathrm{Stab}(q,m,p,a,V) = F^{(8)}(q,m,p,a,V)$$

and

$$\mathrm{Mon}(2,m,p,a,V) = F^{(1k)}(m,p,a,V) \qquad \text{for } k = 1, 2.$$

We carry out these strategies to prove the desired properties of $F^{(8)}$ and $F^{(1k)}$ in Sections 6 and 8, respectively. In Section 6 we also derive a few other versions of formula $F^{(8)}$; more precisely, we show that

$$\begin{aligned} F^{(8)}(q,m,p,a,V) &= F^{(7)}(q,m,p,a,V) && \text{for every } q \\ &= F^{(7*)}(q,m,p,a,V) && \text{for even } q \\ &= F^{(6)}(q,m,p,a,V) && \text{for even } q \end{aligned}$$

$$= F^{(5k)}(q,m,p,a,V) \quad \text{for even } q \text{ and } k = 1, 2, \ldots, q$$

whereas

$$F^{(6)}(q,m,p,a,V) = 0 = F^{(7^*)}(q,m,p,a,V) \quad \text{for odd } q \text{ and } p \geq 2.$$

In Section 7 we transform the formulas deduced in Section 6 in the case of bitableaux (i.e., in the case of $q = 2$) and we derive in this case that

$$F^{(6)}(2,m,p,a,V) = F^{(Lk)}(m,p,a,V)$$

for $L = 1, 1^*, 2, 2^*, 3, 3^*, 4, 4^*$, and $k = 1, 2$.

Finally, all these identities are collated together in Section 9 to show that for every nonnegative integer V, we have

$$\begin{aligned} \mathrm{Stab}(q,m,p,a,V) &= F^{(8)}(q,m,p,a,V) \\ &= F^{(7)}(q,m,p,a,V) \end{aligned}$$

and to show that if q is even then for every nonnegative integer V, we have

$$\begin{aligned} \mathrm{Stab}(q,m,p,a,V) &= F^{(7^*)}(q,m,p,a,V) \\ &= F^{(6)}(q,m,p,a,V) \\ &= F^{(5k)}(q,m,p,a,V) \quad \text{for } k = 1, 2, \ldots, q \end{aligned}$$

and to show that given any $L \in \{1,1^*,2,2^*,3,3^*,4,4^*\}$ and $k \in \{1,2\}$, for every nonnegative integer V we have

$$\mathrm{Stab}(2,m,p,a,V) = F^{(Lk)}(m,p,a,V) = \mathrm{Mon}(2,m,p,a,V)$$

and hence, in particular, for every nonnegative integer V we have

$$\mathrm{Stab}(2,m,p,a,V) = \mathrm{Mon}(2,m,p,a,V).$$

In (5.6''') and (5.8''') we make some Motivational Remarks relevant to the material of the main portion of Section 5. In (6.41) and (6.42) we pose some problems related to the material of Section 6. In (7.33) and (7.41) we pose some problems related to the material of Section 7, and in (7.34) we make a relevant Motivational Remark, and in (7.0) and (7.31) we make some relevant Informational Remarks. In (8.41) and (8.42) we pose some problems related to the material of Section 8. Finally, in (9.10), (9.14), and (9.18) we pose some problems related to the material of Section 9.

In Chapter 3 we prove a certain universal determinantal identity. In Chapter 4 we give several applications of the enumeration formulas above and the universal determinantal identity. These applications will include enumerative proofs of the Straightening Law of Doubilet-Rota-Stein. They will include enumerative proofs of the Second Fundamental Theorem of Invariant Theory and its generalizations. Finally, they will include computations of Hilbert functions of certain polynomial ideals.

(5.2) Summary of General Notation and Terminology In this chapter we freely use the notation and terminology introduced in Section 2 of Chapter 1. To recapitulate the relevant matter from Section 2, let us first note that after fixing some general notation in (2.1), multivectors and tableaux were defined in (2.2) and (2.3), respectively, and then the index of the support of a monomial was defined in (2.4), and then binomial coefficients were introduced in (2.5). In particular, in (2.3) and (2.4) we defined the sets stab(q,m,p,a, V) and mon(q,m,p,a,V) as described in Introduction (5.1); moreover, in (2.3) we also introduced the set stas(q,m,p,a,v) of standard tableaux of given shape (p,v) where v = (v(1),...,v(p)) is a sequence of nonnegative integers and where a tableau T of depth d is said to be of shape (p,v) if for i = 1, 2, ..., p we have that v(i) is the number of those multivectors out of T[1], T[2], ..., T[d] whose length is i; again, Stas(q,m,p,a,v) was defined to be the cardinality of stas(q,m,p,a,v). Now a summation is said to be essentially finite if the number of nonzero summands in it is finite; this definition was introduced in (2.8). In (2.6) the groundwork was prepared for showing that the summations which enter into the definitions of the various determinantal polynomials are essentially finite. The determinantal polynomials

$$F^{(5k)}(q,m,p,a,V) \quad \text{for } k = 1, 2, \ldots, q$$

and

$$F^{(L)}(q,m,p,a,V) \quad \text{for } L = 6, 7, 7^*, 8$$

were defined in (2.7) and the essential finiteness of the summations involved in their definitions was established in (2.9); the definitions of these determinantal polynomials was made in terms of the auxiliary expressions

$$G^{(Lk)}(q,m,p,a,v) \quad \text{and} \quad H^{(L)}(q,m,p,a,v)$$
$$\text{for } L = 6, 7, 8 \text{ and } k = 1, 2, \ldots, q.$$

Finally, the determinantal polynomials

$$F^{(Lk)}(m,p,a,V) \quad \text{for } L = 1, 1^*, 2, 2^*, 3, 3^*, 4, 4^* \text{ and } k = 1, 2$$

and their cognates

$$F^{[Lk]}(m,p,a,V) \quad \text{for } L = 1, 1^*, 2, 2^*, 3, 3^* \text{ and } k = 1, 2$$

were defined in (2 10) and the essential finiteness of the summations involved in their definitions was established in (2.11); the "degree" of these determinantal polynomials was denoted by $C(m,p,a)$ and their "coefficients" were denoted by

$$F_D^{(Lk)}(m,p,a) \quad \text{for } L = 1, 1^*, 2, 2^*, 3, 3^* \text{ and } k = 1, 2$$

and the said "coefficients" were defined in terms of the auxiliary expressions

$$G^{(Lk)} \text{ and } H^{(Lk)} \quad \text{for } L = 1, 1^*, 2, 2^*, 3, 3^* \text{ and } k = 1, 2.$$

(5.3) Summary of Summation Lemmas In Section 3 of Chapter 1 we had proved some lemmas on summations and thereby prepared the groundwork for establishing the conditional equivalence of the determinantal polynomials

$$F^{(5k)}, F^{(6)}, F^{(7^*)}, F^{(7)}, F^{(8)}$$

and the equivalence of the determinantal polynomials

$$F^{(1k)}, F^{(1*k)}, \ldots, F^{(4*k)}.$$

The items from Section 3 that we shall refer to in this chapter are as follows:

Item	Referred to on page	In the proof of
(3.34)	187	(6.1)
(3.32.1)	192	(6.3)
(3.12)	212	(7.21)
(3.38)	225	(7.24)
(3.11)	231	(7.24)

(5.4) Summary of Binomial Lemmas In Section 4 of Chapter 1 we had proved some lemmas on binomial coefficients and thereby prepared the groundwork for establishing the equivalence of the determinantal polynomials

$$F^{(1k)}, F^{(1*k)}, \ldots, F^{(4*k)}$$

and for showing that they satisfy certain recurrence relations. The items from Section 4 that we refer to in this chapter are the following

Item	Referred to on page	In the proof of
(4.21*)	174	(5.8')
(4.12*)	187	(6.1)
(4.35'.3)	233	(7.25)
(4.35*.3)	234	(7.26)
(4.48)	236	(7.27)
(4.59)	239	(7.28)
(4.53)	244	(7.29)
(4.26)	267	(8.10)
(4.46)	268	(8.10)
(4.43)	269	(8.11)

Item	Referred to on page	In the proof of
(4.12*)	271	(8.11)
(4.44)	273	(8.12)
(4.36)	274	(8.14)
(4.11')	277	(8.17)
(4.21*)	307	(9.11)

(5.5) Preview of the Main Portion of Section 5 In the remaining portion of this section, for every $L \in \{6,7,8\}$, we shall state Theorem (5.L*), which is proved in the Demonstrational Remark (L.40) of Section L. Before reading the rest of this section, the reader may find it useful to browse in the Motivational Remarks (5.6''') and (5.8''').

NOTATION (5.6) As the notation to be used in (5.6'), (5.6''), and (5.6*), first for every $p \in N^*$ and $v \in Z(p)$ let $\hat{u}[p,v]$ be the member of $Z(p)$ obtained by putting

$$\hat{u}[p,v](i) = \begin{cases} v(i) & \text{if } i \in [1, p-1] \\ 1 + v(p) & \text{if } i = p \end{cases}$$

and second, for every $p \in N^*\backslash\{1\}$ and $v \in Z(p)$ let $\hat{h}[p,v]$ be the member of $Z(p-1)$ obtained by putting

$$\hat{h}[p,v](i) = v(i) \qquad \text{for all } i \in [1, p-1],$$

and third, for every $q \in N^*$ and $p \in N^*\backslash\{1\}$ and $t \in \text{pre}[q,p]$ let $\tilde{h}[q,p,t]$ be the member of $\text{pre}[q, p-1]$ obtained by putting

$$\tilde{h}[q,p,t](k,i) = t(k,i) \qquad \text{for all } k \in [1,q] \text{ and } i \in [1, p-1],$$

and fourth, for every $p \in N^*\backslash\{1\}$ let

$$\hat{Z}(p) = \{v \in Z(p) : v(p) = 0\}.$$

PROPOSITION (5.6') Given any $q \in N^*$ and $m \in N^*(q)$ and $p \in N^*$ and $a \in \text{vec}(q,m,p)$, we have the following

(5.6'.11) With $Z(p,V,8)$ as in (2.7), for all $V \in Z$ we have

$$\sum_{v \in Z(p,V,8)} \text{Stas}(q,m,p,a,v) = \text{Stab}(q,m,p,a,V).$$

(5.6'.12) With $\hat{u}$ as in (5.6), for all $v \in N(p)$ we have

$$\sum_{b \in vec(q,m,p,a)} Stas(q,m,p,b,v) = Stas(q,m,p,a,\hat{u}[p,v]).$$

(5.6'.13) If $\hat{h}$, $\tilde{h}$, $\hat{Z}$ are as in (5.6) and if $p \neq 1$, then for all $v \in N(p) \cap \hat{Z}(p)$ we have

$$Stas(q,m,p,a,v) = Stas(q,m, p - 1, \tilde{h}[q,p,a],\hat{h}[p,v]).$$

(5.6'.14) If $p = 1$, then for the unique $v \in N(p)$ with $v(1) = 0$, we have $Stas(q,m,p,a,v) = 1$.

Proof: (5.6'.11) follows by noting that given any $V \in Z$, we clearly have

$$\coprod_{v \in Z(p,V,8)} stas(q,m,p,a,v) = stab(q,m,p,a,V).$$

(5.6'.12) follows by noting that given any $v \in N(p)$, we clearly have that

$$\coprod_{b \in vec(q,m,p,a)} \{T \in stas(q,m,p,b,\hat{u}[p,v]) : T[1] = b\}$$
$$= stas(q,m,p,a,\hat{u}[p,v])$$

and for every $b \in vec(q,m,p,a)$ we clearly have that

$T \mapsto T^*$ gives a bijection of $\{T \in stas(q.m,p,b,\hat{u}[p,v]) : T[1] = b\}$ onto $stas(q,m,p,b,v)$ where for every $T \in stas(q,m,p,b,\hat{u}[p,v])$ with $T[1] = b$, we have let T^* be the tableau of width q and depth $dep(T) - 1$ obtained by putting $T^*[j] = T[j + 1]$ for all $j \in [1, dep(T) - 1]$.

(5.6'.13) follows by noting that if $p \neq 1$, then given any $v \in N(p) \cap \hat{Z}(p)$, we clearly have

$$stas(q,m,p,a,v) = stas(q, m, p - 1, \tilde{h}[q,p,a], \hat{h}[p,v]).$$

(5.6'.14) follows by noting that if $p = 1$, then for the unique $v \in N(p)$ with $v(1) = 0$, clearly the unique tableau of width q and depth 0 is the unique member of $stas(q,m,p,a,v)$.

LEMMA (5.6") Let $\hat{u}$, $\hat{h}$, $\tilde{h}$, $\hat{Z}$ be as in (5.6), and recall the definition of $Z(p,V,8)$ given in (2.7), and let there be fixed any $q \in N^*$

and $m \in N^*(q)$. Let there be given any

$$\begin{cases} \hat{S}tas(q,m,p,a,v) \in Q \text{ and } \hat{H}(q,m,p,a,v) \in Q \\ \text{for all } p \in N^* \text{ and } a \in vec(q,m,p) \text{ and } v \in N(p) \end{cases}$$

and

$$\begin{cases} \hat{S}tab(q,m,p,a,V) \in Q \text{ and } \hat{F}(q,m,p,a,V) \in Q \\ \text{for all } p \in N^* \text{ and } a \in vec(q,m,p) \text{ and } V \in Z \end{cases}$$

such that

$$(11) \quad \begin{cases} \text{for all } p \in N^* \text{ and } a \in vec(q,m,p) \text{ and } V \in Z \text{ we have} \\ \sum_{v \in Z(p,V,8)} \hat{S}tas(q,m,p,a,v) = \hat{S}tab(q,m,p,a,V) \text{ and} \\ \sum_{v \in Z(p,V,8)} \hat{H}(q,m,p,a,v) = \hat{F}(q,m,p,a,V) \end{cases}$$

and such that

$$(12) \quad \begin{cases} \text{for all } p \in N^* \text{ and } a \in vec(q,m,p) \text{ and } v \in N(p) \text{ we have} \\ \sum_{b \in vec(q,m,p,a)} \hat{S}tas(q,m,p,b,v) = \hat{S}tas(q,m,p,a,\hat{u}[p,v]) \\ \text{and } \sum_{b \in vec(q,m,p,a)} \hat{H}(q,m,p,b,v) = \hat{H}(q,m,p,a,\hat{u}[p,v]) \end{cases}$$

and such that

$$(13) \quad \begin{cases} \text{for all } p \in N^* \setminus \{1\} \text{ and } a \in vec(q,m,p) \text{ and } v \in N(p) \cap \hat{Z}(p) \\ \text{we have } \hat{S}tas(q,m,p,a,v) = \hat{S}tas(q,\ m,\ p-1,\ \check{h}[q,p,a],\ \hat{h}[p,v]) \\ \text{and } \hat{H}(q,m,p,a,v) = \hat{H}(q,\ m,\ p-1,\ \tilde{h}[q,p,a],\ \hat{h}[p,v]) \end{cases}$$

and such that

$$(14) \quad \begin{cases} \text{if } p = 1, \text{ then for the unique } v \in N(p) \text{ with } v(1) = 0, \\ \text{and for all } a \in vec(q,m,p) \text{ we have} \\ \hat{S}tas(q,m,p,a,v) = \hat{H}(q,m,p,a,v) \end{cases}$$

(where we note that the symbols $\hat{S}tas$, $\hat{S}tab$, $\hat{H}$, and $\hat{F}$ are independent of any previous meaning).

Now

$$(21) \quad \begin{cases} \text{for all } p \in N^* \text{ and } a \in vec(q,m,p) \text{ and } v \in N(p) \\ \text{we have } \hat{S}tas(q,m,p,a,v) = \hat{H}(q,m,p,a,v) \end{cases}$$

and

$$\text{(22)}\quad \left[\begin{array}{l}\text{for all } p \in N^* \text{ and } a \in \text{vec}(q,m,p) \text{ and } V \in Z \text{ we have}\\ \widehat{\text{S}}\text{tab}(q,m,p,a,V) = \hat{F}(q,m,p,a,V).\end{array}\right.$$

Proof: In view of (12), (13), and (14), we get (21) by double induction on p and v(p); in greater detail, for any fixed p, by (12) we can make induction on v(p) provided that we assume the initial case when v(p) = 0; on the other hand, by (13) we see that the general case of p - 1 yields the initial case for p; finally, (14) says that the very initial case (i.e., the initial case for p = 1) is assumed.

By (11) and (21) we get (22).

REMARK (5.6''') In (5.6') we have shown that Stas satisfies certain recursive relations. In (5.6") we have shown that these relations uniquely determine a function. So to find a formula for Stas amounts to finding an expression satisfying these relations. This is an example of a *general counting principle*, which says that *to find a formula for the number of elements in the various members of a family of sets, first try to discover some recursive relations which are satisfied by their cards and which uniquely determine a function, and then construct a formula satisfying the same relations; moreover, if recursive relations of the desired kind cannot be found for the given family, then enlarge the family suitably and try again.*

Thus we have motivated the following two theorems, (5.6*) and (5.7*). Namely, Theorem (5.6*) says that the recursive relations satisfied by Stas are also satisfied by the expression $H^{(8)}$. Moreover, Theorem (5.6*) also says that the expression $H^{(8)}$ can be successively transformed into the expressions $F^{(8)}$ and $F^{(7)}$ and $F^{(6)}$. Finally, Theorem (5.7*) says that the expression $F^{(6)}$ can be transformed into several other expressions. These two theorems, (5.6*) and (5.7*), are proved, respectively, in Remark (6.40) of Section 6 and Remark (7.40) of Section 7.

The general counting principle above also explains why it was necessary to introduce stab(q,m,p,a,V) even if we were only interested in stib(q,m,V) or stab[q,m,p,V]. Namely, recursive relations of the desired kind could not be found for the simpler families of sets stib(q,m,V) or stab[q,m,p,V], so we passed on to the larger family stab(q,m,p,a,V), and then we decomposed this larger family into its constituents stas(q,m,p,a,v) which did satisfy recursive relations of the desired kind.

THEOREM (5.6*) Given any $q \in N^*$ and $m \in N^*(q)$ and $p \in N^*$ and $a \in \mathrm{vec}(q,m,p)$, we have the following.

(5.6*.1) for all $V \in Z$ we have

$$F^{(7)}(q,m,p,a,V) = F^{(8)}(q,m,p,a,V).$$

(5.6*.2) If q is even, then for all $k \in [1,q]$ and $V \in Z$ we have

$$F^{(5k)}(q,m,p,a,V) = F^{(6)}(q,m,p,a,V) = F^{(7*)}(q,m,p,a,V) = F^{(8)}(q,m,p,a,V).$$

(5.6*.3) If $p = 1$, then for all $V \in Z$ we have

$$F^{(6)}(q,m,p,a,V) = F^{(7*)}(q,m,p,a,V) = F^{(8)}(q,m,p,a,V).$$

(5.6*.4) If q is odd and $p \geq 2$, then for all $V \in Z$ we have

$$F^{(6)}(q,m,p,a,V) = 0 = F^{(7*)}(q,m,p,a,V).$$

(5.6*.11) With Z(p,V,8) as in (2.7), for all $V \in Z$ we have

$$\sum_{v \in Z(p,V,8)} H^{(8)}(q,m,p,a,v) = F^{(8)}(q,m,p,a,V)$$

(5.6*.12) With $\hat{u}$ as in (5.6), for all $v \in N(p)$ we have

$$\sum_{b \in \mathrm{vec}(q,m,p,a)} H^{(8)}(q,m,p,b,v) = H^{(8)}(q,m,p,a,\hat{u}[p,v]).$$

(5.6*.13) If $\hat{h}$, $\tilde{h}$, $\hat{Z}$ are as in (5.6) and if $p \neq 1$, then for all $v \in \hat{Z}(p)$ we have

$$H^{(8)}(q,m,p,a,v) = H^{(8)}(q,\ m,\ p - 1,\ \tilde{h}[q,p,a],\ \hat{h}[p,v]).$$

(5.6*.14) If $p = 1$, then for the unique $v \in N(p)$ with $v(1) = 0$, we have $H^{(8)}(q,m,p,a,v) = 1$.

THEOREM (5.7*) Given any $m \in N^*(2)$ and $p \in N^*$ and $a \in vec(2.m,p)$, we have the following.

(5.7*.1) Given any

$$L \in \{1,1^*,2,2^*,3,3^*\} \cup \{3z : z \in W[p]\} \quad \text{and} \quad k \in [1,2]$$

first we have

$$F_D^{(Lk)}(m,p,a) = F_D^{(11)}(m,p,a) \quad \text{for all } D \in Z,$$

and second, we have

$$F^{[Lk]}(m,p,a,V) = F^{[11]}(m,p,a,V) \quad \text{in } Q[V],$$

and third, for every $C' \in Z$ we have

$$F^{[[Lk]]}(m,p,a,C',V) = F^{[[11]]}(m,p,a,C',V) \quad \text{in } Q[V],$$

and fourth, for every $C' \in Z$ we have

$$F^{((Lk))}(m,p,a,C',V) = F^{((11))}(m,p,a,C',V) \quad \text{for all } V \in Z.$$

(5.7*.2) Given any

$$L \in \{1,1^*,2,2^*,3,3^*\} \cup \{3z : z \in W[p]\} \quad \text{and} \quad k \in [1,2]$$

we have

$$F^{[Lk]}(m,p,a,V) = F^{(6)}(2,m,p,a,V) \quad \text{for all } V \in N.$$

(5.7*.3) Given any

$$L \in \{1,1^*,2,2^*,3,3^*,4,4^*\} \cup \{3z : z \in W[p]\} \quad \text{and} \quad k \in [1,2]$$

we have

$$F^{(Lk)}(m,p,a,V) = F^{(6)}(2,m,p,a,V) \quad \text{for all } V \in Z.$$

(5.7*.4) If

$$p = \min(m(1),m(2)) \text{ and } a(1,i) = i = a(2,i) \quad \text{for all } i \in [1,p]$$

and

if $k \in [1,2]$ is such that $m(k) = p$,

then for any given $e \in Z(p)$ we have

$$H^{(2k)}(m,p,a,e) = \begin{cases} 1 & \text{if } e(i) = 0 \text{ for all } i \in [1,p] \\ 0 & \text{if } e(i) \neq 0 \text{ for some } i \in [1,p] \end{cases}$$

and for any given $E \in Z$ we have

$$H_E^{(2k)}(m,p,a) = \begin{cases} 1 & \text{if } E = 0 \\ 0 & \text{if } E \neq 0. \end{cases}$$

(5.7*.5) If

$p = \min(m(1),m(2))$ and $a(1,i) = i = a(2,i)$ for all $i \in [1,p]$,

then for every

$L \in \{1,1^*,2,2^*,3,3^*\} \cup \{3z : z \in W[p]\}$ and $k \in [1,2]$

first we have

$$F_D^{(Lk)}(m,p,a) = \begin{cases} 1 & \text{for } D = 0 \\ 0 & \text{for all } D \in Z\backslash\{0\} \end{cases}$$

and second, we have

$$F^{[Lk]}(m,p,a,V) = \begin{bmatrix} V \\ m(1)m(2) - 1 \end{bmatrix} \text{ in } Q[V].$$

(5.7*.6) If

$p = \min(m(1),m(2))$ and $a(1,i) = i = a(2,i)$ for all $i \in [1,p]$,

then for every

$L \in \{1,1^*,2,2^*,3,3^*,4,4^*\} \cup \{3z : z \in W[p]\}$ and $k \in [1,2]$

we have

$$F^{(Lk)}(m,p,a,V) = F^{(6)}(2,m,p,a,V) = \begin{bmatrix} m(1)m(2) - 1 \\ V \end{bmatrix}$$

for all $V \in Z$.

NOTATION (5.8) As notation to be used in (5.8'), (5.8''), and (5.8*), given any $m \in N^*(2)$ and $p \in N^*$ and $a \in \text{vec}[2,p]$ and $\hat{k} \in [1,2]$ and $b \in \text{vec}[2, p + 1]$, we define

$$C^*(m,p,a,\hat{k},b) = p + m(\hat{k}) - b(\hat{k}, p+1) + \sum_{i=1}^{p} [a(\hat{k},i) - b(\hat{k},i)].$$

As further notation to be used in (5.8'), (5.8''), and (5.8*), given any $p \in N^*$ and $b \in \text{vec}[2, p+1]$ and $k \in [1,2]$, upon letting

$$k' = \begin{cases} 2 & \text{if } k = 1 \\ 1 & \text{if } k = 2, \end{cases}$$

first for every $u \in J(p)$ we let

$$M[p,b,k,u] = \{a \in \text{pre}[2,p] : a(k,i^*) = b(k,i^*) \text{ for all } i^* \in [1,p], \text{ and } a(k',i) \in [b(k',i) + 1, b(k', i+1) - 1] \text{ for all } i \in u, \text{ and } a(k',j) = b(k',j) \text{ for all } j \in [1,p]\backslash u\},$$

and second, for every $U \in [0,p]$ we let

$$M(p,b,k,U) = \bigcup_{u \in J(p,U)} M[p,b,k,u] \tag{5.8.0}$$

and we note that then clearly

$$M(p,b,k,U) \subset \text{vec}[2,p];$$

in (5.8'), (5.8''), and (5.8*), in addition to the quantities p and b and k above, we shall be given $m \in N^*(2)$ and $m^* \in N^*(2)$ and $p^* \in N^*$ and $a^* \in \text{vec}(2,m^*,p^*)$ such that

$$m^*(k') = m(k') \quad \text{and} \quad m^*(k) = m(k) + 1 \tag{5.8.1}$$

and such that either

$$\left[\begin{array}{l} a^*(k,p^*) \neq m^*(k), \text{ and } p^* = p, \text{ and } b(\hat{k},i) = a^*(\hat{k},i) \\ \text{for all } \hat{k} \in [1,2] \text{ and } i \in [1,p], \text{ and } b(\hat{k}, p+1) = \\ m(\hat{k}) + 1 \text{ for all } \hat{k} \in [1,2] \end{array}\right. \tag{5.8.1'}$$

or

$$a^*(k,p^*) = m^*(k) \text{ and } p^* = p + 1 \text{ and } b = a^*; \tag{5.8.1*}$$

let us observe that

$$\left[\begin{array}{l} \text{if (5.8.1) holds and if either (5.8.1') holds or} \\ \text{(5.8.1*) holds, then for all } U \in [0,p] \text{ we have} \\ M(p,b,k,U) \subset \text{vec}(2,m,p). \end{array}\right. \tag{5.8.2}$$

PROPOSITION (5.8') Let C* and M be as in (5.8), and let there be fixed any $k \in [1,2]$, and let us put

$$k' = \begin{cases} 2 & \text{if } k = 1 \\ 1 & \text{if } k = 2. \end{cases}$$

Let there be given any $m \in N^*(2)$ and $p \in N^*$ and $b \in \mathrm{vec}[2, p + 1]$ and $m^* \in N^*(2)$ and $p^* \in N^*$ and $a^* \in \mathrm{vec}(2,m^*,p^*)$ such that (5.8.1) holds and either (5.8.1') holds or (5.8.1*) holds. Let

$$\Sigma^* \text{ stand for } \sum_{U\in[0,p]} \sum_{a\in M(p,b,k,U)} .$$

Now

$$\text{(11)} \quad \left[\begin{array}{l} \text{for every } V \in N \text{ we have} \\ \Sigma^* \displaystyle\sum_{v\in N(2,V)} \mathrm{Mon}(2,m,p,a,v(1)) \begin{bmatrix} C^*(m,p,a,k',b) \\ v(2) - U \end{bmatrix} \\ = \mathrm{Mon}(2,m^*,p^*,a^*,V) \end{array}\right.$$

and

$$\text{(12)} \quad \left[\begin{array}{l} \text{if } p = 1 \text{ and if } a \in \mathrm{vec}(2,m,p) \text{ is such that } m(k) \\ - a(k,p) = 0, \text{ then for every } V \in N \text{ we have} \\ \mathrm{Mon}(2,m,p,a,V) = \begin{bmatrix} m(k') - a(k',1) \\ V \end{bmatrix}. \end{array}\right.$$

Proof: For every finite set X let

$$H(X) = \text{the set of all maps of } X \text{ into } N$$

and for every $A \in Z$ let

$$H[[X,A]] = \left\{h \in H(X) : \sum_{x\in X} h(x) = A\right\}$$

and for every $R \subset X$ and $S \subset X$ let

$$H(X,R,S) = \{h \in H(X) : h(r) = 0 \text{ for all } r \in R \text{ and } h(s) \neq 0 \text{ for all } s \in S\}$$

and

$$H(X,R,S,A) = H(X,R,S) \cap H[[X,A]].$$

By (4.21*) we see that

$$(31)\quad \left[\begin{array}{l}\text{if } X, R, S \text{ are any finite sets such that } R \subset X \text{ and } S \subset X \\ \text{and } R \neq X \text{ and } R \cap S = \emptyset\text{, then for every } A \in Z \text{ we have} \\ \mathrm{card}(H(X,R,S,A)) = \begin{bmatrix}\mathrm{card}(X\backslash R) - 1 \\ A - \mathrm{card}(S)\end{bmatrix}.\end{array}\right.$$

If $p = 1$ and if $a \in \mathrm{vec}(2,m,p)$ is such that $m(k) - a(k,p) = 0$, then given any $V \in N$, upon taking

$$\left[\begin{array}{l}X = \mathrm{cub}(2,m) \text{ and } S = \emptyset \text{ and } A = V \text{ and} \\ R = \{y \in \mathrm{cub}(2,m) : \text{either } y(k) < m(k) \text{ or } y(k') < a(k',1)\}\end{array}\right.$$

we would have

$$\left[\begin{array}{l}\mathrm{mon}(2,m,p,a,V) = H(X,R,S,A) \text{ and} \\ \mathrm{card}(X\backslash R) = m(k') - a(k',1) + 1\end{array}\right.$$

and hence by (31) we would get

$$\mathrm{Mon}(2,m,p,a,V) = \begin{bmatrix}m(k') - a(k',1) \\ V\end{bmatrix}$$

and this proves (12).

To prove (11), let there be fixed any $V \in N$. Now upon letting

$$\left[\begin{array}{l}E = \text{the set of all triples } (u,a,v) \text{ with} \\ u \in J(p) \text{ and } a \in M[p,b,k,u] \text{ and } v \in N(2,V)\end{array}\right.$$

we clearly have

$$(41)\quad [\text{LHS of } (11)] = \sum_{(u,a,v)\in E} \mathrm{Mon}(2,m,p,a,v(1)) \begin{bmatrix}C^*(m,p,a,k',b) \\ v(2) - \mathrm{card}(u)\end{bmatrix}.$$

Let $n\colon [1,m^*(k')] \to \mathrm{cub}(2,m^*)$ be the injective map obtained by putting $n(j)(k) = m^*(k)$ and $n(j)(k') = j$ for all $j \in [1,m^*(k')]$. Now upon letting

$$Y = n([1,m^*(k')])$$

we clearly have

$$(42)\quad \mathrm{cub}(2,m^*) = \mathrm{cub}(2,m) \coprod Y.$$

At this point it may be useful to look at the following picture, in which the big box represents $\mathrm{cub}(2,m^*)$ and in which we have shown $i \in u$ and $i' \in [1,p]\backslash u$.

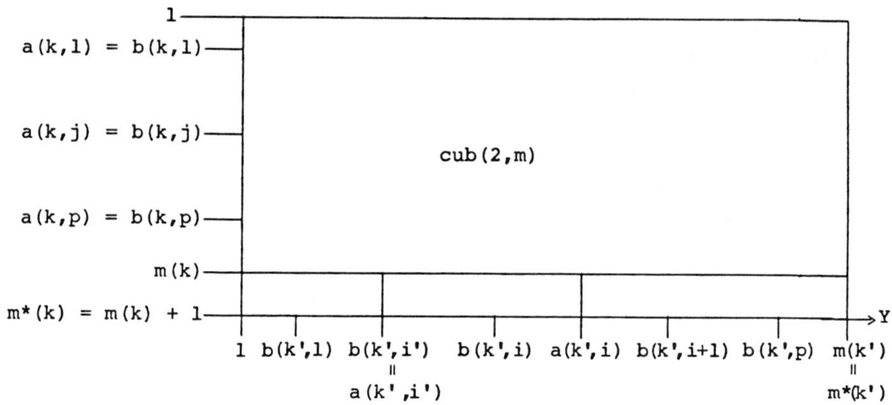

Let

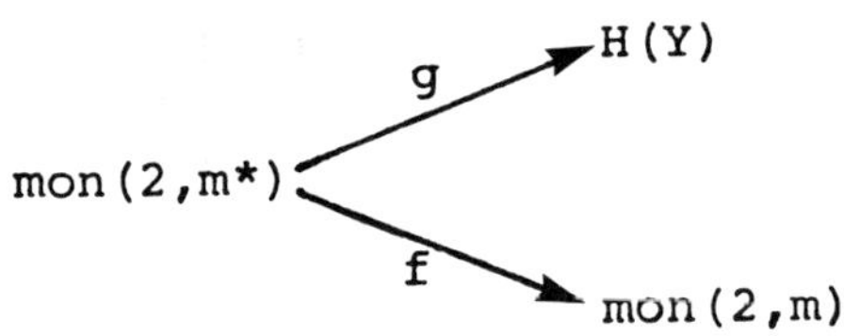

be the maps induced by restrictions, i.e., for every $t \in \text{mon}(2,m^*)$ we let $f(t) \in \text{mon}(2,m)$ and $g(t) \in H(Y)$ be defined by putting

$$f(t)(y) = t(y) \quad \text{for all } y \in \text{cub}(2,m)$$

and

$$g(t)(y) = t(y) \quad \text{for all } y \in Y.$$

By (42) it follows that

(43) $\left[\begin{array}{l} f\colon \text{mon}(2,m^*) \to \text{mon}(2,m) \text{ and } g\colon \text{mon}(2,m^*) \to H(Y) \\ \text{are surjective maps such that for all } t' \neq t^* \text{ in} \\ \text{mon}(2,m^*) \text{ we have either } f(t') \neq f(t^*) \text{ or } g(t') \neq g(t^*) \end{array}\right.$

and briefly speaking this means that mon(2,m*) is the cartesian product of mon(2,m) and H(Y), and we may illustrate it by the following picture.

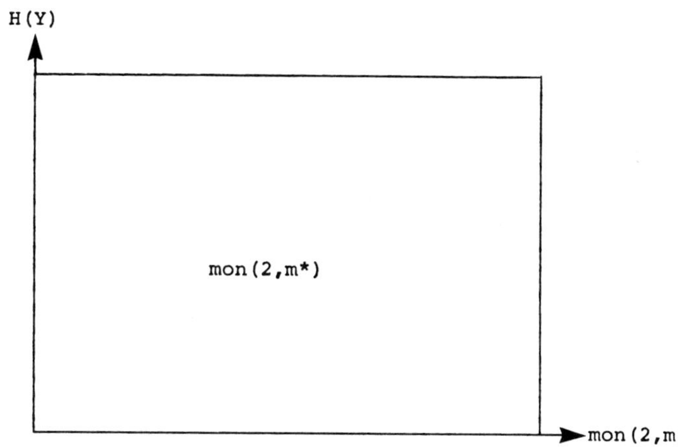

For every $u \in J(p)$ and $a \in M[p,b,k,u]$, let us note that

$$n\left(\bigcup_{i=1}^{p} [a(k',i) + 1,\ b(k',\ i + 1) - 1]\right) \subset Y$$

and

$$n(\{a(k',i) : i \in u\}) \subset Y$$

and let us put

$$\hat{R}(u,a) = n\left(\bigcup_{i=1}^{p} [a(k',i) + 1,\ b(k',\ i + 1) - 1]\right)$$

and

$$S(u,a) = n(\{a(k',i) : i \in u\}).$$

Now upon letting

D = the set of all pairs (u,a) with $u \in J(p)$ and $a \in M[p,b,k,u]$

we clearly have

$$H(Y,\hat{R}(u,a),S(u,a)) \cap H(Y,\hat{R}(\tilde{u},\tilde{a}),S(\tilde{u},\tilde{a})) = \emptyset$$
$$\text{for all } (\tilde{u},\tilde{a}) \neq (u,a) \text{ in } D.$$

Given any $h \in H(Y)$ we get a unique $(u[h],a[h]) \in D$ by putting

$$u[h] = \{i \in [1,p] : h(n(z)) \neq 0 \text{ for some } z \in [b(k',i)+1,\ b(k',i+1)-1]\}$$

and

$$a[h](\hat{k},i) = \begin{cases} b(k,i) \text{ if } \hat{k} = k \text{ and } i \in [1,p] \\ \max\{z \in [b(k',i)+1,\ b(k',i+1)-1] : \\ \qquad h(n(z)) \neq 0\} \text{ if } \hat{k} = k' \text{ and } i \in u[h] \\ b(k',i) \text{ if } \hat{k} = k' \text{ and } i \in [1,p]\backslash u[h] \end{cases}$$

and we clearly have

$$h \in H(Y,\hat{R}(u[h],a[h]),S(u[h],a[h]))$$

and therefore

$$H(Y) \subset \bigcup_{(u,a)\in D} H(Y,\hat{R}(u,a),S(u,a)).$$

Thus we conclude that

$$H(Y) = \coprod_{(u,a)\in D} H(Y,\hat{R}(u,a),S(u,a)). \tag{44}$$

Since $g\colon \mathrm{mon}(2,m^*) \to H(Y)$ is a map, by (44) we get

$$\mathrm{mon}(2,m^*) = \coprod_{(u,a)\in D} g^{-1}(H(Y,\hat{R}(u,a),S(u,a))). \tag{45}$$

Since $\mathrm{mon}(2,m^*,p^*,a^*) \subset \mathrm{mon}(2,m^*)$, by (45) we get

$$\mathrm{mon}(2,m^*,p^*,a^*) = \coprod_{(u,a)\in D} L(u,a) \tag{46}$$

where for every $(u,a) \in D$, we have put

$$L(u,a) = \mathrm{mon}(2,m^*,p^*,a^*) \cap g^{-1}(H(Y,\hat{R}(u,a),S(u,a))).$$

Now, for every $(u,a) \in D$, by letting

$$R(u,a) = \hat{R}(u,a) \cup n([1,\ b(k',1)-1])$$

and by considering the two cases (5.8.1') and (5.8.1*) separately, we see that

$$\left[\begin{array}{l} L(u,a) = \{t \in \mathrm{mon}(2,m^*) : f(t) \in \mathrm{mon}(2,m,p,a) \\ \qquad\qquad \text{and } g(t) \in H(Y,R(u,a),S(u,a))\} \end{array}\right. \tag{47}$$

and briefly speaking, this means that L(u,a) is a cartesian product of mon(2,m,p,a) and H(Y,R(u,a),S(u,a)), and we may illustrate this by the following picture, in which the big box represents mon(2,m*).

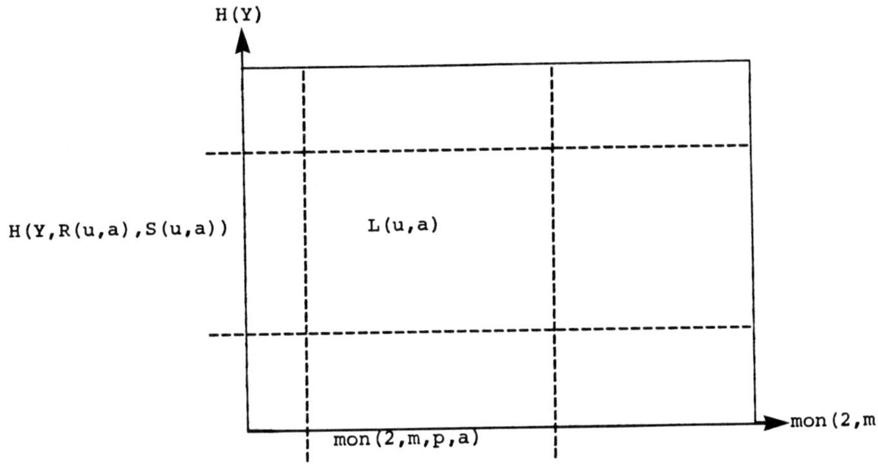

Clearly,

(48) $$\text{mon}[[2,m^*,V]] = \coprod_{v \in N(2,V)} K(v)$$

where for every $v \in N(2,V)$ we have put

(49) $$\left[\begin{array}{l} K(v) = \{t \in \text{mon}(2,m^*) : \quad f(t) \in \text{mon}(2,m,v(1)]] \\ \qquad\qquad\qquad\qquad\qquad\qquad \text{and } g(t) \in H[[Y,v(2)]]\}. \end{array} \right.$$

For every $(u,a,v) \in E$, upon letting

(50) $L(u,a,v) = L(u,a) \cap K(v)$

by (47) and (49) we see that

$$L(u,a,v) = \{t \in \text{mon}(2,m^*) : \\ f(t) \in \text{mon}(2,m,p,a) \cap \text{mon}[[2,m,v(1)]] \text{ and} \\ g(t) \in H(Y,R(u,a),S(u,a)) \cap H[[Y,v(2)]]\}$$

and hence upon noting that by definition we have

$$\text{mon}(2,m,p,a) \cap \text{mon}[[2,m,v(1)]] = \text{mon}(2,m,p,a,v(1))$$

and

$$H(Y,R(u,a),S(u,a)) \cap H[[Y,v(2)]] = H(Y,R(u,a),S(u,a),v(2)),$$

we conclude that

$$(51) \quad \left[\begin{array}{l} L(u,a,v) = \{t \in \mathrm{mon}(2,m^*) : f(t) \in \mathrm{mon}(2,m,p,a,v(1)) \\ \qquad\qquad \text{and } g(t) \in H(Y,R(u,a),S(u,a),v(2))\} \end{array} \right.$$

and briefly speaking this means that L(u,a,v) is a cartesian product of mon(2,m,p,a,v(1)) and H(Y,R(u,a),S(u,a),v(2)) because the intersection of two parallel rectangles is a rectangle, and we may illustrate this by the following picture, in which the big box represents mon(2,m*), whereas the hyphenated box represents L(u,a), and finally, the dotted box represents K(v).

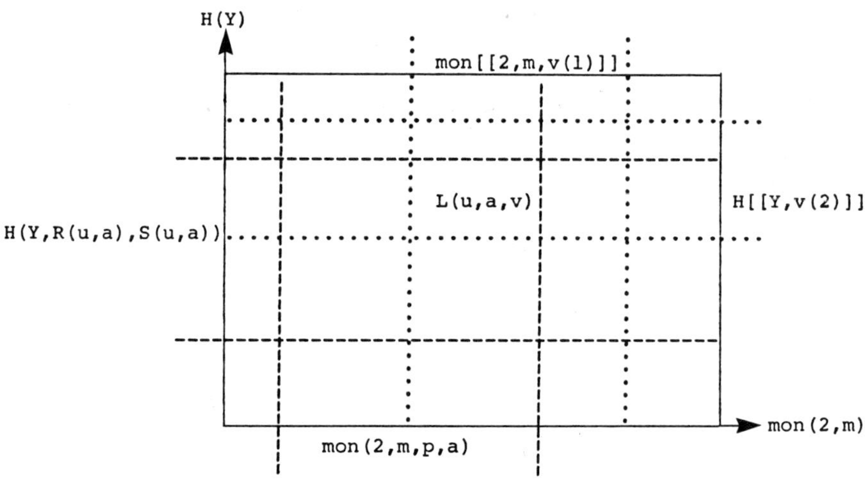

By definition we have

$$\mathrm{mon}(2,m^*,p^*,a^*,V) = \mathrm{mon}(2,m^*,p^*,a^*) \cap \mathrm{mon}[[2,m^*,V]]$$

and hence by (46), (48), and (50) we see that

$$(52) \quad \mathrm{mon}(2,m^*,p^*,a^*,V) = \coprod_{(u,a,v)\in E} L(u,a,v).$$

Given any $(u,a,v) \in E$, by the definition of Mon we have

$$(53) \quad \mathrm{card}(\mathrm{mon}(2,m,p,a,v(1))) = \mathrm{Mon}(2,m,p,a,v(1))$$

and by the definitions of C*, Y, R(u,a), and S(u,a) we also see that

$$\begin{bmatrix} R(u,a) \subset Y \text{ and } S(u,a) \subset Y \\ R(u,a) \neq Y \text{ and } R(u,a) \cap S(u,a) = \emptyset \text{ and} \\ \text{card}(Y \setminus R(u,a)) - 1 = C^*(m,p,a,k',b) \text{ and} \\ \text{card}(S(u,a)) = \text{card}(u) \end{bmatrix}$$

and hence by (31) we conclude that

$$(54) \quad \text{card}(H(Y,R(u,a),S(u,a),v(2))) = \begin{bmatrix} C^*(m,p,a,k',b) \\ v(2) - \text{card}(u) \end{bmatrix}.$$

Given any $(u,a,v) \in E$, by definition we have

$$\text{mon}(2,m,p,a,v(1)) \subset \text{mon}(2,m)$$

and

$$H(Y,R(u,a),S(u,a),v(2)) \subset H(Y)$$

and hence by (43) and (51) we see that $L(u,a,v)$ is a cartesian product of $\text{mon}(2,m,p,a,v(1))$ and $H(Y,R(u,a),S(u,a),v(2))$ and therefore by (53) and (54) we conclude that

$$(55) \quad \text{card}(L(u,a,v)) = \text{Mon}(2,m,p,a,v(1)) \begin{bmatrix} C^*(2,m,p,a,k',b) \\ v(2) - \text{card}(u) \end{bmatrix}.$$

By definition we have

$$\text{card}(\text{mon}(2,m^*,p^*,a^*,V)) = \text{Mon}(2,m^*,p^*,a^*,V)$$

and hence by (52) we get

$$\text{Mon}(2,m^*,p^*,a^*,V) = \sum_{(u,a,v) \in E} \text{card}(L(u,a,v))$$

and therefore by (55) we conclude that

$$(56) \quad [\text{RHS of } (11)] = \sum_{(u,a,v) \in E} \text{Mon}(2,m,p,a,v(1)) \begin{bmatrix} C^*(2,m,p,a,k',b) \\ v(2) - \text{card}(u) \end{bmatrix}.$$

By (41) and (56) we get (11).

LEMMA (5.8'') Let M be as in (5.8), and let there be fixed any $k \in [1,2]$, and let us put

$$k' = \begin{cases} 2 & \text{if } k = 1 \\ 1 & \text{if } k = 2. \end{cases}$$

Let there be given any

$$\begin{cases} \widehat{\text{Mon}}(2,m,p,a,V) \in Q \text{ and } \hat{F}(m,p,a,V) \in Q \text{ for all} \\ m \in N^*(2) \text{ and } p \in N^* \text{ and } a \in \text{vec}(2,m,p) \text{ and } V \in N \end{cases}$$

and

$$\begin{cases} \hat{C}(m,p,a,k',b,V,U) \in Q \text{ for all } m \in N^*(2) \text{ and } p \in N^* \\ \text{and } a \in \text{vec}[2,p] \text{ and } b \in \text{vec}[2,\ p+1] \text{ and } V \in N \text{ and } U \in N \end{cases}$$

(where we note that the symbols $\widehat{\text{Mon}}$, $\hat{F}$, and $\hat{C}$ are independent of any previous meaning).

Assume that (11): if $m \in N^*(2)$ and $p \in N^*$ and $b \in \text{vec}[2,\ p+1]$ and $m^* \in N^*(2)$ and $p^* \in N^*$ and $a^* \in \text{vec}(2,m^*,p^*)$ are such that (5.8.1) holds and either (5.8.1') holds or (5.8.1*) holds, then upon letting

$$\Sigma^* \text{ stand for } \sum_{U\in[0,p]} \sum_{a\in M(p,b,k,U)},$$

for every $V \in N$ we have

$$\Sigma^* \sum_{v\in N(2,V)} \widehat{\text{Mon}}(2,m,p,a,v(1))\hat{C}(m,p,a,k',b,v(2),U)$$
$$= \widehat{\text{Mon}}(2,m^*,p^*,a^*,V)$$

and

$$\Sigma^* \sum_{v\in N(2,V)} \hat{F}(m,p,a,v(1))\hat{C}(m,p,a,k',b,v(2),U) = \hat{F}(m^*,p^*,a^*,V).$$

Also assume that (12): if $m \in N^*(2)$ and $p \in N^*$ and $a \in \text{vec}(2,m,p)$ are such that $p = 1$ and $m(k) - a(k,p) = 0$, then for every $V \in N$ we have $\widehat{\text{Mon}}(2,m,p,a,V) = \hat{F}(m,p,a,V)$.

Now (21): for all $m \in N^*(2)$ and $p \in N^*$ and $a \in \text{vec}(2,m,p)$ and $V \in N$ we have

$$\widehat{\text{Mon}}(2,m,p,a,V) = \hat{F}(m,p,a,V).$$

Proof: Follows by double induction on p and $m(k) - a(k,p)$. In greater detail, for any fixed p, by (5.8.1') we can make induction on $m(k) - a(k,p)$ provided that we assume the initial case when $m(k) - a(k,p) = 0$; on the other hand, by (5.8.1*) we see that the general case of p yields the initial case for $p + 1$; finally, (12) says that the very initial case (i.e., the initial case for p = 1) is assumed.

REMARK (5.8''') In (5.8') we have proved that Mon satisfies certain recursive relations. In (5.8'') we have proved the uniqueness of the function satisfying these relations. So, to find a formula for Mon amounts to finding an expression satisfying these relations. This is another example of the general counting principle discussed in (5.6'''). Thus we have motivated the following Theorem (5.8*), which says that the expression $F^{(1k)}$ satisfies the said relations, and which we shall prove in Remark (8.40) of Section 8. The principle also explains why it was necessary to consider mon(2,m,p,a,V) even if we were interested only in mon[2,m,p,V]; namely, recursive relations of the kind mentioned in the cited principle could not be found for the simpler family of sets mon[2,m,p,V], so we enlarged it to mon(2,m,p,a,V).

Let us also observe that the proofs of (5.6') and (5.8') indicate that all counting methods are eventually based on the following two basic principles: the *additive cardinal principle*, which says that *the card of a disjoint union is the sum of the cards*, and the *multiplicative cardinal principle*, which says that *the card of a cartesian product is the product of the cards*. Of course, these two principles also constitute the foundation of arithmetic.

THEOREM (5.8*) Let C* and M be as in (5.8), and let there be fixed any $k \in [1,2]$, and let us put

$$k' = \begin{cases} 2 & \text{if } k = 1 \\ 1 & \text{if } k = 2. \end{cases}$$

Let there be given any $m \in N^*(2)$ and $p \in N^*$ and $b \in \text{vec}[2, p + 1]$ and $m^* \in N^*(2)$ and $p^* \in N^*$ and $a^* \in \text{vec}(2,m^*,p^*)$ such that (5.8.1) holds and either (5.8.1') holds or (5.8.1*) holds. Finally, let

$$\Sigma^* \text{ stand for } \sum_{U\in[0,p]} \sum_{a\in M(p,b,k,U)} .$$

Now

$$\left[\begin{array}{l}\text{for every } D \in Z \text{ we have} \\ \displaystyle\sum_{I\in[0,p]} \Sigma^*\binom{U}{I}F^{(1k)}_{D-I}(m,p,a) = F^{(1k)}_{D}(m^*,p^*,a^*)\end{array}\right. \tag{5.8*.1}$$

and

$$\left[\begin{array}{l}\text{for every } C' \in Z \text{ in } Q[V] \text{ we have} \\ \Sigma^* F^{[[1k]]}(m,p,a,C',V-U) = F^{[[1k]]}(m^*,p^*,a^*,C',V)\end{array}\right. \tag{5.8*.2}$$

and

$$\left[\begin{array}{l}\text{in } Q[V] \text{ we have} \\ \Sigma^* F^{[[1k]]}(m,p,a,C(m^*,p^*,a^*), V-U) = F^{[1k]}(m^*,p^*,a^*,V)\end{array}\right. \tag{5.8*.3}$$

and

$$\left[\begin{array}{l}\text{for every } V \in N \text{ we have} \\ \Sigma^* F^{((1k))}(m,p,a,C(m^*,p^*,a^*), V-U) = F^{(1k)}(m^*,p^*,a^*,V)\end{array}\right. \tag{5.8*.4}$$

and

$$\left[\begin{array}{l}\text{for every } V \in N \text{ we have} \\ \displaystyle\Sigma^* \sum_{v\in N(2,V)} F^{(1k)}(m,p,a,v(1))\begin{bmatrix}C^*(m,p,a,k',b) \\ v(2)-U\end{bmatrix} \\ = F^{(1k)}(m^*,p^*,a^*,V)\end{array}\right. \tag{5.8*.11}$$

and

$$\left[\begin{array}{l}\text{if } p = 1 \text{ and if } a \in \text{vec}(2,m,p) \text{ is such that} \\ m(k) - a(k,p) = 0, \text{ then for every } V \in N \text{ we have} \\ F^{(1k)}(m,p,a,V) = \begin{bmatrix}m(k') - a(k',1) \\ V\end{bmatrix}.\end{array}\right. \tag{5.8*.12}$$

6. CONDITIONAL EQUIVALENCE OF CERTAIN DETERMINANTAL POLYNOMIALS OF ANY WIDTH

In this section we prove Theorem (5.6*) by establishing the conditional equivalence of the determinantal polynomials $F^{(5k)}$, $F^{(6)}$, ..., $F^{(8)}$ which were introduced in (2.7). The proof is given in the Demonstrational Remark (6.40), and then in (6.41) and (6.42) we pose some related problems.

LEMMA (6.1) Let there be given any $p \in N^*$ and $\hat{D} \in Z$ and $\hat{Y} \in Z$ and $d \in Z(p)$ such that

$$\hat{Y} + \hat{D} \geq 0 \geq \hat{Y} \text{ and } d(i) + p \geq \hat{D} \quad \text{for all } i \in [1,p].$$

Let

$$Y = \{y \in Z(p) : y(1) > y(2) > \cdots > y(p) \geq \hat{Y}\}.$$

For every $n \in [0,p]$ and $y \in Y$ let

$$R(n,y) = \{r \in Y : \begin{array}{ll} r(k) \leq y(k) & \text{for all } k \in [1,n] \quad \text{and} \\ r(j) = y(j) & \text{for all } j \in [n+1, p]\}. \end{array}$$

For every $n \in [1,p]$ and $y \in Y$ let

$$S(n,y) = \{s \in Y : \begin{array}{ll} s(n) \leq y(n) & \text{and} \\ s(j) = y(j) & \text{for all } j \in [1,p]\backslash\{n\}\}. \end{array}$$

For every $n \in [0,p]$ and $s \in Z(p)$ let

$$G_{ij}(n,s) = \begin{cases} \begin{bmatrix} s(j) \\ d(i) + j + 1 \end{bmatrix} & \text{if } i \in [1,p] \text{ and } j \in [1,n] \\ \begin{bmatrix} s(j) \\ d(i) + j \end{bmatrix} & \text{if } i \in [1,p] \text{ and } j \in [n+1, p] \end{cases}$$

and

$$G(n,s) = \text{the } p \text{ by } p \text{ matrix whose } (i,j)\text{th entry is } G_{ij}(n,s).$$

Then we have

$$\sum_{r \in R(0,y)} \det G(0,r) = \det G(0,y) \quad \text{for all } y \in Y \tag{6.1.1}$$

and we have

$$\left[\begin{array}{l} \sum_{r \in R(n,y)} \det G(0,r) = \sum_{s \in S(n,y)} \sum_{r \in R(n-1,s)} \det G(0,r) \\ \text{for all } n \in [1,p] \text{ and } y \in Y \end{array}\right. \tag{6.1.2}$$

and we have

$$\left[\begin{array}{l} \sum_{s \in S(n,y)} \det G(n-1, s) = \det G(n,y) \\ \text{for all } n \in [1,p] \text{ and } y \in Y \end{array}\right. \tag{6.1.3}$$

and we have

$$\left[\begin{array}{l}\sum\limits_{r\in R(n,y)} \det G(0,r) = \det G(n,y) \\ \text{for all } n \in [0,p] \text{ and } y \in Y\end{array}\right. \tag{6.1.4}$$

and we have

$$\sum_{r\in R(p,y)} \det G(0,r) = \det G(p,y) \qquad \text{for all } y \in Y. \tag{6.1.5}$$

Proof: (6.1.1) follows by noting that for every $y \in Y$ we clearly have $R(0,y) = \{y\}$. (6.1.2) follows by noting that for every $n \in [1,p]$ and $y \in Y$ we clearly have

$$R(n,y) = \coprod_{s\in S(n,y)} R(n - 1, s).$$

To prove (6.1.3), let there be fixed any $n \in [1,p]$ and $y \in Y$. Upon letting

$$(1) \quad x = \begin{cases} y(n + 1) & \text{if } n \neq p \\ \hat{Y} - 1 & \text{if } n = p \end{cases}$$

we see that

(2) $x \in Z$ and $y(n) \in Z$ and $x < y(n)$

and we see that

(3) if $n \neq p$, then $x = y(n + 1)$

and we see that

$$(4) \quad \left[\begin{array}{l}\text{if } n = p, \text{ then} \\ d(i) + n + 1 > x + d(i) + n + 1 \geq 0 \qquad \text{for all } i \in [1,p].\end{array}\right.$$

Upon letting

$$(5) \quad \hat{L}_{ij}(s^*) = \begin{cases} \begin{bmatrix} s^* \\ d(i) + j + 1\end{bmatrix} & \text{if } i \in [1,p] \text{ and } j \in [1, n - 1] \text{ and } s^* \in Z \\ \begin{bmatrix} s^* \\ d(i) + j\end{bmatrix} & \text{if } i \in [1,p] \text{ and } j \in [n,p] \text{ and } s^* \in Z \end{cases}$$

we see that

$$(6) \quad \left[\begin{array}{l}\text{for every } s \in S(n,y) \text{ we have that } G(n - 1, s) \text{ is the} \\ p \text{ by } p \text{ matrix whose } (i,j)\text{th entry is } \hat{L}_{ij}(s(j)).\end{array}\right.$$

By (1) we see that

$$(7)\quad \left[\begin{aligned} S(n,y) = \{s \in Z(p) : \ & s(n) \in [x+1, y(n)] \text{ and} \\ & s(j) = y(j) \quad \text{for all } j \in [1,p]\backslash\{n\}\}. \end{aligned}\right.$$

Let

$$(8)\quad \hat{K} = \text{the } p \text{ by } p \text{ matrix whose } (i,j)\text{th entry is } \hat{K}_{ij}$$

where

$$(9)\quad \hat{K}_{ij} = \hat{L}_{ij}(y(j)) \qquad \text{for all } i \in [1,p] \text{ and } j \in [1,p]\backslash\{n\}$$

and where

$$(10)\quad \hat{K}_{in} = \sum_{s^* \in [x+1, y(n)]} \hat{L}_{in}(s^*) \qquad \text{for all } i \in [1,p].$$

In view of (6) to (10), we see that for every $j \in [1,p]\backslash\{n\}$, the jth column of K coincides with the jth column of $G(n-1, s)$ for all $s \in S(n,y)$, whereas the nth column of $\hat{K}$ equals the sum of the nth columns of $G(n-1, s)$ as s ranges over $S(n,y)$, and therefore, by the multilinearity of det as a function of its columns we conclude that

$$(11)\quad \det \hat{K} = \sum_{s \in S(n,y)} \det G(n-1, s).$$

[Equivalently, in view of (6) to (10), we see that (11) follows from (3.34) when we let

$$Y_j = \begin{cases} \{y(j)\} & \text{if } j \in [1,p]\backslash\{n\} \\ [x+1, y(n)] & \text{if } j = n \end{cases}$$

and when, for every $i \in [1,p]$ and $j \in [1,p]$, we let $L_{ij} : Y_j \to Q$ be the map obtained by putting $L_{ij}(s^*) = \hat{L}_{ij}(s^*)$ for all $s^* \in Z$.]

In view of (2), (5), and (10), by (4.12*) we see that

$$(12)\quad \hat{K}_{in} = \begin{bmatrix} y(n) \\ d(i) + n + 1 \end{bmatrix} - \begin{bmatrix} x \\ d(i) + n + 1 \end{bmatrix} \qquad \text{for all } i \in [1,p].$$

In view of (4), by (2.5.6*) we see that

$$(13)\quad \text{if } n = p, \text{ then } \begin{bmatrix} x \\ d(i) + n + 1 \end{bmatrix} = 0 \qquad \text{for all } i \in [1,p].$$

By (5), (9), (12), and (13) we see that

(14) if $n = p$, then $\hat{K} = G(n,y)$.

By (14) we conclude that

(15) if $n = p$, then $\det \hat{K} = \det G(n,y)$.

In view of (3), (5), (9), and (12), for the case of $n \neq p$, we have

$$\hat{K}_{ij} = \begin{cases} G_{ij}(n,y) & \text{if } i \in [1,p] \text{ and } j \in [1,p]\backslash\{n\} \\ G_{in}(n,y) - G_{i,n+1}(n,y) & \text{if } i \in [1,p] \text{ and } j = n \end{cases}$$

which means that $\hat{K}$ can be obtained by subtracting the $(n + 1)$th column of $G(n,y)$ from its nth column, and hence by general properties of dets we deduce that $\det \hat{K} = \det G(n,y)$; thus we conclude that

(16) if $n \neq p$, then $\det \hat{K} = \det G(n,y)$.

By (11), (15), and (16) we get

$$\sum_{s \in S(n,y)} \det G(n - 1, s) = \det G(n,y)$$

and this completes the proof of (6.1.3).

In view of (6.1.1), (6.1.2), and (6.1.3), by induction on n we get (6.1.4). Upon taking $n = p$ in (6.1.4) we get (6.1.5).

LEMMA (6.2) For every $p \in N^*$ and $v \in Z(p)$ let $\hat{u}[p,v]$ be the member of $Z(p)$ obtained by putting

$$\hat{u}[p,v](i) = \begin{cases} v(i) & \text{if } i \in [1, p - 1] \\ 1 + v(p) & \text{if } i = p. \end{cases}$$

For every $p \in N^*\backslash\{1\}$ and $v \in Z(p)$ let $\hat{h}[p,v]$ be the member of $Z(p - 1)$ obtained by putting

$$\hat{h}[p,v](i) = v(i) \quad \text{for all } i \in [1, p - 1].$$

For every $p \in N^*\backslash\{1\}$ let

$$\hat{Z}(p) = \{v \in Z(p) : v(p) = 0\}.$$

For every $p \in N^*$ let

$$Y' = \{y \in Z(p) : y(1) > y(2) > \cdots > y(p) \geq 0\}.$$

For every $p \in N^*$ and $y \in Y'$ let

$$R'(p,y) = \{r \in Y' : r(j) \le y(j) \quad \text{for all } j \in [1,p]\}.$$

For every $p \in N^*$ and $s \in Z(p)$ and $v \in Z(p)$ and $i \in [1,p]$ and $j \in [1,p]$ let

$$G'_{ij}(p,s,v) = \begin{bmatrix} s(j) \\ v(i) + v(i+1) + \cdots + v(p) + j - i \end{bmatrix}.$$

Finally, for every $p \in N^*$ and $s \in Z(p)$ and $v \in Z(p)$ let

$$G'(p,s,v) = \text{the } p \text{ by } p \text{ matrix whose } (i,j)\text{th entry is } G'_{ij}(p,s,v).$$

Now

$$\left[\begin{array}{l}\text{for every } p \in N^* \text{ and } y \in Y' \text{ and } v \in N(p) \text{ we have} \\ \displaystyle\sum_{r \in R'(p,y)} \det G'(p,r,v) = \det G'(p,y,\hat{u}[p,v])\end{array}\right. \tag{6.2.1}$$

and

$$\left[\begin{array}{l}\text{for every } p \in N^*\backslash\{1\} \text{ and } s \in Z(p) \text{ and } v \in \hat{Z}(p) \text{ we have} \\ \det G'(p,s,v) = \det G'(p-1, \hat{h}[p,s], \hat{h}[p,v]).\end{array}\right. \tag{6.2.2}$$

Proof: (6.2.1) follows from (6.1.5) upon taking $\hat{D} = 0 = \hat{Y}$ and upon taking d to be the member of Z(p) obtained by putting

$$d(i) = v(i) + v(i+1) + \cdots + v(p) - i \quad \text{for all } i \in [1,p].$$

To prove (6.2.2), let there be fixed any $p \in N^*\backslash\{1\}$ and $s \in Z(p)$ and $v \in \hat{Z}(p)$. Now clearly

$$\text{(1)} \quad \left[\begin{array}{l}G'_{ij}(p,s,v) = G'_{ij}(p-1, \hat{h}[p,s], \hat{h}[p,v]) \\ \text{for all } i \in [1, p-1] \text{ and } j \in [1, p-1]\end{array}\right.$$

and

$$\text{(2)} \quad G'_{pj}(p,s,v) = \begin{bmatrix} s(j) \\ j - p \end{bmatrix} \quad \text{for all } j \in [1,p].$$

By (1) we see that

$$\text{(3)} \quad \left[\begin{array}{l}\det G'(p-1, \hat{h}[p,s], \hat{h}[p,v]) = \det \text{ of the } p-1 \text{ by } p-1 \\ \text{matrix whose } (i,j)\text{th entry is } G'_{ij}(p,s,v).\end{array}\right.$$

In view of (2), by (2.5.1*) and (2.5.2*) we see that

$$G'_{pj}(p,s,v) = \begin{cases} 0 & \text{if } j \in [1, p - 1] \\ 1 & \text{if } j = p \end{cases}$$

and hence upon expanding det $G'(p,s,v)$ by the pth row we see that

$$\det G'(p,s,v) = \text{det of the } p - 1 \text{ by } p - 1 \text{ matrix whose } (i,j)\text{th entry is } G'_{ij}(p,s,v)$$

and therefore by (3) we get

$$\det G'(p,s,v) = \det G'(p - 1, \hat{h}[p,s], \hat{h}[p,v]).$$

LEMMA (6.3) Let $\hat{u}$, $\hat{h}$, $\hat{Z}$ be as in (6.2). For every $q \in N^*$ and $p \in N^*\backslash\{1\}$ and $t \in \text{pre}[q,p]$ let $\tilde{h}[q,p,t]$ be the member of $\text{pre}[q, p - 1]$ obtained by putting

$$\tilde{h}[q,p,t](k,i) = t(k,i) \quad \text{for all } k \in [1,q] \text{ and } i \in [1, p - 1].$$

For every $q \in N^*$ and $p \in N^*$ let

$$Y(q,p) = \{t \in \text{pre}[q,p] : t(k,1) > t(k,2) > \cdots > t(k,p) \geq 0 \text{ for all } k \in [1,q]\}.$$

For every $q \in N^*$ and $p \in N^*$ and $t \in Y(q,p)$ let

$$R(q,p,t) = \{e \in Y(q,p) : e(k,j) \leq t(k,j) \text{ for all } k \in [1,q] \text{ and } j \in [1,p]\}.$$

For every $q \in N^*$ and $p \in N^*$ and $t \in \text{pre}[q,p]$ and $v \in Z(p)$ and $k \in [1,q]$ and $i \in [1,p]$ and $j \in [1,p]$ let

$$\tilde{G}^{(8k)}_{ij}(q,p,t,v) = \begin{bmatrix} t(k,j) \\ v(i) + v(i + 1) + \cdots + v(p) + j - i \end{bmatrix}.$$

For every $q \in N^*$ and $p \in N^*$ and $t \in \text{pre}[q,p]$ and $v \in Z(p)$ and $k \in [1,q]$ let

$$\tilde{G}^{(8k)}(q,p,t,v) = \text{the } p \text{ by } p \text{ matrix whose } (i,j)\text{th entry is } \tilde{G}^{(8k)}_{ij}(q,p,t,v).$$

Finally, for every $q \in N^*$ and $p \in N^*$ and $t \in \text{pre}[q,p]$ and $v \in Z(p)$ let

$$\tilde{H}^{(8)}(q,p,t,v) = \prod_{k=1}^{q} \det \tilde{G}^{(8k)}(q,p,t,v).$$

Now

$$\left[\begin{array}{l}\text{for every } q \in N^* \text{ and } p \in N^* \text{ and } t \in Y(q,p) \\ \text{and } v \in N(p) \text{ we have} \\ \displaystyle\sum_{e \in R(q,p,t)} \tilde{H}^{(8)}(q,p,e,v) = \tilde{H}^{(8)}(q,p,t,\hat{u}[p,v])\end{array}\right. \tag{6.3.1}$$

and

$$\left[\begin{array}{l}\text{for every } q \in N^* \text{ and } p \in N^*\backslash\{1\} \text{ and } t \in \text{pre}[q,p] \\ \text{and } v \in \hat{Z}(p) \text{ we have} \\ \tilde{H}^{(8)}(q,p,t,v) = \tilde{H}^{(8)}(q,\ p-1,\ \tilde{h}[q,p,t],\ \hat{h}[p,v]).\end{array}\right. \tag{6.3.2}$$

Proof: Let there be fixed any $q \in N^*$ and $p \in N^*$ and $t \in \text{pre}[q,p]$ and $v \in Z(p)$. For every $e \in \text{pre}[q,p]$ and $k \in [1,q]$ let $z[e,k]$ be the member of $Z(p)$ obtained by putting

$$z[e,k](j) = e(k,j) \quad \text{for all } j \in [1,p].$$

Now, in case $p \neq 1$ and $v \in \hat{Z}(p)$, we see that (6.3.2) follows first by taking $s = z[t,k]$ in (6.2.2) and then producting the resulting two sides of (6.2.2) as k ranges over $[1,q]$.

To prove (6.3.1), henceforth assume that $t \in Y(q,p)$ and $v \in N(p)$. Upon letting G' and R' be as in (6.2) and upon letting

(1) $f^{(k)}(r) = \det G'(p,r,v) \quad$ for all $k \in [1,q]$ and $r \in Z(p)$

and

(2) $y_k = z[t,k]$ and $X_k = R'(p,y_k) \quad$ for all $k \in [1,q]$

and

(3) $g^{(k)}(r) = \det G'(p,r,\hat{u}[p,v]) \quad$ for all $k \in [1,q]$ and $r \in Z(p)$

by (6.2.1) we see that

(4) $\displaystyle\sum_{r \in X_k} f^{(k)}(r) = g^{(k)}(y_k) \quad$ for all $k \in [1,q]$.

Now clearly

$$R(q,p,t) = \{e \in \text{pre}[q,p] : z[e,k] \in X_k \text{ for all } k \in [1,q]\}$$

and therefore, upon taking

$$T = \mathrm{pre}[q,p] \text{ and } Y_k = Z(p) \qquad \text{for all } k \in [1,q],$$

by the distributive law (3.32.1) we see that

$$(5) \quad \prod_{k=1}^{q} \sum_{r \in X_k} f^{(k)}(r) = \sum_{e \in R(q,p,t)} \prod_{k=1}^{q} f^{(k)}(z[e,k]).$$

By (1) we see that

$$(6) \quad \prod_{k=1}^{q} f^{(k)}(z[e,k]) = \tilde{H}^{(8)}(q,p,e,v) \qquad \text{for all } e \in \mathrm{pre}[q,p].$$

By (2) and (3) we see that

$$(7) \quad \prod_{k=1}^{q} g^{(k)}(y_k) = \tilde{H}^{(8)}(q,p,t,\hat{u}[p,v]).$$

By (4), (5), (6), and (7) we conclude that

$$\sum_{e \in R(q,p,t)} \tilde{H}^{(8)}(q,p,e,v) = \tilde{H}^{(8)}(q,p,t,\hat{u}[p,v])$$

which completes the proof of (6.3.1).

REMARK (6.40) We shall now prove Theorem (5.6*). So let there be given any $q \in N^*$ and $m \in N^*(q)$ and $p \in N^*$ and $a \in \mathrm{vec}(q,m,p)$. Note that in (5.6*.11) we have simply repeated the definition of $F^{(8)}$. Also, note that (5.6*.14) follows from (2.5.1*). To prove the rest, for every $b \in \mathrm{vec}(q,m,p)$ let $e[b]$ be the member of $\mathrm{pre}[q,p]$ obtained by putting

$$e[b](k,i) = m(k) - b(k,i) \qquad \text{for all } k \in [1,q] \text{ and } i \in [1,p].$$

Also, let $t = e[a]$, and let t' be the member of $\mathrm{pre}[q,p]$ obtained by putting $t'(k,i) = i$ for all $k \in [1,q]$ and $i \in [1,p]$. Now (5.6*.1), (5.6*.2), (5.6*.3), and (5.6*.4) follow from (3.25) and (3.28), whereas (5.6*.13) follows from (6.3.2). Finally, upon letting Y and R be as in (6.3) and upon noting that $b \mapsto e[b]$ gives a bijection of $\mathrm{vec}(q,m,p,a)$ onto $R(q,p,t)$, we see that (5.6*.12) follows from (6.3.1)

PROBLEM (6.41) Having thus proved Theorem (5.6*), we are led to pose the problem: For the interesting identity (5.6*.4), give a direct proof whatever that may mean. As an illustration, if $q = 1$ and $p \geq 2$, then for every $V \in Z$, by the definition of $F^{(6)}$ we have

$$F^{(6)}(1,m,p,a,V) = (1/p!) \sum_{v \in Z(p,V)} \det G^{(61)}(q,m,p,a,v)$$

where $G^{(61)}(a,m,p,a,v)$ is the p by p matrix whose (i,j)th entry is $\begin{bmatrix} m(1) - a(1,j) \\ v(i) + j - i \end{bmatrix}$ and hence upon expanding the det by rows we get

$$\begin{aligned} &F^{(6)}(1,m,p,a,V) \\ &= (1/p!) \sum_{v \in Z(p,V)} \sum_{w \in W(p)} \operatorname{sgn}(w) \prod_{i=1}^{p} \begin{bmatrix} m(1) - a(1,w(i)) \\ v(i) + w(i) - i \end{bmatrix} \\ &= (1/p!) \sum_{w \in W(p)} \operatorname{sgn}(w) \sum_{v \in Z(p,V)} \prod_{i=1}^{p} \begin{bmatrix} m(1) - a(1,w(i)) \\ v(i) + w(i) - i \end{bmatrix} \end{aligned}$$

and upon letting

$$R = \sum_{j=1}^{p} [m(1) - a(1,j)]$$

by (4.35*.3) we also have

$$\sum_{v \in Z(p,V)} \prod_{i=1}^{p} \begin{bmatrix} m(1) - a(1,w(i)) \\ v(i) + w(i) - i \end{bmatrix} = \begin{bmatrix} R + p - 1 \\ V \end{bmatrix} \quad \text{for all } w \in W(p)$$

and therefore we get

$$\begin{aligned} F^{(6)}(1,m,p,a,V) &= (1/p!) \sum_{w \in W(p)} \operatorname{sgn}(w) \begin{bmatrix} R + p - 1 \\ V \end{bmatrix} \\ &= (1/p!) \begin{bmatrix} R + p - 1 \\ V \end{bmatrix} \sum_{w \in W(p)} \operatorname{sgn}(w) \\ &= (1/p!) \begin{bmatrix} R + p - 1 \\ V \end{bmatrix} \text{ times zero} \\ &= 0. \end{aligned}$$

Now this *direct* proof is actually more sophisticated than the *elementary* proof given in (3.25). So perhaps I do not really know what the problem means. Maybe it simply says: Think of the interesting identity and do with it some other interesting thing, which in effect means, do with it what you can!

PROBLEM (6.42) Having proved Theorem (5.6*), we may also pose the problem: Find proper modifications of $F^{(5k)}$ and $F^{(6)}$ and $F^{(7*)}$ so that the equations in (5.6*.2) remain valid for *every* q and not just for even q.

Now given any $q \in N^*$ and $m \in N^*(2)$ and $p \in N^*$ and $a \in \operatorname{vec}(q,m,p)$, in view of (5.6''') and (5.6*.1), by (5.6*.11) to (5.6*.14) we see that

$$\left[\begin{array}{l}\text{for all } V \in N \text{ we have}\\ \operatorname{Stab}(q,m,p,a,V) = F^{(8)}(q,m,p,a,V) = F^{(7)}(q,m,p,a,V)\end{array}\right.$$

and therefore by (5.6*.2) we see that

$$\left[\begin{array}{l}\text{if } q \text{ is even, then for all } k \in [1,q] \text{ and } V \in N, \text{ we have}\\ \operatorname{Stab}(q,m,p,a,V) = F^{(7*)}(q,m,p,a,V) = F^{(6)}(q,m,p,a,V)\\ \qquad\qquad = F^{(5k)}(q,m,p,a,V).\end{array}\right.$$

So the problem above can be paraphrased by saying: Find better formulas for counting standard tableaux of odd width.

7. EQUIVALENCE OF CERTAIN DETERMINANTAL POLYNOMIALS OF WIDTH 2

In this section we prove Theorem (5.7*) by establishing the equivalence of the determinantal polynomials $F^{(1k)}$, $F^{(1*k)}$, ..., $F^{(4*k)}$ which were introduced in (2.10) and which, in a somewhat generalized form, will now be renamed, respectively, as $F^{(10)}$, $F^{(1*0)}$, ..., $F^{(4*0)}$; in (7.33) we pose a problem about these polynomials. Logically speaking, this section is independent of Subsections (2.10) and (2.11) and Sections 5 and 6; the one exception to this is the Demonstrational Remark (7.40), in which the various results of this section will be put together to prove Theorem (5.7*); the only other exception is (7.41), where we pose a related problem. Before reading the rest of this section, the reader may find it useful to first browse in the Motivational Remark (7.34); similarly, while reading this section, Informational Remarks (7.0) and (7.31) should not be omitted.

In this section let p be a fixed positive integer. Also, let there be given any

$$r \in Z(p) \qquad s \in Z(p) \qquad r' \in Z(p) \qquad s' \in Z(p),$$

and let us put

$$R = r(1) + r(2) + \cdots + r(p) \quad \text{and} \quad S = s(1) + s(2) + \cdots + s(p)$$

and

$$R' = r'(1) + r'(2) + \cdots + r'(p)$$

and

$$S' = s'(1) + s'(2) + \cdots + s'(p).$$

Now first, for every $i \in [1,p]$ and $j \in [1,p]$ and $e \in Z(p)$ we define

$$G_{ij}^{(10)}(r,r',s',e) = \begin{Bmatrix} r(j) + r'(j) - s'(i) \\ r(j) - e(i) \end{Bmatrix},$$

and second, for every $e \in Z(p)$ we define

$$G^{(10)}(r,r',s',e) = \text{the p by p matrix where the (i,j)th entry is } G_{ij}^{(10)}(r,r',s',e)$$

and

$$G^{[10]}(s,e) = \prod_{i=1}^{p} \begin{bmatrix} s(i) \\ e(i) \end{bmatrix}$$

and

$$H^{(10)}(r,s,r',s',e) = G^{[10]}(s,e) \det G^{(10)}(r,r',s',e),$$

and third, for every $i \in [1,p]$ and $j \in [1,p]$ and $e \in Z(p)$ we define

$$G_{ij}^{(1*0)}(r,s,r',s',e) = \begin{bmatrix} s(i) \\ e(j) \end{bmatrix} \begin{Bmatrix} r(j) + r'(j) - s'(i) \\ r(j) - e(j) \end{Bmatrix}$$

and fourth, for every $e \in Z(p)$ we define

$$G^{(1*0)}(r,s,r',s',e) = \text{the p by p matrix whose (i,j)th entry is } G_{ij}^{(1*0)}(r,s,r',s',e)$$

and

$$H^{(1*0)}(r,s,r',s',e) = \det G^{(1*0)}(r,s,r',s',e),$$

and fifth, for every $L \in \{1,1^*\}$ and $E \in Z$ we define

$$H_E^{(L0)}(r,s,r',s') = \sum_{e \in Z(p,E)} H^{(L0)}(r,s,r',s',e),$$

and sixth, for every $L \in \{1,1^*\}$ and $D \in Z$ we define

$$F_D^{(L0)}(r,s,r',s') = \sum_{E \in Z} (-1)^{R-E} \binom{E + R' - S'}{D + E - R} H_E^{(L0)}(r,s,r',s'),$$

and seventh, for every $i \in [1,p]$ and $j \in [1,p]$ and $e \in Z(p)$ we define

$$G_{ij}^{(20)}(r,s,r',s',e) = \binom{r(i) + r'(i) - s'(j)}{r(i) - e(i)} \binom{s(j) + s'(j) - r'(i)}{e(i)}$$

and

$$G_{ij}^{(2*0)}(r,s,r',s',e) = \binom{r(i) + r'(i) - s'(j)}{r(i) - e(j)} \binom{s(j) + s'(j) - r'(i)}{e(j)}$$

and eighth, for every $L \in \{2,2^*\}$ and $e \in Z(p)$ we define

$$G^{(L0)}(r,s,r',s',e) = \text{the p by p matrix whose (i,j)th entry is } G_{ij}^{(L0)}(r,s,r',s',e)$$

and

$$H^{(L0)}(r,s,r',s',e) = \det G^{(L0)}(r,s,r',s',e),$$

and ninth, for every $L \in \{2,2^*\}$ and $E \in Z$ we define

$$H_E^{(L0)}(r,s,r',s') = \sum_{e \in Z(p,E)} H^{(L0)}(r,s,r',s',e),$$

and tenth, for every $L \in \{2,2^*\}$ and $D \in Z$ we define

$$F_D^{(L0)}(r,s,r',s') = \sum_{E\in Z} \binom{E}{D} H_E^{(L0)}(r,s,r',s'),$$

and eleventh, for every $i \in [1,p]$ and $j \in [1,p]$ and $d \in Z(p)$ we define

$$G_{ij}^{(30)}(r,s,r',s',d) = \begin{bmatrix} s(i) - d(i) \\ r(j) \end{bmatrix} \binom{r(j) + r'(j) - s'(i)}{d(i)}$$

and

$$G_{ij}^{(3*0)}(r,s,r',s',d) = \begin{bmatrix} s(i) - d(j) \\ r(j) \end{bmatrix} \binom{r(j) + r'(j) - s'(i)}{d(j)}$$

and twelfth, for every $L \in \{3,3^*\}$ and $d \in Z(p)$ we define

$$G^{(L0)}(r,s,r',s',d) = \text{the } p \text{ by } p \text{ matrix whose } (i,j)\text{th entry is } G_{ij}^{(L0)}(r,s,r',s',d)$$

and

$$H^{(L0)}(r,s,r',s',d) = \det G^{(L0)}(r,s,r',s',d),$$

and thirteenth, for every $z \in W[p]$ and $w \in W(p)$ and $d \in Z(p)$ we define

$$G_w^{(3z0)}(r,s,r',s',d) = \prod_{i=1}^{p} \begin{bmatrix} s(i) - d(z(w)(i)) \\ r(w(i)) \end{bmatrix} \times \binom{r(w(i)) + r'(w(i)) - s'(i)}{d(z(w)(i))},$$

and fourteenth, for every $z \in W[p]$ and $d \in Z(p)$ we define

$$H^{(3z0)}(r,s,r',s',d) = \sum_{w\in W(p)} \operatorname{sgn}(w) G_w^{(3z0)}(r,s,r',s',d),$$

and fifteenth, for every $L \in \{3,3^*\} \cup \{3z : z \in W[p]\}$ and $D \in Z$ we define

$$F_D^{(L0)}(r,s,r',s') = \sum_{d\in Z(p,D)} H^{(L0)}(r,s,r',s',d),$$

and sixteenth, for every $i \in [1,p]$ and $j \in [1,p]$ and $v \in Z(p)$ we define

$$G_{ij}^{(40)}(s,s',r',v) = \begin{bmatrix} s(j) \\ v(i) + s'(j) - r'(i) \end{bmatrix},$$

and seventeenth, for every $v \in Z(p)$ we define

$$G^{(40)}(s,s',r',v) = \text{the } p \text{ by } p \text{ matrix whose } (i,j)\text{th entry is } G_{ij}^{(40)}(s,s',r',v)$$

and

$$G^{[40]}(r,v) = \prod_{i=1}^{p} \begin{bmatrix} r(i) \\ v(i) \end{bmatrix}$$

and

$$H^{(40)}(r,s,r',s',v) = G^{[40]}(r,v)\det G^{(40)}(s,s',r',v),$$

and eighteenth, for every $i \in [1,p]$ and $j \in [1,p]$ and $v \in Z(p)$ we define

$$G_{ij}^{(4*0)}(r,s,r',s',v) = \begin{bmatrix} r(i) \\ v(j) \end{bmatrix} \begin{bmatrix} s(j) \\ v(j) + s'(j) - r'(i) \end{bmatrix}$$

and ninteenth, for every $v \in Z(p)$ we define

$$G^{(4*0)}(r,s,r',s',v) = \text{the } p \text{ by } p \text{ matrix whose } (i,j)\text{th entry is } G_{ij}^{(4*0)}(r,s,r',s',v)$$

and

$$H^{(4*0)}(r,s,r',s',v) = \det G^{(4*0)}(r,s,r',s',v),$$

and twentieth, for every $L \in \{4,4*\}$ and $V \in Z$ we define

$$F^{(L0)}(r,s,r',s',V) = \sum_{v \in Z(p,V)} H^{(L0)}(r,s,r',s',v)$$

and twenty-first, for every $v \in Z(p)$ we define

$$H^{(60)}(r,s,r',s',v) = [\det G^{(40)}(r,r',s',v)]\det G^{(40)}(s,s',r',v),$$

and finally, for every $V \in Z$ we define

$$F^{(60)}(r,s,r',s',V) = (1/p!) \sum_{v \in Z(p,V)} H^{(60)}(r,s,r',s',v).$$

Now, *if*

$$R' = S' \tag{*}$$

then for every $L \in \{1,1^*\}$ and $V \in Z$ we define

$$F^{(L0)}(r,s,r',s',V) = \sum_{D \in Z} (-1)^D F_D^{(L0)}(r,s,r',s') \begin{bmatrix} R + S + p - 1 - D \\ V \end{bmatrix}$$

Again, *if* (*), *then* for every $L \in \{1,1^*\}$, in $Q[V]$ we define

$$F^{[L0]}(r,s,r',s',V) = \sum_{D \in Z} (-1)^D F_D^{(L0)}(r,s,r',s') \begin{bmatrix} V \\ R + S + p - 1 - D \end{bmatrix}$$

Moreover, for every $L \in \{2,2^*\}$ and $V \in Z$ we define

$$F^{(L0)}(r,s,r',s',V) = \sum_{D \in Z} (-1)^D F_D^{(L0)}(r,s,r',s') \begin{bmatrix} R + S + p - 1 - D \\ V \end{bmatrix}$$

Again, for every $L \in \{2,2^*\}$, in $Q[V]$ we define

$$F^{[L0]}(r,s,r',s',V) = \sum_{D \in Z} (-1)^D F_D^{(L0)}(r,s,r',s') \begin{bmatrix} V \\ R + S + p - 1 - D \end{bmatrix}$$

Furthermore, *if*

$$\begin{bmatrix} \text{for every } i \in [1,p] \text{ and } j \in [1,p] \text{ we have} \\ s(i) + s'(i) - r'(j) \geq 0 \leq r(j) + r'(j) - s'(i), \end{bmatrix} \tag{**}$$

then for every $L \in \{3,3^*\} \cup \{3z : z \in W[p]\}$ and $V \in Z$ we define

$$F^{(L0)}(r,s,r',s',V) = \sum_{D \in Z} (-1)^D F_D^{(L0)}(r,s,r',s') \begin{bmatrix} R + S + p - 1 - D \\ V \end{bmatrix}$$

Again, *if* (**), *then* for every $L \in \{3,3^*\} \cup \{3z : z \in W[p]\}$, in $Q[V]$ we define

$$F^{[L0]}(r,s,r',s',V) = \sum_{D \in Z} (-1)^D F_D^{(L0)}(r,s,r',s') \begin{bmatrix} V \\ R + S + p - 1 - D \end{bmatrix}$$

REMARK (7.0) In this section, sometimes we shall assume condition (*), and sometimes we shall assume condition (**), and sometimes we shall assume that

$$r'(i) = s'(i) \quad \text{for all } i \in [1,p] \tag{***}$$

and sometimes we shall assume that

$$\text{either } R < 0 \text{ or } S \geq 0. \tag{****}$$

Note that

if (*) and (**) hold, then $S \geq 0 \leq R$ (7.0.0)

and hence in particular

conditions (*) and (**) imply condition (****). (7.0.1)

Also note that

condition (***) implies condition (*). (7.0.2)

In (7.21) to (7.32) we shall establish the equivalence of $F^{(10)}$, $F^{(1*0)}$, ..., $F^{(4*0)}$. Before doing so, in (7.1) to (7.3) we record several alternative characterizations of some of the functions which were introduced above as G and H with different subscripts and superscripts, and in (7.11) to (7.16) we establish the essential finiteness of the summations involved in the definitions above, and in (7.17) to (7.20) we calculate the functions H and F for some special values of the parameters. It may be noted that (7.21) and (7.22) are independent of (7.17) to (7.20). It may also be noted that (7.23) to (7.29) are independent of (7.17) to (7.22). Moreover, it may be noted that (7.27), (7.28), and (7.29) are mutually independent. Finally, it may be noted that (7.30) is independent of (7.17) to (7.29).

REMARK (7.0') In Observations (7.1), (7.2), and (7.3) we use the well-known fact that the determinant of a square matrix equals the determinant of its transpose.

OBSERVATION (7.1) Let there be given any $e \in Z(p)$. Now by (2.5.10') we see that *if* (**), *then* for every $i \in [1,p]$ and $j \in [1,p]$ we have

$$G_{ij}^{(10)}(r,r',s',e) = \begin{pmatrix} r(j) + r'(j) - s'(i) \\ e(i) + r'(j) - s'(i) \end{pmatrix}.$$

In view of (7.0') we also see that

$$H^{(10)}(r,s,r',s',e) = \left(\prod_{j=1}^{p} \begin{bmatrix} s(j) \\ e(j) \end{bmatrix} \right) \text{ times } \left\{ \begin{matrix} \text{det of the transpose} \\ \text{of } G^{(10)}(r,r',s',e) \end{matrix} \right\}$$

and *if* (**), *then* for every $i \in [1,p]$ and $j \in [1,p]$, the (i,j)th entry of the said transpose has two different expressions which are obtained by interchanging i and j in the two expressions of $G_{ij}^{(10)}$ above.

By (2.5.10') we see that *if* (**), *then* for every $i \in [1,p]$ and $j \in [1,p]$ we have

$$G_{ij}^{(1*0)}(r,r',s',e) = \begin{bmatrix} s(i) \\ e(j) \end{bmatrix} \begin{pmatrix} r(j) + r'(j) - s'(i) \\ e(j) + r'(j) - s'(i) \end{pmatrix}.$$

In view of (7.0') we also see that

$$H^{(1*0)}(r,s,r',s',e) = \text{det of the transpose of } G^{(1*0)}(r,s,r',s',e)$$

and *if* (**), *then* for every $i \in [1,p]$ and $j \in [1,p]$, the (i,j)th entry of the transpose has two different expressions which are obtained by interchanging i and j in the two expressions of $G_{ij}^{(1*0)}$ above.

OBSERVATION (7.2) Let there be given any $e \in Z(p)$. Now by (2.5.10') we see that *if* (**), *then* for every $i \in [1,p]$ and $j \in [1,p]$ we have

$$G_{ij}^{(20)}(r,s,r',s',e)$$

$$= \begin{pmatrix} r(i) + r'(i) - s'(j) \\ e(i) + r'(i) - s'(j) \end{pmatrix} \begin{pmatrix} s(j) + s'(j) - r'(i) \\ e(i) \end{pmatrix}$$

$$= \begin{pmatrix} r(i) + r'(i) - s'(j) \\ e(i) + r'(i) - s'(j) \end{pmatrix} \begin{pmatrix} s(j) + s'(j) - r'(i) \\ s(j) + s'(j) - r'(i) - e(i) \end{pmatrix}$$

$$= \begin{pmatrix} r(i) + r'(i) - s'(j) \\ r(i) - e(i) \end{pmatrix} \begin{pmatrix} s(j) + s'(j) - r'(i) \\ s(j) + s'(j) - r'(i) - e(i) \end{pmatrix}$$

In view of (7.0') we also see that

$$H^{(20)}(r,s,r',s',e) = \text{det of the transpose of } G^{(20)}(r,s,r',s',e)$$

and *if* (**), *then* for every $i \in [1,p]$ and $j \in [1,p]$, the (i,j)th entry of the transpose has four different expressions which are obtained by interchanging i and j in the four expressions of $G_{ij}^{(20)}$ above.

By (2.5.10') we see that *if* (**), *then* for every $i \in [1,p]$ and $j \in [1,p]$ we have

$$G_{ij}^{(2*0)}(r,s,r',s',e)$$
$$= \binom{r(i) + r'(i) - s'(j)}{e(j) + r'(i) - s'(j)}\binom{s(j) + s'(j) - r(i)}{e(j)}$$
$$= \binom{r(i) + r'(i) - s'(j)}{e(j) + r'(i) - s'(j)}\binom{s(j) + s'(j) - r'(i)}{s(j) + s'(j) - r'(i) - e(j)}$$
$$= \binom{r(i) + r'(i) - s'(j)}{r(i) - e(j)}\binom{s(j) + s'(j) - r'(i)}{s(j) + s'(j) - r'(i) - e(j)}.$$

In view of (7.0') we also see that

$$H^{(2*0)}(r,s,r',s',e) = \text{det of the transpose of } G^{(2*0)}(r,s,r',s',e)$$

and *if* (**), *then* for every $i \in [1,p]$ and $j \in [1,p]$, the (i,j)th entry of the transpose has four different expressions which are obtained by interchanging i and j in the four expressions of $G_{ij}^{(2*0)}$ above.

OBSERVATION (7.3') By (2.5.23') we see that for every $L \in \{3,3^*\}$ and $d \in Z(p)$ we have

$$H^{(L0)}(r,s,r',s',d) = H^{(3z'[L]0)}(r,s,r',s',d),$$

where $z'[3^*] \in W[p]$ is defined by putting

$$z'[3^*](w) = w \quad \text{for all } w \in W(p),$$

and where $z'[3] \in W[p]$ is defined by putting, for every $w \in W(p)$,

$$z'[3](w) = \text{the identity map of } [1,p] \text{ onto itself},$$

i.e.,

$$z'[3](w)(i) = i \quad \text{for all } i \in [1,p].$$

OBSERVATION (7.3) Let there be given any $d \in Z(p)$. Now, upon taking $d^* = d(i)$ in (2.6.23*) we see that *if* (**), *then* for every $i \in [1,p]$ and $j \in [1,p]$ we have

$$G_{ij}^{(30)}(r,s,r',s',d)$$
$$= \begin{bmatrix} r(j) \\ s(i) - d(i) \end{bmatrix} \begin{pmatrix} r(j) + r'(j) - s'(i) \\ d(i) \end{pmatrix}$$
$$= \begin{bmatrix} r(j) \\ s(i) - d(i) \end{bmatrix} \begin{pmatrix} r(j) + r'(j) - s'(i) \\ r(j) + r'(j) - s'(i) - d(i) \end{pmatrix}$$
$$= \begin{bmatrix} s(i) - d(i) \\ r(j) \end{bmatrix} \begin{pmatrix} r(j) + r'(j) - s'(i) \\ r(j) + r'(j) - s'(i) - d(i) \end{pmatrix}.$$

In view of (7.0') we also see that

$$H^{(30)}(r,s,r',s',d) = \text{det of the transpose of } G^{(30)}(r,s,r',s',d)$$

and *if* (**), *then* for every $i \in [1,p]$ and $j \in [1,p]$, the (i,j)th entry of the transpose has four different expressions which are obtained by interchanging i and j in the four expressions of $G_{ij}^{(30)}$ above.

Upon taking $d^* = d(j)$ in (2.6.23*) we see that *if* (**), *then* for every $i \in [1,p]$ and $j \in [1,p]$ we have

$$G_{ij}^{(3^*0)}(r,s,r',s',d)$$
$$= \begin{bmatrix} r(j) \\ s(i) - d(j) \end{bmatrix} \begin{pmatrix} r(j) + r'(j) - s'(i) \\ d(j) \end{pmatrix}$$
$$= \begin{bmatrix} r(j) \\ s(i) - d(j) \end{bmatrix} \begin{pmatrix} r(j) + r'(j) - s'(i) \\ r(j) + r'(j) - s'(i) - d(j) \end{pmatrix}$$
$$= \begin{bmatrix} s(i) - d(j) \\ r(j) \end{bmatrix} \begin{pmatrix} r(j) + r'(j) - s'(i) \\ r(j) + r'(j) - s'(i) - d(j) \end{pmatrix}.$$

In view of (7.0') we also see that

$$H^{(3^*0)}(r,s,r',s',d) = \text{det of the transpose of } G^{(3^*0)}(r,s,r',s',d)$$

and *if* (**), *then* for every $i \in [1,p]$ and $j \in [1,p]$, the (i,j)th entry of the transpose has four different expressions which are obtained by interchanging i and j in the four expressions of $G_{ij}^{(3^*0)}$ above.

Given any $z \in W[p]$ and $w \in W(p)$, upon taking $j = w(i)$ and $d^* = d(z(w)(i))$ in (2.6.23*) and then producting as i varies over $[1,p]$, we see that *if* (**), *then*

$$G_{w}^{(3z0)}(r,s,r',s',d)$$

$$= \prod_{i=1}^{p} \begin{bmatrix} r(w(i)) \\ s(i) - d(z(w)(i)) \end{bmatrix} \begin{pmatrix} r(w(i)) + r'(w(i)) - s'(i) \\ d(z(w)(i)) \end{pmatrix}$$

$$= \prod_{i=1}^{p} \begin{bmatrix} r(w(i)) \\ s(i) - d(z(w)(i)) \end{bmatrix} \begin{pmatrix} r(w(i)) + r'(w(i)) - s'(i) \\ r(w(i)) + r'(w(i)) - s'(i) - d(z(w)(i)) \end{pmatrix}$$

$$= \prod_{i=1}^{p} \begin{bmatrix} s(i) - d(z(w)(i)) \\ r(w(i)) \end{bmatrix} \begin{pmatrix} r(w(i)) + r'(w(i)) - s'(i) \\ r(w(i)) + r'(w(i)) - s'(i) - d(z(w)(i)) \end{pmatrix}$$

OBSERVATION (7.11) Let there be given any $L \in \{1,1^*\}$. Now by (2.6.21) we see that

(1) $\{e \in Z(p) : H^{(L0)}(r,s,r',s',e) \neq 0\} \subset N[p,R]$

and by (2.6.13) we see that

(2) $\left[\begin{array}{l} \textit{if } (*), \textit{ then} \\ \left\{D \in Z : \left\{E \in [0,R] : \begin{pmatrix} E + R' - S' \\ D + E - R \end{pmatrix} \neq 0\right\} \neq \emptyset\right\} \subset [0,R] \end{array}\right.$

and clearly we see that

(3) $\left[\begin{array}{l} \textit{if } (****), \textit{ then} \text{ for every } D \in [0,R] \text{ we have} \\ R + S + p - 1 - D \geq 0. \end{array}\right.$

By (1), first we see that for every $E \in Z$, the summation in the definition of $H_{E}^{(L0)}(r,s,r',s')$ is essentially finite, and second, we see that

(4) $\{E \in Z : H_{E}^{(L0)}(r,s,r',s') \neq 0\} \subset [0,R]$.

By (4), first we see that for every $D \in Z$, the summation in the definition of $F_{D}^{(L0)}(r,s,r',s')$ is essentially finite, and second, in view of (2), we see that

(5) $\left[\begin{array}{l} \textit{if } (*), \textit{ then} \\ \{D \in Z : F_{D}^{(L0)}(r,s,r',s') \neq 0\} \subset [0,R]. \end{array}\right.$

By (5), first we see that

(6) $\left[\begin{array}{l}\textit{if } (*), \textit{ then the summations in the definitions of} \\ F^{(L0)}(r,s,r',s',V) \text{ and } F^{[L0]}(r,s,r',s',V) \text{ are} \\ \text{essentially finite,}\end{array}\right.$

and second, in view of (3) and (2.5.7*), we see that

(7) $\left[\begin{array}{l}\textit{if } (*) \textit{ and } (****), \textit{ then for every } V \in N \text{ we have} \\ F^{(L0)}(r,s,r',s',V) = F^{[L0]}(r,s,r',s',V).\end{array}\right.$

OBSERVATION (7.12) Let there be given any $L \in \{2,2^*\}$. Now by (2.6.22) we see that

(1) $\{e \in Z(p) : H^{(L0)}(r,s,r',s',e) \neq 0\} \subset N[p,R]$

and by (2.6.14) we see that

(2) $\left\{D \in Z : \left\{E \in [0,R] : \begin{pmatrix}E\\D\end{pmatrix} \neq 0\right\} \neq \emptyset\right\} \subset [0,R]$

and clearly we see that

(3) $\left[\begin{array}{l}\textit{if } (****), \textit{ then for every } D \in [0,R] \text{ we have} \\ R + S + p - 1 - D \geq 0.\end{array}\right.$

By (1), first we see that for every $E \in Z$, the summation in the definitions of $H_E^{(L0)}(r,s,r',s')$ is essentially finite, and second, we see that

(4) $\{E \in Z : H_E^{(L0)}(r,s,r',s') \neq 0\} \subset [0,R]$.

By (4), first we see that for every $D \in Z$, the summation in the definition of $F_D^{(L0)}(r,s,r',s')$ is essentially finite, and second, in view of (2), we see that

(5) $\{D \in Z : F_D^{(L0)}(r,s,r',s') \neq 0\} \subset [0,R]$.

By (5), first we see that

$\left[\begin{array}{l}\text{the summations in the definitions of } F^{(L0)}(r,s,r',s',V) \\ \text{and } F^{[L0]}(r,s,r',s',V) \text{ are essentially finite}\end{array}\right.$

and second, in view of (3) and (2.5.7*), we see that

(7) $\left[\begin{array}{l}\textit{if } (****), \textit{ then} \text{ for every } V \in N \text{ we have} \\ F^{(L0)}(r,s,r',s',V) = F^{[L0]}(r,s,r',s',V).\end{array}\right.$

OBSERVATION (7.13) Let there be given any $L \in \{3,3^*\} \cup \{3z : z \in W[p]\}$. Now by (2.6.23) we see that

(1) $\{d \in Z(p) : H^{(L0)}(r,s,r',s',d) \neq 0\} \subset N(p)$

and by (2.6.23) we also see that

(1*) $\left[\begin{array}{l}\textit{if } (**), \textit{ then} \\ \{d \in Z(p) : H^{(L0)}(r,s,r',s',d) \neq 0\} \subset N[p, R + R' - S']\end{array}\right.$

and clearly we see that

(2) $\left[\begin{array}{l}\textit{if } (*) \textit{ and } (****), \textit{ then} \text{ for every } D \in [0, R + R' - S'] \\ \text{we have } R + S + p - 1 - D \geq 0.\end{array}\right.$

By (1) we see that for every $D \in Z$, the summation in the definition of $F_D^{(L0)}(r,s,r',s')$ is essentially finite. By (1*) we see that

(3) $\left[\begin{array}{l}\textit{if } (**), \textit{ then} \\ \{D \in Z : F_D^{(L0)}(r,s,r',s') \neq 0\} \subset [0, R + R' - S'].\end{array}\right.$

By (3), first we see that

(4) $\left[\begin{array}{l}\textit{if } (**), \textit{ then} \text{ the summations in the definitions of} \\ F^{(L0)}(r,s,r',s',V) \text{ and } F^{[L0]}(r,s,r',s',V) \text{ are} \\ \text{essentially finite}\end{array}\right.$

and second, in view of (2) and (2.5.7*) and (7.0.1), we see that

(5) $\left[\begin{array}{l}\textit{if } (*) \textit{ and } (**), \textit{ then} \text{ for every } V \in N \text{ we have} \\ F^{(L0)}(r,s,r',s',V) = F^{[L0]}(r,s,r',s',V).\end{array}\right.$

OBSERVATION (7.14) Let there be given any $L \in \{4,4^*\}$. Now by (2.6.24) we see that

$$\{v \in Z(p) : H^{(L0)}(r,s,r',s',v) \neq 0\} \subset N(p)$$

and hence the summation in the definition of $F^{(L0)}(r,s,r',s',V)$ is essentially finite.

OBSERVATION (7.16) Let

$$\hat{R} = \max\{r'(j) : j \in [1,p]\} \quad \text{and} \quad \hat{S} = \max\{s'(j) : j \in [1,p]\}.$$

Now by (2.6.24") we see that

$$\begin{aligned}&\{v \in Z(p) : H^{(60)}(r,s,r',s',v) \neq 0\}\\ &\subset \{v \in Z(p) : v(i) \geq \max(s'(i) - \hat{R},\ r'(i) - \hat{S})\\ &\qquad \text{for all } i \in [1,p]\}\end{aligned}$$

and hence the summation in the definition of $F^{(60)}(r,s,r',s',V)$ is essentially finite.

LEMMA (7.17) Given any $t \in Z(p)$ and $u \in Z(p)$, for every $q \in [0,p]$ let $T^{(q)}$ be the p by p matrix whose (i,j)th entry is $T_{ij}^{(q)}$ where

$$T_{ij}^{(q)} = \begin{cases} \begin{pmatrix} t(i) - j - q \\ u(i) - j \end{pmatrix} & \text{if } i \in [1,p] \text{ and } j \in [1, p - q] \\[2ex] \begin{pmatrix} t(i) - p \\ u(i) - j \end{pmatrix} & \text{if } i \in [1,p] \text{ and } j \in [p - q + 1, p]. \end{cases}$$

Then we have

$$\det T^{(q)} = \det T^{(p)} \qquad \text{for all } q \in [0,p].$$

Proof: Given any $q \in [0, p - 1]$, we obviously have

$$T_{ij}^{(q+1)} = T_{ij}^{(q)} \qquad \text{for all } i \in [1,p] \text{ and } j \in [p - q, p]$$

and by (4.11') we also see that

$$T_{i,j}^{(q+1)} = T_{i,j}^{(q)} - T_{i,j+1}^{(q)} \qquad \text{for all } i \in [1,p] \text{ and } j \in [1, p - q - 1]$$

and thus

$$[j\text{th column of } T^{(q+1)}]$$

$$= \begin{cases} [j\text{th column of } T^{(q)}] & \text{if } j \in [p - q, p] \\ [j\text{th column of } T^{(q)}] - [(j + 1)\text{th column of } T^{(q)}] & \\ \qquad\qquad\qquad\qquad\qquad \text{if } j \in [1, p - q - 1] & \end{cases}$$

and therefore, because det is an alternating multilinear function of its columns, we get

$$\det T^{(q+1)} = \det T^{(q)}.$$

This being so for all $q \in [0, p - 1]$, by an obvious (decreasing) induction on q we see that

$$\det T^{(q)} = \det T^{(p)} \qquad \text{for all } q \in [0,p].$$

LEMMA (7.18) For every $q \in [1,p]$ and $e \in Z(p)$ let $\hat{T}^{(q)}(e)$ be the q by q matrix whose (i,j)th entry is $\hat{T}^{(q)}_{ij}(e)$ where

$$\hat{T}^{(q)}_{ij}(e) = \binom{0}{e(i) + i - j}.$$

Then given any $q \in [1,p]$ and $e \in Z(p)$ we have that

$$\text{(1')} \quad \det \hat{T}^{(q)}(e) = \begin{cases} 1 & \text{if } e(i) = 0 \text{ for all } i \in [1,q] \\ 0 & \text{if } e(i) > 0 \text{ for some } i \in [1,q]. \end{cases}$$

Proof: Given any $q \in [2,p]$ and $e \in Z[p]$, obviously we have

$$\hat{T}^{(q)}_{ij}(e) = \hat{T}^{(q-1)}_{ij}(e) \qquad \text{for all } i \in [1, q-1] \text{ and } j \in [1, q-1]$$

and by (2.5.1') and (2.5.6') we also have

$$\hat{T}^{(q)}_{qj}(e) = \begin{cases} 1 & \text{if } e(q) = 0 \text{ and } j = q \\ 0 & \text{if } e(q) = 0 \text{ and } j \in [1, q-1] \\ 0 & \text{if } e(q) > 0 \text{ and } j \in [1,q] \end{cases}$$

and therefore upon expanding $\det \hat{T}^{(q)}(e)$ by the last row we see that

$$\text{(11')} \quad \det \hat{T}^{(q)}(e) = \begin{cases} \det \hat{T}^{(q-1)}(e) & \text{if } e(q) = 0 \\ 0 & \text{if } e(q) > 0. \end{cases}$$

By (2.5.1') and (2.5.6') we also see that for all $e \in Z(p)$ we have

$$\text{(12')} \quad \det \hat{T}^{(1)}(e) = \begin{cases} 1 & \text{if } e(1) = 0 \\ 0 & \text{if } e(1) > 0. \end{cases}$$

In view of (12'), and because (11') is true for all $q \in [2,p]$, by an obvious induction we see that given any $q \in [1,p]$ and $e \in Z(p)$ we have (1').

LEMMA (7.19) Given any $e \in Z(p)$, upon letting $\tilde{T}(e)$ be the p by p matrix whose (i,j)th entry is

$$\binom{p - j}{e(i) + i - j}$$

we have that

$$(1^*) \quad \det \tilde{T}(e) = \begin{cases} 1 & \text{if } e(i) = 0 \text{ for all } i \in [1,p] \\ 0 & \text{if } e(i) > 0 \text{ for some } i \in [1,p]. \end{cases}$$

Proof: Using the notation of (7.17) and (7.18), and upon taking t and u to be the members of Z(p) obtained by putting

$$t(i) = p \text{ and } u(i) = e(i) + i \qquad \text{for all } i \in [1,p],$$

we get

$$\tilde{T}(e) = T^{(0)} \qquad \text{and} \qquad \hat{T}^{(p)}(e) = T^{(p)}$$

and hence upon taking $q = 0$ in (7.17) we get

$$\det \tilde{T}(e) = \det \hat{T}^{(p)}(e)$$

and therefore by taking $q = p$ in (7.18) we get (1*).

LEMMA (7.20) If for some $n \in Z$, we have

$$r(i) = p - i \text{ and } s(i) = n - i \text{ and } r'(i) = s'(i) = i$$
$$\text{for all } i \in [1,p],$$

then first, for any given $e \in Z(p)$ we have

$$(1) \quad H^{(20)}(r,s,r',s',e) = \begin{cases} 1 & \text{if } e(i) = 0 \text{ for all } i \in [1,p] \\ 0 & \text{if } e(i) \neq 0 \text{ for some } i \in [1,p], \end{cases}$$

and second, for any given $E \in Z$ we have

$$(2) \quad H_E^{(20)}(r,s,r',s') = \begin{cases} 1 & \text{if } E = 0 \\ 0 & \text{if } E \neq 0, \end{cases}$$

and third, for any given $D \in Z$ we have

(3) $$F_D^{(20)}(r,s,r',s') = \begin{cases} 1 & \text{if } D = 0 \\ 0 & \text{if } D \neq 0, \end{cases}$$

and fourth, for all $V \in Z$ we have

(4) $$F^{(20)}(r,s,r',s',V) = \begin{bmatrix} pn - 1 \\ V \end{bmatrix},$$

and finally, in $Q[V]$ we have

(5) $$F^{[20]}(r,s,r',s',V) = \begin{bmatrix} V \\ pn - 1 \end{bmatrix}.$$

Proof: Let there be given any $n \in Z$ and let us assume that

$r(i) = p - i$ and $s(i) = n - i$ and $r'(i) = s'(i) = i$ for all $i \in [1,p]$.

Now given any $e \in Z(p)$, we clearly have

$$G_{ij}^{(20)}(r,s,r',s',e) = \binom{p - j}{p - i - e(i)}\binom{n - i}{e(i)}$$
for all $i \in [1,p]$ and $j \in [1,p]$

and by (2.5.10') we also have

$$\binom{p - j}{p - i - e(i)} = \binom{p - j}{e(i) + i - j} \quad \text{for all } i \in [1,p] \text{ and } j \in [1,p]$$

and hence we have

$$G_{ij}^{(20)}(r,s,r',s',e) = \binom{p - j}{e(i) + i - j}\binom{n - i}{e(i)}$$
for all $i \in [1,p]$ and $j \in [1,p]$

and therefore, we get

(11) $$H^{(20)}(r,s,r',s',e) = \left[\prod_{i=1}^{p}\binom{n - i}{e(i)}\right]\det \tilde{T}(e)$$

with $\tilde{T}(e)$ as in (7.19) and by (2.5.1') and (2.5.2') we see that

(12) $$\prod_{i=1}^{p}\binom{n - i}{e(i)} = \begin{cases} 1 & \text{if } e(i) = 0 \text{ for all } i \in [1,p] \\ 0 & \text{if } e(i) < 0 \text{ for some } i \in [1,p], \end{cases}$$

and finally, by (11), (12), and (7.19) we get (1).

Clearly, (2) follows from (1). By (2.5.1'), (2.5.2'), and (2.5.6') we see that

$$\binom{0}{D} = \begin{cases} 1 & \text{if } D = 0 \\ 0 & \text{if } D \in Z\setminus\{0\} \end{cases}$$

and therefore (3) follows from (2). Now

$$R = \sum_{i=1}^{p} (p - i) = \sum_{i=1}^{p-1} i$$

and hence

$$R + p = \sum_{i=1}^{p} i$$

and therefore

$$R + p + S = R + p + \sum_{i=1}^{p} (n - i) = \sum_{i=1}^{p} i + \sum_{i=1}^{p} (n - i) = pn$$

and hence

$$R + S + p - 1 = pn - 1$$

and therefore by (3) we get (4) and (5).

LEMMA (7.21) Let there be given any

$$K_{ij}(d^*) \in Q \qquad \text{for all } i \in [1,p] \text{ and } j \in [1,p] \text{ and } d^* \in Z.$$

Then for every $z \in W[p]$ and $D \in Z$ we have

$$\left[\begin{aligned} &\sum_{d\in N(p,D)} \sum_{w\in W(p)} \operatorname{sgn}(w) \prod_{i=1}^{p} K_{iw(i)}(d(z(w)(i))) \\ &= \sum_{d\in N(p,D)} \det\left([K_{ij}(d(i))]_{1\le j\le p}^{1\le i\le p}\right) \end{aligned}\right. \tag{7.21.1}$$

and for every $D \in Z$ we have

$$\left[\begin{aligned} &\sum_{d\in N(p,D)} \det\left([K_{ij}(d(j))]_{1\le j\le p}^{1\le i\le p}\right) \\ &= \sum_{d\in N(p,D)} \det\left([K_{ij}(d(i))]_{1\le j\le p}^{1\le i\le p}\right) \end{aligned}\right. \tag{7.21.2}$$

where, as usual,

$$\left([K_{ij}(d(i))]_{1\le j\le p}^{1\le i\le p}\right) \qquad \text{and} \qquad \left([K_{ij}(d(j))]_{1\le j\le p}^{1\le i\le p}\right)$$

denote the p by p matrices whose (i,j)th entries, respectively, are

$$K_{ij}(d(i)) \qquad \text{and} \qquad K_{ij}(d(j)).$$

Proof: Let there be fixed any $D \in Z$. For every $w \in W(p)$, let f_w: $Z(p) \to Q$ be the map defined by putting

$$(1) \quad f_w(d) = \text{sgn}(w) \prod_{i=1}^{p} K_{iw(i)}(d(i)) \qquad \text{for all } d \in Z(p).$$

For every $d \in Z(p)$ and $g \in W(p)$, let $h[d,g] \in Z(p)$ be defined by putting

$$(2) \quad h[d,g](i) = d(g(i)) \qquad \text{for all } i \in [1,p].$$

Now by (3.12) we see that

$$(3) \quad \left[\begin{array}{l}\text{for every } w \in W(p) \text{ and } g \in W(p) \text{ we have} \\ \displaystyle\sum_{d\in N(p,D)} f_w(d) = \sum_{d\in N(p,D)} f_w(h[d,g]).\end{array}\right.$$

Upon taking $g = z(w)$ in (3) we see that

$$(4) \quad \left[\begin{array}{l}\text{for every } z \in W[p] \text{ and } w \in W(p) \text{ we have} \\ \displaystyle\sum_{d\in N(p,D)} f_w(d) = \sum_{d\in N(p,D)} f_w(h[d,z(w)]).\end{array}\right.$$

Upon summing both sides of (4) as w varies over W(p) we see that

$$(5) \quad \left[\begin{array}{l}\text{for every } z \in W[p] \text{ we have} \\ \displaystyle\sum_{w\in W(p)} \sum_{d\in N(p,D)} f_w(d) = \sum_{w\in W(p)} \sum_{d\in N(p,D)} f_w(h[d,z(w)]).\end{array}\right.$$

Upon interchanging the orders of summation in (5) we see that

$$(6) \quad \left[\begin{array}{l}\text{for every } z \in W[p] \text{ we have} \\ \displaystyle\sum_{d\in N(p,D)} \sum_{w\in W(p)} f_w(d) = \sum_{d\in N(p,D)} \sum_{w\in W(p)} f_w(h[d,z(w)]).\end{array}\right.$$

By (1) and (2) we see that

(7) $\left[\begin{array}{l}\text{for every } z \in W[p] \text{ and } w \in W(p) \text{ and } d \in Z(p) \text{ we have} \\ f_w(h[d,z(w)]) = \operatorname{sgn}(w) \prod_{i=1}^{p} K_{iw(i)}(d(z(w)(i))).\end{array}\right.$

In view of (1), upon expanding the det on the RHS of (7.21.1) by rows, we see that

(8) $\left[\begin{array}{l}\text{for every } d \in Z(p) \text{ we have} \\ \det[K_{ij}(d(i))]_{1\le j\le p}^{1\le i\le p} = \sum_{w\in W(p)} f_w(d).\end{array}\right.$

By (7) we see that

(9) $\left[\begin{array}{l}\text{for every } z \in W[p] \text{ we have} \\ [\text{LHS of } (7.21.1)] = [\text{RHS of } (6)]\end{array}\right.$

and by (8) we see that

(10) [RHS of (7.21.1)] = [LHS of (6)].

By (6), (9), and (10) we get (7.21.1).

Let $z'[3^*] \in W[p]$ be defined by putting

$$z'[3^*](w) = w \quad \text{for all } w \in W(p).$$

Now by the second equation of (2.6.23") we see that for every $d \in Z(p)$ we have

$$\det\left([K_{ij}(d(j))]_{1\le j\le p}^{1\le i\le p}\right) = \sum_{w\in W(p)} \operatorname{sgn}(w) \prod_{i=1}^{p} K_{iw(i)}(d(z'[3^*](w)(i)))$$

and upon summing both sides of the equation above as d varies over N(p,D), we get

(11) [LHS of (7.21.2)] = [LHS of (7.21.1) for $z = z'[3^*]$].

Upon taking $z = z'[3^*]$, by (7.21.1) we get

(12) [LHS of (7.21.1) for $z = z'[3^*]$] = [RHS of (7.21.1)].

Clearly, the RHS of (7.21.1) coincides with the RHS of (7.21.2), and hence by (12) we get

(13) [LHS of (7.21.1) for $z = z'[3^*]$] = [RHS of (7.21.2)].

Finally, by (11) and (13) we get (7.21.2).

PROPOSITION (7.22) For every $z \in W[p]$ and $D \in Z$ we have

$$F_D^{(3z0)}(r,s,r',s') = F_D^{(30)}(r,s,r',s') \tag{7.22.1}$$

and for every $D \in Z$ we have

$$F_D^{(3*0)}(r,s,r',s') = F_D^{(30)}(r,s,r',s'). \tag{7.22.2}$$

Proof: For every $i \in [1,p]$ and $j \in [1,p]$ and $d^* \in Z$ let

$$K_{ij}(d^*) = \begin{bmatrix} s(i) - d^* \\ r(j) \end{bmatrix} \begin{pmatrix} r(j) + r'(j) - s'(i) \\ d^* \end{pmatrix}.$$

Now by (7.13) and the definition of $F_D^{(3z0)}$ we see that for every $z \in W[p]$ and $D \in Z$ we have

$$F_D^{(3z0)}(r,s,r',s')$$

$$= \sum_{d \in N(p,D)} \sum_{w \in W(p)} \operatorname{sgn}(w) \prod_{i=1}^{p} K_{ij}(d(z(w)(i)))$$

and by (7.13) and the definitions of $F_D^{(30)}$ and $F_D^{(3*0)}$ we see that for every $D \in Z$ we have

$$F_D^{(30)}(r,s,r',s') = \sum_{d \in N(p,D)} \det G^{(30)}(r,s,r',s')$$

and

$$F_D^{(3*0)}(r,s,r',s') = \sum_{d \in N(p,D)} \det G^{(3*0)}(r,s,r',s')$$

where $G^{(30)}(r,s,r',s')$ and $G^{(3*0)}(r,s,r',s')$ are the p by p matrices whose (i,j)th entries, respectively, are

$$K_{ij}(d(i)) \quad \text{and} \quad K_{ij}(d(j))$$

and therefore (7.22.1) and (7.22.2) follow, respectively, from (7.21.1) and (7.21.2).

REMARK (7.23) We shall now discuss some rules for changing the order of summation in a multiple sum; these rules will be used in the proof of Lemma (7.24); *the reader may wish to skip this entire Remark and return to it if and when necessary for reference.*

Let W*, D*, E*, and Y be any sets, and

let there be given any $y(E) \subset E^*$ for every $E \in Y$, such that for all $E_1 \neq E_2$ in Y we have $y(E_1) \cap y(E_2) = \emptyset$, and such that $E^* = \bigcup_{E \in Y} y(E)$.

Also, let there be given any

$$h(E) \in Q \quad \text{for all } E \in Y$$

and let there be given any

$$g(w,e) \in Q \quad \text{for all } w \in W^* \text{ and } e \in E^*$$

and let us put

$$\hat{g} = \text{card}(\{(w,e) \text{ with } w \in W^* \text{ and } e \in E^* : g(w,e) \neq 0\})$$

and let there be given any

$$f(w,d,e) \in Q \quad \text{for all } w \in W^* \text{ and } d \in D^* \text{ and } e \in E^*$$

and let us put

$$\hat{f} = \text{card}(\{(w,d,e) \text{ with } w \in W^* \text{ and } d \in D^* \text{ and } e \in E^* : f(w,d,e) \neq 0\}).$$

Now the said rules follow from two basic, but obvious principles, the first of which says that

if $G: E^* \to Q$ is any map such that $\text{card}(\{e \in E^* : G(e) \neq 0\}) < \infty$, then $\sum_{e \in E^*} G(e) = \sum_{E \in Y} \sum_{e \in y(E)} G(e)$ where all the summations are essentially finite, (7.23.1)

and the second of which says that

if $F: D^* \to Q$ is any map such that $\text{card}(\{d \in D^* : F(d) \neq 0\}) < \infty$ and if H is any fixed element of Q, then $\sum_{d \in D^*} F(d)H = H \sum_{d \in D^*} F(d)$ where both the summations are essentially finite. (7.23.2)

As an application of the principles above, first we claim that

$$\begin{cases} \text{if } \hat{g} < \infty, \text{ then } \sum_{w \in W^*} \sum_{e \in E^*} g(w,e) = \sum_{e \in E^*} \sum_{w \in W^*} g(w,e) \\ \text{where all the summations are essentially finite.} \end{cases} \tag{7.23.3}$$

To prove this, let Σ' stand for summation over the product of W^* and E^*; moreover, in (7.23.1), let E^* be replaced by the said product, now on the one hand, in (7.23.1), upon replacing Y by W^* and for every $w \in W^*$, upon letting $y(w)$ be the set of all pairs (w',e') with $w' = w$ and $e' \in E^*$, by (7.23.1) we see that

$$(1) \quad \begin{cases} \text{if } \hat{g} < \infty, \text{ then } \Sigma' g(w,e) = \sum_{w \in W^*} \sum_{e \in E^*} g(w,e) \\ \text{where all the summations are essentially finite;} \end{cases}$$

on the other hand, in (7.23.1), upon replacing Y by E^* and for every $e \in E^*$, upon letting $y(e)$ be the set of all pairs (w',e') with $w' \in W^*$ and $e' = e$, by (7.23.1) we see that

$$(2) \quad \begin{cases} \text{if } \hat{g} < \infty, \text{ then } \Sigma' g(w,e) = \sum_{e \in E^*} \sum_{w \in W^*} g(w,e) \\ \text{where all the summations are essentially finite.} \end{cases}$$

By (1) and (2) we get (7.23.3).

[Clearly, (7.23.3) is a *discrete incarnation* of Fubini's Theorem

$$\int_w \int_e = \int_e \int_w$$

and hence (7.23.3) may be called *Finite Fubini Rule*; note that in this connection, the integral analog of (7.23.1) is the statement which says that a multiple integral equals a repeated integral.]

Second, we claim that

$$\begin{cases} \text{if } \hat{g} < \infty, \text{ then } \sum_{w \in W^*} \sum_{e \in E^*} g(w,e) = \sum_{E \in Y} \sum_{e \in y(E)} \sum_{w \in W^*} g(w,e) \\ \text{where all the summations are essentially finite.} \end{cases} \tag{7.23.4}$$

Namely, if $\hat{g} < \infty$, then upon letting

$$(1) \quad G'(e) = \sum_{w \in W^*} g(w,e) \qquad \text{for all } e \in E^*,$$

by (7.23.3) we see that the summation above is essentially finite and $\operatorname{card}(\{e \in E^* : G'(e) \neq 0\}) < \infty$, and hence by (7.23.1) we see that

$$(2) \quad \sum_{e \in E^*} G'(e) = \sum_{E \in Y} \sum_{e \in y(E)} G'(e)$$

where all the summations are essentially finite; now by substituting (1) in (2) we see that

$$(3) \quad \left[\begin{array}{l} \text{if } \hat{g} < \infty, \text{ then} \\ \displaystyle\sum_{e \in E^*} \sum_{w \in W^*} g(w,e) = \sum_{E \in Y} \sum_{e \in y(E)} \sum_{w \in W^*} g(w,e) \\ \text{where all the summations are essentially finite.} \end{array}\right.$$

By (3) and (7.23.3) we get (7.23.4).

Third, we claim that

$$\left[\begin{array}{l} \text{if } \hat{g} < \infty, \text{ then} \\ \displaystyle\sum_{E \in Y} \sum_{e \in y(E)} \sum_{w \in W^*} g(w,e) = \sum_{w \in W^*} \sum_{E \in Y} \sum_{e \in y(E)} g(w,e) \\ \text{where all the summations are essentially finite.} \end{array}\right. \qquad (7.23.5)$$

Namely, if $\hat{g} < \infty$, then upon fixing any $w \in W^*$ and upon letting

$$(11) \quad G''(e) = g(w,e) \qquad \text{for all } e \in E^*$$

we clearly have $\operatorname{card}(\{e \in E^* : G''(e) \neq 0\}) < \infty$ and hence by (7.23.1) we see that

$$(12) \quad \sum_{e \in E^*} G''(e) = \sum_{E \in Y} \sum_{e \in y(E)} G''(e)$$

where all the summations are essentially finite; by substituting (11) in (12) we deduce that

$$(13) \quad \left[\begin{array}{l} \text{if } \hat{g} < \infty, \text{ then for every } w \in W^* \text{ we have} \\ \displaystyle\sum_{e \in E^*} g(w,e) = \sum_{E \in Y} \sum_{e \in y(E)} g(w,e) \\ \text{where all the summations are essentially finite.} \end{array}\right.$$

Now

$$(14)\quad\begin{cases}\text{upon letting } G^*(w) = \sum_{e\in E^*} g(w,e) \text{ for all } w \in W^*,\\ \text{by (7.23.3) we see that if } \hat{g} < \infty, \text{ then the}\\ \text{summation is essentially finite and}\\ \text{card}(\{w \in W^* : G^*(w) \neq 0\}) < \infty.\end{cases}$$

In view of (14), upon summing both sides of (13) as w varies over W^*, we conclude that

$$(15)\quad\begin{cases}\text{if } \hat{g} < \infty, \text{ then } \sum_{w\in W^*} \sum_{e\in E^*} g(w,e) = \sum_{w\in W^*} \sum_{E\in Y} \sum_{e\in y(E)} g(w,e)\\ \text{where all the summations are essentially finite.}\end{cases}$$

By (15) and (7.23.4) we get (7.23.5).

Fourth, we claim that

$$\begin{cases}\text{if } \hat{g} < \infty, \text{ then}\\ \sum_{w\in W^*} \sum_{E\in Y} \sum_{e\in y(E)} g(w,e)h(E) = \sum_{E\in Y} h(E) \sum_{e\in y(E)} \sum_{w\in W^*} g(w,e)\\ \text{where all the summations are essentially finite.}\end{cases} \qquad (7.23.6)$$

Namely, upon letting

$$(21)\quad\begin{cases}g'(w,e) = g(w,e)h(E)\\ \text{for all } w \in W^* \text{ and } E \in Y \text{ and } e \in y(E)\end{cases}$$

it is obvious that

$$\begin{cases}\text{if } \hat{g} < \infty, \text{ then}\\ \text{card}(\{(w,e) \text{ with } w \in W^* \text{ and } e \in E^* : g'(w,e) \neq 0\}) < \infty\end{cases}$$

and hence by (7.23.5) we see that

$$(22)\quad\begin{cases}\text{if } \hat{g} < \infty, \text{ then}\\ \sum_{E\in Y} \sum_{e\in y(E)} \sum_{w\in W^*} g'(w,e) = \sum_{w\in W^*} \sum_{E\in Y} \sum_{e\in y(E)} g'(w,e)\\ \text{where all the summations are essentially finite.}\end{cases}$$

By substituting (21) in (22) we see that

$$(23)\quad\begin{cases}\text{if } \hat{g} < \infty, \text{ then}\\ \sum_{w\in W^*} \sum_{E\in Y} \sum_{e\in y(E)} g(w,e)h(E) = \sum_{E\in Y} \sum_{e\in y(E)} \sum_{w\in W^*} g(w,e)h(E)\\ \text{where all the summations are essentially finite.}\end{cases}$$

If $\hat{g} < \infty$, then upon fixing any $E \in Y$ and $e \in y(E)$ and

$$(24)\quad \begin{cases} \text{upon letting } H' = h(E) \\ \text{and upon letting } F'(w) = g(w,e) \text{ for all } w \in W^*, \end{cases}$$

we clearly have $\operatorname{card}(\{w \in W^* : F'(w) \neq 0\}) < \infty$ and hence by (7.23.2) we deduce that

$$(25)\quad \sum_{w \in W^*} F'(w)H' = H' \sum_{w \in W^*} F'(w)$$

where both the summations are essentially finite; now by substituting (24) in (25) we conclude that

$$(26)\quad \begin{cases} \text{if } \hat{g} < \infty, \text{ then for every } E \in Y \text{ and } e \in y(E) \text{ we have} \\ \displaystyle\sum_{w \in W^*} g(w,e)h(E) = h(E) \sum_{w \in W^*} g(w,e) \\ \text{where both the summations are essentially finite.} \end{cases}$$

Again, if $\hat{g} < \infty$, then upon fixing any $E \in Y$ and

$$(27)\quad \begin{cases} \text{upon letting } H^* = h(E) \text{ and upon letting} \\ F^*(e) = \displaystyle\sum_{w \in W^*} g(w,e) \text{ for all } e \in y(E), \end{cases}$$

by (7.23.3) we see that the summation in (27) is essentially finite and $\operatorname{card}(\{e \in y(E) : F^*(e) \neq 0\}) < \infty$ and hence by (7.23.2) we deduce that

$$(28)\quad \sum_{e \in y(E)} H^*F^*(e) = H^* \sum_{e \in y(E)} F^*(e)$$

where both the summations are essentially finite; now by substituting (27) in (28) we conclude that

$$(29)\quad \begin{cases} \text{if } \hat{g} < \infty, \text{ then for every } E \in Y \text{ we have} \\ \displaystyle\sum_{e \in y(E)} h(E) \sum_{w \in W^*} g(w,e) = h(E) \sum_{e \in y(E)} \sum_{w \in W^*} g(w,e) \\ \text{where all the summations are essentially finite.} \end{cases}$$

By (26) and (29) we see that

$$(30)\quad \begin{cases} \text{if } \hat{g} < \infty, \text{ then} \\ \displaystyle\sum_{e \in y(E)} \sum_{w \in W^*} g(w,e)h(E) = h(E) \sum_{e \in y(E)} \sum_{w \in W^*} g(w,e) \\ \text{where all the summations are essentially finite.} \end{cases}$$

By (23) we see that

$$(31)\quad \left[\begin{array}{l}\text{if } \hat{g} < \infty, \text{ then in the expression}\\ \sum_{E\in Y}\ \sum_{e\in y(E)}\ \sum_{w\in W^*} g(w,e)h(E)\\ \text{all the summations are essentially finite.}\end{array}\right.$$

In view of (31), upon summing both sides of (30) as E varies over Y, we conclude that

$$(32)\quad \left[\begin{array}{l}\text{if } \hat{g} < \infty, \text{ then}\\ \sum_{E\in Y}\ \sum_{e\in y(E)}\ \sum_{w\in W^*} g(w,e)h(E) = \sum_{E\in Y} h(E) \sum_{e\in y(E)}\ \sum_{w\in W^*} g(w,e)\\ \text{where all the summations are essentially finite.}\end{array}\right.$$

Now by (23) and (32) we get (7.23.6).

Fifth, we claim that

$$\left[\begin{array}{l}\text{if } \hat{f} < \infty, \text{ then}\\ \sum_{d\in D^*}\ \sum_{e\in E^*}\ \sum_{w\in W^*} f(w,d,e) = \sum_{w\in W^*}\ \sum_{d\in D^*}\ \sum_{e\in E^*} f(w,d,e)\\ \text{where all the summations are essentially finite.}\end{array}\right. \tag{7.23.7}$$

This follows from (7.23.5) upon replacing E* by the product of D* and E*, and upon replacing Y by D*, and for every $d \in D^*$, upon letting y(d) be the set of all pairs (d',e') with $d' = d$ and $e' \in E^*$.

Concerning the condition on $\hat{g}$ used in (7.23.3) to (7.23.6), we note that

$$\left[\begin{array}{l}\hat{g} < \infty \text{ iff for every } w \in W^* \text{ we have}\\ \mathrm{card}(\{e \in E^* : g(w,e) \neq 0\}) < \infty, \text{ and we have}\\ \mathrm{card}(\{w \in W^* : \text{there exists } e \in E^* \text{ for which}\\ g(w,e) \neq 0\}) < \infty.\end{array}\right. \tag{7.23.8}$$

Concerning the condition on $\hat{g}$ used in (7.23.3) to (7.23.6), we also note that

$$\left[\begin{array}{l}\text{if } \mathrm{card}(W^*) < \infty, \text{ and if for every } w \in W^* \text{ we have}\\ \mathrm{card}(\{e \in E^* : g(w,e) \neq 0\}) < \infty, \text{ then we have } \hat{g} < \infty.\end{array}\right. \tag{7.23.9}$$

Concerning the condition on $\hat{f}$ used in (7.23.7), we note that

$$\left[\begin{array}{l}\text{if } \operatorname{card}(W^*) < \infty, \text{ and if for every } w \in W^* \text{ and } d \in D^* \\ \text{we have } \operatorname{card}(\{e \in E^* : f(w,d,e) \neq 0\}) < \infty, \text{ and if} \\ \text{for every } w \in W^* \text{ we have } \operatorname{card}(\{d \in D^* : \text{there exists} \\ e \in E^* \text{ for which } f(w,d,e) \neq 0\}) < \infty \text{ then we have} \\ \hat{f} < \infty. \end{array}\right. \qquad (7.23.10)$$

We shall now restate the rules in a form suitable for the proof of (7.24). First, by (7.23.4) and (7.23.8), upon changing W* to D*, we deduce that

$$\left[\begin{array}{l}\text{if } w \in W^* \text{ is such that for every } d \in D^* \text{ we have} \\ \operatorname{card}(\{e \in E^* : f(w,d,e) \neq 0\}) < \infty \text{ and we have} \\ \operatorname{card}(\{d \in D^* : \text{there exists } e \in E^* \text{ for which} \\ f(w,d,e) = 0\}) < \infty, \text{ then we have} \\ \sum_{d \in D^*} \sum_{e \in E^*} f(w,d,e) = \sum_{E \in Y} \sum_{e \in y(E)} \sum_{d \in D^*} f(w,d,e) \\ \text{where all the summations are essentially finite} \end{array}\right.$$

and now, upon summing both sides of the equation above as w varies over W*, we conclude that

$$\left[\begin{array}{l}\text{if } \operatorname{card}(W^*) < \infty, \text{ and if for every } w \in W^* \text{ and } d \in D^* \\ \text{we have } \operatorname{card}(\{e \in E^* : f(w,d,e) \neq 0\}) < \infty, \text{ and if} \\ \text{for every } w \in W^* \text{ we have } \operatorname{card}(\{d \in D^* : \text{there} \\ \text{exists } e \in E^* \text{ for which } f(w,d,e) \neq 0\}) < \infty, \text{ then} \\ \sum_{w \in W^*} \sum_{d \in D^*} \sum_{e \in E^*} f(w,d,e) = \sum_{w \in W^*} \sum_{E \in Y} \sum_{e \in y(E)} \sum_{d \in D^*} f(w,d,e) \\ \text{where all the summations are essentially finite.} \end{array}\right. \qquad (7.23.11)$$

Second, by (7.23.6) and (7.23.9) we see that

$$\left[\begin{array}{l}\text{if } \operatorname{card}(W^*) < \infty, \text{ and if for every } w \in W^* \text{ we have} \\ \operatorname{card}(\{e \in E^* : g(w,e) \neq 0\}) < \infty, \text{ then we have} \\ \sum_{w \in W^*} \sum_{E \in Y} \sum_{e \in y(E)} g(w,e)h(E) = \sum_{E \in Y} h(E) \sum_{e \in y(E)} \sum_{w \in W^*} g(w,e) \\ \text{where all the summations are essentially finite.} \end{array}\right. \qquad (7.23.12)$$

Third, by (7.23.7) and (7.23.10) we see that

$$\left[\begin{array}{l}\text{if } \operatorname{card}(W^*) < \infty, \text{ and if for every } w \in W^* \text{ and } d \in D^* \\ \text{we have } \operatorname{card}(\{e \in E^* : f(w,d,e) \neq 0\}) < \infty, \text{ and if} \end{array}\right.$$

$$\left[\begin{array}{l}\text{for every } w \in W^* \text{ we have } \operatorname{card}(\{d \in D^* : \text{there} \\ \text{exists } e \in E^* \text{ for which } f(w,d,e) \neq 0\}) < \infty, \\ \text{then we have} \\ \sum_{d \in D^*} \sum_{e \in E^*} \sum_{w \in W^*} f(w,d,e) = \sum_{w \in W^*} \sum_{d \in D^*} \sum_{e \in E^*} f(w,d,e) \\ \text{where all the summations are essentially finite.}\end{array}\right. \tag{7.23.13}$$

Finally, we note that

$$\left[\begin{array}{l}\text{in the proof of (7.24), while using (7.23.11) and} \\ \text{(7.23.12) and (7.23.13), we shall have } W^* = W(p) \\ \text{and } D^* = Z(p,D) \text{ and } E^* = Z(p) \text{ and } Y = Z \text{ and} \\ y(E) = Z(p,E).\end{array}\right. \tag{7.23.14}$$

LEMMA (7.24) Let there be given any $A_{ij}(e^*) \in Q$ for all $i \in [1,p]$ and $j \in [1,p]$ and $e^* \in Z$ such that

(1) $\operatorname{card}(\{e^* \in Z : A_{ij}(e^*) \neq 0\}) < \infty$ for all $i \in [1,p]$ and $j \in [1,p]$.

Also, let there be given any

$$\left[\begin{array}{ll}B^*(d^*,e^*) \in Q & \text{for all } d^* \in Z \text{ and } e^* \in Z \text{ and} \\ h^*(d^*,e^*) \in Q & \text{for all } d^* \in Z \text{ and } e^* \in Z\end{array}\right.$$

such that

(2) for every $d^* \in Z \setminus N$ and $e^* \in Z$ we have $B^*(d^*,e^*) = 0$

and such that

$$\text{(3)}\quad \left[\begin{array}{l}\text{for every } D \in Z \text{ and } D' \in Z \text{ and } d' \in Z(p,D') \text{ and} \\ E' \in Z \text{ and } e' \in Z(p,E') \text{ we have} \\ \displaystyle\sum_{d \in Z(p,D)} \prod_{k=1}^{p} B^*(d(k) + d'(k), e'(k)) = h^*(D + D', E')\end{array}\right.$$

[note that as a consequence of (2),

$$\text{(4)}\quad \left[\begin{array}{l}\text{for every } D \in Z \text{ and } d' \in Z(p) \text{ and } e' \in Z(p) \text{ we have} \\ \operatorname{card}\left(\left\{d \in Z(p,D) : \displaystyle\prod_{k=1}^{p} B^*(d(k) + d'(k), e'(k)) \neq 0\right\}\right) < \infty\end{array}\right.$$

and hence the summation in (3) is essentially finite].

Moreover, let there be given any

$$q^* \in Z \quad \text{and} \quad q^{**} \in Z \quad \text{and} \quad q'' \in Z,$$

and let there be given any

$$\begin{cases} r^*[k] \in Z,\ r^{**}[k] \in Z,\ r''[k] \in Z, \\ s^*[k] \in Z,\ s^{**}[k] \in Z, \text{ and } s''[k] \in Z, \\ \text{for all } k \in [1,p], \end{cases}$$

and let us put

$$(5) \quad R^* = \sum_{k=1}^{p} r^*[k] \quad \text{and} \quad S^* = \sum_{k=1}^{p} s^*[k]$$

and let us put

$$(6) \quad \begin{cases} R^{**} = \sum_{k=1}^{p} r^{**}[k] \quad \text{and} \quad S^{**} = \sum_{k=1}^{p} s^{**}[k] \\ \text{and} \quad R'' = \sum_{k=1}^{p} r''[k] \quad \text{and} \quad S'' = \sum_{k=1}^{p} s''[k] \end{cases}$$

and let us put

$$(7) \quad \begin{cases} B_{ij}(d^*,e^*) = (-1)^{q^*e^*+r^*[j]+s^*[i]} B^*(d^* + q^{**}e^* \\ \qquad\qquad + r^{**}[j] + s^{**}[i],\ q''e^* + r''[j] + s''[i]) \\ \text{for all } i \in [1,p] \text{ and } j \in [1,p] \text{ and } d^* \in Z \text{ and } e^* \in Z \end{cases}$$

and let us put

$$(8) \quad \begin{cases} h(D,E) = (-1)^{q^*E+R^*+S^*} h^*(D + q^{**}E + R^{**} + S^{**},\ q''E + R'' + S'') \\ \text{for all } D \in Z \text{ and } E \in Z. \end{cases}$$

Finally, let there be fixed any $D \in Z$. Now we have that

$$\begin{cases} \sum_{d \in Z(p,D)} \det\left(\left[\sum_{e^* \in Z} A_{ij}(e^*) B_{ij}(d(i),e^*)\right]_{1 \le j \le p}^{1 \le i \le p}\right) \\ = \sum_{E \in Z} h(D,E) \sum_{e \in Z(p,E)} \det\left([A_{ij}(e(i))]_{1 \le j \le p}^{1 \le i \le p}\right) \end{cases} \tag{7.24.1}$$

and we have that

$$\left[\begin{aligned}&\sum_{d\in Z(p,D)} \det\left(\left[\sum_{e^*\in Z} A_{ij}(e^*)B_{ij}(d(j),e^*)\right]_{1\le j\le p}^{1\le i\le p}\right)\\ &= \sum_{E\in Z} h(D,E) \sum_{e\in Z(p,E)} \det\left([A_{ij}(e(j))]_{1\le j\le p}^{1\le i\le p}\right)\end{aligned}\right. \tag{7.24.2}$$

and we have that

$$\left[\begin{aligned}&\sum_{E\in Z} h(D,E) \sum_{e\in Z(p,E)} \det\left([A_{ij}(e(i))]_{1\le j\le p}^{1\le i\le p}\right)\\ &= \sum_{E\in Z} h(D,E) \sum_{e\in Z(p,E)} \det\left([A_{ij}(e(j))]_{1\le j\le p}^{1\le i\le p}\right)\end{aligned}\right. \tag{7.24.3}$$

and we have that

$$\left[\begin{aligned}&\textit{if } (*),\ \textit{then}\\ &\sum_{E\in Z} h(D,E) \sum_{e\in Z(p,E)} \det\left([A_{ij}(e(i) + r'(j) - s'(i))]_{1\le j\le p}^{1\le i\le p}\right)\\ &= \sum_{E\in Z} h(D,E) \sum_{e\in Z(p,E)} \det\left([A_{ij}(e(i))]_{1\le j\le p}^{1\le i\le p}\right)\end{aligned}\right. \tag{7.24.4}$$

and we have that

$$\left[\begin{aligned}&\textit{if } (*),\ \textit{then}\\ &\sum_{E\in Z} h(D,E) \sum_{e\in Z(p,E)} \det\left([A_{ij}(e(j) - s'(i) + r'(j))]_{1\le j\le p}^{1\le i\le p}\right)\\ &= \sum_{E\in Z} h(D,E) \sum_{e\in Z(p,E)} \det\left([A_{ij}(e(j))]_{1\le j\le p}^{1\le i\le p}\right)\end{aligned}\right. \tag{7.24.5}$$

where all the summations are essentially finite and where, as usual,

$$\left(\left[\sum_{e^*\in Z} A_{ij}(e^*)B_{ij}(d(i),e^*)\right]_{1\le j\le p}^{1\le i\le p}\right)$$

and

$$\left([A_{ij}(e(i))]_{1\le j\le p}^{1\le i\le p}\right) \quad \text{and so on}$$

denote the p by p matrices whose (i,j)th entries, respectively, are

$$\sum_{e^*\in Z} A_{ij}(e^*)B_{ij}(d(i),e^*) \text{ and } A_{ij}(e(i)) \text{ and so on.}$$

Proof: By (1) we see that

$$(11)\quad \left[\begin{array}{l}\text{for every } d \in Z(p) \text{ and } i \in [1,p] \text{ and } j \in [1,p] \text{ we have}\\ \operatorname{card}(\{e^* \in Z : A_{ij}(e^*)B_{ij}(d(i),e^*) \neq 0\}) < \infty.\end{array}\right.$$

By (1), (2), and (7) we see that

$$(12)\quad \left[\begin{array}{l}\operatorname{card}(\{d \in Z(p,D) : \text{there exists } j \in [1,p] \text{ and } e^* \in Z\\ \qquad\qquad \text{such that for every } i \in [1,p]\\ \qquad\qquad \text{we have } A_{ij}(e^*)B_{ij}(d(i)e^*) \neq 0\}) < \infty.\end{array}\right.$$

In view of (11) and (12),

$$(13)\quad \left[\begin{array}{l}\text{upon letting } \hat{L}_{ij}(d) = \sum_{e^* \in Z} A_{ij}(e^*)B_{ij}(d(i),e^*)\\ \text{we see that for every } d \in Z(p) \text{ and } i \in [1,p] \text{ and } j \in [1,p],\\ \text{the summation above is essentially finite, and}\\ \operatorname{card}(\{d \in Z(p,D) : \text{there exists } j \in [1,p] \text{ such that}\\ \qquad \text{for every } i \in [1,p] \text{ we have } \hat{L}_{ij}(d) \neq 0\}) < \infty.\end{array}\right.$$

Let us note that

$$(14)\quad \left[\begin{array}{l}\text{if all the entries either in a row or a column of}\\ \text{a } p \text{ by } p \text{ matrix are zero, then its det is zero.}\end{array}\right.$$

In view of (13) and (14)

$$(15)\quad \left[\begin{array}{l}\text{upon letting } \hat{L}(d) = \det\left(\left[\sum_{e^* \in Z} A_{ij}(e^*)B_{ij}(d(i),e^*)\right]_{1\le j\le p}^{1\le i\le p}\right)\\ \text{we see that for every } d \in Z(p) \text{ all the summations above}\\ \text{are essentially finite, and}\\ \operatorname{card}(\{d \in Z(p,D) : \hat{L}(d) \neq 0\}) < \infty.\end{array}\right.$$

Given any $d \in Z(p)$, upon letting

$$K_{ij}(e^*) = A_{ij}(e^*)B_{ij}(d(i),e^*)$$
$$\text{for all } i \in [1,p] \text{ and } j \in [1,p] \text{ and } e^* \in Z,$$

in view of (11), by taking $X = Z$ in (3.38) we see that

$$\det\left(\left[\sum_{e^* \in Z} A_{ij}(e^*)B_{ij}(d(i),e^*)\right]_{1\le j\le p}^{1\le i\le p}\right)$$
$$= \sum_{e \in Z(p)} \det\left([A_{ij}(e(i))B_{ij}(d(i),e(i))]_{1\le j\le p}^{1\le i\le p}\right)$$

and

$$\det\left(\left[\sum_{e^*\in Z} A_{ij}(e^*)B_{ij}(d(i),e^*)\right]_{1\le j\le p}^{1\le i\le p}\right)$$

$$= \sum_{e\in Z(p)} \det\left([A_{ij}(e(j))B_{ij}(d(i),e(j))]_{1\le j\le p}^{1\le i\le p}\right)$$

where all the summations are essentially finite, and now, in view of (15), upon summing both sides of each of the two equations above as d varies over Z(p,D), we conclude that

(16) $\left[\begin{array}{l}\text{LHS of (7.24.1)}\\ = \displaystyle\sum_{d\in Z(p,D)}\sum_{e\in Z(p)} \det\left([A_{ij}(e(i))B_{ij}(d(i),e(i))]_{1\le j\le p}^{1\le i\le p}\right)\\ \text{where all the summations are essentially finite}\end{array}\right.$

and

(17) $\left[\begin{array}{l}\text{LHS of (7.24.1)}\\ = \displaystyle\sum_{d\in Z(p,D)}\sum_{e\in Z(p)} \det\left([A_{ij}(e(j))B_{ij}(d(i),e(j))]_{1\le j\le p}^{1\le i\le p}\right)\\ \text{where all the summations are essentially finite.}\end{array}\right.$

Upon expanding determinants we see that

(21) $\left[\begin{array}{l}\text{for every } e \in Z(p) \text{ we have}\\ \det\left([A_{ij}(e(i))]_{1\le j\le p}^{1\le i\le p}\right) = \displaystyle\sum_{w\in W(p)} g(w,e)\end{array}\right.$

and

(22) $\left[\begin{array}{l}\text{for every } d \in Z(p) \text{ and } e \in Z(p) \text{ we have}\\ \det\left([A_{ij}(e(i))B_{ij}(d(i),e(i))]_{1\le j\le p}^{1\le i\le p}\right) = \displaystyle\sum_{w\in W(p)} f(w,d,e)\end{array}\right.$

where

(23) $\left[\begin{array}{l}\text{for every } w \in W(p) \text{ and } e \in Z(p) \text{ we have put}\\ g(w,e) = \operatorname{sgn}(w) \displaystyle\prod_{i=1}^{p} A_{iw(i)}(e(i))\end{array}\right.$

and where

$$(24)\quad \left[\begin{array}{l}\text{for every } w \in W(p) \text{ and } d \in Z(p) \text{ and } e \in Z(p) \text{ we have put} \\ f(w,d,e) = g(w,e) \prod_{i=1}^{p} B_{iw(i)}(d(i),e(i)).\end{array}\right.$$

By (1) and (23) we see that

$$(25)\quad \left[\begin{array}{l}\text{for every } w \in W(p) \text{ we have} \\ \mathrm{card}(\{e \in Z(p) : g(w,e) \neq 0\}) < \infty.\end{array}\right.$$

By (11), (23), and (24) we see that

$$(26)\quad \left[\begin{array}{l}\text{for every } w \in W(p) \text{ and } d \in Z(p,D) \text{ we have} \\ \mathrm{card}(\{e \in Z(p) : f(w,d,e) \neq 0\}) < \infty.\end{array}\right.$$

By (12), (23), and (24) we see that

$$(27)\quad \left[\begin{array}{l}\text{for every } w \in W(p) \text{ we have } \mathrm{card}(\{d \in Z(p,D) : \text{there} \\ \qquad\qquad \text{exists } e \in Z(p) \text{ for which } f(w,d,e) \neq 0\}) < \infty.\end{array}\right.$$

In view of (7.23.13) and (7.23.14), by (22), (26), and (27) we see that

$$(28)\quad \left[\begin{array}{l}[\text{RHS of (16)}] = \sum_{w\in W(p)} \sum_{d\in Z(p,D)} \sum_{e\in Z(p)} f(w,d,e) \\ \text{where all the summations are essentially finite.}\end{array}\right.$$

In view of (7.23.11) and (7.23.14), by (26) and (27) we see that

$$(29)\quad \left[\begin{array}{l}[\text{RHS of (28)}] = \sum_{w\in W(p)} \sum_{E\in Z} \sum_{e\in Z(p,E)} \sum_{d\in Z(p,D)} f(w,d,e) \\ \text{where all the summations are essentially finite.}\end{array}\right.$$

By (5), (7), and (24) we see that

$$(30)\quad \left[\begin{array}{l}\text{for every } w \in W(p) \text{ and } E \in Z \text{ and } e \in Z(p,E) \text{ we have} \\ f(w,d,e) = g(w,e)(-1)^{q^*E+R^*+S^*} \prod_{i=1}^{p} B^*(d(i) + \hat{d}[e,w](i), \hat{e}[w](i))\end{array}\right.$$

where

$$(31)\quad \left[\begin{array}{l}\text{for every } w \in W(p) \text{ and } e \in Z(p) \text{ we have let } \hat{d}[e,w] \text{ and } \hat{e}[w] \\ \text{be the members of } Z(p) \text{ obtained by putting for all } i \in [1,p], \\ \hat{d}[e,w](i) = q^{**}e(i) + r^{**}[w(i)] + s^{**}[i] \text{ and} \\ \hat{e}[w](i) = q''e(i) + r''[w(i)] + s''[i].\end{array}\right.$$

By (6) and (31) we see that

$$(32)\quad \left[\begin{array}{l}\text{for every } w \in W(p) \text{ and } E \in Z \text{ and } e \in Z(p,E) \text{ we have}\\ \hat{d}[e,w] \in Z(p,\ q^{**}E + R^{**} + S^{**})\\ \text{and } \hat{e}[w] \in Z(p,\ q''E + R'' + S'').\end{array}\right.$$

By (3), (4), (8), (30), and (32) we see that

$$(33)\quad \left[\begin{array}{l}\text{for every } w \in W(p) \text{ and } E \in Z \text{ and } e \in Z(p,E) \text{ we have}\\ \displaystyle\sum_{d\in Z(p,D)} f(w,d,e) = g(w,e)h(D,E)\\ \text{where the summation is essentially finite.}\end{array}\right.$$

By (29) and (33) we see that

$$(34)\quad \left[\begin{array}{l}[\text{RHS of } (29)] = \displaystyle\sum_{w\in W(p)}\ \sum_{E\in Z}\ \sum_{e\in Z(p,E)} g(w,e)h(D,E)\\ \text{where all the summations are essentially finite.}\end{array}\right.$$

In view of (7.23.12) and (7.23.14), by (21) and (25) we see that

$$(35)\quad \left[\begin{array}{l}[\text{RHS of } (34)] = [\text{RHS of } (7.24.1)]\\ \text{where all the summations are essentially finite.}\end{array}\right.$$

Now by (16), (28), (29), (34), and (35) we conclude that

$$(36)\quad \left[\begin{array}{l}[\text{LHS of } (7.24.1)] = [\text{RHS of } (7.24.1)]\\ \text{where all the summations are essentially finite.}\end{array}\right.$$

Upon expanding determinants we see that

$$(41)\quad \left[\begin{array}{l}\text{for every } e \in Z(p) \text{ we have}\\ \det\left\{[A_{ij}(e(j))]_{1\le j\le p}^{1\le i\le p}\right\} = \displaystyle\sum_{w\in W(p)} g'(w,e)\end{array}\right.$$

and

$$(42)\quad \left[\begin{array}{l}\text{for every } d \in Z(p) \text{ and } e \in Z(p) \text{ we have}\\ \det\left\{[A_{ij}(e(j))B_{ij}(d(i),e(j))]_{1\le j\le p}^{1\le i\le p}\right\} = \displaystyle\sum_{w\in W(p)} f'(w,d,e)\end{array}\right.$$

where

$$(43)\quad \left[\begin{array}{l}\text{for every } w \in W(p) \text{ and } e \in Z(p) \text{ we have put}\\ g'(w,e) = \operatorname{sgn}(w) \displaystyle\prod_{i=1}^{p} A_{iw(i)}(e(w(i))\end{array}\right.$$

and where

$$(44)\quad \left[\begin{array}{l}\text{for every } w \in W(p) \text{ and } d \in Z(p) \text{ and } e \in Z(p) \text{ we have put} \\ f'(w,d,e) = g'(w,e) \prod_{i=1}^{p} B_{iw(i)}(d(i),e(w(i))).\end{array}\right.$$

By (1) and (43) we see that

$$(45)\quad \left[\begin{array}{l}\text{for every } w \in W(p) \text{ we have} \\ \mathrm{card}(\{e \in Z(p) : g'(w,e) \neq 0\}) < \infty.\end{array}\right.$$

By (11), (43), and (44) we see that

$$(46)\quad \left[\begin{array}{l}\text{for every } w \in W(p) \text{ and } d \in Z(p,D) \text{ we have} \\ \mathrm{card}(\{e \in Z(p) : f'(w,d,e) \neq 0\}) < \infty.\end{array}\right.$$

By (12), (43), and (44) we see that

$$(47)\quad \left[\begin{array}{l}\text{for every } w \in W(p) \text{ we have} \\ \mathrm{card}(\{d \in Z(p,D) : \text{there exists } e \in Z(p) \text{ for which} \\ \qquad\qquad\qquad\qquad f'(w,d,e) \neq 0\}) < \infty.\end{array}\right.$$

In view of (7.23.13) and (7.23.14), by (42), (46), and (47) we see that

$$(48)\quad \left[\begin{array}{l}[\text{RHS of } (17)] = \sum_{w\in W(p)} \sum_{d\in Z(p,D)} \sum_{e\in Z(p)} f'(w,d,e) \\ \text{where all the summations are essentially finite.}\end{array}\right.$$

In view of (7.23.11) and (7.23.14), by (46) and (47) we see that

$$(49)\quad \left[\begin{array}{l}[\text{RHS of } (48)] = \sum_{w\in W(p)} \sum_{E\in Z} \sum_{e\in Z(p,E)} \sum_{d\in Z(p,D)} f'(w,d,e) \\ \text{where all the summations are essentially finite.}\end{array}\right.$$

By (5), (7), and (44) we see that

$$(50)\quad \left[\begin{array}{l}\text{for every } w \in W(p) \text{ and } E \in Z \text{ and } e \in Z(p,E) \text{ we have} \\ f'(w,d,e) \\ = g'(w,e)(-1)^{q^*E+R^*+S^*} \prod_{i=1}^{p} B^*(d(i) + d'[e,w](i), e'[w](i))\end{array}\right.$$

where

$$\left[\begin{array}{l}\text{for every } w \in W(p) \text{ and } e \in Z(p) \text{ we have let } d'[e,w] \text{ and} \\ e'[w] \text{ be the members of } Z(p) \text{ obtained by putting,}\end{array}\right.$$

$$(51)\quad \begin{cases} \text{for all } i \in [1,p], \\ d'[e,w](i) = q^{**}e(w(i)) + r^{**}[w(i)] + s^{**}[i] \text{ and} \\ e'[w](i) = q''e(w(i)) + r''[w(i)] + s''[i]. \end{cases}$$

By (6) and (51) we see that

$$(52)\quad \begin{cases} \text{for every } w \in W(p) \text{ and } E \in Z \text{ and } e \in Z(p,E) \text{ we have} \\ d'[e,w] \in Z(p,\ q^{**}E + R^{**} + S^{**}) \text{ and} \\ e'[w] \in Z(p,\ q''E + R'' + S''). \end{cases}$$

By (3), (4), (8), (50), and (52) we see that

$$(53)\quad \begin{cases} \text{for every } w \in W(p) \text{ and } E \in Z \text{ and } e \in Z(p,E) \text{ we have} \\ \sum_{d \in Z(p,D)} f'(w,d,e) = g'(w,e)h(D,E) \\ \text{where the summation is essentially finite.} \end{cases}$$

By (49) and (53) we see that

$$(54)\quad \begin{cases} [\text{RHS of (49)}] = \sum_{w \in W(p)} \sum_{E \in Z} \sum_{e \in Z(p,E)} g'(w,e)h(D,E) \\ \text{where all the summations are essentially finite.} \end{cases}$$

In view of (7.23.12) and (7.23.14), by (41) and (45) we see that

$$(55)\quad \begin{cases} [\text{RHS of (54)}] = [\text{RHS of (7.24.2)}] \\ \text{where all the summations are essentially finite.} \end{cases}$$

Now by (17), (36), (48), (49), (54), and (55) we conclude that

$$(56)\quad \begin{cases} [\text{RHS of (7.24.1)}] = [\text{RHS of (7.24.2)}] \\ \text{where all the summations are essentially finite.} \end{cases}$$

In view of (23), upon expanding the determinant we see that

$$(61)\quad \begin{cases} \text{for every } e \in Z(p) \text{ we have} \\ \det\left([A_{ij}(e(i) + r'(j) - s'(i))]_{1 \le j \le p}^{1 \le i \le p}\right) = \sum_{w \in W(p)} g(w,u[e,t[w]]) \end{cases}$$

where

$$(62)\quad \begin{cases} \text{for every } e \in Z(p) \text{ and } T \in Z(p) \text{ we have let } u[e,T] \text{ be} \\ \text{the member of } Z(p) \text{ obtained by putting} \\ u[e,T](k) = e(k) + T(k) \text{ for all } k \in [1,p] \end{cases}$$

and where

(63) $\left[\begin{array}{l}\text{for every } w \in W(p) \text{ we have let } t[w] \text{ be the member of } Z(p) \\ \text{obtained by putting } t[w](i) = r'(w(i)) - s'(i) \text{ for all} \\ i \in [1,p].\end{array}\right.$

By (1), (23), (62), and (63) we see that

(64) $\left[\begin{array}{l}\text{for every } w \in W(p) \text{ we have} \\ \text{card}(\{e \in Z(p) : g(w,u[e,t[w]]) \neq 0\}) < \infty.\end{array}\right.$

By (11), (23), (24), (62), and (63) we see that

(65) $\left[\begin{array}{l}\text{for every } w \in W(p) \text{ and } d \in Z(p,D) \text{ we have} \\ \text{card}(\{e \in Z(p) : f(w,d,u[e,t[w]]) \neq 0\}) < \infty.\end{array}\right.$

By (12), (23), (24), (62), and (63) we see that

(66) $\left[\begin{array}{l}\text{for every } w \in W(p) \text{ we have} \\ \text{card}(\{d \in Z(p,D) : \text{there exists } e \in Z(p) \\ \qquad\qquad \text{for which } f(w,d,u[e,t[w]]) \neq 0\}) < \infty.\end{array}\right.$

In view of (3.11), by (26), (62), and (63) we see that

(67) $\left[\begin{array}{l}\text{for every } w \in W(p) \text{ and } d \in Z(p,D) \text{ we have} \\ \displaystyle\sum_{e\in Z(p)} f(w,d,e) = \sum_{e\in Z(p)} f(w,d,u[e,t[w]]) \\ \text{where both the summations are essentially finite.}\end{array}\right.$

By (28) and (67) we conclude that

(68) $\left[\begin{array}{l}[\text{RHS of } (16)] = \displaystyle\sum_{w\in W(p)} \sum_{d\in Z(p,D)} \sum_{e\in Z(p)} f(w,d,u[e,t[w]]) \\ \text{where all the summations are essentially finite.}\end{array}\right.$

In view of (7.23.11) and (7.23.14), by (65) and (66) we see that

(69) $\left[\begin{array}{l}[\text{RHS of } (68)] = \displaystyle\sum_{w\in W(p)} \sum_{E\in Z} \sum_{e\in Z(p,E)} \sum_{d\in Z(p,D)} f(w,d,u[e,t[w]]) \\ \text{where all the summations are essentially finite.}\end{array}\right.$

By (5), (7), (24), (62), and (63) we see that

(70) $\left[\begin{array}{l}\text{for every } w \in W(p) \text{ and } E \in Z \text{ and } e \in Z(p,E) \text{ we have} \\ f(w,d,u[e,t[w]]) = g(w,u[e,t[w]])(-1)^{q^*E+R^*+S^*} \displaystyle\prod_{i=1}^{p} B^*(d(i) \\ \qquad\qquad + \tilde{d}[e,w](i), \tilde{e}[w](i))\end{array}\right.$

where

(71) $\left[\begin{array}{l}\text{for every } w \in W(p) \text{ and } e \in Z(p) \text{ we have let } \tilde{d}[e,w] \text{ and } \tilde{e}[w] \\ \text{be the members of } Z(p) \text{ obtained by putting for all } i \in [1,p], \\ \tilde{d}[e,w](i) = q^{**}(e(i) + r'(w(i)) - s'(i)) + r^{**}[w(i)] + s^{**}[i] \\ \text{and } \tilde{e}[w](i) = q''(e(i) + r'(w(i)) - s'(i)) + r''[w(i)] + s''[i].\end{array}\right.$

By (6) and (71) we see that

(72) $\left[\begin{array}{l}\textit{if } (*), \textit{ then} \text{ for every } w \in W(p) \text{ and } E \in Z \text{ and } e \in Z(p,E) \\ \text{we have } \tilde{d}[e,w] \in Z(p,\ q^{**}E + R^{**} + S^{**}) \\ \text{and } \tilde{e}[w] \in Z(p,\ q''E + R'' + S'').\end{array}\right.$

By (3), (4), (8), (70), and (72) we see that

(73) $\left[\begin{array}{l}\textit{if } (*), \textit{ then} \text{ for every } w \in W(p) \text{ and } E \in Z \text{ and } e \in Z(p,E) \\ \text{we have } \displaystyle\sum_{d\in Z(p,D)} f(w,d,u[e,t[w]]) = g(w,u[e,t[w]])h(D,E) \\ \text{where the summation is essentially finite.}\end{array}\right.$

By (69) and (73) we see that

(74) $\left[\begin{array}{l}\textit{if } (*), \textit{ then} \\ [\text{RHS of } (69)] = \displaystyle\sum_{w\in W(p)} \sum_{E\in Z} \sum_{e\in Z(p,E)} g(w,u[e,t[w]])h(D,E) \\ \text{where all the summations are essentially finite.}\end{array}\right.$

In view of (7.23.12) and (7.23.14), by (61) and (64) we see that

(75) $\left[\begin{array}{l}[\text{RHS of } (74)] = [\text{LHS of } (7.24.4)] \\ \text{where all the summations are essentially finite.}\end{array}\right.$

Now the RHS of (7.24.1) coincides with the RHS of (7.24.4) and therefore by (16), (36), (68), (69), (74), and (75) we conclude that

(76) $\left[\begin{array}{l}\textit{if } (*), \textit{ then} \\ [\text{LHS of } (7.24.4)] = [\text{RHS of } (7.24.4)] \\ \text{where all the summations are essentially finite.}\end{array}\right.$

The proof above remains verbatim valid when in it everywhere we change

$\left[\begin{array}{l}(7.24.1),\ (7.24.2),\ (7.24.4),\ i \in [1,p],\ j \in [1,p],\ \Pi_{i=1}^{p}, \\ d(i),\ e(i),\ e(j),\ e(w(i)),\ A_{iw(i)},\ B_{iw(i)},\ r^{**}[w(i)],\end{array}\right.$

$\left| \begin{array}{l} s^{**}[i],\ r''[w(i)],\ s''[i],\ t[w](i),\ s'(i),\ r'(j),\ r'(w(i)), \\ \hat{d}[e,w](i),\ \hat{e}[w](i),\ d'[e,w](i),\ e'[w](i),\ \tilde{d}[e,w](i),\ \tilde{e}[w](i), \end{array} \right.$

to

$\left[\begin{array}{l} (7.24.2),\ (7.24.1),\ (7.24.5),\ j \in [1,p],\ i \in [1,p],\ \Pi_{j=1}^{p}, \\ d(j),\ e(j),\ e(i),\ e(w(j)),\ A_{w(j)j},\ B_{w(j)j'},\ s^{**}[w(j)], \\ r^{**}[j],\ s''[w(j)],\ r''[j],\ t[w](j),\ -r'(j),\ -s'(i),\ -s'(w(j)), \\ \hat{d}[e,w](j),\ \hat{e}[w](j),\ d'[e,w](j),\ e'[w](j),\ \tilde{d}[e,w](j),\ \tilde{e}[w](j), \end{array} \right.$

respectively. Let (36*), (56*), and (76*) be the changed versions of (36), (56), and (76), respectively. Then these changed versions are thus:

(36*) $\left[\begin{array}{l} \text{[LHS of (7.24.2)] = [RHS of (7.24.2)]} \\ \text{where all the summations are essentially finite.} \end{array} \right.$

(56*) $\left[\begin{array}{l} \text{[RHS of (7.24.2)] = [RHS of (7.24.1)]} \\ \text{where all the summations are essentially finite.} \end{array} \right.$

(76*) $\left[\begin{array}{l} \textit{If } (*),\ \textit{then} \\ \text{[LHS of (7.24.5)] = [RHS of (7.24.5)]} \\ \text{where all the summations are essentially finite.} \end{array} \right.$

Now (7.24.1), (7.24.2), (7.24.4), and (7.24.5) follow from (36), (36*), (76), and (76*), respectively. Finally, (56) and (56*) are mutually equivalent and each of them yields (7.24.3) because clearly RHS of (7.24.1) is the same as the LHS of (7.24.3), while RHS of (7.24.2) is the same as the RHS of (7.24.3).

REMARK (7.25) Upon letting

$$\left[\begin{array}{l} B^*(d^*,e^*) = \binom{e^*}{d^*} \text{ for all } d^* \in Z \text{ and } e^* \in Z \\ \text{and} \\ h^*(d^*,e^*) = \binom{e^*}{d^*} \text{ for all } d^* \in Z \text{ and } e^* \in Z \end{array} \right.$$

by (2.5.2') and (4.35'.3) we see that conditions (2) and (3) of (7.24) are satisfied.

REMARK (7.26) Upon letting

$$\begin{cases} B^*(d^*,e^*) = \begin{bmatrix} e^* \\ d^* \end{bmatrix} \text{ for all } d^* \in Z \text{ and } e^* \in Z \\ \text{and} \\ h^*(d^*,e^*) = \begin{bmatrix} e^* + p - 1 \\ d^* \end{bmatrix} \text{ for all } d^* \in Z \text{ and } e^* \in Z \end{cases}$$

by (2.5.2*), and by taking $t = \emptyset$ in (4.35*.3), we see that conditions (2) and (3) of (7.24) are satisfied.

PROPOSITION (7.27) For every $D \in Z$ we have

$$F_D^{(30)}(r,s,r',s') = F_D^{(10)}(r,s,r',s') \tag{7.27.1}$$

and

$$F_D^{(3^*0)}(r,s,r',s') = F_D^{(1^*0)}(r,s,r',s') \tag{7.27.2}$$

and

$$F_D^{(10)}(r,s,r',s') = F_D^{(1^*0)}(r,s,r',s'). \tag{7.27.3}$$

Proof: Let

$$(1) \quad \begin{cases} A_{ij}(e^*) = \begin{bmatrix} s(i) \\ e^* \end{bmatrix} \begin{pmatrix} r(j) + r'(j) - s'(i) \\ r(j) - e^* \end{pmatrix} \\ \text{for all } i \in [1,p] \text{ and } j \in [1,p] \text{ and } e^* \in Z \end{cases}$$

and let

$$(2) \quad B^*(d^*,e^*) = \begin{pmatrix} e^* \\ d^* \end{pmatrix} \quad \text{for all } d^* \in Z \text{ and } e^* \in Z$$

and let

$$(3) \quad h^*(d^*,e^*) = \begin{pmatrix} e^* \\ d^* \end{pmatrix} \quad \text{for all } d^* \in Z \text{ and } e^* \in Z$$

and let

$$(4) \quad \begin{cases} q^* = -1 \text{ and } q^{**} = 1 \text{ and } q'' = 1, \\ \text{and } r^*[k] = r(k) \text{ and } s^*[k] = 0 \text{ for all } k \in [1,p], \\ \text{and } r^{**}[k] = -r(k) \text{ and } s^{**}[k] = 0 \text{ for all } k \in [1,p], \\ \text{and } r''[k] = r'(k) \text{ and } s''[k] = -s'(k) \text{ for all } k \in [1,p], \end{cases}$$

and let (5) to (8) be the same as (5) to (8) of (7.24), respectively.

By (2.5.2*) we see that for every $i \in [1,p]$ and $j \in [1,p]$ we have

$$\left\{e^* \in Z : \begin{bmatrix} s(i) \\ e^* \end{bmatrix} \neq 0\right\} \subset N$$

and by (2.5.2') we see that for every $i \in [1,p]$ and $j \in [1,p]$ we have

$$\left\{e^* \in Z : \binom{r(j) + r'(j) - s'(i)}{r(j) - e^*} \neq 0\right\} \subset \{e^* \in Z : e^* \leq r(j)\}$$

and hence by (1) we see that for every $i \in [1,p]$ and $j \in [1,p]$ we have

$$\{e^* \in Z : A_{ij}(e^*) \neq 0\} \subset [0,r(j)]$$

and thus we conclude that

(9) $\operatorname{card}(\{e^* \in Z : A_{ij}(e^*) \neq 0\}) < \infty$ for all $i \in [1,p]$ and $j \in [1,p]$.

Henceforth let there be fixed any $D \in Z$. Now in view of (1), (3), (4), (5), (6), and (8), by (7.11) and the definitions of $F_D^{(10)}$ and $F_D^{(1*0)}$ we see that

(10) $$F_D^{(10)}(r,s,r',s') = \sum_{E \in Z} h(D,E) \sum_{e \in Z(p,E)} \det\left([A_{ij}(e(i))]_{1\leq j\leq p}^{1\leq i\leq p}\right)$$

and

(10*) $$F_D^{(1*0)}(r,s,r',s') = \sum_{E \in Z} h(D,E) \sum_{e \in Z(p,E)} \det\left([A_{ij}(e(j))]_{1\leq j\leq p}^{1\leq i\leq p}\right)$$

where all the summations are essentially finite. In view of (1) to (10) and (10*), by (7.24.3) and (7.25) we get (7.27.3).

By (7.13) and the definitions of $F_D^{(30)}$ and $F_D^{(3*0)}$ we see that

(11) $$F_D^{(30)}(r,s,r',s') = \sum_{d \in Z(p,D)} \det\left([K_{ij}(d(i))]_{1\leq j\leq p}^{1\leq i\leq p}\right)$$

and

(11*) $$F_D^{(3*0)}(r,s,r',s') = \sum_{d \in Z(p,D)} \det\left([K_{ij}(d(j))]_{1\leq j\leq p}^{1\leq i\leq p}\right)$$

where all the summations are essentially finite and where, for all $i \in [1,p]$ and $j \in [1,p]$ and $d^* \in Z$, we have put

$$K_{ij}(d^*) = \begin{bmatrix} s(i) - d^* \\ r(j) \end{bmatrix} \begin{pmatrix} r(j) + r'(j) - s'(i) \\ d^* \end{pmatrix}.$$

In view of (1), (2), (4), (7), and the equation above, by (4.48) we see that

$$(12)\quad \begin{cases} \text{for every } i \in [1,p] \text{ and } j \in [1,p] \text{ and } d^* \in Z \text{ we have} \\ K_{ij}(d^*) = \sum_{e^* \in Z} A_{ij}(e^*) B_{ij}(d^*,e^*) \\ \text{where the summation is essentially finite.} \end{cases}$$

In view of (1) to (9) and (10) and (11) and (12), by (7.24.1) and (7.25) we get (7.27.1). In view of (1) to (9) and (10*) and (11*) and (12), by (7.24.2) and (7.25) we get (7.27.2).

PROPOSITION (7.28) Assume (**). Then for every $V \in Z$ we have

$$F^{(4^*0)}(r,s,r',s',V) = F^{(30)}(r,s,r',s',V) \tag{7.28.1}$$

and

$$F^{(40)}(r,s,r',s',V) = F^{(3^*0)}(r,s,r',s',V) \tag{7.28.2}$$

and

$$F^{(30)}(r,s,r',s',V) = F^{(3^*0)}(r,s,r',s',V) \tag{7.28.3}$$

Proof: Let

$$(1)\quad \begin{cases} A_{ij}(e^*) = \begin{bmatrix} s(i) - e^* \\ r(j) \end{bmatrix} \begin{pmatrix} r(j) + r'(j) - s'(i) \\ e^* \end{pmatrix} \\ \text{for all } i \in [1,p] \text{ and } j \in [1,p] \text{ and } e^* \in Z \end{cases}$$

and let

$$(2)\quad B^*(d^*,e^*) = \begin{bmatrix} e^* \\ d^* \end{bmatrix} \qquad \text{for all } d^* \in Z \text{ and } e^* \in Z$$

and let

(3) $$h^*(d^*,e^*) = \begin{bmatrix} e^* + p - 1 \\ d^* \end{bmatrix} \qquad \text{for all } d^* \in Z \text{ and } e^* \in Z$$

and let

(4) $$\begin{cases} q^* = 1 \text{ and } q^{**} = 0 \text{ and } q'' = -1, \\ \text{and } r^*[k] = 0 \text{ and } s^*[k] = 0 \text{ for all } k \in [1,p], \\ \text{and } r^{**}[k] = 0 \text{ and } s^{**}[k] = 0 \text{ for all } k \in [1,p], \\ \text{and } r''[k] = r(k) \text{ and } s''[k] = s(k) \text{ for all } k \in [1,p], \end{cases}$$

and let (5) to (8) be the same as (5) to (8) of (7.24), respectively. By (2.5.2') and (2.5.6') we see that for every $i \in [1,p]$ and $j \in [1,p]$ we have

$$\left\{e^* \in Z : \begin{pmatrix} r(j) + r'(j) - s'(i) \\ e^* \end{pmatrix} \neq 0\right\}$$
$$\subset [0, r(j) + r'(j) - s'(i)]$$

and hence by (1) we see that for every $i \in [1,p]$ and $j \in [1,p]$ we have

$$\{e^* \in Z : A_{ij}(e^*) \neq 0\} \subset [0, r(j) + r'(j) - s'(i)]$$

amd thus we conclude that

(9) $$\text{card}(\{e^* \in Z : A_{ij}(e^*) \neq 0\}) < \infty \qquad \text{for all } i \in [1,p] \text{ and } j \in [1,p].$$

Henceforth let there be fixed any $D \in Z$. Now in view of (1), (3), (4), (5), (6), and (8), by (7.13) and the definitions of $F^{(30)}$ and $F^{(3*0)}$ we see that

(10) $$F^{(30)}(r,s,r',s',D) = \sum_{E \in Z} h(D,E) \sum_{e \in Z(p,E)} \det\left([A_{ij}(e(i))]_{1 \le j \le p}^{1 \le i \le p}\right)$$

and

(10*) $$F^{(3*0)}(r,s,r',s',D) = \sum_{E \in Z} h(D,E) \sum_{e \in Z(p,E)} \det\left([A_{ij}(e(j))]_{1 \le j \le p}^{1 \le i \le p}\right)$$

where all the summations are essentially finite. In view of (1) to

(10) and (10*), by (7.24.3) and (7.26) we see that

(10') $F^{(30)}(r,s,r',s',D) = F^{(3*0)}(r,s,r',s',D)$.

By (7.14) and the definitions of $F^{(4*0)}$ and $F^{(40)}$ we see that

(11) $$F^{(4*0)}(r,s,r',s',D) = \sum_{d\in Z(p,D)} \det\left([\hat{K}_{ij}(d(j))]_{1\le j\le p}^{1\le i\le p}\right)$$

and

(11*) $$F^{(40)}(r,s,r',s',D) = \sum_{d\in Z(p,D)} \det\left([\hat{K}_{ij}(d(i))]_{1\le j\le p}^{1\le i\le p}\right)$$

where all the summations are essentially finite and where, for all $i \in [1,p]$ and $j \in [1,p]$ and $d^* \in Z$, we have put

$$\hat{K}_{ij}(d^*) = \begin{bmatrix} r(i) \\ d^* \end{bmatrix}\begin{bmatrix} s(j) \\ d^* + s'(j) - r'(i) \end{bmatrix}.$$

For all $i \in [1,p]$ and $j \in [1,p]$ and $d^* \in Z$, let us put

(11') $$K_{ij}(d^*) = \begin{bmatrix} r(j) \\ d^* \end{bmatrix}\begin{bmatrix} s(i) \\ d^* + s'(i) - r'(j) \end{bmatrix}.$$

Now for all $d^* \in Z$, clearly we have

(12) $$[K_{ij}(d(i))]_{1\le j\le p}^{1\le i\le p} = \text{transpose of } [\hat{K}_{ij}(d(j))]_{1\le j\le p}^{1\le i\le p}$$

and

(12*) $$[K_{ij}(d(j))]_{1\le j\le p}^{1\le i\le p} = \text{transpose of } [\hat{K}_{ij}(d(i))]_{1\le j\le p}^{1\le i\le p}$$

and by general properties of determinants we have

(13) $$\det\left([\hat{K}_{ij}(d(j))]_{1\le j\le p}^{1\le i\le p}\right) = \det \text{ of transpose of } [\hat{K}_{ij}(d(j))]_{1\le j\le p}^{1\le i\le p}$$

and

(13*) $\det\left([\hat{K}_{ij}(d(i))]_{1\le j\le p}^{1\le i\le p}\right)$ = det of transpose of $[\hat{K}_{ij}(d(i))]_{1\le j\le p}^{1\le i\le p}$.

By (11), (12), and (13) we see that

(14) $\begin{cases} F^{(4*0)}(r,s,r',s',D) = \sum\limits_{d\in Z(p,D)} \det\left([K_{ij}(d(i))]_{1\le j\le p}^{1\le i\le p}\right) \\ \text{where the summation is essentially finite.}\end{cases}$

By (11*), (12*), and (13*) we see that

(14*) $\begin{cases} F^{(40)}(r,s,r',s',D) = \sum\limits_{d\in Z(p,D)} \det\left([K_{ij}(d(j))]_{1\le j\le p}^{1\le i\le p}\right) \\ \text{where the summation is essentially finite.}\end{cases}$

In view of (1), (2), (4), (7), and (11'), by (4.59) we see that

(15) $\begin{cases} \text{for every } i \in [1,p] \text{ and } j \in [1,p] \text{ and } d^* \in Z \text{ we have} \\ K_{ij}(d^*) = \sum\limits_{e^*\in Z} A_{ij}(e^*)B_{ij}(d^*,e^*) \\ \text{where the summation is essentially finite.}\end{cases}$

In view of (1) to (9) and (10) and (14) and (15), by (7.24.1) and (7.26) we see that

(16) $F^{(4*0)}(r,s,r',s',D) = F^{(30)}(r,s,r',s',D)$.

In view of (1) to (9) and (10*) and (14*) and (15), by (7.24.2) and (7.26) we see that

(16*) $F^{(40)}(r,s,r',s',D) = F^{(3*0)}(r,s,r',s',D)$.

Now upon writing V for D, by (10'), (16), and (16*) we get (7.28.3), (7.28.1), and (7.28.2), respectively.

PROPOSITION (7.29) Given any $D \in Z$, we have that

$$\textit{if } (**), \textit{ then } F_D^{(30)}(r,s,r',s') = F_D^{(20)}(s,r,s',r') \qquad (7.29.1)$$

and we have that

$$\textit{if } (**), \textit{ then } F_D^{(3*0)}(r,s,r',s') = F_D^{(2*0)}(s,r,s',r') \qquad (7.29.2)$$

and we have that

$$F_D^{(20)}(s,r,s',r') = F_D^{(2*0)}(s,r,s',r') \tag{7.29.3}$$

and we have that

if (*) *and* (**), *then*

$$F_D^{(2*0)}(r,s,r',s') = F_D^{(20)}(s,r,s',r') \tag{7.29.4}$$

and we have that

if (*) *and* (**), *then*

$$F_D^{(20)}(r,s,r',s') = F_D^{(2*0)}(s,r,s',r'). \tag{7.29.5}$$

[Note that upon taking s, r, s', r' for r, s, r', s', respectively, by (7.12) we know that the summations in the definitions of $F_D^{(20)}(s,r,s',r')$ and $F_D^{(2*0)}(s,r,s',r')$ are essentially finite.]

Proof: Let

(1) $\begin{cases} A_{ij}(e^*) = \binom{s(i) + s'(i) - r'(j)}{s(i) - e^*}\binom{r(j) + r'(j) - s'(i)}{e^*} \\ \text{for all } i \in [1,p] \text{ and } j \in [1,p] \text{ and } e^* \in Z \end{cases}$

and let

(2) $B^*(d^*,e^*) = \binom{e^*}{d^*}$ for all $d^* \in Z$ and $e^* \in Z$

and let

(3) $h^*(d^*,e^*) = \binom{e^*}{d^*}$ for all $d^* \in Z$ and $e^* \in Z$

and let

(4) $\begin{cases} q^* = 0 \text{ and } q^{**} = 0 \text{ and } q'' = 1, \\ \text{and } r^*[k] = 0 \text{ and } s^*[k] = 0 \text{ for all } k \in [1,p], \\ \text{and } r^{**}[k] = 0 \text{ and } s^{**}[k] = 0 \text{ for all } k \in [1,p], \\ \text{and } r''[k] = 0 \text{ and } s''[k] = 0 \text{ for all } k \in [1,p], \end{cases}$

and let (5) to (8) be the same as (5) to (8) of (7.24), respectively. By (2.5.2') we see that for every $i \in [1,p]$ and $j \in [1,p]$ we have

$$\left\{e^* \in Z : \binom{s(i) + s'(i) - r'(j)}{s(i) - e^*} \neq 0\right\} \subset \{e^* \in Z : e^* \leq s(i)\}$$

and

$$\left\{e^* \in Z : \binom{r(j) + r'(j) - s'(i)}{e^*} \neq 0\right\} \subset N$$

and hence by (1) we see that for every $i \in [1,p]$ and $j \in [1,p]$ we have

$$\{e^* \in Z : A_{ij}(e^*) \neq 0\} \subset [0,s(i)]$$

and thus we conclude that

(9) $\quad \operatorname{card}(\{e^* \in Z : A_{ij}(e^*) \neq 0\}) < \infty$
for all $i \in [1,p]$ and $j \in [1,p]$.

Henceforth let there be fixed any $D \in Z$. Now in view of (1), (3), (4), (5), (6), and (8), by (7.12) and the definitions of $F_D^{(20)}$ and $F_D^{(2*0)}$ we see that

$$(10) \qquad F_D^{(20)}(s,r,s',r') = \sum_{E \in Z} h(D,E) \sum_{e \in Z(p,E)} \det\left([A_{ij}(e(i))]_{1\leq j\leq p}^{1\leq i\leq p}\right)$$

and

$$(10^*) \qquad F_D^{(2*0)}(s,r,s',r') = \sum_{E \in Z} h(D,E) \sum_{e \in Z(p,E)} \det\left([A_{ij}(e(j))]_{1\leq j\leq p}^{1\leq i\leq p}\right)$$

where all the summations are essentially finite. In view of (1) to (10) and (10*), by (7.24.3) and (7.25) we get (7.29.3).

In view of (3), (4), (5), (6), and (8), by (7.12) and the definitions of $F_D^{(2*0)}$ and $F_D^{(20)}$ we see that

$$(11) \qquad F_D^{(2*0)}(r,s,r',s') = \sum_{E \in Z} h(D,E) \sum_{e \in Z(p,E)} \det G^{(2*0)}(r,s,r',s',e)$$

and

$$(11^*) \qquad F_D^{(20)}(r,s,r',s') = \sum_{E \in Z} h(D,E) \sum_{e \in Z(p,E)} \det G^{(20)}(r,s,r',s',e)$$

where all the summations are essentially finite and where for every $e \in Z(p)$ we have

(12) $$\begin{cases} G^{(2*0)}(r,s,r',s',e) = \text{the } p \text{ by } p \text{ matrix whose } (i,j)\text{th} \\ \qquad \text{entry is } G_{ij}^{(2*0)}(r,s,r',s',e) \end{cases}$$

and

(12*) $$\begin{cases} G^{(20)}(r,s,r',s',e) = \text{the } p \text{ by } p \text{ matrix whose } (i,j)\text{th} \\ \qquad \text{entry is } G_{ij}^{(20)}(r,s,r',s',e) \end{cases}$$

By general properties of determinants we know that for every $e \in Z(p)$ we have

(13) $$\begin{cases} \det G^{(2*0)}(r,s,r',s',e) \\ = \det \text{ of the transpose of } G^{(2*0)}(r,s,r',s',e) \end{cases}$$

and

(13*) $$\begin{cases} \det G^{(20)}(r,s,r',s',e) \\ = \det \text{ of the transpose of } G^{(20)}(r,s,r',s',e). \end{cases}$$

By (7.2) we know that

$$\begin{cases} \textit{if } (**), \textit{ then} \text{ for every } i \in [1,p] \text{ and } j \in [1,p] \text{ and } e \in Z(p) \\ \text{we have that} \\ G_{ij}^{(2*0)}(r,s,r',s',e) = \binom{r(i) + r'(i) - s'(j)}{e(j) + r'(i) - s'(j)} \\ \qquad \times \binom{s(j) + s'(j) - r'(i)}{s(j) + s'(j) - r'(i) - e(j)} \end{cases}$$

and therefore, upon interchanging the two factors on the RHS of the equations above and upon taking the transpose, in view of (12) we deduce that

$$\begin{cases} \textit{if } (**), \textit{ then} \text{ for every } i \in [1,p] \text{ and } j \in [1,p] \text{ and} \\ e \in Z(p) \text{ we have that} \\ \binom{s(i) + s'(i) - r'(j)}{s(i) + s'(i) - r'(j) - e(i)}\binom{r(j) + r'(j) - s'(i)}{e(i) + r'(j) - s'(i)} \\ = \text{the } (i,j)\text{th entry in the transpose of } G^{(2*0)}(r,s,r',s',e) \end{cases}$$

and by (1) we see that

$$A_{ij}(e(i) + r'(j) - s'(i)) = \text{LHS of the equation above}$$

so we conclude that

$$\left[\begin{array}{l} \textit{if } (**), \textit{ then for every } i \in [1,p] \text{ and } j \in [1,p] \text{ and} \\ e \in Z(p) \text{ we have that} \\ A_{ij}(e(i) + r'(j) - s'(i)) \\ = \text{the } (i,j)\text{th entry in the transpose of } G^{(2*0)}(r,s,r',s',e) \end{array}\right.$$

and therefore in view of (13) we see that

$$(14)\quad \left[\begin{array}{l} \textit{if } (**), \textit{ then for every } e \in Z(p) \text{ we have} \\ \det\left([A_{ij}(e(i) + r'(j) - s'(i))]_{1\le j\le p}^{1\le i\le p}\right) \\ = \det G^{(2*0)}(r,s,r',s',e). \end{array}\right.$$

Again by (7.2) we know that

$$\left[\begin{array}{l} \textit{if } (**), \textit{ then for every } i \in [1,p] \text{ and } j \in [1,p] \text{ and} \\ e \in Z(p) \text{ we have that} \\ G_{ij}^{(20)}(r,s,r',s',e) = \binom{r(i) + r'(i) - s'(j)}{e(i) + r'(i) - s'(j)} \\ \qquad \times \binom{s(j) + s'(j) - r'(i)}{s(j) + s'(j) - r'(i) - e(i)} \end{array}\right.$$

and therefore, upon interchanging the two factors on the RHS of the equation above and upon taking the transpose, in view of (12*) we deduce that

$$\left[\begin{array}{l} \textit{if } (**), \textit{ then for every } i \in [1,p] \text{ and } j \in [1,p] \text{ and} \\ e \in Z(p) \text{ we have that} \\ \binom{s(i) + s'(i) - r'(j)}{s(i) + s'(i) - r'(j) - e(j)}\binom{r(j) + r'(j) - s'(i)}{e(j) + r'(j) - s'(i)} \\ = \text{the } (i,j)\text{th entry in the transpose of } G^{(20)}(r,s,r',s',e) \end{array}\right.$$

and by (1) we see that

$$A_{ij}(e(j) - s'(i) + r'(j)) = \text{LHS of the equation above}$$

so we conclude that

$$\left[\begin{array}{l}\textit{if } (**), \textit{ then} \text{ for every } i \in [1,p] \text{ and } j \in [1,p] \text{ and} \\ e \in Z(p) \text{ we have that} \\ A_{ij}(e(j) - s'(i) + r'(j)) \\ = \text{the } (i,j)\text{th entry in the transpose of } G^{(20)}(r,s,r',s',e)\end{array}\right.$$

and therefore in view of (13*) we see that

$$(14^*)\quad \left[\begin{array}{l}\textit{if } (**), \textit{ then} \text{ for every } e \in Z(p) \text{ we have} \\ \det\left([A_{ij}(e(j) - s'(i) + r'(j))]_{1\le j\le p}^{1\le i\le p}\right) = \det G^{(20)}(r,s,r',s',e)\end{array}\right.$$

In view of (1) to (9) and (10) and (11) and (14), by (7.24.4) and (7.25) we get (7.29.4). In view of (1) to (9) and (10*) and (11*) and (14*), by (7.24.5) and (7.25) we get (7.29.5).

By (7.3) and (7.13) and the definitions of $F_D^{(30)}$ and $F_D^{(3*0)}$, we see that

$$(15)\quad \left[\begin{array}{l}\textit{if } (**), \textit{ then} \\ F_D^{(30)}(r,s,r',s') = \sum_{d\in Z(p,D)} \det\left([K_{ij}(d(i))]_{1\le j\le p}^{1\le i\le p}\right)\end{array}\right.$$

and

$$(15^*)\quad \left[\begin{array}{l}\textit{if } (**), \textit{ then} \\ F_D^{(3*0)}(r,s,r',s') = \sum_{d\in Z(p,D)} \det\left([K_{ij}(d(j))]_{1\le j\le p}^{1\le i\le p}\right)\end{array}\right.$$

where all the summations are essentially finite and where for all $i \in [1,p]$ and $j \in [1,p]$ and $d^* \in Z$, we have put

$$K_{ij}(d^*) = \begin{bmatrix} r(j) \\ s(i) - d^* \end{bmatrix} \begin{pmatrix} r(j) + r'(j) - s'(i) \\ d^* \end{pmatrix}.$$

In view of (1), (2), (4), (7), and the equation above, by (4.53) we see that

$$(16)\quad \left[\begin{array}{l}\text{for every } i \in [1,p] \text{ and } j \in [1,p] \text{ and } d^* \in Z \text{ we have} \\ K_{ij}(d^*) = \sum_{e^*\in Z} A_{ij}(e^*)B_{ij}(d^*,e^*) \\ \text{where the summation is essentially finite.}\end{array}\right.$$

In view of (1) to (9) and (10) and (15) and (16), by (7.24.1) and (7.25) we set (7.29.1). In view of (1) to (9) and (10*) and (15*) and (16), by (7.24.2) and (7.25) we get (7.29.2).

PROPOSITION (7.30) Assume (***). Then for every $V \in Z$ we have

$$F^{(40)}(r,s,r',s',V) = F^{(60)}(r,s,r',s',V)$$
$$= F^{(40)}(s,r,s',r',V)$$

Proof: Let $q = 2$ and let $t \in \mathrm{pre}[q,p]$ and $t' \in \mathrm{pre}[q,p]$ be defined by putting

$$\begin{cases} t(1,i) = r(i) \text{ and } t(2,i) = s(i) \text{ and } t'(1,i) = t'(2,i) = r'(i) \\ \text{for all } i \in [1,p]. \end{cases}$$

Now by taking $k = 1$ in (3.25) we see that

$$F^{(60)}(r,s,r',s',V) = F^{(40)}(r,s,r',s',V) \quad \text{for all } V \in Z$$

whereas by taking $k = 2$ in (3.25) we see that

$$F^{(60)}(r,s,r',s',V) = F^{(40)}(s,r,s',r',V) \quad \text{for all } V \in Z.$$

REMARK (7.31) By (7.27) we get a *second* proof of (7.22.2); moreover, assuming (**), by (7.29.1), (7.29.2), and (7.29.3) we get a *third* proof of (7.22.2). Likewise, by (7.27.1) and (7.27.2) we get a second proof of (7.27.3). And so on. In other words, by (7.22) and (7.27) we get several proofs of the fact that

$$\begin{cases} \text{given any } L \text{ and } L' \text{ in } \{1,1^*,3,3^*\} \cup \{3z : z \in W[p]\} \\ \text{we have } F_D^{(L0)}(r,s,r',s') = F_D^{(L'0)}(r,s,r',s') \\ \text{for all } D \in Z \end{cases} \tag{7.31.0}$$

and moreover, if condition (**) is satisfied, then by (7.22), (7.27), (7.29.1), (7.29.2), and (7.29.3) we get some more proofs of (7.31.0). Similarly, by (7.22), (7.27), and (7.29) we get several proofs of the fact that

$$(1) \quad \begin{cases} \text{if } (*) \text{ and } (**), \text{ then for every} \\ L \in \{1,1^*,2,2^*,3,3^*\} \cup \{3z : z \in W[p]\} \text{ we have} \\ F_D^{(L0)}(r,s,r',s') = F_D^{(20)}(r,s,r',s') \\ \qquad = F_D^{(20)}(s,r,s',r') \text{ for all } D \in Z. \end{cases}$$

Upon interchanging (r,r') with (s,s'), by (1) we see that

(2) $$\begin{cases}\text{if } (*) \text{ and } (**), \text{ then for every} \\ L \in \{1,1^*,2,2^*,3,3^*\} \cup \{3z : z \in W[p]\} \\ \text{we have} \\ F_D^{(L0)}(s,r,s',r') = F_D^{(20)}(s,r,s',r') = F_D^{(20)}(r,s,r',s') \\ \text{for all } D \in Z.\end{cases}$$

By (1) and (2) we see that

$$\begin{cases}\text{if } (*) \text{ and } (**), \text{ then for every} \\ L \in \{1,1^*,2,2^*,3,3^*\} \cup \{3z : z \in W[p]\} \\ \text{we have} \\ F_D^{(L0)}(r,s,r',s') = F_D^{(L0)}(s,r,s',r') \\ \qquad = F_D^{(10)}(r,s,r',s') \text{ for all } D \in Z.\end{cases} \tag{7.31.1}$$

Note that the first equation in (7.31.1) is an *equation of symmetry* and its primary source in the argument above was the case of L = 2 which was proved in (7.29.4) and (7.29.5).

By (7.31.1) we see that

$$\begin{cases}\text{if } (*) \text{ and } (**), \text{ then for every} \\ L \in \{1,1^*,2,2^*,3,3^*\} \cup \{3z : z \in W[p]\} \\ \text{we have} \\ F^{[L0]}(r,s,r',s',V) = F^{[L0]}(s,r,s',r',V) \\ \qquad = F^{[10]}(r,s,r',s',V) \text{ in } Q[V].\end{cases} \tag{7.31.1*}$$

By (7.31.1) we also see that

$$\begin{cases}\text{if } (*) \text{ and } (**), \text{ then for every} \\ L \in \{1,1^*,2,2^*,3,3^*\} \cup \{3z : z \in W[p]\} \\ \text{we have} \\ F^{(L0)}(r,s,r',s',V) = F^{(L0)}(s,r,s',r',V) \\ \qquad = F^{(30)}(r,s,r',s',V) \text{ for all } V \in Z.\end{cases} \tag{7.31.2'}$$

In view of (5) of (7.13), by (7.31.1*) we see that

$$\left[\begin{array}{l}\text{if } (*) \text{ and } (**), \text{ then for every} \\ L \in \{1,1^*,2,2^*,3,3^*\} \cup \{3z : z \in W[p]\} \\ \text{we have} \\ F^{[L0]}(r,s,r',s',V) = F^{[L0]}(s,r,s',r',V) \\ \qquad = F^{(30)}(r,s,r',s',V) \text{ for all } V \in N.\end{array}\right. \qquad (7.31.2'')$$

In view of (5) of (7.13), by (7.28.3) we see that

$$\left[\begin{array}{l}\text{if } (*) \text{ and } (**), \text{ then for all } V \in N, \\ F^{[3^*0]}(r,s,r',s',V) = F^{[30]}(r,s,r',s',V)\end{array}\right.$$

and therefore in view of (4.33) and (4.65.2) we conclude that

$$\left[\begin{array}{l}\text{if } (*) \text{ and } (**), \text{ then for all } D \in Z, \\ F_D^{(3^*0)}(r,s,r',s') = F_D^{(30)}(r,s,r',s');\end{array}\right.$$

thus, assuming (*) and (**), by (7.28.3) we get *fourth*, although *indirect*, proof of (7.22.2).

By (7.28) we see that

$$(3) \quad \left[\begin{array}{l}\text{if } (**), \text{ then for all } V \in Z, \\ F^{(4^*0)}(r,s,r',s',V) = F^{(40)}(r,s,r',s',V)\end{array}\right.$$

and by interchanging (r,r') with (s,s') in (3) we see that

$$(4) \quad \left[\begin{array}{l}\text{if } (**), \text{ then for all } V \in Z, \\ F^{(4^*0)}(s,r,s',r',V) = F^{(40)}(s,r,s',r',V).\end{array}\right.$$

By (7.28) we also see that

$$(5) \quad \left[\begin{array}{l}\text{if } (**), \text{ then for all } V \in Z, \\ F^{(40)}(r,s,r',s',V) = F^{(30)}(r,s,r',s',V)\end{array}\right.$$

and by interchanging (r,r') with (s,s') in (5) we see that

$$(6) \quad \left[\begin{array}{l}\text{if } (**), \text{ then for all } V \in Z, \\ F^{(40)}(s,r,s',r',V) = F^{(30)}(s,r,s',r',V).\end{array}\right.$$

In view of (7.0.2), by (7.30), (7.31.2''), and (5) we conclude that

$$\left[\begin{array}{l}\text{if } (**) \text{ and } (***), \text{ then for every} \\ L \in \{1,1^*,2,2^*,3,3^*\} \cup \{3z : z \in W[p]\} \\ \text{we have} \\ F^{[L0]}(r,s,r',s',V) = F^{[L0]}(s,r,s',r',V) \\ \qquad\qquad = F^{(60)}(r,s,r',s',V) \text{ for all } V \in N.\end{array}\right. \tag{7.31.2}$$

In view of (7.0.2), by (7.30), (7.31.2'), (3), (4), and (5) we conclude that

$$\left[\begin{array}{l}\text{if } (**) \text{ and } (***), \text{ then for every} \\ L \in \{1,1^*,2,2^*,3,3^*,4,4^*\} \cup \{3z : z \in W[p]\} \\ \text{we have} \\ F^{(L0)}(r,s,r',s',V) = F^{(L0)}(s,r,s',r',V) \\ \qquad\qquad = F^{(60)}(r,s,r',s',V) \text{ for all } V \in Z.\end{array}\right. \tag{7.31.3}$$

By (7.29.1) we see that

(7) $$\left[\begin{array}{l}\text{if } (**), \text{ then for all } V \in Z \\ F^{(30)}(r,s,r',s',V) = F^{(20)}(r,s,r',s',V)\end{array}\right.$$

and by interchanging (r,r') with (s,s') in (7) we see that

(8) $$\left[\begin{array}{l}\text{if } (**), \text{ then for all } V \in Z \\ F^{(30)}(s,r,s',r',V) = F^{(20)}(s,r,s',r',V).\end{array}\right.$$

By (5), (6), (7), (8), and (7.30) we see that

(9) $$\left[\begin{array}{l}\text{if } (**) \text{ and } (***), \text{ then for all } V \in Z \\ F^{(20)}(r,s,r',s',V) = F^{(20)}(s,r,s',r',V).\end{array}\right.$$

In view of (7.0.1) and (7.0.2) and in view of (7) of (7.12), by (9) we see that

$$\left[\begin{array}{l}\text{if } (**) \text{ and } (***), \text{ then for all } V \in N \\ F^{[20]}(r,s,r',s',V) = F^{[20]}(s,r,s',r',V)\end{array}\right.$$

and therefore in view of (4.33) and (4.65.2) we conclude that

$$\left[\begin{array}{l}\text{if } (**) \text{ and } (***), \text{ then for all } D \in Z \\ F_D^{(20)}(r,s,r',s') = F_D^{(20)}(s,r,s',r').\end{array}\right.$$

This is the indirect argument, based on (7.28), (7.29.1), *and* (7.30), *which first led us to the equation of symmetry*

$$F_D^{(20)}(r,s,r',s') = F_D^{(20)}(s,r,s',r').$$

The direct argument, which consisted of (7.29.4) *and* (7.29.5), *and which was mentioned at the end of the first paragraph of this Remark, was an afterthought.*

THEOREM (7.32) Assuming (**) and (***), we have the following:

(7.32.1) Given any

$$L \in \{1,1^*,2,2^*,3,3^*\} \cup \{3z : z \in W[p]\}$$

first we have

$$F_D^{(L0)}(r,s,r',s') = F_D^{(L0)}(s,r,s',r') = F_D^{(10)}(r,s,r',s')$$

for all $D \in \mathbf{Z}$,

and second, we have

$$F^{[L0]}(r,s,r',s',V) = F^{[L0]}(s,r,s',r',V) = F^{[10]}(r,s,r',s',V)$$

in $Q[V]$.

(7.32.2) Given any

$$L \in \{1,1^*,2,2^*,3,3^*\} \cup \{3z : z \in W[p]\}$$

we have

$$F^{[L0]}(r,s,r',s',V) = F^{[L0]}(s,r,s',r',V) = F^{(60)}(r,s,r',s',V)$$

for all $V \in N$.

(7.32.3) Given any

$$L \in \{1,1^*,2,2^*,3,3^*,4,4^*\} \cup \{3z : z \in W[p]\}$$

we have

$$F^{(L0)}(r,s,r',s',V) = F^{(L0)}(s,r,s',r',V) = F^{(60)}(r,s,r',s',V)$$

for all $V \in Z$.

(7.32.4) If for some $n \in Z$, we have

$r(i) = p - i$ and $s(i) = n - i$ and $r'(i) = s'(i) = i$
for all $i \in [1,p]$,

then for any given $e \in Z(p)$ we have

$$H^{(20)}(r,s,r',s',e) = \begin{cases} 1 & \text{if } e(i) = 0 \text{ for all } i \in [1,p] \\ 0 & \text{if } e(i) \neq 0 \text{ for some } i \in [1,p] \end{cases}$$

and for any given $E \in Z$ we have

$$H_E^{(20)}(r,s,r',s') = \begin{cases} 1 & \text{if } E = 0 \\ 0 & \text{if } E \neq 0. \end{cases}$$

(7.32.5) If either

[for some $n \in Z$ we have
$r(i) = p - i$ and $s(i) = n - i$ and $r'(i) = s'(i) = i$
for all $i \in [1,p]$]

or

[for some $n \in Z$ we have
$r(i) = n - i$ and $s(i) = p - i$ and $r'(i) = s'(i) = i$
for all $i \in [1,p]$]

then for every

$$L \in \{1,1^*,2,2^*,3,3^*\} \cup \{3z : z \in W[p]\}$$

first, we have

$$F_D^{(L0)}(r,s,r',s') = F_D^{(L0)}(s,r,s',r') = \begin{cases} 1 & \text{for } D = 0 \\ 0 & \text{for all } D \in Z\backslash\{0\}, \end{cases}$$

and second, we have

$$F^{[L0]}(r,s,r',s',V) = F^{[L0]}(s,r,s',r',V) = \begin{bmatrix} V \\ pn - 1 \end{bmatrix} \text{ in } Q[V].$$

(7.32.6) If either

[for some $n \in Z$ we have
$r(i) = p - i$ and $s(i) = n - i$ and $r'(i) = s'(i) = i$
for all $i \in [,p]$

or

$$\left[\begin{array}{l}\text{for some } n \in Z \text{ we have}\\ r(i) = n - i \text{ and } s(i) = p - i \text{ and } r'(i) = s'(i) = i\\ \text{for all } i \in [1,p]\end{array}\right.$$

then for every

$$L \in \{1,1^*,2,2^*,3,3^*,4,4^*\}$$

we have

$$\begin{aligned} F^{(L0)}(r,s,r',s',V) &= F^{(L0)}(s,r,s',r',V)\\ &= F^{(60)}(r,s,r',s',V)\\ &= \begin{bmatrix} pn - 1\\ V \end{bmatrix} \quad \text{for all } V \in Z. \end{aligned}$$

Proof: As in (7.31.1) and (7.31.1*) and in view of (7.0.2), by (7.22), (7.27), and (7.29) we get (7.32.1). As in (7.31.2) and (7.31.3), by (7.22), (7.27), (7.28), (7.29), and (7.30) we get (7.32.2) and (7.32.3). By (7.20) we get (7.32.4). By (7.20) and (7.32.1) we get (7.32.5). Finally, by (7.20) and (7.32.3) we get (7.32.6).

PROBLEM (7.33) Assuming that $r'(i) = s'(i) = i$ for all $i \in [1,p]$ and upon restricting L to vary in $\{1,1^*,2,2^*,3,3^*\}$, by (7.32.1) we see that the function F_D has 12 different expressions, and we see that *the function is symmetric in the pair* (r,s); i.e., it remains unchanged when we interchange r and s. Moreover, by (7.1) we know that $F_D^{(10)}$ has four variations and so does $F_D^{(1^*0)}$; by (7.2) we know that $F_D^{(20)}$ has eight variations and so does $F_D^{(2^*0)}$; by (7.3) we know $F_D^{(30)}$ has eight variations and so does $F_D^{(3^*0)}$. Thus, in a sense, we have found 80 different expressions for the function F_D. However, *none* of these 80 expressions *look* symmetric. So *the problem is to find an expression for* F_D *whose symmetry in* (r,s) *should be obvious by the very looks of it.*

For example, the function

$$\sum_{i=1}^{p} \{[(r(i) - s(i)]^2 + 4r(i)s(i)\}$$

does not look symmetric in (r,s), but it equals

$$\sum_{i=1}^{p} [r(i) + s(i)]^2$$

which is symmetric by the very looks of it.

REMARK (7.34) We now motivate the various results that have been given in this section.

Concerning the function F_D discussed in (7.33), originally in our Nice lectures, which appeared as "Combinatoire des tableaux de Young, varietes determinantielles et calcul de fontions de Hilbert" in Volume 42 of *Rendiconti del Seminario Matematico of Universita et Polytecnico di Torino* for the year 1984, we had only found the expression $F_D^{(20)}$. A secondary motivation for transforming $F_D^{(20)}$ into the 12 (or 80) different expressions was the search for a symmetric expression.

To explain the primary motivation, let us note that in $F_D^{(20)}$ the variables r and s are intertwined and do not appear in a nice shape. So we try to write $F_D^{(20)}$ as a polynomial in s whose coefficients are functions of r. However, we do not do this in the usual sense, but we do it in the sense of the twisted binomial basis (4.65.1*). In other words, we try to express F_D as a linear combination of the twisted monomials

$$\left(\prod_{i=1}^{p} \begin{bmatrix} s(i) \\ e(i) \end{bmatrix}\right)_{e \in N(p)}$$

with coefficients depending on r. In this manner $F_D^{(20)}$ *is converted into* $F_D^{(10)}$.

A further motivation for this conversion is to facilitate the establishment of the recursive property of F_D discussed in Section 8. Very approximately speaking, the recursive property relates the values of F_D at two different values of certain parameters. Comparing the two values of F_D is similar to proving a polynomial identity in the two sets of variables r and s. So, as usual, we write the sides as polynomials in s whose coefficients are polynomials in r and then we equate the corresponding coefficients. But we do this in the twisted binomial sense. And so on.

A final motivation for transforming $F_D^{(20)}$ into as may expressions as possible is simply the desire to do so—like climbing a mountain because it is there!

Carrying the above-mentioned idea of converting $F_D^{(20)}$ into $F_D^{(10)}$ one step further (i.e., backwards), we can say that in our Nice lectures we really started with the function $F^{(60)}$ and, as indicated in the last paragraph of (7.31), by expressing $F^{(60)}$ as a twisted polynomial in V, we obtained $F^{[20]}$; the degree of $F^{[20]}$ in V is R + S + p - 1, and the coefficient of $\begin{bmatrix} V \\ R + S + p - 1 - D \end{bmatrix}$ in $F^{[20]}$ is $F_D^{(20)}$. For the reasons outlined above, we have now expressed $F_D^{(20)}$ as a twisted polynomial in s and this has given us $F_D^{(10)}$. And so on.

REMARK (7.40) We shall now prove Theorem (5.7*). So let there be given any $m \in N^*(2)$ and $p \in N^*$ and $a \in vec(2,m,p)$ and $k \in [1,2]$, and let us put

$$k' = \begin{cases} 2 & \text{if } k = 1 \\ 1 & \text{if } k = 2. \end{cases}$$

Now, in view of (2.2.2), we see that (5.7*.1), (5.7*.2), and (5.7*.3) follow, respectively, from (7.32.1), (7.32.2), and (7.32.3) by taking

$$\left[\begin{array}{l} r(i) = m(k) - a(k,i) \text{ and } s(i) = m(k') - a(k',i) \text{ and} \\ r'(i) = s'(i) = i \text{ for all } i \in [1,p]. \end{array}\right.$$

Similarly, in view of (2.2.2), we see that (5.7*.4), (5.7*.5), and (5.7*.6) follow, respectively, from (7.32.4), (7.32.5), and (7.32.6) by taking

$$n = \max(m(1),m(2))$$

and

$$\begin{cases} r(i) = m(k) - a(k,i) \text{ and } s(i) = m(k') - a(k',i) \text{ and} \\ r'(i) = s'(i) = i \text{ for all } i \in [1,p]. \end{cases}$$

PROBLEM (7.41) Let there be given any $q \in N^*$ and $m \in N^*(q)$ and $p \in N^*$ and $a \in \text{vec}(q,m,p)$. Now, in view of (5.6'''), by (5.6*.11) to (5.6*.14) we see that

$$\text{Stab}(q,m,p,a,V) = F^{(8)}(q,m,p,a,V) \qquad \text{for all } V \in N$$

and hence by (5.6*.2) and (5.7*.2) we see that

$$\begin{cases} \text{if } q = 2, \text{ then for every } L \in \{1,1^*,2,2^*,3,3^*\} \text{ and} \\ k \in [1,2] \text{ we have } \text{Stab}(2\ m,p,a,V) = F^{[Lk]}(m,p,a,V) \\ \text{for all } V \in N \text{ where we recall that } F^{[Lk]}(m,p,a,V) \in Q[V]. \end{cases}$$

So we may pose the problem: Find better formulas to express Stab(q,m,p,a,V) for *every* q and not just for q = 2; in other words, find better formulas to count standard tableaux of any width and not just standard bitableaux. As a guideline, we may ask the concrete question: For every q is Stab(q,m,p,a,V) a polynomial in V; i.e., does there exist $F(q,m,p,a,V) \in Q[V]$ such that Stab(q,m,p,a,V) = F(q,m,p,a,V) for all $V \in N$?

8. RECURSIVE RELATIONS

In this section we prove Theorem (5.8*) by showing that certain recursive relations are satisfied by the determinantal polynomial $F^{(1k)}$ which was introduced in (2.10) and which, in a somewhat generalized form, will now be renamed as $F^{(91)}$; certain variations of $F^{(91)}$ will be called $F^{(92)}, \ldots, F^{(96)}$. Logically speaking, this section is independent of Subsections (2.10) and (2.11) and Sections 5 to 7; the one exception to this is Demonstrational Remark (8.40), in which the various results of this section are put together to prove Theorem (5.8*); the only other exceptions are (8.41) and (8.42), where we pose some related problems.

In this section let p be any fixed positive integer and also let any

$$x \in Z(p) \qquad y \in Z(p) \qquad r \in Z(p) \qquad r' \in Z(p) \qquad s' \in Z(p)$$

be fixed and let

$$R = \sum_{i=1}^{p} r(i) \quad \text{and} \quad R' = \sum_{i=1}^{p} r'(i) \quad \text{and} \quad S' = \sum_{i=1}^{p} s'(i)$$

and

$$C_0 = R + 2p + x(p) + \sum_{i=1}^{p} y(i)$$

and finally for every $s \in Z(p)$, let

$$C((s)) = R + p - 1 + \sum_{i=1}^{p} s(i)$$

and

$$C^*((s)) = p + x(p) + \sum_{i=1}^{p} [y(i) - s(i)].$$

Now first, for every $i \in [1,p]$ and $j \in [1,p]$ and $e \in Z(p)$ we define

$$G_{ij}^{(91)}(e) = \begin{Bmatrix} r(j) + r'(j) - s'(i) \\ r(j) - e(i) \end{Bmatrix},$$

and second, for every $e \in Z(p)$ we define

$$G^{(91)}(e) = \text{the } p \text{ by } p \text{ matrix whose } (i,j)\text{th entry is } G_{ij}^{(91)}(e),$$

and third, for every $s \in Z(p)$ and $e \in Z(p)$ we define

$$G^{[91]}(s,e) = \prod_{i=1}^{p} \begin{bmatrix} s(i) \\ e(i) \end{bmatrix}$$

and

$$H^{(91)}(s,e) = G^{[91]}(s,e) \det G^{(91)}(e),$$

and fourth, for every $s \in Z(p)$ and $E \in Z$ we define

$$H_E^{(91)}(s) = \sum_{e \in Z(p,E)} H^{(91)}(s,e),$$

and fifth, for every $s \in Z(p)$ and $D \in Z$ we define

$$F_D^{(91)}(s) = \sum_{E \in Z} (-1)^{R-E}\binom{E + R' - S'}{D + E - R} H_E^{(91)}(s).$$

Again, first for every $u \in J(p)$ and $e \in Z(p)$ we define

$$G_{ij}^{(92)}(u,e) = \begin{cases} \dbinom{r(j) + r'(j) - s'(i)}{1 + r(j) - e(i)} & \text{if } j \in [1,p] \text{ and } i \in u \\ \dbinom{r(j) + r'(j) - s'(i)}{r(j) - e(i)} & \text{if } j \in [1,p] \text{ and } i \in [1,p]\backslash u \end{cases}$$

and

$$G^{(92)}(u,e) = \text{the } p \text{ by } p \text{ matrix whose } (i,j)\text{th entry is } G_{ij}^{(92)}(u,e)$$

and

$$G_i^{[92]}(u,e) = \begin{cases} \begin{bmatrix} y(i) \\ e(i) \end{bmatrix} - \begin{bmatrix} x(i) \\ e(i) \end{bmatrix} & \text{if } i \in u \\ \begin{bmatrix} y(i) \\ e(i) \end{bmatrix} & \text{if } i \in [1,p]\backslash u \end{cases}$$

and

$$G^{[92]}(u,e) = \prod_{i=1}^{p} G_i^{[92]}(u,e)$$

and

$$H^{(92)}(u,e) = G^{[92]}(u,e) \det G^{(92)}(u,e),$$

and second, for every $E \in Z$ we define

$$H_E^{(92)} = \sum_{e \in Z(p,E)} \sum_{u \in J(p)} H^{(92)}(u,e),$$

and third, for every $t \in J(p)$ and $e \in Z(p)$ we define

$$G_{ij}^{(93)}(t,e) = \begin{cases} \dbinom{1 + r(j) + r'(j) - s'(i)}{1 + r(j) - e(i)} & \text{if } j \in [1,p] \text{ and } i \in [1,p]\backslash t \\ \dbinom{r(j) + r'(j) - s'(i)}{1 + r(j) - e(i)} & \text{if } j \in [1,p] \text{ and } i \in t \end{cases}$$

and

$$G^{(93)}(t,e) = \text{the p by p matrix whose (i,j)th entry is } G_{ij}^{(93)}(t,e)$$

and

$$G_i^{[93]}(t,e) = \begin{cases} \begin{bmatrix} y(i) \\ e(i) \end{bmatrix} & \text{if } i \in [1,p]\backslash t \\ -\begin{bmatrix} x(i) \\ e(i) \end{bmatrix} & \text{if } i \in t \end{cases}$$

and

$$G^{[93]}(t,e) = \prod_{i=1}^{p} G_i^{[93]}(t,e)$$

and

$$H^{(93)}(t,e) = G^{[93]}(t,e) \det G^{(93)}(t,e),$$

and fourth, for every $E \in Z$ we define

$$H_E^{(93)} = \sum_{e \in Z(p,E)} \sum_{t \in J(p)} H^{(93)}(t,e),$$

and fifth, for every $I \in [0,p]$ and $e \in Z(p)$ we define

$$G_{ij}^{(94)}(I,e) = \begin{cases} \binom{1 + r(j) + r'(j) - s'(i)}{1 + r(j) - e(i)} & \text{if } j \in [1,p] \text{ and } i \in [1,I] \\ \binom{r(j) + r'(j) - s'(i)}{1 + r(j) - e(i)} & \text{if } j \in [1,p] \text{ and } i \in [I + 1, p] \end{cases}$$

and

$$G^{(94)}(I,e) = \text{the p by p matrix whose (i,j)th entry is } G_{ij}^{(94)}(I,e)$$

and

$$G_i^{[94]}(I,e) = \begin{cases} \begin{bmatrix} y(i) \\ e(i) \end{bmatrix} & \text{if } i \in [1,I] \\ -\begin{bmatrix} x(i) \\ e(i) \end{bmatrix} & \text{if } i \in [I + 1, p] \end{cases}$$

and

$$G^{[94]}(I,e) = \prod_{i=1}^{p} G_i^{[94]}(I,e)$$

and

$$H^{(94)}(I,e) = G^{[94]}(I,e) \det G^{(94)}(I,e)$$

and sixth, for every $E \in Z$ we define

$$H_E^{(94)} = \sum_{e \in Z(p,E)} \sum_{I \in [0,p]} H^{(94)}(I,e)$$

and seventh, for every $e \in Z(p)$ we define

$$G_{ij}^{(95)}(e) = \binom{1 + r(j) + r'(j) - s'(i)}{1 + r(j) - e(i)} \qquad \text{if } j \in [1,p] \text{ and } i \in [1,p]$$

and

$$G^{(95)}(e) = \text{the p by p matrix whose (i,j)th entry is } G_{ij}^{(95)}(e)$$

and

$$G^{[95]}(e) = \prod_{i=1}^{p} \begin{bmatrix} y(i) \\ e(i) \end{bmatrix}$$

and

$$H^{(95)}(e) = G^{[95]}(e) \det G^{(95)}(e),$$

and eighth, for every $E \in Z$ we define

$$H_E^{(95)} = \sum_{e \in Z(p,E)} H^{(95)}(e),$$

and ninth, for every $e \in Z(p + 1)$ we define

$$G_{ij}^{(96)}(e) = \begin{cases} \binom{1 + r(j) + r'(j) - s'(i)}{1 + r(j) - e(i)} & \text{if } j \in [1,p] \text{ and } i \in [1,p] \\ \binom{r(j) + r'(j) - s'(p)}{1 + r(j) - e(p + 1)} & \text{if } j \in [1,p] \text{ and } i = p + 1 \\ \binom{1 + r'(p) - s'(i)}{-e(i)} & \text{if } j = p + 1 \text{ and } i \in [1,p] \\ \binom{r'(p) - s'(p)}{-e(p + 1)} & \text{if } j = p + 1 \text{ and } i = p + 1 \end{cases}$$

and

$$G^{(96)}(e) = \text{the } p + 1 \text{ by } p + 1 \text{ matrix whose } (i,j)\text{th entry is } G^{(96)}_{ij}(e)$$

and

$$G^{[96]}(e) = \begin{bmatrix} x(p) \\ e(p+1) \end{bmatrix} \prod_{i=1}^{p} \begin{bmatrix} y(i) \\ e(i) \end{bmatrix}$$

and

$$H^{(96)}(e) = G^{[96]}(e) \det G^{(96)}(e),$$

and tenth, for every $E \in Z$ we define

$$H^{(96)}_E = \sum_{e \in Z(p+1,E)} H^{(96)}(e),$$

and finally, for every $L \in [2,6]$ and $D \in Z$ we define

$$F^{(9L)}_D = \sum_{E \in Z} (-1)^{p+R-E} \binom{E + R' - S'}{D + E - p - R} H^{(9L)}_E .$$

Now, *if*

$$R' = S' \qquad [*]$$

then for every $s \in Z(p)$ and $C' \in Z$ and $V \in Z$ we define

$$F^{((91))}(s,C',V) = \sum_{D \in Z} (-1)^D F^{(91)}_D(s) \begin{bmatrix} C' - D \\ V \end{bmatrix}.$$

Again, *if* [*], *then* for every $s \in Z(p)$ and $C' \in Z$, in $Q[V]$ we define

$$F^{[[91]]}(s,C',V) = \sum_{D \in Z} (-1)^D F^{(91)}_D(s) \begin{bmatrix} V \\ C' - D \end{bmatrix}.$$

Again, *if* [*], *then* for every $L \in [2,6]$ and $C' \in Z$ and $V \in Z$ we define

$$F^{((9L))}(C',V) = \sum_{D \in Z} (-1)^D F^{(9L)}_D \begin{bmatrix} C' - D \\ V \end{bmatrix}.$$

Finally, *if* [*], *then* for every $L \in [2,6]$ and $C' \in Z$, in $Q[V]$ we define

$$F^{[[9L]]}(C',V) = \sum_{D \in Z} (-1)^D F_D^{(9L)} \begin{bmatrix} & V \\ C' & D \end{bmatrix}.$$

REMARK (8.0) In this section, sometimes we shall assume condition [*], and sometimes we shall assume that

$$\text{for all } i \in [1,p] \text{ we have } x(i) < y(i) \qquad [**]$$

and sometimes we shall assume that

$$p + x(p) + \sum_{i=1}^{p} y(i) \geq 0 \qquad [***]$$

and sometimes we shall assume that

$$\begin{matrix} \text{for all } i \in [1, p-1] \text{ we have} \\ y(i+1) = x(i) \text{ and } s'(i+1) = 1 + s'(i) \end{matrix} \qquad [****]$$

and sometimes we shall assume that

$$\begin{bmatrix} x(p) = -1 \text{ and for every } j \in [1,p] \text{ we have} \\ r(j) + r'(j) - s'(p) \geq 0 \leq s'(p) - r'(j) \end{bmatrix} \qquad (8.0.1')$$

and sometimes we shall assume that

$$\begin{bmatrix} \text{for all } i \in [1,p] \text{ we have} \\ -1 \leq x(i) < y(i) \text{ and } r'(i) = s'(i) = i, \\ \text{and for all } i \in [1, p-1] \text{ we have } y(i+1) = x(i). \end{bmatrix} \qquad (8.0.2)$$

In (8.10) to (8.40) we prove certain recurrence relations concerning the various functions $F^{(91)}$, $F^{(92)}$, ..., etc., defined above. Before doing so, in (8.1) to (8.6) we establish the essential finiteness of the summations involved in the definitions of these functions above.

OBSERVATION (8.1) Let there be given any

$x^* \in Z(p)$, $y^* \in Z(p)$, $x' \in Z(p)$, $y' \in Z(p)$, $r^{**} \in Z(p)$, and $r^* \in Z(p)$

such that

$$r^{**}(j) \leq r^*(j) \qquad \text{for all } j \in [1,p]$$

and let us put

$$R^* = \sum_{i=1}^{p} r^*(j).$$

Moreover, for every $i \in [1,p]$ and $j \in [1,p]$, let there be given any

$$g[i,j] \in Z.$$

For every $i \in [1,p]$ and $j \in [1,p]$ and $e \in Z(p)$ let us put

$$G^*_{ij}(e) = \binom{g[i,j]}{r^{**}(j) - e(i)}.$$

Finally, for every $e \in Z(p)$, let us put

$$G^*(e) = \text{the } p \text{ by } p \text{ matrix whose } (i,j)\text{th entry is } G^*_{ij}(e)$$

and

$$G^{[*]}(e) = \prod_{i=1}^{p} \left(x'(i)\begin{bmatrix}x^*(i)\\ e(i)\end{bmatrix} + y'(i)\begin{bmatrix}y^*(i)\\ e(i)\end{bmatrix}\right)$$

and

$$H^*(e) = G^{[*]}(e) \det G^*(e).$$

Then we have

$$\{e \in Z(p) : H^*(e) \neq 0\} \subset N[p,R^*].$$

Proof: By (2.5.2*) and the definition of $G^{[*]}(e)$ we see that

$$(1) \quad \{e \in Z(p) : G^{[*]}(e) \neq 0\} \subset N(p).$$

Upon taking r^{**} for r in (2.6.20) we get

$$\{e \in Z(p) : \det G^*(e) \neq 0\} \subset Z[p, r^{**}(1) + r^{**}(2) + \cdots + r^{**}(p)]$$

and clearly, we have

$$Z[p, r^{**}(1) + r^{**}(2) + \cdots + r^{**}(p)] \subset Z[p,R^*]$$

and therefore in view of (1) we get

$$\{e \in Z(p) : H^*(e) \neq 0\} \subset N[p,R^*].$$

OBSERVATION (8.2) Given any $R^* \in Z$, we have that

$$\left\{D \in Z : \left\{E \in [0,R^*] : \binom{E}{D + E - R^*} \neq 0\right\} \neq \emptyset\right\} \subset [0,R^*].$$

Proof: Follows by taking R^* for R in (2.6.13).

OBSERVATION (8.3) Given any $s \in Z(p)$, we have

$$\{e \in Z(p) : H^{(91)}(s,e) \neq 0\} \subset N[p,R]. \tag{8.3.1}$$

Given any $u \in J(p)$, we have

$$\{e \in Z(p) : H^{(92)}(u,e) \neq 0\} \subset N[p, R + p]. \tag{8.3.2}$$

Given any $t \in J(p)$, we have

$$\{e \in Z(p) : H^{(93)}(t,e) \neq 0\} \subset N[p, R + p]. \tag{8.3.3}$$

Given any $I \in [0,p]$, we have

$$\{e \in Z(p) : H^{(94)}(I,e) \neq 0\} \subset N[p, R + p]. \tag{8.3.4}$$

Also, we have

$$\{e \in Z(p) : H^{(95)}(e) \neq 0\} \subset N[p, R + p]. \tag{8.3.5}$$

Finally, we have

$$\{e \in Z(p + 1) : H^{(96)}(e) \neq 0\} \subset N[p, R + p]. \tag{8.3.6}$$

Proof: (8.3.1) follows by taking $L = 1$ in (2.6.21). Given any $L \in [2,5]$, (8.3.L) follows from (8.1) by letting r^* be the member of $Z(p)$ obtained by putting

$$r^*(j) = 1 + r(j) \qquad \text{for all } j \in [1,p]$$

and by letting x^*, y^*, x', y', r^{**}, and $g[i,j]$ have appropriate values which are obvious from the context. Finally, (8.3.6) follows from (8.1) by changing p to $p + 1$ and by letting r^* be the member of $Z(p + 1)$ obtained by putting

$$r^*(j) = \begin{cases} 1 + r(j) & \text{if } j \in [1,p] \\ 0 & \text{if } j = p + 1 \end{cases}$$

and by letting x*, y*, x', y', r**, and g[i,j] have appropriate values which are obvious from the context.

OBSERVATION (8.4) We have that

$$\left[\begin{array}{l} \textit{if } [*], \textit{ then} \\ \left\{D \in Z : \left\{E \in [0,R] : \binom{E + R' - S'}{D + E - R} \neq 0\right\} \neq \emptyset\right\} \subset [0,R]. \end{array}\right. \tag{8.4.1}$$

Moreover,

$$\left[\begin{array}{l} \textit{if } [***], \textit{ then for every } D \in [0,R] \text{ and } V \in N \text{ and} \\ U \in [0,p] \text{ we have } (V - U) + (C_0 - D) \geq 0. \end{array}\right. \tag{8.4.1'}$$

Also, we have that

$$\left[\begin{array}{l} \textit{if } [*], \textit{ then} \\ \left\{D \in Z : \left\{E \in [0, R + p] : \binom{E + R' - S'}{D + E - p - R} \neq 0\right\} \neq \emptyset\right\} \\ \subset [0, R + p]. \end{array}\right. \tag{8.4.2}$$

Moreover,

$$\left[\begin{array}{l} \textit{if } [***], \textit{ then for every } D \in [0, R + p] \\ \text{we have } C_0 - D \geq 0. \end{array}\right. \tag{8.4.2'}$$

Proof: (8.4.1) follows from (2.6.13). (8.4.1') is obvious. (8.4.2) follows from (8.2) by taking R* = R + p. (8.4.2') is obvious.

OBSERVATION (8.5) Let there be given any $s \in Z(p)$. Then by (8.3.1), first we see that for every $E \in Z$, the summation in the definition of $H_E^{(91)}(s)$ is essentially finite, and second, we see that

$$\{E \in Z : H_E^{(91)}(s) \neq 0\} \subset [0,R]. \tag{8.5.1}$$

By (8.5.1), first we see that for every $D \in Z$, the summation in the definition of $F_D^{(91)}(s)$ is essentially finite, and second, in view of (8.4.1) we see that

$$\left[\begin{array}{l} \textit{if } [*], \textit{ then} \\ \{D \in Z : F_D^{(91)}(s) \neq 0\} \subset [0,R]. \end{array}\right. \tag{8.5.2}$$

By (8.5.2), first we see that

$$\left[\begin{array}{l}\textit{if } [*], \textit{ then the summations in the definitions of} \\ F^{((91))}(s,C',V) \text{ and } F^{[[91]]}(s,C',V) \text{ are essentially} \\ \text{finite}\end{array}\right. \quad (8.5.3)$$

and second, in view of (8.4.1') and (2.5.10*), we see that

$$\left[\begin{array}{l}\textit{if } [*] \textit{ and } [***], \textit{ then for every } V \in N \text{ and } U \in [0,p] \\ \text{we have } F^{((91))}(s,\ C_0,\ V - U) = F^{[[91]]}(s,\ C_0,\ V - U).\end{array}\right. \quad (8.5.4)$$

Proof: The proofs of all the assertions are embedded in those assertions.

OBSERVATION (8.6) Let there be given any $L \in [2,6]$. Then by (8.3.L), first we see that for every $E \in Z$, the summations in the definition of $H_E^{(9L)}$ are essentially finite, and second, we see that

$$\{E \in Z : H_E^{(9L)} \neq 0\} \subset [0,\ R + p]. \quad (8.6.1)$$

By (8.6.1), first we see that for every $D \in Z$, the summation in the definition of $F_D^{(9L)}$ is essentially finite, and second, in view of (8.4.2) we see that

$$\left[\begin{array}{l}\textit{if } [*], \textit{ then} \\ \{D \in Z : F_D^{(9L)} \neq 0\} \subset [0,\ R + p].\end{array}\right. \quad (8.6.2)$$

By (8.6.2), first we see that

$$\left[\begin{array}{l}\textit{if } [*], \textit{ then the summations in the definitions of} \\ F^{((9L))}(c',V) \text{ and } F^{[[9L]]}(C',V) \text{ are essentially} \\ \text{finite,}\end{array}\right. \quad (8.6.3)$$

and second, in view of (8.4.2') and (2.5.7*), we see that

$$\left[\begin{array}{l}\textit{if } [*] \textit{ and } [***], \textit{ then for every } V \in N \text{ we have} \\ F^{((9L))}(C_0,V) = F^{[[9L]]}(C_0,V).\end{array}\right. \quad (8.6.4)$$

Proof: The proofs of all the assertions are embedded in those assertions.

(8.10) Notation for Summation Over the Faces of a Box Let us consider the p-dimensional box bounded by x and y, and let us consider the open faces and the closed faces of this box defined thus. For every u ∈ J(p), let X[u] be the card(u)-dimensional open face obtained by putting

$$\left[\begin{array}{l} X[u] = \{s \in Z(p) : s(i) \in [x(i) + 1, y(i) - 1] \text{ for} \\ \qquad \text{all } i \in u, \text{ and } s(j) = y(j) \text{ for all } j \in [1,p]\backslash u\} \end{array}\right. \tag{8.10.1}$$

and let Y[u] be the card(u)-dimensional closed face obtained by putting

$$\left[\begin{array}{l} Y[u] = \{s \in Z(p) : s(i) \in [x(i) + 1, y(i)] \text{ for} \\ \qquad \text{all } i \in u, \text{ and } s(j) = y(j) \text{ for all } j \in [1,p]\backslash u\}. \end{array}\right. \tag{8.10.2}$$

For every U ∈ [0,p], let us also consider the U-dimensional open skeleton (i.e., the union of all U-dimensional open faces) X(U) of our box obtained by putting

$$X(U) = \bigcup_{u \in J(p,U)} X[u]. \tag{8.10.3}$$

(Note that Y[[1,p]] is the box itself. Also note that the dimensionality descriptions above make only heuristic sense and then only if condition [**] is satisfied. Finally, note that we are considering only faces emanating from the top vertex. For example, in the case p = 2 we consider the two open edges marked with an arrow

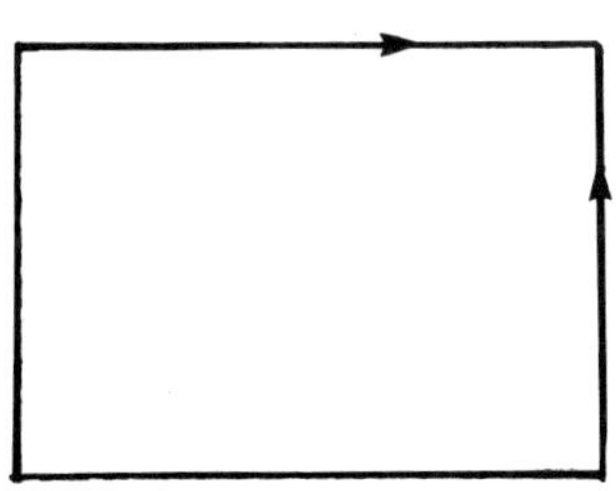

and disregard the two unmarked edges; similarly, we consider only the two corresponding edges.)

Now clearly

$$\left[\begin{array}{l} \text{given any } u \in J(p) \text{ and } s \in X[u], \text{ we have} \\ \{j \in [1,p] : s(j) \neq y(j)\} = u \end{array}\right.$$

and hence

$$\begin{bmatrix}\text{if } u' \in J(p) \text{ and } u^* \in J(p) \text{ are such that } u' \neq u^*, \\ \text{then } X[u'] \cap X[u^*] = \emptyset.\end{bmatrix} \tag{8.10.4}$$

Also, clearly,

$$\begin{bmatrix}\text{given any } u \in J(p) \text{ and } s \in Y[u], \text{ upon letting} \\ t = \{j \in [1,p] : s(j) \neq y(j)\}, \text{ we have } t \subset u \text{ and } s \in X[t]\end{bmatrix}$$

and clearly,

$$\begin{bmatrix}\textit{if } [**], \textit{ then } \text{for every } u \in J(p) \text{ and } t \subset u \\ \text{we have } X[t] \subset Y[u]\end{bmatrix}$$

and therefore

$$\begin{bmatrix}\textit{if } [**], \textit{ then } \text{for every } u \in J(p) \text{ we have} \\ Y[u] = \bigcup_{t \in P(u)} X[t].\end{bmatrix} \tag{8.10.5}$$

By (8.10.4) we see that the union in the RHS of (8.10.3) is pairwise disjoint and hence

$$\begin{bmatrix}\text{given any } U \in [0,p], \text{ for every map } f\colon Z(p) \to Q \\ \text{we have } \sum_{s \in X(U)} f(s) = \sum_{u \in J(p,U)} \sum_{s \in X[u]} f(s).\end{bmatrix} \tag{8.10.6}$$

However, the same does not hold for closed faces (namely, if condition [**] is satisfied, then for every $u \in J(p)$ we have $y \in Y[u]$ and hence for every $u' \in J(p)$ and $u^* \in J(p)$ we have $Y[u'] \cap Y[u^*] \neq \emptyset$). So we proceed as follows.

For every set $\hat{Z}$, let $\mathrm{Hom}(\hat{Z},Q)$ denote the set of all maps of $\hat{Z}$ into Q; note that then $\mathrm{Hom}(\hat{Z},Q)$ is an additive abelian group; namely, for any f and g in $\mathrm{Hom}(\hat{Z},Q)$ we define $f + g$ in $\mathrm{Hom}(\hat{Z},Q)$ by putting

$$(f + g)(s) = f(s) + g(s) \qquad \text{for all } s \in \hat{Z}.$$

Now for every $u \in J(p)$ we define

$$X^*[u] \text{ and } Y^*[u] \text{ in } \mathrm{Hom}(\mathrm{Hom}(Z(p),Q),Q)$$

by putting

$$X^*[u](f) = \sum_{s \in X[u]} f(s) \qquad \text{for all } f \in \mathrm{Hom}(Z(p),Q)$$

and

$$Y^*[u](f) = \sum_{s \in Y[u]} f(s) \qquad \text{for all } f \in \mathrm{Hom}(Z(p),Q).$$

Finally, for every $U \in [0,p]$ we define

$$X^*(U) \text{ and } Y^*(U) \text{ in } \mathrm{Hom}(\mathrm{Hom}(Z(p),Q),Q)$$

by putting

$$X^*(U) = \sum_{u \in J(p,U)} X^*[u] \quad \text{and} \quad Y^*(U) = \sum_{u \in J(p,U)} Y^*[u];$$

note that then, for all $f \in \mathrm{Hom}(Z(p),Q)$, we have

$$X^*(U)(f) = \sum_{u \in J(p,U)} X^*[u](f) = \sum_{u \in J(p,U)} \sum_{s \in X[u]} f(s)$$

and

$$Y^*(U)(f) = \sum_{u \in J(p,U)} Y^*[u](f) = \sum_{u \in J(p,U)} \sum_{s \in Y[u]} f(s).$$

(A more suggestive notation for $X^*[u]$ could have been

$$\sum_{X[u]} \qquad \text{or} \qquad \int_{X[u]}$$

and similarly for $Y^*[u]$ and so on; at any rate, summation and integration are essentially the same operation, the former being restricted to discrete objects.)

Now by (8.10.4) and (8.10.5) we see that

$$\left[\begin{array}{l} \textit{if } [**], \textit{ then} \text{ for every } u \in J(p) \text{ we have} \\ Y^*[u] = \sum_{t \in P(u)} X^*[t]. \end{array}\right. \tag{8.10.7}$$

In view of (8.10.7), by (4.26) we see that

$$\left[\begin{array}{l} \textit{if } [**], \textit{ then} \text{ for every } U \in [0,p] \text{ we have} \\ Y^*(U) = \sum_{T \in [0,p]} \binom{p - T}{U - T} X^*(T). \end{array}\right. \tag{8.10.8}$$

Now *if* [**], *then* given any $I \in Z$ we get

$$\sum_{U \in [0,p]} \binom{U}{I} X^*(U)$$

$$= \sum_{T\in[0,p]} X^*(T)\binom{T}{I} \qquad \text{(by putting } U = T)$$

$$= \sum_{T\in[0,p]} X^*(T) \sum_{U\in[0,p]} (-1)^{p-U}\binom{U}{p-I}\binom{p-T}{U-T} \qquad \text{[by (4.46)]}$$

$$= \sum_{U\in[0,p]} (-1)^{p-U}\binom{U}{p-I} \sum_{T\in[0,p]} \binom{p-T}{U-T} X^*(T)$$

$$= \sum_{U\in[0,p]} (-1)^{p-U}\binom{U}{p-I} Y^*(U) \qquad \text{[by (8.10.8)]}$$

and thus we see that

if [**], *then* for every $I \in Z$ we have

$$\sum_{U\in[0,p]} \binom{U}{I} X^*(U) = \sum_{U\in[0,p]} (-1)^{p-U}\binom{U}{p-I} Y^*(U). \tag{8.10.9}$$

Finally, by (8.10.6) and (8.10.9) we get the

Box Lemma, which says that *if* [**], *then* given any map $f\colon Z(p) \to Q$ and given any $I \in Z$, we have

$$\sum_{U\in[0,p]} \sum_{s\in X(U)} \binom{U}{I} f(s) = \sum_{U\in[0,p]} \sum_{u\in J(p,U)} \sum_{s\in Y[u]} (-1)^{p-U}\binom{U}{p-I} f(s). \tag{8.10.10}$$

REMARK *From this subsection we shall explicitly use only* (8.10.1), (8.10.2), (8.10.3), (8.10.4), and (8.10.10). Note that in view of (4.27), the discussion of this subsection (8.10) remains verbatim valid if we replace Q by any additive abelian group.

PROPOSITION (8.11) Assume [**]. Then for every $D \in Z$ we have

$$\sum_{I\in[0,p]} \sum_{U\in[0,p]} \sum_{s\in X(U)} \binom{U}{I} F^{(91)}_{D-I}(s) = F^{(92)}_{D}.$$

Proof: Let there be fixed any $D \in Z$. Now upon letting

$$\Sigma^{**} \text{ stand for } \sum_{U\in[0,p]} \sum_{u\in J(p,U)} \sum_{s\in Y[u]}$$

by (8.10.10) we see that

(1) $$\left[\begin{array}{l}\text{for every } I \in Z \text{ we have} \\ \displaystyle\sum_{U\in[0,p]} \sum_{s\in X(U)} \binom{U}{I} F_{D-I}^{(91)}(s) = \Sigma^{**} (-1)^{p-U}\binom{U}{p-I} F_{D-I}^{(91)}(s).\end{array}\right.$$

Next we have

$$\begin{aligned}
&\sum_{I\in[0,p]} \sum_{U\in[0,p]} \sum_{s\in X(U)} \binom{U}{I} F_{D-I}^{(91)}(s) \\
&= \sum_{I\in[0,p]} \Sigma^{**} (-1)^{p-U}\binom{U}{p-I} F_{D-I}^{(91)}(s) \qquad \text{[by (1)]} \\
&= \Sigma^{**} \sum_{I\in[0,p]} (-1)^{p-U}\binom{U}{p-I} F_{D-I}^{(91)}(s) \\
&= \Sigma^{**} \sum_{I\in[0,p]} (-1)^{p-U}\binom{U}{p-I} \sum_{E\in Z} (-1)^{R-E}\binom{E+R'-S'}{D-I+E-R} H_E^{(91)}(s) \\
&\qquad\qquad (\text{by definition of } F_D^{(91)}) \\
&= \Sigma^{**} \sum_{E\in Z} (-1)^{p+R-E-U}\left[\sum_{I\in[0,p]} \binom{U}{p-I}\binom{E+R'-S'}{D-I+E-R}\right] H_E^{(91)}(s) \\
&= \Sigma^{**} \sum_{E\in Z} (-1)^{p+R-E-U}\binom{E+U+R'-S'}{D+E+U-p-R} H_E^{(91)}(s) \qquad \text{[by (4.43)]} \\
&= \Sigma^{**} \sum_{T\in Z} (-1)^{p+R-T}\binom{T+R'-S'}{D+T-p-R} H_{T-U}^{(91)}(s) \qquad (\text{by putting } E+U=T) \\
&= \Sigma^{**} \sum_{E\in Z} (-1)^{p+R-E}\binom{E+R'-S'}{D+E-p-R} H_{E-U}^{(91)}(s) \qquad (\text{by putting } T=E) \\
&= \sum_{E\in Z} \left[(-1)^{p+R-E}\binom{E+R'-S'}{D+E-p-R} \Sigma^{**} H_{E-U}^{(91)}(s)\right]
\end{aligned}$$

where in view of (8.5.1), the one occurrence of the summation $\Sigma_{T\in Z}$ and the five occurrences of the summation $\Sigma_{E\in Z}$ are all essentially finite, and therefore we conclude that

(2) $$\left[\begin{array}{l}\displaystyle\sum_{I\in[0,p]} \sum_{U\in[0,p]} \sum_{s\in X(U)} \binom{U}{I} F_{D-I}^{(91)}(s) \\ \displaystyle= \sum_{E\in Z} \left[(-1)^{p+R-E}\binom{E+R'-S'}{D+E-p-R} \Sigma^{**} H_{E-U}^{(91)}(s)\right] \\ \text{where the summation } \Sigma_{E\in Z} \text{ is essentially finite.}\end{array}\right.$$

Now for every $E \in Z$ we have

$$\Sigma^{**} H^{(91)}_{E-U}(s)$$
$$= \sum_{U\in[0,p]} \sum_{u\in J(p,U)} \sum_{s\in Y[u]} H^{(91)}_{E-U}(s) \qquad \text{(by definition of } \Sigma^{**})$$
$$= \sum_{u\in J(p)} \sum_{s\in Y[u]} H^{(91)}_{E-\mathrm{card}(u)}(s)$$

and thus we see that

$$(3) \quad \left[\begin{array}{l}\text{for every } E \in Z \text{ we have} \\ \Sigma^{**} H^{(91)}_{E-U}(s) = \sum_{u\in J(p)} \sum_{s\in Y[u]} H^{(91)}_{E-\mathrm{card}(u)}(s).\end{array}\right.$$

Given any $u \in J(p)$ and $e \in Z(p)$, upon letting

$d(u,e)$ be the member of $Z(p)$

obtained by putting

$$d(u,e)(i) = \begin{cases} e(i) - 1 & \text{if } i \in u \\ e(i) & \text{if } i \in [1,p]\backslash u, \end{cases}$$

by the definitions of $G^{(91)}_{ij}$ and $G^{(92)}_{ij}$ we see that

$$G^{(91)}_{ij}(d(u,e)) = G^{(92)}_{ij}(u,e) \qquad \text{for all } i \in [1,p] \text{ and } j \in [1,p]$$

and hence by the definitions of $G^{(91)}$ and $G^{(92)}$ we see that

$$\det G^{(91)}(d(u,e)) = \det G^{(92)}(u,e)$$

and upon multiplying both sides of the equation above by $G^{[91]}(s, d(u,e))$, and upon noting that by the definition of $H^{(91)}$ we have

$$H^{(91)}(s,d(u,e)) = G^{[91]}(s,d(u,e)) \det G^{(91)}(d(u,e)),$$

we see that for every $s \in Z(p)$ we have

$$H^{(91)}(s,d(u,e)) = G^{[91]}(s,d(u,e)) \det G^{(92)}(u,e)$$

and now, upon summing both sides of the equation above as s varies over $Y[u]$, we conclude that

$$(4)\quad \left[\begin{array}{l}\text{for every } u \in J(p) \text{ and } e \in Z(p) \text{ we have}\\ \sum_{s\in Y[u]} H^{(91)}(s,d(u,e)) = \left[\sum_{s\in Y[u]} G^{[91]}(s,d(u\ e))\right]\det G^{(92)}(u,e).\end{array}\right.$$

Given any $u \in J(p)$ and $e \in Z(p)$ we have

$$\begin{aligned}
&\sum_{s\in Y[u]} G^{[91]}(s,d(u,e))\\
&= \sum_{s\in Y[u]} \left(\prod_{i\in u}\begin{bmatrix} s(i)\\ e(i)-1\end{bmatrix}\right)\left(\prod_{i\in[1,p]\backslash u}\begin{bmatrix} s(i)\\ e(i)\end{bmatrix}\right) && \text{(by definitions of } G^{[91]} \text{ and d)}\\
&= \left(\prod_{i\in u}\ \sum_{s(i)\in[x(i)+1,y(i)]}\begin{bmatrix} s(i)\\ e(i)-1\end{bmatrix}\right)\left(\prod_{i\in[1,p]\backslash u}\begin{bmatrix} y(i)\\ e(i)\end{bmatrix}\right) && \text{(by definition of Y)}\\
&= \left(\prod_{i\in u}\left\{\begin{bmatrix} y(i)\\ e(i)\end{bmatrix} - \begin{bmatrix} x(i)\\ e(i)\end{bmatrix}\right\}\right)\left(\prod_{i\in[1,p]\backslash u}\begin{bmatrix} y(i)\\ e(i)\end{bmatrix}\right) && \text{[by (4.12*)]}\\
&= G^{[92]}(e) && \text{(by definition of } G^{[92]})
\end{aligned}$$

and therefore, in view of (4) and upon remembering the defining equation

$$H^{(92)}(u,e) = G^{[92]}(e)\ \det G^{(92)}(e),$$

we conclude that

$$(5)\quad \left[\begin{array}{l}\text{for every } u \in J(p) \text{ and } e \in Z(p) \text{ we have}\\ \sum_{s\in Y[u]} H^{(91)}(s,d(u,e)) = H^{(92)}(u,e).\end{array}\right.$$

Given any $u \in J(p)$ and $s \in Z(p)$ and $E \in Z$, we have

$$\begin{aligned}
&H^{(91)}_{E-\mathrm{card}(u)}(s)\\
&= \sum_{e\in Z(p,E-\mathrm{card}(u))} H^{(91)}(s,e) && \text{(by definition of } H^{(91)}_E)\\
&= \sum_{e\in Z(p,E)} H^{(91)}(s,d(u,e)) && \text{(by definition of d)}
\end{aligned}$$

where by (8.3.1) we know that both the summations are essentially finite, and hence we see that

$$(6)\quad \left[\begin{array}{l}\text{for every } u \in J(p) \text{ and } s \in Z(p) \text{ and } E \in Z \text{ we have} \\ H^{(91)}_{E-\mathrm{card}(u)}(s) = \sum_{e\in Z(p,E)} H^{(91)}(s,d(u,e)) \\ \text{where summation in the RHS is essentially finite.}\end{array}\right.$$

Given any $E \in Z$ we have

$$\begin{aligned}
&\Sigma^{**}\, H^{(91)}_{E-U}(s) \\
&= \sum_{u\in J(p)} \sum_{s\in Y[u]} H^{(91)}_{E-\mathrm{card}(u)}(s) && \text{[by (3)]} \\
&= \sum_{u\in J(p)} \sum_{s\in Y[u]} \sum_{e\in Z(p,E)} H^{(91)}(s,d(u,e)) && \text{[by (6)]} \\
&= \sum_{e\in Z(p,E)} \sum_{u\in J(p)} \left[\sum_{s\in Y[u]} H^{(91)}(s,d(u,e))\right] \\
&= \sum_{e\in Z(p,E)} \sum_{u\in J(p)} H^{(92)}(u,e) && \text{[by (5)]} \\
&= H^{(92)}_E && \text{(by definition of } H^{(92)}_E\text{)}
\end{aligned}$$

where the first two occurrences of the summation

$$\sum_{e\in Z(p,E)}$$

are essentially finite by (8.3.1), whereas its third occurrence is essentially finite by (8.3.2), and thus we conclude that

(7) for every $E \in Z$ we have $\Sigma^{**}\, H^{(91)}_{E-U}(s) = H^{(92)}_E$.

Now in view of (8.5) and the definition of $F^{(92)}_D$, by (2) and (7) we conclude that

$$\sum_{I\in[0,p]} \sum_{U\in[0,p]} \sum_{s\in X(U)} \binom{U}{I} F^{(91)}_{D-I}(s) = F^{(92)}_D.$$

PROPOSITION (8.12) Assume [*] and [**]. Then for every $C' \in Z$, in $Q[V]$ we have

$$\sum_{U\in[0,p]} \sum_{s\in X(U)} F^{[[91]]}(s, C', V - U) = F^{[[92]]}(C',V).$$

Proof: Let there be given any $C' \in Z$. Now upon letting

$$\Sigma' \text{ stand for } \sum_{U\in[0,p]} \sum_{s\in X(U)}$$

in $Q[V]$ we have

$$\begin{aligned}
&\Sigma' F^{[[91]]}(s, C', V - U) \\
&= \Sigma' \sum_{D\in Z} (-1)^D F_D^{(91)}(s) \begin{bmatrix} V - U \\ C' - D \end{bmatrix} && \text{(by definition of } F^{[[91]]}\text{)} \\
&= \Sigma' \sum_{D\in Z} \left\{ (-1)^D F_D^{(91)}(s) \sum_{I\in[0,p]} (-1)^I \binom{U}{I} \begin{bmatrix} V \\ C' - D - I \end{bmatrix} \right\} && \text{[by (4.44)]} \\
&= \Sigma' \sum_{I\in[0,p]} \left\{ \sum_{D\in Z} (-1)^{D+I} \binom{U}{I} F_D^{(91)}(s) \begin{bmatrix} V \\ C' - D - I \end{bmatrix} \right\} \\
&= \Sigma' \sum_{I\in[0,p]} \left\{ \sum_{T\in Z} (-1)^T \binom{U}{I} F_{T-I}^{(91)}(s) \begin{bmatrix} V \\ C' - T \end{bmatrix} \right\} && \text{(by putting } D + I = T\text{)} \\
&= \Sigma' \sum_{I\in[0,p]} \left\{ \sum_{D\in Z} (-1)^D \binom{U}{I} F_{D-I}^{(91)}(s) \begin{bmatrix} V \\ C' - D \end{bmatrix} \right\} && \text{(by putting } T = D\text{)} \\
&= \sum_{D\in Z} (-1)^D \left\{ \sum_{I\in[0,p]} \Sigma' \binom{U}{I} F_{D-I}^{(91)}(s) \right\} \begin{bmatrix} V \\ C' - D \end{bmatrix} \\
&= \sum_{D\in Z} (-1)^D F_D^{(92)} \begin{bmatrix} V \\ C' - D \end{bmatrix} && \text{[by (8.11)]} \\
&= F^{[[92]]}(C', V) && \text{(by definition of } F^{[[92]]}\text{)}
\end{aligned}$$

where, by (8.5.2), we see that the one occurrence of $\Sigma_{T\in Z}$ and the first five occurrences of $\Sigma_{D\in Z}$ are essentially finite, and by (8.6.2) we see that the last occurrence of $\Sigma_{D\in Z}$ is also essentially finite. Thus we conclude that in $Q[V]$ we have

$$\sum_{U\in[0,p]} \sum_{s\in X(U)} F^{[[91]]}(s, C', V - U) = F^{[[92]]}(C', V).$$

PROPOSITION (8.13) Assume [*] and [**] and [***]. Then for every $V \in N$ we have

$$\sum_{U\in[0,p]} \sum_{s\in X(U)} F^{((91))}(s, C_0, V - U) = F^{((92))}(C_0, V).$$

Proof: Follows from (8.5.4), (8.6.4), and (8.12).

PROPOSITION (8.14) Assume [*]. Then for every $s \in Z(p)$ and $V \in N$ and $U \in N$ we have

$$\sum_{v \in N(2,V)} F^{((91))}(s, C((s)), v(1)) \begin{bmatrix} C^*((s)) \\ V(2) - U \end{bmatrix}$$
$$= F^{((91))}(s, C_0, V - U).$$

Proof: Let there be fixed any $s \in Z(p)$ and $V \in N$ and $U \in N$. Now given any $D \in Z$, we clearly have

$$(C((s)) - D) + C^*((s)) + 1 = C_0 - D$$

and hence upon taking

$$A = C((s)) - D \quad \text{and} \quad B = C^*((s))$$

in (4.36) we get

$$(1) \quad \sum_{v \in N(2,V)} \begin{bmatrix} C((s)) - D \\ v(1) \end{bmatrix} \begin{bmatrix} C^*((s)) \\ v(2) - U \end{bmatrix} = \begin{bmatrix} C_0 - D \\ V - U \end{bmatrix}.$$

In view of (8.5.2), by the definition of $F^{((91))}$ we see that

$$(2) \quad F^{((91))}(s, C_0, V) = \sum_{D \in [0,R]} (-1)^D F_D^{(91)}(s) \begin{bmatrix} C_0 - D \\ V \end{bmatrix}$$

and

$$(3) \quad \left[\begin{array}{l} \text{for every } v(1) \in N \text{ we have} \\ F^{((91))}(s, C((s)), v(1)) = \displaystyle\sum_{D \in [0,R]} (-1)^D F_D^{(91)}(s) \begin{bmatrix} C((s)) - D \\ v(1) \end{bmatrix}. \end{array} \right.$$

Now

$$\sum_{v \in N(2,V)} F^{((91))}(s, C((s)), v(1)) \begin{bmatrix} C^*((s)) \\ v(2) - U \end{bmatrix}$$
$$= \sum_{v \in N(2,V)} \left\{ \left(\sum_{D \in [0,R]} (-1)^D F_D^{(91)}(s) \begin{bmatrix} C((s)) - D \\ v(1) \end{bmatrix} \right) \begin{bmatrix} C^*((s)) \\ v(2) - U \end{bmatrix} \right\}$$

[by (3)]

$$= \sum_{D \in [0,R]} \left\{ (-1)^D F_D^{(91)}(s) \left(\sum_{v \in N(2,V)} \begin{bmatrix} C((s)) - D \\ v(1) \end{bmatrix} \begin{bmatrix} C^*((s)) \\ v(2) - U \end{bmatrix} \right) \right\}$$

$$= \sum_{D \in [0,R]} (-1)^D F_D^{(91)}(s) \begin{bmatrix} C_0 - D \\ V - U \end{bmatrix} \qquad \text{[by (1)]}$$

$$= F^{((91))}(s, C_0, V - U) \qquad \text{[by (2)]}$$

and therefore

$$\sum_{v \in N(2,V)} F^{((91))}(s,C((s)),v(1)) \begin{bmatrix} C^*((s)) \\ v(2) - U \end{bmatrix} = F^{((91))}(s, C_0, V - U).$$

THEOREM (8.15) Assume [*] and [**] and [***]. Then for every $V \in N$ we have

$$\sum_{U \in [0,p]} \sum_{s \in X(U)} \sum_{v \in N(2,V)} F^{((91))}(s,C((s)),v(1)) \begin{bmatrix} C^*((s)) \\ v(2) - U \end{bmatrix}$$

$$= F^{((92))}(C_0,V).$$

Proof: Follows from (8.13) and (8.14).

LEMMA (8.16) Given any $t \in J(p)$, and given any

$$A'_{ij} \in Q \text{ and } A^*_{ij} \in Q \qquad \text{for all } i \in [1,p]\backslash t \text{ and } j \in [1,p]$$

and given any

$$A_{ij} \in Q \qquad \text{for all } i \in t \text{ and } j \in [1,p],$$

upon letting

$$B_{ij} = \begin{cases} A'_{ij} + A^*_{ij} & \text{if } i \in [1,p]\backslash t \text{ and } j \in [1,p] \\ A_{ij} & \text{if } i \in t \text{ and } j \in [1,p] \end{cases}$$

and upon letting

B = the p by p matrix whose (i,j)th entry is B_{ij},

we have

$$\det B = \sum_{u \in J^*(p,t)} \det B^{(u)}$$

where for every $u \in J^*(p,t)$, we have put

$$B_{ij}^{(u)} = \begin{cases} A_{ij}^{*} & \text{if } i \in [1,p]\backslash u \text{ and } j \in [1,p] \\ A_{ij}' & \text{if } i \in u\backslash t \text{ and } j \in [1,p] \\ A_{ij} & \text{if } i \in t \text{ and } j \in [1,p] \end{cases}$$

and

$$B^{(u)} = \text{the } p \text{ by } p \text{ matrix whose } (i,j)\text{th entry is } B_{ij}^{(u)}.$$

Proof: Follows from the multilinearity of det as a function of its rows.

PROPOSITION (8.17) For every $e \in Z(p)$, we have

$$\sum_{u \in J(p)} H^{(92)}(u,e) = \sum_{t \in J(p)} H^{(93)}(t,e).$$

Proof: Given any

$$\hat{x}_i \in Q \text{ and } \hat{y}_i \in Q \qquad \text{for } 1 \le i \le p,$$

for every $u \in J(p)$ we clearly have

$$\prod_{i \in u} (\hat{y}_i - \hat{x}_i) = \sum_{t \in P(u)} \left[\prod_{i \in t} (-\hat{x}_i)\right]\left[\prod_{i \in u\backslash t} \hat{y}_i\right]$$

and upon multiplying both sides of this equation by

$$\prod_{i \in [1,p]\backslash u} \hat{y}_i$$

we get

$$\left[\prod_{i \in u} (\hat{y}_i - \hat{x}_i)\right]\left[\prod_{i \in [1,p]\backslash u} \hat{y}_i\right]$$
$$= \sum_{t \in P(u)} \left[\prod_{i \in t} (-\hat{x}_i)\right]\left[\prod_{i \in [1,p]\backslash t} \hat{y}_i\right]$$

and upon substituting in this equation,

$$\begin{bmatrix} x(i) \\ e(i) \end{bmatrix} \text{ and } \begin{bmatrix} y(i) \\ e(i) \end{bmatrix} \qquad \text{for } \hat{x}_i \text{ and } \hat{y}_i, \text{ respectively,}$$

we see that for every $e \in Z(p)$ we have

$$G^{[92]}(u,e) = \sum_{t \in P(u)} G^{[93]}(t,e).$$

Given any $e \in Z(p)$, upon multiplying both sides of the equation above by $\det G^{(92)}(u,e)$ and then summing over all u in $J(p)$, we get

$$\sum_{u \in J(p)} G^{[92]}(u,e) \det G^{(92)}(u,e)$$
$$= \sum_{u \in J(p)} \left[\sum_{t \in P(u)} G^{[93]}(t,e) \det G^{(92)}(u,e) \right].$$

Upon letting $J[p]$ be the set of all pairs (t,u) with $u \in J(p)$ and $t \in P(u)$, we see that the RHS of the equation above clearly equals

$$\sum_{(t,u) \in J[p]} G^{[93]}(t,e) \det G^{(92)}(u,e)$$

and this, in turn, is clearly equal to

$$\sum_{t \in J(p)} G^{[93]}(t,e) \sum_{u \in J^*(p,t)} \det G^{(92)}(u,e)$$

and therefore we see that for every $e \in Z(p)$ we have

$$(1) \quad \left[\begin{array}{l} \displaystyle\sum_{u \in J(p)} G^{[92]}(u,e) \det G^{(92)}(u,e) \\ \displaystyle= \sum_{t \in J(p)} \left[G^{[93]}(t,e) \sum_{u \in J^*(p,t)} \det G^{(92)}(u,e) \right]. \end{array} \right.$$

Given any $e \in Z(p)$ and $t \in J(p)$, by (4.11'), for all $i \in [1,p]\backslash t$ and $j \in [1,p]$ we have

$$\binom{1 + r(j) + r'(j) - s'(i)}{1 + r(j) - e(i)} = \binom{r(j) + r'(j) - s'(i)}{1 + r(j) - e(i)} + \binom{r(j) + r'(j) - s'(i)}{r(j) - e(i)}$$

and hence upon letting

$$A'_{ij} = \binom{r(j) + r'(j) - s'(i)}{1 + r(j) - e(i)} \quad \text{for all } i \in [1,p]\backslash t \text{ and } j \in [1,p]$$

and upon letting

$$A^*_{ij} = \binom{r(j) + r'(j) - s'(i)}{r(j) - e(i)} \quad \text{for all } i \in [1,p]\backslash t \text{ and } j \in [1,p]$$

and upon letting

$$A_{ij} = \binom{r(j) + r'(j) - s'(i)}{1 + r(j) - e(i)} \qquad \text{for all } i \in t \text{ and } j \in [1,p]$$

by the definition of $G^{(93)}$ we see that

$$G_{ij}^{(93)}(t,e) = \begin{cases} A'_{ij} + A^*_{ij} & \text{if } i \in [1,p]\backslash t \text{ and } j \in [1,p] \\ A_{ij} & \text{if } i \in t \text{ and } j \in [1,p] \end{cases}$$

and therefore by (8.16) we get

(2) $$\det G^{(93)}(t,e) = \sum_{u \in J^*(p,t)} \det B^{(u)}$$

where for every $u \in J^*(p,t)$, we have put

$$B_{ij}^{(u)} = \begin{cases} A^*_{ij} & \text{if } i \in [1,p]\backslash u \text{ and } j \in [1,p] \\ A'_{ij} & \text{if } i \in u\backslash t \text{ and } j \in [1,p] \\ A_{ij} & \text{if } i \in t \text{ and } j \in [1,p] \end{cases}$$

and

$B^{(u)}$ = the p by p matrix whose (i,j)th entry is $B_{ij}^{(u)}$.

Now by the definition of $G^{(92)}(u,e)$ we see that for every $e \in Z(p)$ and $t \in J(p)$ and $u \in J^*(p,t)$ we have

$$G^{(92)}(u,e) = B^{(u)}$$

and therefore by (1) and (2) we see that for every $e \in Z(p)$ we have

$$\sum_{u \in J(p)} G^{[92]}(u,e) \det G^{(92)}(u,e)$$

$$= \sum_{t \in J(p)} G^{[93]}(t,e) \det G^{(93)}(t,e)$$

and hence upon remembering the defining equations

$$H^{(92)}(u,e) = G^{[92]}(u,e) \det G^{(92)}(u,e)$$

and

$$H^{(93)}(t,e) = G^{[93]}(t,e) \det G^{(93)}(t,e)$$

we conclude that for every $e \in Z(p)$ we have

$$\sum_{u \in J(p)} H^{(92)}(u,e) = \sum_{t \in J(p)} H^{(93)}(t,e).$$

PROPOSITION (8.18) For every $E \in Z$ we have

$$H_E^{(92)} = H_E^{(93)}. \tag{8.18.2}$$

Moreover, for every $D \in Z$ we have

$$F_D^{(92)} = F_D^{(93)}.$$

Proof: In view of (8.3.2) and (8.3.3) and the definitions of $H_E^{(92)}$ and $H_E^{(93)}$, by (8.17) we see that for every $E \in Z$ we have (8.18.1). In view of (8.6.1) and the definitions of $F_D^{(92)}$ and $F_D^{(93)}$, by (8.18.1) we see that for every $D \in Z$ we have (8.18.2).

PROPOSITION (8.19) Assume [*]. Then for every $C' \in Z$, in $Q[V]$ we have

$$F^{[[92]]}(C',V) = F^{[[93]]}(C',V).$$

Proof: In view of (8.6.2) and the definitions of $F^{[[92]]}$ and $F^{[[93]]}$, our assertion follows from (8.18.2).

THEOREM (8.20) Assume [*]. Then for every $C' \in Z$ and $V \in Z$ we have

$$F^{((92))}(C',V) = F^{((93))}(C',V).$$

Proof: In view of (8.6.2) and the definitions of $F^{((92))}$ and $F^{((93))}$, our assertion follows from (8.18.2).

LEMMA (8.21) Let there be given any $t \in J(p)$ such that there exists $I \in [1, p - 1]$ for which we have $I \in t$ and $I + 1 \notin t$ and $y(I + 1) = x(I)$ and $s'(I + 1) = 1 + s'(I)$. Then for every $E \in Z$ we have

$$\sum_{e \in N(p,E)} H^{(93)}(t,e) = 0.$$

Proof: For every $e \in Z(p)$, let $e^* \in Z(p)$ be defined by putting

$$e^*(i) = \begin{cases} e(i) & \text{if } i \in [1,p]\backslash\{I, I+1\} \\ e(I+1) & \text{if } i = I \\ e(I) & \text{if } i = I+1. \end{cases}$$

Now given any $e \in Z(p)$, for every $j \in [1,p]$ we clearly have

$$G_{ij}^{(93)}(t,e^*) = G_{ij}^{(93)}(t,e) \qquad \text{if } i \in [1,p]\backslash\{I, I+1\}$$

and we have

$$G_{Ij}^{(93)}(t,e^*)$$

$$= \binom{r(j) + r'(j) - s'(I)}{1 + r(j) - e^*(I)} \qquad \text{(because } I \in t\text{)}$$

$$= \binom{1 + r(j) + r'(j) - s'(I+1)}{1 + r(j) - e(I+1)} \qquad [\text{because } s'(I+1) = 1 + s'(1) \text{ and } e^*(I) = e(I+1)]$$

$$= G_{I+1,j}^{(93)}(t,e) \qquad \text{(because } I + 1 \in [1,p]\backslash t\text{)}$$

$$G_{I+1,j}^{(93)}(t,e^*)$$

$$= \binom{1 + r(j) + r'(j) - s'(I+1)}{1 + r(j) - e^*(I+1)} \qquad \text{(because } I + 1 \in [1,p]\backslash t\text{)}$$

$$= \binom{r(j) + r'(j) - s'(I)}{1 + r(j) - e(I)} \qquad [\text{because } s'(I+1) = 1 + s'(I) \text{ and } e^*(I+1) = e(I)]$$

$$= G_{Ij}^{(93)}(t,e) \qquad \text{(because } I \in t\text{);}$$

thus the matrix $G^{(93)}(t,e^*)$ is obtained by interchanging the Ith and (I + 1)th rows of the matrix $G^{(93)}(t,e)$ and hence

(1) $\det G^{(93)}(t,e^*) = -\det G^{(93)}(t,e)$.

Given any $e \in Z(p)$, we clearly have

$$G_i^{[93]}(t,e^*) = G_i^{[93]}(t,e) \qquad \text{if } i \in [1,p]\backslash\{I, I+1\}$$

and we have

$$G_I^{[93]}(t,e^*) = -\begin{bmatrix} x(I) \\ e^*(I) \end{bmatrix} \qquad \text{(because } I \in t\text{)}$$

$$= -\begin{bmatrix} y(I+1) \\ e(I+1) \end{bmatrix} \qquad [\text{because } x(I) = y(I+1) \text{ and } e^*(I) = e(I+1)]$$

$$= -G_{I+1}^{[93]}(t,e) \qquad \text{(because } I+1 \in [1,p]\backslash t\text{)}$$

and we also have

$$G_{I+1}^{[93]}(t,e^*) = \begin{bmatrix} y(I+1) \\ e^*(I+1) \end{bmatrix} \qquad \text{(because } I+1 \in [1,p]\backslash t\text{)}$$

$$= \begin{bmatrix} x(I) \\ e(I) \end{bmatrix} \qquad [\text{because } y(I+1) = X(I) \text{ and } e^*(I+1) = e(I)]$$

$$= (-1)\left(-\begin{bmatrix} x(I) \\ e(I) \end{bmatrix}\right)$$

$$= -G_I^{[93]}(t,e) \qquad \text{(because } I \in t\text{);}$$

consequently,

$$\prod_{i=1}^{p} G_i^{[93]}(t,e^*) = \prod_{i=1}^{p} G_i^{[93]}(t,e)$$

and hence

(2) $G^{[93]}(t,e^*) = G^{[93]}(t,e)$.

Upon remembering the defining equation

$$H^{(93)}(t,e) = G^{[93]}(t,e) \det G^{(93)}(t,e),$$

by (1) and (2) we see that for every $e \in Z(p)$ we have

(3) $H^{(93)}(t,e^*) = -H^{(93)}(t,e)$.

Given any $E \in Z$, upon summing both sides of (3) as e varies over $N(p,E)$, we get

$$\sum_{e \in N(p,E)} H^{(93)}(t,e^*) = -\sum_{e \in N(p,E)} H^{(93)}(t,e);$$

now $e \mapsto e^*$ clearly gives a bijection of $N(p,E)$ onto itself and hence we get

$$\sum_{e\in N(p,E)} H^{(93)}(t,e^*) = \sum_{e\in N(p,E)} H^{(93)}(t,e);$$

finally, the two equations above yield

$$\sum_{e\in N(p,E)} H^{(93)}(t,e) = 0.$$

PROPOSITION (8.22) Assume [****]. Then for every $E \in Z$ we have

$$H_E^{(93)} = H_E^{(94)}.$$

Proof: For every $t \in J(p)$, upon letting

$$t^* = \begin{cases} 0 & \text{if } [1,p]\backslash t = \emptyset \\ \max\{i : i \in [1,p]\backslash t\} & \text{if } [1,p]\backslash t \neq \emptyset \end{cases}$$

we see that

(1) $t^* \in [0,p]$ and $[t^* + 1, p] \subset t$ and $t^* \notin t$

and then upon letting

$$t' = \begin{cases} 0 & \text{if } [1,t^*] \cap t = \emptyset \\ \max\{j : j \in [1,t^*] \cap t\} & \text{if } [1,t^*] \cap t \neq \emptyset \end{cases}$$

we see that

(2) if $t' = 0$, then $t = [t^* + 1, p]$

whereas

(3) if $t' \neq 0$, then $t' \in t$ and $t' + 1 \in [1,p]\backslash t$.

Upon taking t' for I in (8.21), in view of (3) we see that

(4) $\left[\begin{array}{l} \text{if } t \in J(p) \text{ is such that } t' \neq 0, \text{ then} \\ \text{for every } E \in Z \text{ we have} \\ \displaystyle\sum_{e\in N(p,E)} H^{(93)}(t,e) = 0. \end{array}\right.$

By (1), (2), and (3) we see that

(5) $\left[\begin{array}{l} I \mapsto [I + 1, p] \text{ gives} \\ \text{a bijection of } [0,p] \text{ onto } \{t \in J(p) : t' = 0\}. \end{array}\right.$

By (4) and (5) we see that for every $E \in Z$ we have

$$(6)\quad \left[\begin{array}{l} \sum_{e\in N(p,E)} \sum_{t\in J(p)} H^{(93)}(t,e) \\ = \sum_{e\in N(p,E)} \sum_{I\in[0,p]} H^{(93)}([I+1,\ p],\ e). \end{array}\right.$$

By the definitions of $H^{(93)}$ and $H^{(94)}$ we see that

$$(7)\quad \left[\begin{array}{l} \text{for every } e \in Z(p) \text{ and } I \in [0,p] \text{ we have} \\ H^{(93)}([I+1,\ p],\ e) = H^{(94)}(I,e). \end{array}\right.$$

In view of (8.3.3) and (8.3.4), by the definitions of $H_E^{(93)}$ and $H_E^{(94)}$ we see that for every $E \in Z$ we have

$$(8)\quad H_E^{(93)} = \sum_{e\in N(p,E)} \sum_{t\in J(p)} H^{(93)}(t,e)$$

and

$$(9)\quad H_E^{(94)} = \sum_{e\in N(p,E)} \sum_{I\in[0,p]} H^{(94)}(I,e).$$

By (6), (7), (8), and (9) we see that for every $E \in Z$ we have

$$H_E^{(93)} = H_E^{(94)}.$$

PROPOSITION (8.23) Assume [****]. Then for every $D \in Z$ we have

$$F_D^{(93)} = F_D^{(94)}.$$

Proof: In view of (8.6.1) and the definitions of $F_D^{(93)}$ and $F_D^{(94)}$, our assertion follows from (8.22).

PROPOSITION (8.24) Assume [*] and [****]. Then for every $C' \in Z$, in $Q[V]$ we have

$$F^{[[93]]}(C',V) = F^{[[94]]}(C',V).$$

Proof: In view of (8.6.2) and the definitions of $F^{[[93]]}$ and $F^{[[94]]}$, our assertion follows from (8.23).

THEOREM (8.25) Assume [*] and [****]. Then for every $C' \in Z$ and $V \in Z$ we have

$$F^{((93))}(C',V) = F^{((94))}(C',V).$$

Proof: In view of (8.6.2) and the definitions of $F^{((93))}$ and $F^{((94))}$, our assertion follows from (8.23).

LEMMA (8.26) Assume (8.0.1'). Then for every $I \in [0, p-1]$ and $e \in Z(p)$ we have

$$H^{(94)}(I,e) = 0.$$

Proof: Let there be given any $I \in [0. p-1]$ and $e \in Z(p)$. Now by definition we have

$$G^{(94)}_{pj}(I,e) = \begin{bmatrix} r(j) + r'(j) - s'(p) \\ 1 + r(j) - e(p) \end{bmatrix} \quad \text{for all } j \in [1,p]$$

and by assumption we have

$$r(j) + r'(j) - s'(p) \geq 0 \quad \text{for all } j \in [1,p]$$

and hence in view of (2.5.6') we see that

(1) $\left[\begin{array}{l}\text{if } j \in [1,p] \text{ is such that} \\ 1 + r(j) - e(p) > r(j) + r'(j) - s'(p), \text{ then } G^{(94)}_{pj}(I,e) = 0.\end{array}\right.$

By assumption

$$0 \leq s'(p) - r'(j) \quad \text{for all } j \in [1,p]$$

and hence we see that

(2) $\left[\begin{array}{l}\text{if } e(p) < 1, \text{ then for all } j \in [1,p] \text{ we have} \\ 1 + r(j) - e(p) > r(j) + r'(j) - s'(p).\end{array}\right.$

In view of the definition of $G^{(94)}$, by (1) and (2) we see that

$\left[\begin{array}{l}\text{if } e(p) < 1, \text{ then the pth row of} \\ G^{(94)}(I,e) \text{ consists entirely of zeros}\end{array}\right.$

and hence we get that

(3) if $e(p) < 1$, then $\det G^{(94)}(I,e) = 0$.

By definition we have

$$G_p^{[94]}(I,e) = -\begin{bmatrix} x(p) \\ e(p) \end{bmatrix}$$

and hence in view of (2.5.6*) we see that

(4) if $e(p) > x(p) + e(p) \geq 0$, then $G_p^{[94]}(I,e) = 0$.

By assumption we have

$$x(p) = -1$$

and hence we see that

(5) if $e(p) \geq 1$, then $e(p) > x(p) + e(p) \geq 0$.

By (4) and (5) we see that

(6) if $e(p) \geq 1$, then $G_p^{[94]}(I,e) = 0$.

In view of the definition of $G^{[94]}$, by (6) we see that

(7) if $e(p) \geq 1$, then $G^{[94]}(I,e) = 0$.

By definition we have

$$H^{(94)}(I,e) = G^{[94]}(I,e) \det G^{(94)}(I,e)$$

and hence by (3) and (7) we conclude that

$$H^{(94)}(I,e) = 0.$$

PROPOSITION (8.27) Assume (8.0.1'). Then for every $E \in Z$ we have

$$H_E^{(94)} = H_E^{(95)}.$$

Proof: Let there be given any $E \in Z$. Now by (8.3.4) and the definition of $H_E^{(94)}$ we see that

(1) $$H_E^{(94)} = \sum_{e \in N(p,E)} \sum_{I \in [0,p]} H^{(94)}(I,e)$$

and by (8.3.5) and the definition of $H_E^{(95)}$ we see that

(2) $H_E^{(95)} = \sum_{e \in N(p,E)} H^{(95)}(e).$

By the definitions of $H^{(94)}$ and $H^{(95)}$ we also see that

(3) $H^{(94)}(p,e) = H^{(95)}(e)$ for all $e \in Z(p)$.

Finally, by (8.26) we see that

(4) $H^{(94)}(I,e) = 0$ for all $I \in [0, p - 1]$ and $e \in Z(p)$.

By (1) to (4) we see that

$$H_E^{(94)} = H_E^{(95)}.$$

PROPOSITION (8.28) Assume (8.0.1'). Then for every $D \in Z$ we have

$$F_D^{(94)} = F_D^{(95)}.$$

Proof: In view of (8.6.1) and the definitions of $F_D^{(94)}$ and $F_D^{(95)}$, our assertion follows from (8.27).

PROPOSITION (8.29) Assume [*] and (8.0.1'). Then for every $C' \in Z$, in $Q[V]$ we have

$$F^{[[94]]}(C',V) = F^{[[95]]}(C',V).$$

Proof: In view of (8.6.2) and the definitions of $F^{[[94]]}$ and $F^{[[95]]}$, our assertion follows from (8.28).

THEOREM (8.30) Assume [*] and (8.0.1'). Then for every $C' \in Z$ and $V \in Z$ we have

$$F^{((94))}(C',V) = F^{((95))}(C',V).$$

Proof: In view of (8.6.2) and the definitions of $F^{((94))}$ and $F^{((95))}$, our assertion follows from (8.28).

PROPOSITION (8.31) Assume [****]. Then for every $E \in Z$ we have

$$H_E^{(94)} = H_E^{(96)}.$$

Proof: For every $e \in Z(p+1)$ and $I \in [0,p]$, let $G^{(94I)}(e)$ = the p by p matrix whose (i,j)th entry is $G_{ij}^{(94I)}(e)$ where

$$G_{ij}^{(94I)}(e) = \begin{cases} G_{ij}^{(96)}(e) & \text{if } j \in [1,p] \text{ and } i \in [1,I] \\ G_{i+1,j}^{(96)}(e) & \text{if } j \in [1,p] \text{ and } i \in [I+1, p]. \end{cases}$$

Now, given any $e \in Z(p+1)$, upon expanding $\det G^{(96)}(e)$ by the $(p+1)$th column we get

$$\det G^{(96)}(e) = \sum_{I \in [0,p]} (-1)^{p-I} G_{I+1,p+1}^{(96)}(e) \det G^{(94I)}(e)$$

and upon multiplying both sides of the equation above by $G^{[96]}(e)$ we get

$$G^{[96]}(e) \det G^{(96)}(e)$$

$$= \sum_{I \in [0,p]} G^{[96]}(e)(-1)^{p-I} G_{I+1,p+1}^{(96)}(e) \det G^{(94I)}(e);$$

therefore, upon remembering the defining equation

$$H^{(96)}(e) = G^{[96]}(e) \det G^{(96)}(e)$$

we see that

$$(1)\quad \begin{cases} \text{for every } e \in Z(p+1) \text{ we have} \\ H^{(96)}(e) = \sum_{I \in [0,p]} G^{[96]}(e)(-1)^{p-I} G_{I+1,p+1}^{(96)}(e) \det G^{(94I)}(e). \end{cases}$$

For every $e \in Z(p+1)$ and $I \in [0,p]$, upon letting

$d[I,e]$ be the member of $Z(p)$

obtained by putting

$$d[I,e](i) = \begin{cases} e(i) & \text{if } i \in [1,I] \\ e(i+1) & \text{if } i \in [I+1, p] \end{cases}$$

we see that for every $i \in [1,I]$ and $j \in [1,p]$ we have

$$G_{ij}^{(94I)}(e)$$

$= G_{ij}^{(96)}(e)$ (by definition of $G^{(94I)}$)

$= \begin{Bmatrix} 1 + r(j) + r'(j) - s'(i) \\ 1 + r(j) - e(i) \end{Bmatrix}$ (by definition of $G^{(96)}$)

$= \begin{Bmatrix} 1 + r(j) + r'(j) - s'(i) \\ 1 + r(j) - d[I,e](i) \end{Bmatrix}$ (by definition of $d[I,e]$)

$= G_{ij}^{(94)}(I,d[I,e])$ (by definition of $G^{(94)}$)

and we see that for every $i \in [I = 1, p - 1]$ and $j \in [1,p]$ we have

$G_{ij}^{(94I)}(e)$

$= G_{i+1,j}^{(96)}(e)$ (by definition of $G^{(94I)}$)

$= \begin{Bmatrix} 1 + r(j) + r'(j) - s'(i + 1) \\ 1 + r(j) - e(i + 1) \end{Bmatrix}$ (by definition of $G^{(96)}$)

$= \begin{Bmatrix} r(j) + r'(j) - s'(i) \\ 1 + r(j) - d[I,e](i) \end{Bmatrix}$ [by definition of d and because $s'(i + 1) = 1 + s'(i)$]

$= G_{ij}^{(94)}(I,d[I,e])$ (by definition of $G^{(94)}$)

and we see that if $I + 1 \le p$, then for every $j \in [1,p]$ we have

$G_{pj}^{(94I)}(e)$

$= G_{p+1,j}^{(96)}(e)$ (by definition of $G^{(94I)}$)

$= \begin{Bmatrix} r(j) + r'(j) - s'(p) \\ 1 + r(j) - e(p + 1) \end{Bmatrix}$ (by definition of $G^{(96)}$)

$= \begin{Bmatrix} r(j) + r'(j) - s'(p) \\ 1 + r(j) - d[I,e](p) \end{Bmatrix}$ (by definition of d)

$= G_{pj}^{(94)}(I,d[I,e])$ (by definition of $G^{(94)}$);

thus we conclude that

(2) $\left[\begin{array}{l} \text{for every } e \in Z(p + 1) \text{ and } I \in [0,p] \text{ we have} \\ G^{(94I)}(e) = G^{(94)}(I,d[I,e]). \end{array}\right.$

Given any $e \in Z(p+1)$ and $I \in [0, p-1]$, we see that

$$G^{(96)}_{I+1,p+1}(e) = \left\{\genfrac{}{}{0pt}{}{1 + r'(p) - s'(I+1)}{-\ e(I+1)}\right\} \qquad \text{(by definition of } G^{(96)}\text{)}$$

$$= \begin{cases} 1 & \text{if } e(I+1) = 0 \qquad \text{[by (2.5.1')]} \\ 0 & \text{if } e(I+1) > 0 \qquad \text{[by (2.5.2')]} \end{cases}$$

and we see that

$$G^{[96]}(e) = \left[\genfrac{}{}{0pt}{}{x(p)}{e(p+1)}\right] \prod_{i=1}^{p} \left[\genfrac{}{}{0pt}{}{y(i)}{e(i)}\right] \qquad \text{(by definition of } G^{[96]}\text{)}$$

$$= \begin{cases} \left[\genfrac{}{}{0pt}{}{x(p)}{e(p+1)}\right] \displaystyle\prod_{i \in [1,p]\backslash\{I+1\}} \left[\genfrac{}{}{0pt}{}{y(i)}{e(i)}\right] & \text{if } e(I+1) = 0 \ \text{[by (2.5.1*)]} \\ 0 & \text{if } e(I+1) < 0 \ \text{[by (2.5.2*)]} \end{cases}$$

and we see that

$$G^{[94]}(I, d[I,e])$$

$$= \left(\prod_{i \in [1,I]} \left[\genfrac{}{}{0pt}{}{y(i)}{d[I,e](i)}\right]\right)\left(\prod_{i \in [I+1,p]} - \left[\genfrac{}{}{0pt}{}{x(i)}{d[I,e](i)}\right]\right) \qquad \text{(by definition of } G^{[94]}\text{)}$$

$$= (-1)^{p-I}\left[\genfrac{}{}{0pt}{}{x(p)}{e(p+1)}\right]\left(\prod_{i \in [1,p]\backslash\{I+1\}} \left[\genfrac{}{}{0pt}{}{y(i)}{e(i)}\right]\right)$$

(by definition of d and because $y(i+1) = x(i)$ for all $i \in [1, p-1]$)

and hence we conclude that

$$(3) \qquad \left[\begin{array}{l} \text{for every } e \in Z(p+1) \text{ and } I \in [0, p-1] \text{ we have} \\ G^{[96]}(e)(-1)^{p-I} G^{(96)}_{I+1,p+1}(e) = \begin{cases} G^{[94]}(I, d[I,e]) & \text{if } e(I+1) = 0 \\ 0 & \text{if } e(I+1) \neq 0. \end{cases} \end{array}\right.$$

Given any $e \in Z(p+1)$, we see that

$$G^{(96)}_{p+1,p+1}(e) = \left\{\genfrac{}{}{0pt}{}{r'(p) - s'(p)}{-e(p+1)}\right\} \qquad \text{(by definition of } G^{(96)}\text{)}$$

$$= \begin{cases} 1 & \text{if } e(p+1) = 0 \qquad \text{[by (2.5.1')]} \\ 0 & \text{if } e(p+1) > 0 \qquad \text{[by (2.5.2')]} \end{cases}$$

and we see that

$$G^{[96]}(e) = \begin{bmatrix} x(p) \\ e(p+1) \end{bmatrix} \prod_{i=1}^{p} \begin{bmatrix} y(i) \\ e(i) \end{bmatrix} \qquad \text{(by definition of } G^{[96]}\text{)}$$

$$= \begin{cases} \prod_{i=1}^{p} \begin{bmatrix} y(i) \\ e(i) \end{bmatrix} & \text{if } e(p+1) = 0 \qquad \text{[by (2.5.1*)]} \\ 0 & \text{if } e(p+1) < 0 \qquad \text{[by (2.5.2*)]} \end{cases}$$

and we see that

$$G^{[94]}(p,d[p,e]) = \prod_{i=1}^{p} \begin{bmatrix} y(i) \\ d[p,e](i) \end{bmatrix} \qquad \text{(by definition of } G^{[94]}\text{)}$$

$$= (-1)^{p-p} \prod_{i=1}^{p} \begin{bmatrix} y(i) \\ e(i) \end{bmatrix} \qquad \text{(by definition of d)}$$

and hence we conclude that

$$(4) \quad \left[\begin{array}{l} \text{for every } e \in Z(p+1) \text{ we have} \\ G^{[96]}(e)(-1)^{p-p} G^{(96)}_{p+1,p+1}(e) = \begin{cases} G^{[94]}(p,d[p,e]) & \text{if } e(p+1) = 0 \\ 0 & \text{if } e(p+1) \neq 0. \end{cases} \end{array}\right.$$

By (3) and (4) we see that

$$(5) \quad \left[\begin{array}{l} \text{for every } e \in Z(p+1) \text{ and } I \in [0,p] \text{ we have} \\ G^{[96]}(e)(-1)^{p-I} G^{(96)}_{I+1,p+1}(e) = \begin{cases} G^{[94]}(I,d[I,e]) & \text{if } e(I+1) = 0 \\ 0 & \text{if } e(I+1) \neq 0. \end{cases} \end{array}\right.$$

By definition we know that for every $e' \in Z(p)$ and $I \in [0,p]$ we have

$$H^{(94)}(I,e') = G^{[94]}(I,e') \det G^{(94)}(I,e')$$

and therefore by (2) and (5) we conclude that

$$(6) \quad \left[\begin{array}{l} \text{for every } e \in Z(p+1) \text{ and } I \in [0,p] \text{ we have} \\ G^{[96]}(e)(-1)^{p-I} G^{(96)}_{I+1,p+1}(e) \det G^{(94I)}(e) \\ = \begin{cases} H^{(94)}(I,d[I,e]) & \text{if } e(I+1) = 0 \\ 0 & \text{if } e(I+1) \neq 0 \end{cases} \end{array}\right.$$

In view of (8.3.6), by the definition of $H_E^{(96)}$ we see that for every $E \in Z$ we have

$$(7) \quad H_E^{(96)} = \sum_{e \in N(p+1,E)} H^{(96)}(e).$$

In view of (8.3.4), by the definition of $H_E^{(94)}$ we see that for every $E \in Z$ we have

$$(8) \quad H_E^{(94)} = \sum_{e' \in N(p,E)} \sum_{I \in [0,p]} H^{(94)}(I,e').$$

In view of the definition of d, by (1), (6), (7), and (8) we see that for every $E \in Z$ we have

$$H_E^{(94)} = H_E^{(96)}.$$

PROPOSITION (8.32) Assume [****]. Then for every $D \in Z$ we have

$$F_D^{(94)} = F_D^{(96)}.$$

Proof: In view of (8.6.1) and the definitions of $F_D^{(94)}$ and $F_D^{(96)}$, our assertion follows from (8.31).

PROPOSITION (8.33) Assume [*] and [****]. Then for every $C' \in Z$, in $Q[V]$ we have

$$F^{[[94]]}(C',V) = F^{[[96]]}(C',V).$$

Proof: In view of (8.6.2) and the definitions of $F^{[[94]]}$ and $F^{[[96]]}$, our assertion follows from (8.32).

THEOREM (8.34) Assume [*] and [****]. Then for every $C' \in Z$ and $V \in Z$ we have

$$F^{((94))}(C',V) = F^{((96))}(C',V).$$

Proof: In view of (8.6.2) and the definitions of $F^{((94))}$ and $F^{((96))}$, our assertion follows from (8.32).

LEMMA (8.35) Assume (8.0.2). Then we have [*] and [**] and [***] and [****].

Proof: Obvious.

THEOREM (8.35') Assume (8.0.1') and (8.0.2). Let

Σ' stand for $\sum_{U\in[0,p]}\ \sum_{s\in X(U)}$.

Then

$$\left[\begin{array}{l}\text{for every } D \in Z \text{ we have}\\ \sum_{I\in[0,p]} \Sigma' \binom{U}{I} F^{(91)}_{D-I}(s) = F^{(95)}_{D}\end{array}\right. \tag{8.35'.1}$$

and

$$\left[\begin{array}{l}\text{for every } C' \in Z, \text{ in } Q[V] \text{ we have}\\ \Sigma'\ F^{[[91]]}(s,\ C',\ V-U) = F^{[[95]]}(C',V)\end{array}\right. \tag{8.35'.2}$$

and

$$\left[\begin{array}{l}\text{in } Q[V] \text{ we have}\\ \Sigma'\ F^{[[91]]}(s,\ C_0,\ V-U) = F^{[[95]]}(C_0,V)\end{array}\right. \tag{8.35'.3}$$

and

$$\left[\begin{array}{l}\text{for every } V \in N \text{ we have}\\ \Sigma'\ F^{((91))}(s,\ C_0,\ V-U) = F^{((95))}(C_0,V)\end{array}\right. \tag{8.35'.4}$$

and

$$\left[\begin{array}{l}\text{for every } V \in N \text{ we have}\\ \Sigma' \sum_{v\in N(2,V)} F^{((91))}(s,C((s)),v(1))\begin{bmatrix}C^*((s))\\ v(2)-U\end{bmatrix} = F^{((95))}(C_0,V).\end{array}\right. \tag{8.35'.5}$$

Proof: By (8.11), (8.18.2), (8.23), (8.28), and (8.35) we get (8.35'.1). By (8.12), (8.19), (8.24), (8.29), and (8.35) we get (8.35'.2). By (8.35'.2) we get (8.35'.3). By (8.13), (8.20), (8.25), (8.30), and (8.35) we get (8.35'.4). By (8.15), (8.20), (8.25), (8.30), and (8.35) we get (8.35'.5).

THEOREM (8.36*) Assume (8.0.2). Let

Σ' stand for $\sum_{U\in[0,p]}\ \sum_{s\in X(U)}$.

Then

$$\left[\begin{array}{l}\text{for every } D \in Z \text{ we have}\\ \sum_{I\in[0,p]} \Sigma' \binom{U}{I} F^{(91)}_{D-I}(s) = F^{(96)}_{D}\end{array}\right. \tag{8.36*.1}$$

and

$$\left[\begin{array}{l}\text{for every } C' \in Z, \text{ in } Q[V] \text{ we have}\\ \Sigma' \; F^{[[91]]}(s, C', V - U) = F^{[[96]]}(C',V)\end{array}\right. \tag{8.36*.2}$$

and

$$\left[\begin{array}{l}\text{in } Q[V] \text{ we have}\\ \Sigma' \; F^{[[91]]}(s, C_0, V - U) = F^{[[96]]}(C_0,V)\end{array}\right. \tag{8.36*.3}$$

and

$$\left[\begin{array}{l}\text{for every } V \in N \text{ we have}\\ \Sigma' \; F^{((91))}(s, C_0, V - U) = F^{((96))}(C_0,V)\end{array}\right. \tag{8.36*.4}$$

and

$$\left[\begin{array}{l}\text{for every } V \in N \text{ we have}\\ \Sigma' \sum_{v\in N(2,V)} F^{((91))}(s,C((s)),v(1)) \begin{bmatrix} C^*((s)) \\ v(2) - U\end{bmatrix} = F^{((96))}(C_0,V).\end{array}\right. \tag{8.36*.5}$$

Proof: By (8.11), (8.18.2), (8.23), (8.32), and (8.35) we get (8.36*.1). By (8.12), (8.19), (8.24), (8.33), and (8.35) we get (8.36*.2). By (8.26*.2) we get (8.36*.3). By (8.13), (8.20), (8.25), (8.34), and (8.35) we get (8.36*.4). By (8.15), (8.20), (8.25), (8.34), and (8.35) we get (8.36*.5).

REMARK (8.40) We shall now prove Theorem (5.8*); so let the notation and assumptions be as in that theorem. Now remembering that x, y, r, r', s' have been arbitrary members of Z(p), we may henceforth assume that

(1) $\left[\begin{array}{l}\text{for all } i \in [1,p] \text{ we have}\\ x(i) = m(k') - b(k', i + 1) \text{ and } y(i) = m(k') - b(k',i)\end{array}\right.$

and

(2) $\begin{cases} \text{for all } j \in [1,p] \text{ we have} \\ r(j) = m(k) - b(k,j) \text{ and } r'(j) = j = s'(j). \end{cases}$

By (1) we see that

(3) for all $i \in [1, p - 1]$ we have $y(i + 1) = x(i)$.

By assumption $b \in \text{vec}[2, p + 1]$ and hence by (1) we see that

(4) for all $i \in [1,p]$ we have $x(p) \le x(i) < y(i)$.

By assumption $a^* \in \text{vec}(2,m^*,p^*)$ and $m^*(k') = m(k')$ and hence by (1) we see that

(5*) if condition (5.8.1*) is satisfied, then $x(p) \ge 0$.

By (1) we see that

(5') if condition (5.8.1') is satisfied, then $x(p) = -1$.

By (2) we know that for all $i \in [1,p]$ we have $r'(i) = s'(i) = i$, and hence by (3), (4), (5'), and (5*) we see that

(5) condition (8.0.2) is satisfied.

By (2) we see that

(6) $R' - S' = 0$.

By (2) we see that

(7') for all $j \in [1,p]$ we have $0 \le s'(p) - r'(j)$.

By (2) we also see that

(8') $\begin{cases} \text{if condition (5.8.1') is satisfied, then for all } j \in [1,p] \\ \text{we have } r(j) + r'(j) - s'(p) = m(k) - a^*(k,j) + j - p. \end{cases}$

By assumption $a^* \in \text{vec}(2,m^*,p^*)$ and $m^*(k) = m(k) + 1$ and hence we see that

(9') if condition (5.8.1') is satisfied, then $a^* \in \text{vec}(2,m,p)$.

In view of (9'), by (2.2.2) we see that

(10') $\begin{cases} \text{if condition (5.8.1') is satisfied, then for all } j \in [1,p] \\ \text{we have } m(k) - a^*(k,j) + j - p \ge 0. \end{cases}$

By (5'), (7'), (8'), and (10') we see that

(11') if condition (5.8.1') is satisfied, then condition (8.0.1') is satisfied.

By assumption $m^*(k) = m(k) + 1$ and hence by (2) we see that

(12') $\left[\begin{array}{l}\text{if condition (5.8.1') is satisfied, then } p^* = p, \text{ and} \\ \text{for all } j \in [1,p] \text{ we have } m^*(k) - a^*(k,j) = 1 + r(j), \\ \text{and for all } j \in [1,p] \text{ and } i \in [1,p] \text{ we have} \\ m^*(k) - a^*(k,j) + j - i = 1 + r(j) + r'(j) - s'(i).\end{array}\right.$

By assumption $m^*(k') = m(k')$ and hence by (1) we see that

(13') if condition (5.8.1') is satisfied, then $p^* = p$ and for all $i \in [1,p]$ we have $m^*(k') - a^*(k',i) = y(i)$.

By assumption $m^*(k) = m(k) + 1$ and hence by (2) we see that

(14') if condition (5.8.1') is satisfied, then $C[m^*,p^*,a^*,k] = p + R$.

By assumption $m^*(k') = m(k')$ and $m^*(k) = m(k) + 1$ and hence by (1), (2), (5'), (13'), and (14') we see that

(15') if condition (5.8.1') is satisfied, then $C(m^*,p^*,a^*) = C_0$.

In view of the definitions of $F_D^{(1k)}$ and $F_D^{(95)}$, by (6), (12'), (13'), and (14') we see that

(16') $\left[\begin{array}{l}\text{if condition (5.8.1') is satisfied, then for every } D \in Z \\ \text{we have } F_D^{(1k)}(m^*,p^*,a^*) = F_D^{(95)}.\end{array}\right.$

In view of (2.11.11) and (8.6) and the definitions of $F^{[[1k]]}$ and $F^{[[95]]}$, by (16') we see that

(17') $\left[\begin{array}{l}\text{if condition (5.8.1') is satisfied, then for every } C' \in Z, \\ \text{in } Q[V] \text{ we have } F^{[[1k]]}(m^*,p^*,a^*,C',V) = F^{[[95]]}(C',V).\end{array}\right.$

In view of (2.11.11) and (8.6) and the definitions of $F^{[1k]}$ and $F^{[[95]]}$, by (15') and (16') we see that

(18') $\left[\begin{array}{l}\text{if condition (5.8.1') is satisfied, then in } Q[V] \text{ we have} \\ F^{[1k]}(m^*,p^*,a^*,V) = F^{[[95]]}(C_0,V).\end{array}\right.$

In view of (2.11.11) and (8.6) and the definitions of $F^{(1k)}$ and $F^{((95))}$, by (15') and (16') we see that

(19') $\left[\begin{array}{l}\text{if condition (5.8.1') is satisfied, then for every } V \in Z \\ \text{we have } F^{(1k)}(m^*,p^*,a^*,V) = F^{((95))}(C_0,V).\end{array}\right.$

By assumption $m^*(k) = m(k) + 1$ and hence by (2) we see that

(20*) $$\left[\begin{array}{l}\text{if condition (5.8.1*) is satisfied, then } p^* = p + 1 \text{ and for} \\ \text{every } j \in [1,p] \text{ we have } m^*(k) - a^*(k,j) = 1 + r(j) \text{ and} \\ m^*(k) - a^*(k,j) + j - i \\ = \begin{cases} 1 + r(j) + r'(j) - s'(i) & \text{for all } i \in [1,p] \\ r(j) + r'(j) - s'(p) & \text{for } i = p + 1.\end{cases}\end{array}\right.$$

By assumption $m^*(k) = m(k) + 1$ and hence by (2) we see that

(21*) $$\left[\begin{array}{l}\text{if condition (5.8.1*) is satisfied, then } p^* = p + 1 \\ \text{and for } j = p + 1 \text{ we have } m^*(k) - a^*(k,j) = 0 \text{ and} \\ m^*(k) - a^*(k,j) + j - i \\ = \begin{cases} 1 + r'(p) - s'(i) & \text{for all } i \in [1,p] \\ r'(p) - s'(p) & \text{for } i = p + 1.\end{cases}\end{array}\right.$$

By assumption $m^*(k') = m(k')$ and hence by (1) we see that

(22*) $$\left[\begin{array}{l}\text{if condition (5.8.1*) is satisfied, then } p^* = p + 1 \\ \text{and we have} \\ m^*(k') - a^*(k',i) = \begin{cases} y(i) & \text{for all } i \in [1,p] \\ x(p) & \text{for } i = p + 1.\end{cases}\end{array}\right.$$

By (20*) and (21*) we see that

(23*) if condition (5.8.1*) is satisfied, then $C[m^*,p^*,a^*,k] = p + R$.

By assumption $m^*(k') = m(k')$ and $m^*(k) = m(k) + 1$ and hence by (22*) and (23*) we see that

(24*) if condition (5.8.1*) is satisfied, then $C(m^*,p^*,a^*) = C_0$.

In view of the definitions of $F_D^{(1k)}$ and $F_D^{(96)}$, by (6), (20*), (21*), (22*), and (23*) we see that

$$(25^*)\quad \left[\begin{array}{l}\text{if condition (5.8.1*) is satisfied, then for every } D \in Z\\ \text{we have } F_D^{(1k)}(m^*,p^*,a^*) = F_D^{(96)}.\end{array}\right.$$

In view of (2.11.11) and (8.6) and the definitions of $F^{[[1k]]}$ and $F^{[[96]]}$, by (25*) we see that

$$(26^*)\quad \left[\begin{array}{l}\text{if condition (5.8.1*) is satisfied, then for every } C' \in Z,\\ \text{in } Q[V] \text{ we have } F^{[[1k]]}(m^*,p^*,a^*,C',V) = F^{[[96]]}(C',V).\end{array}\right.$$

In view of (2.11.11) and (8.6) and definitions of $F^{[1k]}$ and $F^{[[96]]}$, by (24*) and (25*) we see that

$$(27^*)\quad \left[\begin{array}{l}\text{if condition (5.8.1*) is satisfied, then in } Q[V]\\ \text{we have } F^{[1k]}(m^*,p^*,a^*,V) = F^{[[96]]}(C_0,V).\end{array}\right.$$

In view of (2.11.11) and (8.6) and the definitions of $F^{(1k)}$ and $F^{((96))}$, by (24*) and (25*) we see that

$$(28^*)\quad \left[\begin{array}{l}\text{if condition (5.8.1*) is satisfied, then for every } V \in Z\\ \text{we have } F^{(1k)}(m^*,p^*,a^*,V) = F^{((96))}(C_0,V).\end{array}\right.$$

Let

$$(29)\quad \hat{M} = \{a \in \text{pre}[2,p] : a(k,i) = b(k,i) \text{ for all } i \in [1,p]\}.$$

Then by (2) we see that

$$(30)\quad \left[\begin{array}{l}\text{given any } a \in \hat{M} \cap \text{vec}(2,m,p), \text{ we have } C[m,p,a,k] = R,\\ \text{and for all } j \in [1,p] \text{ we have } m(k) - a(k,j) = r(j),\\ \text{and for all } j \in [1,p] \text{ and } i \in [1,p] \text{ we have}\\ m(k) - a(k,j) + j - i = r(j) + r'(j) - s'(i).\end{array}\right.$$

Now

$$(31)\quad \left[\begin{array}{l}\text{for every } a \in \hat{M} \text{ let } S[a] \in Z(p) \text{ be defined by putting}\\ S[a](i) = m(k') - a(k',i) \text{ for all } i \in [1,p].\end{array}\right.$$

By (1) and (31) we see that

$$(32)\quad \text{for all } a \in \hat{M} \cap \text{vec}(2,m,p) \text{ we have } C^*(m,p,a,k',b) = C^*((S[a])).$$

By (30) and (31) we see that

(33) for all $a \in \hat{M} \cap \mathrm{vec}(2,m,p)$ we have $C(m,p,a) = C((S[a]))$.

In view of the definitions of $F_D^{(1k)}$ and $F_D^{(91)}$, by (6), (30), and (31) we see that

(34) $\left[\begin{array}{l}\text{for all } a \in \hat{M} \cap \mathrm{vec}(2,m,p) \text{ and } D \in Z \text{ we have} \\ F_D^{(1k)}(m,p,a) = F_D^{(91)}(S[a]).\end{array}\right.$

In view of (2.11.11) and (8.5) and the definitions of $F^{[[1k]]}$ and $F^{[[91]]}$, by (34) we see that

(35) $\left[\begin{array}{l}\text{for all } a \in \hat{M} \cap \mathrm{vec}(2,m,p) \text{ and } C' \in Z \text{ and } U \in Z, \text{ in } Q[V] \\ \text{we have } F^{[[1k]]}(m, p, a, C', V - U) = F^{[[91]]}(S[a], C', V - U).\end{array}\right.$

By (15'), (24*), and (35) we see that

(36) $\left[\begin{array}{l}\text{for all } a \in \hat{M} \cap \mathrm{vec}(2,m,p) \text{ and } U \in Z, \text{ in } Q[V] \text{ we have} \\ F^{[[1k]]}(m,p,a,C(m^*,p^*,a^*), V - U) = F^{[[91]]}(S[a], C_0, V - U).\end{array}\right.$

In view of (2.11.11) and (8.5) and the definitions of $F^{((1k))}$ and $F^{((91))}$, by (15'), (24*), and (34) we see that

(37) $\left[\begin{array}{l}\text{for all } a \in \hat{M} \cap \mathrm{vec}(2,m,p) \text{ and } U \in Z \text{ and } V \in Z \text{ we have} \\ F^{((1k))}(m,p,a,C(m^*,p^*,a^*), V - U) = F^{((91))}(S[a], C_0, V - U).\end{array}\right.$

In view of (2.11.11) and (8.5) and the definitions of $F^{(1k)}$ and $F^{((91))}$, by (33) and (34) we see that

(38) $\left[\begin{array}{l}\text{for all } a \in \hat{M} \cap \mathrm{vec}(2,m,p) \text{ and } v^* \in Z \text{ we have} \\ F^{(1k)}(m,p,a,v^*) = F^{((91))}(S[a],C((S[a])),v^*).\end{array}\right.$

By the definition of M as given in (5.8) we see that

(39) $\left[\begin{array}{l}\text{for every } u \in J(p) \text{ we have } M[p,b,k,u] = \{a \in \hat{M} : \\ \quad a(k',i) \in [b(k',i) + 1,\ b(k', i + 1) - 1] \text{ for all} \\ \quad i \in u \text{ and } a(k',j) = b(k',j) \text{ for all } j \in [1,p]\backslash u\}.\end{array}\right.$

By (1), (31), and (39) we see that

(40) $\left[\begin{array}{l}\text{for every } u \in J(p),\ a \mapsto S[a] \text{ gives a bijection} \\ \text{of } M[p,b,k,u] \text{ onto } X[u].\end{array}\right.$

Now in view of (5.8.0) and (8.10.3) and (8.10.4), by (40) we see that

$$(41)\quad \left| \begin{array}{l} \text{for every } U \in [0,p],\ a \mapsto S[a] \text{ gives a} \\ \text{bijection of } M(p,b,k,U) \text{ onto } X(U). \end{array} \right.$$

By (5.8.2) we see that

(42) for every $U \in [0,p]$ we have $M(p,b,k,U) \subset \mathrm{vec}(2,m,p)$.

Let

Σ' stand for $\sum_{U\in[0,p]} \sum_{s\in X(U)}$

and recall that we have let

Σ^* stand for $\sum_{U\in[0,p]} \sum_{a\in M(p,b,k,U)}$

Now first, by (34), (41), and (42) we see that for every $I \in Z$ and $D \in Z$ we have

$$\Sigma^* \binom{U}{I} F_{D-I}^{(1k)}(m,p,a) = \Sigma' \binom{U}{I} F_{D-I}^{(91)}(s)$$

and hence by (5), (11'), (16'), (25*), (8.35'.1), and (8.36*.1) we get (5.8*.1). Second, by (35), (41), and (42) we see that for every $C' \in Z$, in $Q[V]$ we have

$$\Sigma^* F^{[[1k]]}(m,\ p,\ a,\ C',\ V-U) = \Sigma' F^{[[91]]}(s,\ C',\ V-U)$$

and hence by (5), (11'), (17'), (26*), (8.35'.2), and (8.36*.2) we get (5.8*.2). Third, by (36), (41), and (42) we see that in $Q[V]$ we have

$$\Sigma^* F^{[[1k]]}(m,p,a,C(m^*,p^*,a^*),\ V-U)$$
$$= \Sigma' F^{[[91]]}(s,\ C_0,\ V-U)$$

and hence by (5), (11'), (18'), (27*), (8.35'.3), and (8.36*.3) we get (5.8*.3). Fourth, by (37), (41), and (42) we see that for every $C' \in Z$ we have

$$\Sigma^* F^{((1k))}(m,p,a,C(m^*,p^*,a^*),\ V-U) = \Sigma' F^{((91))}(s,\ C_0,\ V-U)$$

and hence by (5), (11'), (19'), (28*), (8.35'.4), and (8.36*.4) we get (5.8*.4). Fifth, by (32), (38), (41), and (42) we see that for every $V \in N$ we have

$$\left[\begin{array}{l} \Sigma^* \sum_{v\in N(2,V)} F^{(1k)}(m,p,a,v(1))\begin{bmatrix}C^*(m,p,a,k',b)\\ v(2)-U\end{bmatrix} \\ = \Sigma' \sum_{v\in N(2,V)} F^{(91)}(s,C((s)),v(1))\begin{bmatrix}C^*((s))\\ v(2)-U\end{bmatrix}\end{array}\right.$$

and hence by (5), (11'), (19'), (28*), (8.35'.5), and (8.36*.5) we get (5.8*.11). Finally (5.8*.12) is obvious.

PROBLEM (8.41) Let there be given any $q \in N^*$ and $m \in N^*(q)$ and $p \in N^*$ and $a \in \text{vec}(q,m,p)$. Now, in view of (5.8'''), by (5.8*.11) and (5.8*.12) we see that

$$(1) \quad \left[\begin{array}{l}\text{if } q = 2, \text{ then} \\ \text{Mon}(2,m,p,a,V) = F^{(11)}(m,p,a,V) \text{ for all } V \in N\end{array}\right.$$

and hence by (5.7*.2) and (5.7*.3) we see that

$$(1') \quad \left[\begin{array}{l}\text{if } q = 2, \text{ then for every } L \in \{1,1^*,2,2^*,3,3^*\} \text{ and } k \in [1,2] \\ \text{we have Mon}(2,m,p,a,V) = F^{[Lk]}(m,p,a,V) \text{ for all } V \in N \\ \text{where we recall that } F^{[Lk]}(m,p,a,V) \in Q[V].\end{array}\right.$$

So we may pose the problem: Find a formula to express Mon(q,m,p,a,V) for *every* q and not just for q = 2. As a guideline, we may ask the concrete question: For every q, is Mon(q,m,p,a,V) a polynomial in V; i.e., does there exist $\tilde{F}(q,m,p,a,V) \in Q[V]$ such that Mon(q,m,p,a,V) = $\tilde{F}(q,m,p,a,V)$ for all $V \in N$?

PROBLEM (8.42) Let there be given any $q \in N^*$ and $m \in N^*(q)$ and $p \in N^*$ and $a \in \text{vec}(q,m,p)$. In view of (5.8'''), by (5.8*.11) and (5.8*.12) we see that

$$(1) \quad \left[\begin{array}{l}\text{if } q = 2, \text{ then} \\ \text{Mon}(2,m,p,a,V) = F^{(11)}(m,p,a,V) \text{ for all } V \in N\end{array}\right.$$

and in view of (5.6''') and (5.6*.2) and (5.7*.3), by (5.6*.11) to (5.6*.14) we see that

(2) $\begin{cases} \text{if } q = 2, \text{ then} \\ \mathrm{Stab}(2,m,p,a,V) = F^{(11)}(m,p,a,V) \text{ for all } V \in N. \end{cases}$

By (1) and (2) we see that *if* $q = 2$, *then*

(3*) $\mathrm{Mon}(q,m,p,a,V) = \mathrm{Stab}(q,m,p,a,V) \quad \text{for all } V \in N.$

So we are led to ask: For what values of q is (3*) true? Moreover, given any q, is there a suitable modification of (3*) which is true?

9. COUNTING TABLEAUX

Having proved Theorems (5.6*), (5.7*), and (5.8*) in Sections 6, 7, and 8, respectively, we are now ready to complete the discussion of Section 5 by establishing the main results of this chapter. In Remarks (9.10), (9.14), and (9.18) we pose some related problems.

THEOREM (9.6*) Given any $q \in N^*$ and $m \in N^*(q)$ and $p \in N^*$ and $a \in \mathrm{vec}(q,m,p)$, we have the following.

(9.6*.1) For all $V \in Z$ we have

$$F^{(7)}(q,m,p,a,V) = F^{(8)}(q,m,p,a,V).$$

(9.6*.2) If q is even, then for all $k \in [1,q]$ and $V \in Z$ we have

$$F^{(5k)}(q,m,p,a,V) = F^{(6)}(q,m,p,a,V) = F^{(7^*)}(q,m,p,a,V) = F^{(8)}(q,m,p,a,V).$$

(9.6*.3) If $p = 1$, then for all $V \in Z$ we have

$$F^{(6)}(q,m,p,a,V) = F^{(7^*)}(q,m,p,a,V) = F^{(8)}(q,m,p,a,V).$$

(9.6*.4) If q is odd and $p \geq 2$, then for all $V \in Z$ we have

$$F^{(6)}(q,m,p,a,V) = 0 = F^{(7^*)}(q,m,p,a,V).$$

(9.6*.11) With $Z(p,V,8)$ as in (2.7), for all $V \in Z$ we have

$$\sum_{v \in Z(p,V,8)} H^{(8)}(q,m,p,a,v) = F^{(8)}(q,m,p,a,V).$$

(9.6*.12) With $\hat{u}$ as in (5.6), for all $v \in N(p)$ we have

$$\sum_{b \in vec(q,m,p,a)} H^{(8)}(q,m,p,b,v) = H^{(8)}(q,m,p,a,\hat{u}[p,v]).$$

(9.6*.13) If $\hat{h}$, $\tilde{h}$, $\hat{Z}$ are as in (5.6) and if $p \neq 1$, then for all $v \in \hat{Z}(p)$ we have

$$H^{(8)}(q,m,p,a,v) = H^{(8)}(q, m, p - 1, \tilde{h}[q,p,a], \hat{h}[p,v]).$$

(9.6*.14) If $p = 1$, then for the unique $v \in N(p)$ with $v(1) = 0$, we have $H^{(8)}(q,m,p,a,v) = 1$.

Proof: This is a verbatim restatement of Theorem (5.6*) which was asserted in Section 5 and which was proved in Remark (6.40) of Section 6.

THEOREM (9.6) Let there be given any $q \in N^*$ and $m \in N^*(q)$ and $p \in N^*$ and $a \in vec(q,m,p)$. Then for every $v \in N(p)$ we have

$$Stas(q,m,p,a,v) = H^{(8)}(q,m,p,a,v)$$

and for every $V \in Z$ we have

$$\begin{aligned} Stab(q,m,p,a,V) &= F^{(8)}(q,m,p,a,V) \\ &= F^{(7)}(q,m,p,a,V). \end{aligned}$$

Moreover, if q is even, then for every $V \in Z$ we have

$$\begin{aligned} Stab(q,m,p,a,V) &= F^{(7*)}(q,m,p,a,V) \\ &= F^{(6)}(q,m,p,a,V) \\ &= F^{(5k)}(q,m,p,a,V) \quad \text{for all } k \in [1,q]. \end{aligned}$$

Proof: In view of (5.6') and in view of (9.6*.11) to (9.6*.14), upon taking

$$Stas,\ Stab,\ H^{(8)},\ F^{(8)}, \text{ respectively, for } \hat{Stas},\ \hat{Stab},\ \hat{H},\ \hat{F}$$

in (5.6") we see that for every $v \in N(p)$ we have

$$Stas(q,m,p,a,v) = H^{(8)}(q,m,p,a,v)$$

and for every $V \in Z$ we have

$$\mathrm{Stab}(q,m,p,a,V) = F^{(8)}(q,m,p,a,V).$$

The rest now follows from (9.6*.1) and (9.6*.2).

THEOREM (9.7*) Given any $m \in N^*(2)$ and $p \in N^*$ and $a \in \mathrm{vec}(2,m,p)$, we have the following.

(9.7*.1) Given any

$$L \in \{1,1^*,2,2^*,3,3^*\} \cup \{3z : z \in W[p]\} \quad \text{and} \quad k \in [1,2]$$

first, we have

$$F_D^{(Lk)}(m,p,a) = F_D^{(11)}(m,p,a) \quad \text{for all } D \in Z,$$

and second, we have

$$F^{[Lk]}(m,p,a,V) = F^{[11]}(m,p,a,V) \text{ in } Q[V],$$

and third, for every $C' \in Z$ we have

$$F^{[[Lk]]}(m,p,a,C',V) = F^{[[11]]}(m,p,a,C',V) \text{ in } Q[V],$$

and fourth, for every $C' \in Z$ we have

$$F^{((Lk))}(m,p,a,C',V) = F^{((11))}(m,p,a,C',V) \quad \text{for all } V \in Z.$$

(9.7*.2) Given any

$$L \in \{1,1^*,2,2^*,3,3^*\} \cup \{3z : z \in W[p]\} \quad \text{and} \quad k \in [1,2]$$

we have

$$F^{[Lk]}(m,p,a,V) = F^{(6)}(2,m,p,a,V) \quad \text{for all } V \in N.$$

(9.7*.3) Given any

$$L \in \{1,1^*,2,2^*,3,3^*,4,4^*\} \cup \{3z : z \in W[p]\} \quad \text{and} \quad k \in [1,2]$$

we have

$$F^{(Lk)}(m,p,a,V) = F^{(6)}(2,m,p,a,V) \quad \text{for all } V \in Z.$$

(9.7*.4) If

$$p = \min(m(1),m(2)) \text{ and } a(1,i) = i = a(2,i) \quad \text{for all } i \in [1,p]$$

and

if $k \in [1,2]$ is such that $m(k) = p$

then for any given $e \in Z(p)$ we have

$$H^{(2k)}(m,p,a,e) = \begin{cases} 1 & \text{if } e(i) = 0 \text{ for all } i \in [1,p] \\ 0 & \text{if } e(i) \neq 0 \text{ for some } i \in [1,p] \end{cases}$$

and for any given $E \in Z$ we have

$$H_E^{(2k)}(m,p,a) = \begin{cases} 1 & \text{if } E = 0 \\ 0 & \text{if } E \neq 0. \end{cases}$$

(9.7*.5) if

$p = \min(m(1),m(2))$ and $a(1,i) = i = a(2,i)$ for all $i \in [1,p]$

then for every

$L \in \{1,1^*,2,2^*,3,3^*\} \cup \{3z : z \in W[p]\}$ and $k \in [1,2]$

first, we have

$$F_D^{(Lk)}(m,p,a) = \begin{cases} 1 & \text{for } D = 0 \\ 0 & \text{for all } D \in Z\backslash\{0\}, \end{cases}$$

and second, we have

$$F^{[Lk]}(m,p,a,V) = \begin{bmatrix} V \\ m(1)m(2) - 1 \end{bmatrix} \text{ in } Q[V].$$

(9.7*.6) if

$p = \min(m(1),m(2))$ and $a(1,i) = i = a(2,i)$ for all $i \in [1,p]$

then for every

$L \in \{1,1^*,2,2^*,3,3^*,4,4^*\} \cup \{3z : z \in W[p]\}$ and $k \in [1,2]$

we have

$$F^{(Lk)}(m,p,a,V) = F^{(6)}(2,m,p,a,V) = \begin{bmatrix} m(1)m(2) - 1 \\ V \end{bmatrix}$$

for all $V \in Z$.

Proof: This is a verbatim restatement of Theorem (5.7*) which was asserted in Section 5 and which was proved in Remark (7.40) of Section 7.

THEOREM (9.7) Let there be given any $m \in N^*(2)$ and $p \in N^*$ and $a \in \text{vec}(2,m,p)$ and

$$L \in \{1,1^*,2,2^*,3,3^*\} \cup \{3z : z \in W[p]\} \quad \text{and} \quad k \in [1,2].$$

Then for every $V \in Z$ we have

$$\text{Stab}(2,m,p,a,V) = F^{(Lk)}(m,p,a,V).$$

Proof: Follows from (9.6) and (9.7*.3).

THEOREM (9.8*) Let there be given any

$$L \in \{1,1^*,2,2^*,3,3^*\} \cup \{3z : z \in W[p]\} \quad \text{and} \quad Y \in [1,2]$$

and

$$\ell \in \{1,1^*,2,2^*,3,3^*\} \cup \{3z : z \in W[p]\} \quad \text{and} \quad y \in [1,2].$$

Let C^* and M be as in (5.8), and let there be fixed any $k \in [1,2]$, and let us put

$$k' = \begin{cases} 2 & \text{if } k = 1 \\ 1 & \text{if } k = 2. \end{cases}$$

Let there be given any $m \in N^*(2)$ and $p \in N^*$ and $b \in \text{vec}[2, p + 1]$ and $m^* \in N^*(2)$ and $p^* \in N^*$ and $a^* \in \text{vec}(2,m^*,p^*)$ such that (5.8.1) holds and either (5.8.1') holds or (5.8.1*) holds. Finally, let

$$\Sigma^* \text{ stand for } \sum_{U \in [0,p]} \sum_{a \in M(p,b,k,U)}$$

Now

$$\left[\begin{array}{l} \text{for every } D \in Z \text{ we have} \\ \displaystyle\sum_{I \in [0,p]} \Sigma^* \binom{U}{I} F^{(\ell y)}_{D-I}(m,p,a) = F^{(LY)}_{D}(m^*,p^*,a^*) \end{array}\right. \tag{9.8*.1}$$

and

$$\left[\begin{array}{l} \text{for every } C' \in Z \text{ in } Q[V] \text{ we have} \\ \Sigma^* F^{[[\ell y]]}(m, p, a, C', V - U) \\ = F^{[[LY]]}(m^*,p^*,a^*,C',V) \end{array}\right. \tag{9.8*.2}$$

and

$$\left[\begin{array}{l}\text{in } Q[V] \text{ we have}\\ \Sigma^* F^{[[\ell y]]}(m,p,a,C(m^*,p^*,a^*),\ V - U)\\ = F^{[LY]}(m^*,p^*,a^*,V)\end{array}\right. \qquad (9.8^*.3)$$

and

$$\left[\begin{array}{l}\text{for every } V \in N \text{ we have}\\ \Sigma^* F^{((\ell y))}(m,p,a,C(m^*,p^*,a^*),\ V - U)\\ = F^{(LY)}(m^*,p^*,a^*,V)\end{array}\right. \qquad (9.8^*.4)$$

and

$$\left[\begin{array}{l}\text{for every } V \in N \text{ we have}\\ \Sigma^* \sum_{v\in N(2,V)} F^{(\ell y)}(m,p,a,v(1)) \begin{bmatrix} C^*(m,p,a,k',b) \\ v(2) - U \end{bmatrix}\\ = F^{(LY)}(m^*,p^*,a^*,V)\end{array}\right. \qquad (9.8^*.11)$$

and

$$\left[\begin{array}{l}\text{if } p = 1 \text{ and if } a \in vec(2,m,p) \text{ is such that}\\ m(k) - a(k,p) = 0, \text{ then for every } V \in N \text{ we have}\\ F^{(LY)}(m,p,a,V) = \begin{bmatrix} m(k') - a(k',1) \\ V \end{bmatrix}.\end{array}\right. \qquad (9.8^*.12)$$

Proof: In view of (9.7*.1) to (9.7*.3), our assertion follows from Theorem (5.8*) which was stated in Section 5 and which was proved in Remark (8.40) of Section 8.

THEOREM (9.8) Let there be given any $m \in N^*(2)$ and $p \in N^*$ and $a \in vec(2,m,p)$ and

$$L \in \{1,1^*,2,2^*,3,3^*\} \cup \{3z : z \in W[p]\} \quad \text{and} \quad k \in [1,2].$$

Then for every $V \in Z$ we have

$$\mathrm{Mon}(2,m,p,a,V) = F^{(Lk)}(m,p,a,V).$$

Proof: In view of (5.8') and in view of assertions (5.8*.11) and (5.8*.12) which were made in Section 5 and which were proved in Remark (8.40) of Section 8, upon taking

$$\text{Mon}(2,m,p,a,V) \text{ and } F^{(1k)}(m,p,a,V) \text{ and } \begin{bmatrix} C^*(m,p,a,k',b) \\ V - U \end{bmatrix},$$

respectively, for

$$\widehat{\text{Mon}}(2,m,p,a,V) \text{ and } \hat{F}(m,p,a,V) \text{ and } \hat{C}(m,p,a,k',b,V,U)$$

in (5.8") we see that

$$\text{Mon}(2,m,p,a,V) = F^{(1k)}(m,p,a,V) \quad \text{for all } V \in N.$$

By the definitions of Mon and $F^{(1k)}$ as given in (2.4) and (2.10), respectively, we see that

$$\text{Mon}(2,m,p,a,V) = 0 = F^{(1k)}(m,p,a,V) \quad \text{for all } V \in Z\backslash N.$$

Therefore,

$$\text{Mon}(2,m,p,a,V) = F^{(1k)}(m,p,a,V) \quad \text{for all } V \in Z.$$

The rest now follows from (9.7*.3).

THEOREM (9.9) Let there be given any $m \in N^*(2)$ and $p \in N^*$ and $a \in \text{vec}(2,m,p)$. Then for every $V \in Z$ we have

$$\text{Stab}(2,m,p,a,V) = \text{Mon}(2,m,p,a,V).$$

Proof: Follows from (9.7) and (9.8).

REMARK (9.10) Given any $m \in N^*(2)$ and $p \in N^*$ and $a \in \text{vec}(2,m,p)$ and $V \in Z$, in view of (9.9) we may ask whether there exists a "natural" one-to-one correspondence between stab(2,m,p,a,V) and mon(2,m,p,a,V).

THEOREM (9.11) Let there be given any $m \in N^*(2)$ and $V \in Z$. Then we have

$$\text{Stib}(2,m,V) = \begin{bmatrix} m(1)m(2) - 1 \\ V \end{bmatrix} = \text{Mon}[[2,m,V]].$$

NOTE We shall give two proofs of (9.11). For both of them we note that by (4.21*) we have

$$\text{Mon}[[2,m,V]] = \begin{bmatrix} m(1)m(2) - 1 \\ V \end{bmatrix} \tag{*}$$

and we put $p = \min(m(1),m(2))$ and we fix any $k \in [1,2]$ and we let $a \in \mathrm{vec}(2,m,p)$ be defined by putting $a(1,i) = i = a(2,i)$ for all $i \in [1,p]$.

First Proof: In view of (*) our assertion follows by noting that by (9.7*.6) we have

$$F^{(2k)}(m,p,a,V) = \begin{bmatrix} m(1)m(2) - 1 \\ V \end{bmatrix}$$

and by (9.7) we have

$$\mathrm{Stab}(2,m,p,a,V) = F^{(2k)}(m,p,a,V)$$

and obviously, we have

$$\mathrm{Stib}(2,m,V) = \mathrm{Stab}(2,m,p,a,V).$$

Second Proof: In view of (*) our assertion follows by noting that by (9.9) we have

$$\mathrm{Stab}(2,m,p,a,V) = \mathrm{Mon}(2,m,p,a,V)$$

and obviously, we have

$$\mathrm{Stib}(2,m,V) = \mathrm{Stab}(2,m,p,a,V)$$

and

$$\mathrm{Mon}[[2,m,V]] = \mathrm{Mon}(2,m,p,a,V).$$

LEMMA (9.12) Let there be given any $m \in N^*(2)$ and $p \in N^*$ and $a \in \mathrm{vec}(2,m,p)$ such that $p = 1$ and such that for some $k \in [1,2]$ we have $m(k) - a(k,1) = 0$. Then for every

$$L \in \{1,1^*,2,2^*,3,3^*\} \cup \{3z : z \in W[p]\} \quad \text{and} \quad Y \in [1,2]$$

we have that

$$F_D^{(LY)}(m,p,a) = \begin{cases} 1 & \text{if } D = 0 \\ 0 & \text{if } D \in Z\setminus\{0\}. \end{cases}$$

Proof: The case of $L = 3$ and $Y = k$ follows from the definition of $F_D^{(3k)}$ given in (2.10), and then in view of (9.7*.1), the general case follows from this special case.

THEOREM (9.13) Let there be given any $m \in N^*(2)$ and $p \in N^*$ and $a \in \text{vec}(2,m,p)$ and

$$L \in \{1,1^*,2,2^*,3,3^*\} \cup \{3z : z \in W[p]\} \quad \text{and} \quad Y \in [1,2].$$

Then $F_D^{(LY)}(m,p,a)$ is a nonnegative integer for all $D \in Z$.

Proof: Now a sum of products of nonnegative integers is again a nonnegative integer, and for every $I \in [0,p]$ and $U \in [0,p]$ we have that $\binom{U}{I}$ is a nonnegative integer. Therefore, in view of (9.8*.1), our assertion follows by double induction on p and $m(k) - a(k,p)$. In greater detail, for any fixed p, by (5.8.1') we can make induction on $m(k) - a(k,p)$ provided that we assume the initial case when $m(k) - a(k,p) = 0$; on the other hand, by (5.8.1*) we see that the general case of p yields the initial case for $p + 1$; finally, by (9.12) we get the initial case for $p = 1$.

REMARK (9.14) In view of (9.13) we may ask whether $F_D^{(LY)}(m,p,a)$ is the cardinality of some "natural" family of finite sets parametrized by m, p, a, D.

LEMMA (9.15) Let there be given any $m \in N^*(2)$ and $p \in N^*$ and $a \in \text{vec}(2,m,p)$ and

$$L \in \{1,1^*,2,2^*,3,3^*\} \cup \{3z : z \in W[p]\} \quad \text{and} \quad Y \in [1,2]$$

and let us put

$$c(m,p,a) = \min(C[m,p,a,1], C[m,p,a,2])$$

where we recall that

$$C[m,p,a,k] = \sum_{i=1}^{p} [m(k) - a(k,i)] \quad \text{for all } k \in [1,2].$$

Then we have

$$\{D \in Z : F_D^{(LY)}(m,p,a) \neq 0\} \subset [0,c(m,p,a)].$$

Proof: Follows from (9.7*.1) and part (5) of (2.11.11).

PROPOSITION (9.16) Let there be given any

$$L \in \{1,1^*,2,2^*,3,3^*\} \cup \{3z : z \in W[p]\} \quad \text{and} \quad Y \in [1,2].$$

Let there be fixed any $k \in [1,2]$, and let us put

$$k' = \begin{cases} 2 & \text{if } k = 1 \\ 1 & \text{if } k = 2. \end{cases}$$

Let there be given any $m \in N^*(2)$ and $p \in N^*$ and $b \in \mathrm{vec}[2, p + 1]$ and $m^* \in N^*(2)$ and $p^* \in N^*$ and $a^* \in \mathrm{vec}(2,m^*,p^*)$ such that (5.8.1) holds and either (5.8.1') holds or (5.8.1*) holds. Finally, let

$$\begin{aligned} M^*(p,b,k) = \{a \in \mathrm{pre}[2,p] : {} & a(k,i^*) = b(k,i^*) \text{ for all } i^* \in [1,p] \\ & \text{and } a(k',i) \in [b(k',i), b(k', i + 1) - 1] \\ & \text{for all } i \in [1,p]\}. \end{aligned}$$

Then for every $D \in Z$ we have

$$\sum_{a \in M^*(p,b,k)} F_D^{(LY)}(m,p,a) = F_D^{(LY)}(m^*,p^*,a^*).$$

Proof: Follows from (9.8*.1) and (9.15) by noting that

$$\binom{U}{0} = 1 \quad \text{for all } U \in [0,p]$$

and by noting that with M as in (5.8), we clearly have

$$M^*(p,b,k) = \coprod_{U \in [0,p]} M(p,b,k,U).$$

THEOREM (9.17) Let there be given any $m \in N^*(2)$ and $p \in N^*$ and $a \in \mathrm{vec}(2,m,p)$ and

$$L \in \{1,1^*,2,2^*,3,3^*\} \cup \{3z : z \in W[p]\} \quad \text{and} \quad Y \in [1,2].$$

Then we have that $F_0^{(LY)}(m,p,a)$ is a positive integer.

Proof: Now a nonempty sum of products of positive integers is again a positive integer. Therefore, in view of (9.16), our assertion follows by double induction on p and $m(k) - a(k,p)$. In greater detail, for any fixed p, by (5.8.1') we can make induction on $m(k) - a(k,p)$ provided that we assume the initial case when $m(k) - a(k,p) = 0$;

on the other hand, by (5.8.1*) we see that the general case of p yields the initial case for p + 1; finally, by (9.12) we get the initial case for p = 1.

REMARK (9.18) Given any $m \in N^*(2)$ and $p \in N^*$ and $a \in \text{vec}(2,m,p)$, in view of (9.7*.1), (9.13), (9.15), and (9.17), we may ask whether there exists a nonnegative integer $c^*(m,p,a)$ such that for every

$$L \in \{1,1^*,2,2^*,3,3^*\} \cup \{3z : z \in W[p]\} \quad \text{and} \quad Y \in [1,2]$$

we have

$$\{D \in Z : F_D^{(LY)}(m,p,a) \neq 0\} = [0,c^*(m,p,a)]$$

and if so, then whether it is true that

$$c^*(m,p,a) = c(m,p,a)$$

where $c(m,p,a)$ is as in (9.15).

3
Universal Determinantal Identity

10. INFORMAL DISCUSSION

(10.1) Preamble In this chapter we prove a universal identity satisfied by the minors of any matrix. In the general case of this identity, the minors may be of different sizes, so we call it the mixed size case. We prove the mixed size case by reducing it to the maximal size case, by which we mean the case of p by p minors of a p by q matrix. In turn, we prove the maximal size case by reducing it to the basic case, by which we mean the maximal size case when the minors are disjoint. The proof of the basic case depends only on the Laplace development of a determinant. From the mixed size case we deduce what may be called the full depth case. The full depth case subsumes the existence part of the Straightening Law of Doubilet-Rota-Stein. In other words, as a corollary of the mixed size case we get a proof of the existence part of the Straightening Law.

In the rest of this section we discuss matters informally. Specifically, in (10.2) we briefly state the mixed size case, and in (10.3) we shall give an informal discussion of the basic case. Then in (10.4) we informally discuss the maximal size case, and in (10.5) we give an informal discussion of the mixed size case. Finally, in (10.6) we informally discuss the full depth case.

In Section 11 we give a preview of the formal part of this chapter by relating it to the informal discussion of Section 10. In greater detail, in (11.1) we introduce some notation to be used in the rest of this chapter, and in (11.2) we give a preview of Laplace development, and in (11.3) we preview the basic case. Then in (11.4) we give a preview of the maximal size case, and in (11.5) we preview the mixed size case. Finally, in (11.6) we give a preview of the full depth case.

The formal part of this chapter begins with Section 12. In other words, Section 10 and parts (11.2) to (11.6) of Section 11 do not form a logical part of this chapter.

In Section 12 we give a self-contained treatment of Laplace development, and in Section 13 we deal formally with the basic case. Then in Section 14 we formalize the maximal size case, and in Section 15 we deal formally with the mixed size case. Next in Section 16 we formalize the full depth case, and in Sections 17 and 18 we give some extra information concerning this case, and in Remark (18.7) of Section 18 we pose some related problems. Finally, in Section 19 we outline an alternative treatment of the full depth case.

To avoid repetition, as *temporary* notation to be used in the rest of this chapter, for $k = 1, 2$, we put

$$k' = \begin{cases} 2 & \text{if } k = 1 \\ 1 & \text{if } k = 2 \end{cases}$$

and for $e = 1, 2$, we put

$$e' = \begin{cases} 2 & \text{if } e = 1 \\ 1 & \text{if } e = 2. \end{cases}$$

(10.2) Brief Statement of the Mixed Size Case Let x be a p by p matrix. Let T[1] and T[2] be two square submatrices of x whose sizes may be different and which are allowed to overlap. For e = 1, 2, let T[e](1) and T[e](2) be the rows and columns of T[e], by which we actually mean that T[e](1) and T[e](2) are those rows and columns of x whose portions constitute the rows and columns of T[e]. For e = 1, 2, let us mark some rows and columns B[e](1) and B[e](2) of T[e]; i.e., let there be given any

$$B[e](1) \subset T[e](1) \quad \text{and} \quad B[e](2) \subset T[e](2).$$

Let us assume that

$$B[1](1) \cap B[2](1) = \emptyset. \tag{1*}$$

Let K(T,B) be the set of all "exchangeable subsets" A of B; i.e., let K(T,B) be the set of all quartets

$$A = (A[1](1), A[1](2), A[2](1), A[2](2))$$

such that for e = 1, 2, we have

$$A[e](1) \subset B[e](1) \quad \text{and} \quad A[e](2) \subset B[e](2)$$

and such that upon exchanging A[1] and A[2] between T[1] and T[2] we get two square submatrices S(T,A)[1] and S(T,A)[2] of x whereby, for e = 1, 2, the rows of S(T,A)[e] are

$$S(T,A)[e](1) = (T[e](1) \setminus A[e](1)) \cup A[e'](1)$$

and the columns of S(T,A)[e] are

$$S(T,A)[e](2) = (T[e](2) \setminus A[e](2)) \cup A[e'](2).$$

Also, let L(T,B) be the set obtained by deleting the empty quartet from K(T,B).

We shall prove that certain entries of x, together with some zeros, can be arranged to form a 2p by 2p matrix y such that

$$\det y = \sum_{A \in K(T,B)} \pm (\det S(T,A)[1])(\det S(T,A)[2]). \tag{10.2.1}$$

In the summation above we may transform to the LHS the term which corresponds to the empty quartet, and then we would get that

$$\begin{cases} (\det T[1])(\det T[2]) + \det y \\ = \displaystyle\sum_{A \in L(T,B)} \pm (\det S(T,A)[1])(\det S(T,A)[2]). \end{cases} \tag{10.2.2}$$

We shall also discover a certain cardinality condition (*) on B such that

$$\text{if } (*) \text{ is satisfied, then } \det y = 0 \tag{10.2.3}$$

and now by (10.2.1) and (10.2.3) we would get that

$$\left[\begin{array}{l}\text{if } (*) \text{ is satisfied, then} \\ \displaystyle\sum_{A \in K(T,B)} \pm (\det S(T,A)[1])(\det S(T,A)[2]) = 0\end{array}\right. \tag{10.2.4}$$

and finally, by (10.2.2) and (10.2.3) we would get that

$$\left[\begin{array}{l}\text{if } (*) \text{ is satisfied, then} \\ (\det T[1])(\det T[2]) \\ \displaystyle = \sum_{A \in L(T,B)} \pm (\det S(T,A)[1])(\det S(T,A)[2]).\end{array}\right. \tag{10.2.5}$$

(10.3) Informal Discussion of the Basic Case Let x be a p by 2p matrix. Let T[1] be the first p columns of x. Let T[2] be the last p columns of x. Thus

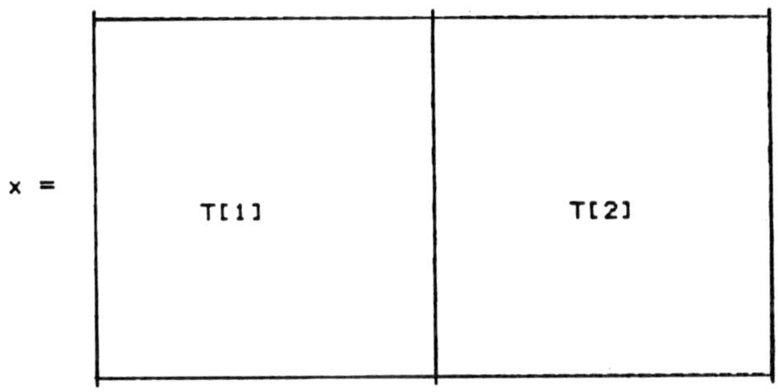

Also, let there be given any

$$B[e] \subset T[e] \qquad \text{for } e = 1, 2$$

and let us consider the 2p by 2p matrix in the following picture where $\hat{B}[e]$ is the p by p matrix formed by augmenting B[e] with zeros in T[e] \ B[e].

Now let us take the Laplace development of y by cutting it into the upper half and the lower half. So, at a time, from the upper half we have to choose p columns. Among them we must choose T[1] \ B[1] because under them we have zeros. This means that the columns A[1] which we do not take from T[1] must be from B[1]. In place of them we may take any columns A[2] from B[2], the only proviso being that

y =

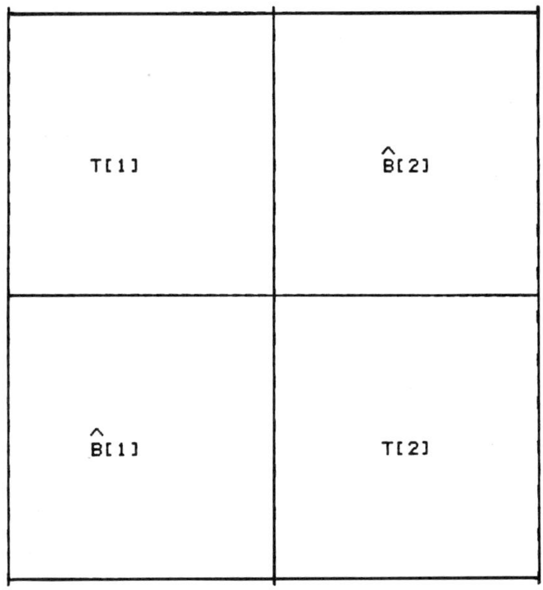

$$\text{card}(A[1]) = \text{card}(A[2]).$$

This gives

$$S(T,A)[1] = (T[1] \setminus A[1]) \cup A[2]$$

as the set of columns of the typical minor from the upper half. The set of columns of the corresponding complementary minor from the lower half is

$$S(T,A)[2] = (T[2] \setminus A[2]) \cup A[1].$$

Thus

$$\det y = \sum_{A \in K(T,B)} \pm (\det S(T,A)[1])(\det S(T,A)[2]) \qquad (10.3.1)$$

where $K(T,B)$ is the set of all pairs $A = (A[1],A[2])$ such that

$$A[1] \subset B[1] \quad \text{and} \quad A[2] \subset B[2] \quad \text{and}$$
$$\text{card}(A[1]) = \text{card}(A[2]).$$

If from the summation above we transfer to the LHS the term which corresponds to the empty pair A, i.e., to the pair A for which $A[1] = \emptyset = A[2]$, then we get

$$\left[\begin{array}{l}(\det T[1])(\det T[2]) + \det y \\ = \sum\limits_{A \in L(T,B)} \pm (\det S(T,A)[1])(\det S(T,A)[2])\end{array}\right. \qquad (10.3.2)$$

where L(T,B) is the set obtained by deleting the empty pair from K(T,B).

Consider the condition

$$\mathrm{card}(B[1]) + \mathrm{card}(B[2]) + \mathrm{card}(b^*) > p \qquad (*)$$

where b* is a certain set of columns which are common to

$$T[1] \setminus B[1] \quad \text{and} \quad T[2] \setminus B[2]$$

Now subtract the bottom half of y from its top half, and note that the resulting new version of y is again a 2p by 2p matrix and the det of this new version of y clearly equals det y. Considering any p by p minor of the top half of the new version of y, we see that if (*) is satisfied then either the said minor has a zero column or two of its columns are negatives of each other, and hence the minor equals zero. Consequently, if (*) is satisfied, then the det of the new version of y is zero. Therefore,

if (*) is satisfied, then det y = 0. (10.3.3)

By (10.3.1) and (10.3.3) we see that

$$\left[\begin{array}{l}\text{if } (*) \text{ is satisfied, then} \\ \sum\limits_{A \in K(T,B)} \pm (\det S(T,A)[1])(\det S(T,A)[2]) = 0.\end{array}\right. \qquad (10.3.4)$$

By (10.3.2) and (10.3.3) we see that

$$\left[\begin{array}{l}\text{if } (*) \text{ is satisfied, then } (\det T[1])(\det T[2]) \\ = \sum\limits_{A \in L(T,B)} \pm (\det S(T,A)[1])(\det S(T,A)[2]).\end{array}\right. \qquad (10.3.5)$$

(10.4) Informal Discussion of the Maximal Size Case Let x* be a p by q* matrix. For e = 1, 2, let T*[e] be any p columns of x* and let there be given any $B^*[e] \subset T^*[e]$, and let us depict the situation by the following picture.

x* =

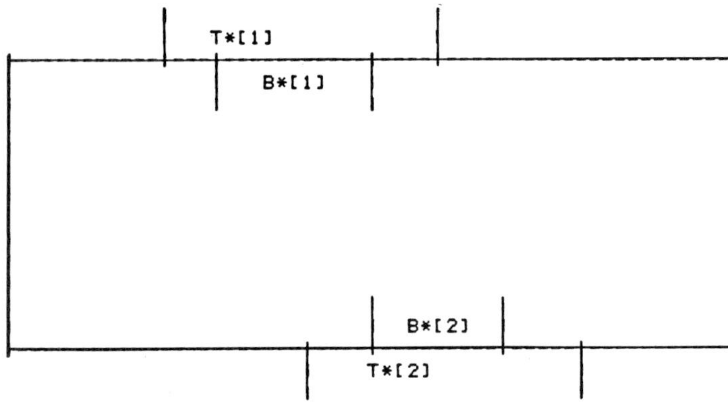

Upon taking

x =

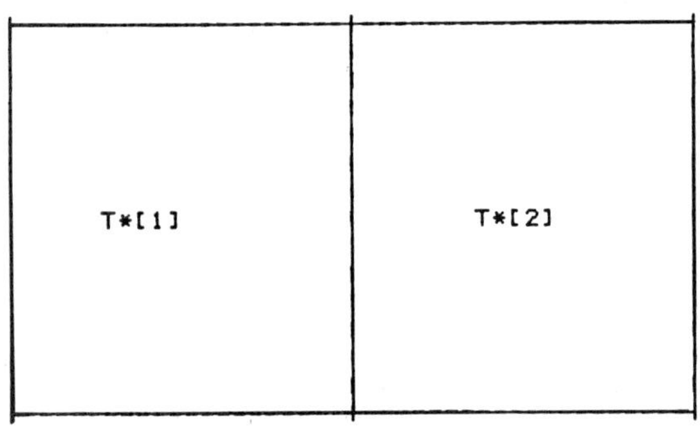

by (10.3.1) we get

$$\det y = \sum_{A^* \in K(T^*,B^*)} \pm (\det S(T^*,A^*)[1])(\det S(T^*,A^*)[2]) \tag{10.4.1}$$

where $K(T^*,B^*)$ is the set of all pairs $A^* = (A^*[1],A^*[2])$ such that

$$A^*[1] \subset B^*[1] \quad \text{and} \quad A^*[2] \subset B^*[2] \quad \text{and}$$
$$\operatorname{card}(A^*[1]) = \operatorname{card}(A^*[2])$$

and where

$$S(T^*,A^*)[1] = (T^*[1] \setminus A^*[1]) \cup A^*[2]$$

and

$$S(T^*,A^*)[2] = (T^*[2] \setminus A^*[2]) \cup A^*[1]$$

and where y is the 2p by 2p matrix obtained by putting

$$y = \begin{bmatrix} T^*[1] & \hat{B}^*[2] \\ \hat{B}^*[1] & T^*[2] \end{bmatrix}$$

where $\hat{B}^*[e]$ is the p by p matrix formed by augmenting $B^*[e]$ with zeros in $T^*[e] \setminus B^*[e]$, and by (10.3.2) we get

$$\left[\begin{aligned} &(\det T^*[1])(\det T^*[2]) + \det y \\ &= \sum_{A^* \in L(T^*,B^*)} \pm (\det S(T^*,A^*)[1])(\det S(T^*,A^*)[2]) \end{aligned}\right. \qquad (10.4.2)$$

where $L(T^*,B^*)$ is the set obtained by deleting the empty pair from $K(T^*,B^*)$, and upon considering the condition

$$\operatorname{card}(B^*[1]) + \operatorname{card}(B^*[2]) + \operatorname{card}(b^*) > p \qquad (*)$$

where

$$b^* = (T^*[1] \setminus B^*[1]) \cap (T^*[2] \setminus B^*[2])$$

by (10.3.3) we see that

$$\text{if } (*) \text{ is satisfied, then } \det y = 0 \qquad (10.4.3)$$

and by (10.3.4) we see that

$$\left[\begin{aligned} &\text{if } (*) \text{ is satisfied, then} \\ &\sum_{A^* \in K(T^*,B^*)} \pm (\det S(T^*,A^*)[1])(\det S(T^*,A^*)[2]) = 0 \end{aligned}\right. \qquad (10.4.4)$$

and by (10.3.5) we see that

$$\left[\begin{array}{l}\text{if } (*) \text{ is satisfied, then} \\ (\det T^*[1])(\det T^*[2]) \\ = \sum_{A^*\in L(T^*,B^*)} \pm (\det S(T^*,A^*)[1])(\det S(T^*,A^*)[2]).\end{array}\right. \tag{10.4.5}$$

(10.5) Informal Discussion of the Mixed Size Case Let x be a p by q matrix. Let T[1] and T[2] be two square submatrices of x whose sizes may be different and which are allowed to overlap. For e = 1, 2, let T[e](1) and T[e](2) be the rows and columns of T[e], by which we actually mean that T[e](1) and T[e](2) are those rows and columns of x whose portions constitute the rows and columns of T[e]. For e = 1, 2, let us mark some rows and columns B[e](1) and B[e](2) of T[e], i.e., let there be given any

$$B[e](1) \subset T[e](1) \quad \text{and} \quad B[e](2) \subset T[e](2).$$

Let us assume that

$$B[1](1) \cap B[2](1) = \emptyset. \tag{1*}$$

Let K(T,B) be the set of all "exchangeable subsets" A of B; i.e., let K(T,B) be the set of all quartets

$$A = (A[1](1),A[1](2),A[2](1),A[2](2))$$

such that for e = 1, 2, we have

$$A[e](1) \subset B[e](1) \quad \text{and} \quad A[e](2) \subset B[e](2)$$

and such that upon exchanging A[1] and A[2] between T[1] and T[2] we get two square submatrices S(T,A)[1] and S(T,A)[2] of x whereby, for e = 1, 2, the rows of S(T,A)[e] are

$$S(T,A)[e](1) = (T[e](1) \setminus A[e](1)) \cup A[e'](1)$$

and the columns of S(T,A)[e] are

$$S(T,A)[e](2) = (T[e](2) \setminus A[e](2)) \cup A[e'](2).$$

Also, let L(T,B) be the set obtained by deleting the empty quartet from K(T,B). Let us depict the situation by the following picture of x.

x =
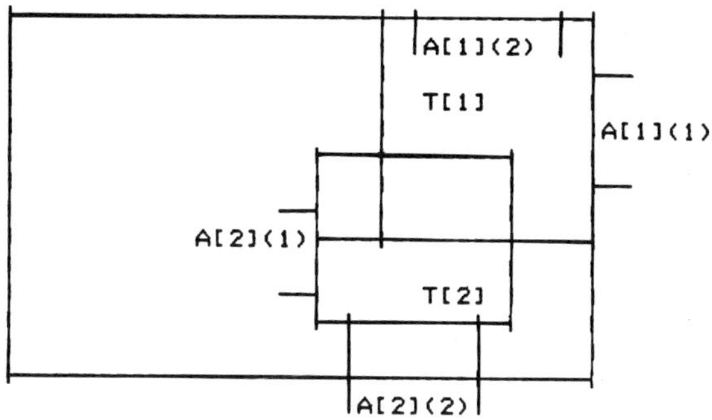

To extend the discussion of (10.4) to include the present case of minors of any size, we proceed thus. Let q* = p + q and let x* be the p by q* matrix which is obtained when we augment x by the p by p anti-identity matrix as depicted below.

x* =
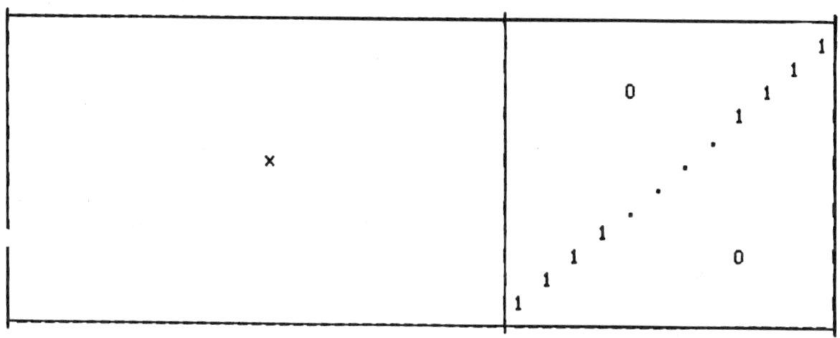

Let V[e] be the set of those rows of x which are not among the rows of T[e]. To T[e] let us associate the p by p submatrix of x* whose columns T*[e] consist of the columns of T[e] together with those columns of x* whose column numbers are obtained when from q* + 1 we subtract the row numbers of V[e]. It can be seen that this sets up a one-to-one correspondence between all square submatrices of x and all p by p submatrices of x*. Moreover, det of the p by p submatrix equals either det T[e] or its negative; this follows by noting that the p by p submatrix can be obtained by suitably scrambling the p by p matrix, which is depicted below and in which the bottom right-hand block is the p - r[e] by p - r[e] anti-identity matrix, where r[e] is the number of rows of T[e], and in

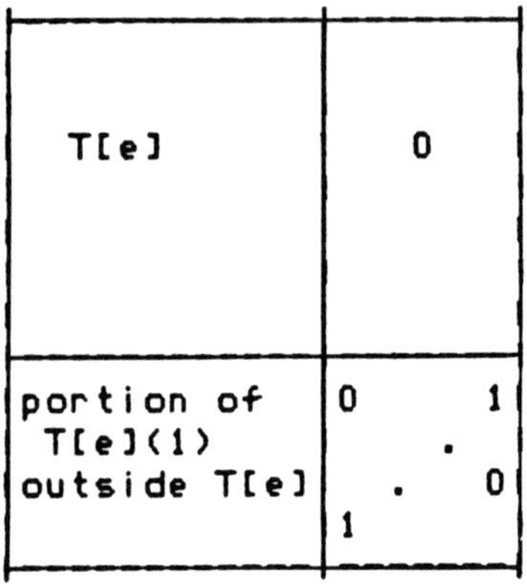

which the bottom left-hand block is as suggested by the picture

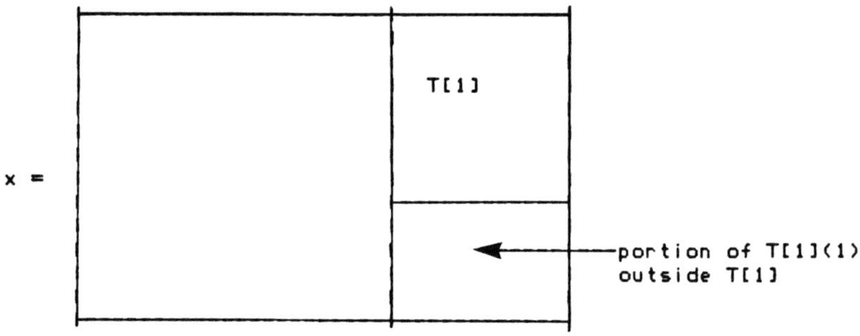

and the picture

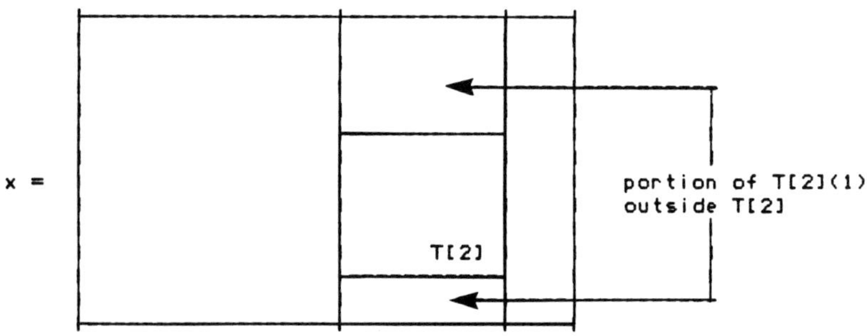

Let U[e] be the set of those rows of B[e'] which are not among the rows of T[e]. Let B*[e] be the subset of T*[e] consisting of the columns of B[e] together with those columns of x* whose column numbers are obtained when from $q^* + 1$ we subtract the row numbers of the rows of $U[e]$. Let y be the $2p$ by $2p$ matrix obtained by putting

$$y = \begin{bmatrix} T^*[1] & \hat{B}^*[2] \\ \hat{B}^*[1] & T^*[2] \end{bmatrix}$$

where $\hat{B}^*[e]$ is the p by p matrix formed by augmenting $B^*[e]$ with zeros in $T^*[e] \setminus B^*[e]$.

In view of the one-to-one correspondence above, (10.4.1) can be converted to the assertion which says that

$$\det y = \sum_{A \in K(T,B)} \pm (\det S(T,A)[1])(\det S(T,A)[2]), \qquad (10.5.1)$$

whereas (10.4.2) can be converted to the assertion which says that

$$\left[\begin{array}{l} (\det T[1])(\det T[2]) + \det y \\ = \displaystyle\sum_{A \in L(T,B)} \pm (\det S(T,A)[1])(\det S(T,A)[2]) \end{array}\right. \qquad (10.5.2)$$

and upon considering the condition

$$E(T,B) > p \qquad (*)$$

where

$$E(T,B) = \operatorname{card}(B[1](2)) + \operatorname{card}(B[2](2)) + \operatorname{card}\left(\bigcap_{e=1,2} [T[e](2) \setminus B[e](2)]\right) + \operatorname{card}(U[1]) + \operatorname{card}(U[2]) + \operatorname{card}\left(\bigcap_{e=1,2} [V[e] \setminus U[e]]\right),$$

(10.4.3) can be converted to the assertion which says that

$$\text{if } (*) \text{ is satisfied, then } \det y = 0 \tag{10.5.3}$$

and now by (10.5.1) and (10.5.3) we see that

$$\left[\begin{array}{l}\text{if } (*) \text{ is satisfied, then} \\ \displaystyle\sum_{A\in K(T,B)} \pm (\det S(T,A)[1])(\det S(T,A)[2]) = 0\end{array}\right. \tag{10.5.4}$$

and finally, by (10.5.2) and (10.5.3) we see that

$$\left[\begin{array}{l}\text{if } (*) \text{ is satisfied, then} \\ (\det T[1])(\det T[2]) \\ \displaystyle = \sum_{A\in L(T,B)} \pm (\det S(T,A)[1])(\det S(T,A)[2]).\end{array}\right. \tag{10.5.5}$$

(10.6) Informal Discussion of the Full Depth Case Let x be a p by q matrix. Let

$$T = T[e]_{1\le e\le d}$$

be any sequence where T[e] is a square submatrix of x, and let T[e](1) and T[e](2) be the rows and columns of T[e]. Let us write the members of T[e](1) as an increasing sequence

$$T[e](1,1) < T[e](1,2) < \cdots < T[e](1,r[e])$$

and let us write the members of T[e](2) as an increasing sequence

$$T[e](2,1) < T[e](2,2) < \cdots < T[e](2,r[e]).$$

For any $e \in [1, d - 1]$, let us define $T[e] \le T[e + 1]$ to mean that $r[e] \ge r[e + 1]$ and for all $i \in [1, r[e + 1]]$ we have

$$T[e](1,i) \le T[e + 1](1,i) \quad \text{and} \quad T[e](2,i) \le T[e + 1](2,i)$$

where, for any integers a and b, by [a,b] we denote the set of all integers c with $a \le c \le b$. Now either

(1) $T[1] \le T[2] \le \cdots \le T[d]$

or there exists a unique $e^* \in [1, d - 1]$ such that

(2) $T[1] \le T[2] \le \cdots \le T[e^* - 1] \le T[e^*] \not\le T[e^* + 1]$.

For a while, suppose that we are in case (2). Now we must have that either

$$(21)\quad \left[\begin{array}{l} r[e^*] \ge r[e^* + 1] \text{ and} \\ T[e^*](1,i) \le T[e^* + 1](1,i) \text{ for all } i \in [1, r[e^* + 1]] \\ \text{and there exists } v \in [1, r[e^* + 1]] \text{ such that} \\ T[e^*](2,v) > T[e^* + 1](2,v) \text{ and} \\ T[e^*](2,j) \le T[e^* + 1](2,j) \text{ for all } j \in [1, v - 1] \end{array}\right.$$

or

$$(22)\quad \left[\begin{array}{l} r[e^*] \ge r[e^* + 1] \text{ and} \\ \text{there exists } w \in [1, r[e^* + 1]] \text{ such that} \\ T[e^*](1,w) > T[e^* + 1](1,w) \text{ and} \\ T[e^*](1,j) \le T[e^* + 1](1,j) \text{ for all } j \in [1, w - 1] \end{array}\right.$$

or

(23) $r[e^*] < r[e^* + 1]$.

Let us consider the pair $t = (t[1],t[2])$ where $t[1] = T[e^*]$ and $t[2] = T[e^* + 1]$. In case of (21), let us define the quartet

$$B = (B[1](1),B[1](2),B[2](1),B[2](2))$$

by putting

$$B[1](1) = \emptyset$$

and

$$B[2](1) = t[2](1) \setminus t[1](1)$$

and

$$B[1](2) = \{t[1](2,i) : i \in [v,r[e^*]]\}$$

and

$$B[2](2) = \{t[2](2,i) : i \in [1,v]\};$$

now clearly,

$$B[e](1) \subset t[e](1) \text{ and } B[e](2) \subset t[e](2) \quad \text{for all } e \in [1,2]$$

and it can be seen that

$$\begin{bmatrix} B[1](1) \cap B[2](1) = \emptyset \\ \text{and } E(t,B) > p \text{ with } E \text{ as in } (10.5) \end{bmatrix} \tag{10.6.1}$$

and hence by (10.5.5) we get that

$$\begin{bmatrix} (\det t[1])(\det t[2]) \\ = \sum_{A \in L(t,B)} \pm (\det S(t,A)[1])(\det S(t,A)[2]). \end{bmatrix} \tag{10.6.2}$$

The definition of B and the proof of equation (10.6.2) can be extended to include cases (22) and (23). Moreover, it can also be shown that

$$\text{for every } A \in L(t,B) \text{ we have } S(t,A)[1] < t[1]. \tag{10.6.3}$$

Let us suppose that all this has been done.

Let $Y(d)$ be the set of all sequences $S = (S[1],S[2],\ldots,S[d])$ where $S[1]$, $S[2]$, ..., $S[d]$ are square submatrices of x. For every $S \in Y(d)$ let us define $g(T,S,1)$ by saying that if for some $A \in L(t,B)$ we have

$$S[e] = \begin{cases} T[e] & \text{for all } e \in [1,d] \setminus [e^*, e^* + 1] \\ S(t,A)[e + 1 - e^*] & \text{for all } e \in [e^*, e^* + 1] \end{cases}$$

then $g(T,S,1) = 1$ or -1 according as the sign in (10.6.2) is + or -, whereas if there is no such A, then $g(T,S,1) = 0$. Let us extend this definition to get

$$g(T,S,n) \in Z \quad \text{for all } n \in N^* \setminus \{1\} \text{ and } S \in Y(d)$$

by the recurrence relation

$$g(T, S, n + 1) = \sum_{U \in Y(d)} g(T,U,n)g(U,S,1)$$

where Z is the set of all integers and N^* is the set of all positive integers. Let $Y^*(d)$ be the set of all S in $Y(d)$ such that

$$S[1] \le S[2] \le \cdots \le S[d].$$

Because of (10.6.3) there exists $n^* \in N^*$ such that for all $n \in N^*$ with $n \geq n^*$ we have

$$g(T,S,n) = \begin{cases} g(T,S,n^*) & \text{if } S \in Y^*(d) \\ 0 & \text{if } S \in Y(d) \setminus Y^*(d). \end{cases}$$

Let $M^*(d)$ be the set of all S in $Y^*(d)$ such that S[d] is not the 0 by 0 matrix. Let

$$M(d) = M^*(0) \cup M^*(1) \cup \cdots \cup M^*(d).$$

For every S in M(d) let S^* be the unique member of $Y^*(d)$ such that for some $d^* \in [0,d]$ we have

$$S^*[e] = \begin{cases} S[e] & \text{if } e \in [1,d^*] \\ \text{the 0 by 0 matrix} & \text{if } e \in [d^* + 1, d] \end{cases}$$

and let

$$f(T,S) = g(T,S^*,n^*).$$

Now by repeated applications of (10.6.2) we get

$$\prod_{e\in[1,d]} \det T[e] = \sum_{S\in M(d)} f(T,S) \prod_{e\in[1,d]} \det S[e] \qquad (10.6.4)$$

which may be called the existence part of the Straightening Law of Doubilet-Rota-Stein.

11. PREVIEW

(11.1) Notation In this chapter we use the notation and terminology introduced in Subsections (2.1) to (2.3) of Chapter 1. Moreover, in this subsection we introduce some more terminology which will be used in the rest of this chapter.

Given any $p \in N$, we introduce the *duplication* of p, which we denote by dup(p) and which we define by saying that dup(p) is the unique member of N(2) such that

$$\text{dup}(p)(1) = p = \text{dup}(p)(2).$$

Given any ring R and given any $m \in N(2)$ and given any m(1) by m(2) matrix x with entries in R, the (i,j)th entry of x is denoted by x_{ij} or $x_{i,j}$; note that here the row index i ranges over [1,m(1)], whereas the column index j ranges over [1,m(2)]; in other words, x is a mapping rec(m(1),m(2)) → R; we may say that x *is of size* m, and we may also say that m *is a size of* x; finally, we introduce *the size of* x, which we denote by siz(x) and which we define by putting

$$\mathrm{siz}(x) = \begin{cases} m & \text{if } m \in N^*(2) \\ \mathrm{dup}(0) & \text{if } m \in N(2) \setminus N^*(2). \end{cases}$$

Given any ring R, first we define

mat(R) = the set of all matrices with entries in R,

and second, for every m ∈ N(2) we define

mat(R,m) = the set of all m(1) by m(2) matrices with entries in R,

and third, we note that

$$\mathrm{mat}(R) = \coprod_{m \in N^*(2) \cup \{\mathrm{dup}(0)\}} \mathrm{mat}(R,m)$$

and fourth, we note that

$$\mathrm{mat}(R,m) = \mathrm{mat}(R,\mathrm{dup}(0)) \qquad \text{for all } m \in N(2) \setminus N^*(2),$$

and fifth, we note that

$$\mathrm{card}(\mathrm{mat}(R,\mathrm{dup}(0))) = 1.$$

Given any ring R and given any p ∈ N and y ∈ mat(R,dup(p)), we recall that by definition

$$\det y = \sum_{w \in W(p)} \mathrm{sgn}(w) \prod_{i \in [1,p]} y_{i,w(i)} \tag{11.1.1}$$

or equivalently,

$$\det y = \sum_{w \in W(p)} \mathrm{sgn}(w) \prod_{j \in [1,p]} y_{w(j),j}; \tag{11.1.2}$$

let us note that (11.1.1) may be called *the expansion of* det y *by rows*, whereas (11.1.2) may be called *the expansion of* det y *by columns*; let us also note that by convention

if p = 0, then det y = 1.

Given any ring R, and given any $m \in N(2)$ and $x \in mat(R,m)$ and $k \in [1,2]$, and given any map $g: [1,m(k')] \to [1,m(k)]$, we introduce the (k,g)th *maximal square submatrix* of x which we denote by mas(x,k,g) and which we define by saying that mas(x,k,g) is the unique member of mat(R,dup(m(k'))) such that for all $i \in [1,m(k')]$ and $j \in [1,m(k')]$ we have

$$mas(x,k,g)_{ij} = \begin{cases} x_{i,g(j)} & \text{if } k = 2 \\ x_{g(i),j} & \text{if } k = 1. \end{cases}$$

We note that as an application of this notation, the usual alternating property of det can be stated by saying that given any ring R, and given any $n \in N(2)$ and $n(1) = n(2)$, and given any $y \in mat(R,n)$, and given any map $w: [1,n(k')] \to [1,n(k)]$, we have

$$\det mas(y,k,w) = \begin{cases} sgn(w) \det y & \text{if } w \text{ is injective} \\ 0 & \text{if } w \text{ is not injective.} \end{cases} \tag{11.1.3}$$

Given any ring R and given any $m \in N(2)$ and $x \in mat(R,m)$, first for every $a \in popre(2,m)$ we introduce the ath *square submatrix* of x, which we denote by sub(x,a) and which we define by saying that sub(x,a) is the unique member of mat(R,dup(len(a))) such that for all $i \in [1,len(a)]$ and $j \in [1,len(a)]$ we have

$$sub(x,a)_{ij} = x_{a(1,i),a(2,j)},$$

and second, for every $a \in popre(2,m)$ we introduce the ath *minor* of x, which we denote by mor(x,a) and which we define by putting

$$mor(x,a) = \det sub(x,a)$$

and we note that by convention

$$\text{if } len(a) = 0, \text{ then } mor(x,a) = 1,$$

and third, for every $T \in popab(2,m)$ we introduce the Tth *monomial in the minors* of x, which we denote by mom(x,T) and which we define by putting

$$mom(x,T) = \prod_{e\in[1,dep(T)]} mor(x,T[e])$$

and we note that by convention

if dep(T) = 0, then mom(x,T) = 1.

Given any q ∈ N, first by a *vectroid of width* q we mean a map [1,q] → P(Z) and we put

veco(q) = the set of all vectroids of width q,

and second, for every d ∈ N, by a *tabloid of width* q *and depth* d we mean a mapping A: [1,d] → veco(q) which, to each e ∈ [1,d], associates A[e] ∈ veco(q), and we put

tabo(q,d) = the set of all tabloids of width q and depth d

and third, by a *tabloid of width* q we mean a tabloid of width q and depth d for some d ∈ N, and we put

tabo(q) = the set of all tabloids of width q.

By a *univectroid* (resp., *unitabloid*) we mean a vectroid (resp., tabloid) of width 1. By a *bivectroid* (resp., *bitabloid*) we mean a vectroid (resp., tabloid) of width 2. For every d ∈ N we put

bid(d) = tabo(2,d)

i.e.,

bid(d) = the set of all bitabloids of depth d.

Given any q ∈ N, by a *protovector of width* q we mean a mapping t: rec(q) → N which, to each k ∈ [1,q] and j ∈ Z, associates t(k,j) ∈ N, and we put

proc(q) = the set of all protovectors of width q.

Given any sets X and Y, and given any map f: X → Y, by im(f) we denote the *image* of f, i.e.,

im(f) = {f(x) : x ∈ X}.

Given any q ∈ N and a ∈ pre(q), first we introduce the *vectorial semiimage* of a, which we denote by sim[a] and which we define by saying that sim[a] is the unique member of veco(q) such that for all k ∈ [1,q] we have

$$\mathrm{sim}[a](k) = \{a(k,i) : i \in [1,\mathrm{len}(a)]\}$$

and second, we introduce the *vectorial content* of a, which we denote by con[a] and which we define by saying that con[a] is the unique member of proc(q) such that for all $k \in [1,q]$ and $j \in Z$ we have

$$\mathrm{con}[a](k,j) = \mathrm{card}(\{i \in [1,\mathrm{len}(a)] : a(k,i) = j\}).$$

Given any $q \in N$ and $T \in \mathrm{pab}(q)$, first we introduce the *semiimage* of T, which we denote by sim(T) and which we define by saying that sim(T) is the unique member of tabo(q,dep(T)) such that for all $e \in [1,\mathrm{dep}(T)]$ and $k \in [1,q]$ we have

$$\mathrm{sim}(T)[e](k) = \mathrm{sim}[T[e]](k),$$

and second, we introduce the *content* of T, which we denote by con(T) and which we define by saying that con(T) is the unique member of proc(q) such that for all $k \in [1,q]$ and $j \in Z$ we have

$$\mathrm{con}(T)(k,j) = \sum_{e\in[1,\mathrm{dep}(T)]} \mathrm{con}[T[e]](k,j).$$

Given any $m \in N(2)$, first we define

$$\mathrm{bic}(m) = \mathrm{vec}(2,m)$$

and we recall that a member of bic(m) is a *bivector bounded by* m, and second, for every $k \in [1,2]$ we define

$$\mathrm{bic}(m,k) = \mathrm{vec}(2,m,m(k'))$$

and we recall that a member of bic(m,k) is a bivector which is bounded by m and which is of length m(k'), and we remark that a member of bic(m,k) may also be called a *bivector of type* (m,k).

Given any $d \in N$, first we define

$$\mathrm{bit}(d) = \mathrm{tid}[2,d]$$

and we recall that a member of bit(d) is a *bitableau of depth* d, and second, for every $m \in N(2)$ we define

$$\mathrm{bit}(d,m) = \mathrm{tid}(2,m,d)$$

and we recall that a member of bit(d,m) is a bitableau which is of depth d and which is bounded by m, and we remark that a member of bit(d,m) may also be called a *bitableau of type* (d,m), and third, for every m ∈ N(2) and k ∈ [1,2] we define

bit(d,m,k) = tip[2,m,m(k'),d]

and we recall that a member of bit(d,m,k) is a bitableau which is of depth d and which is bounded by m and which is such that len(T[e]) = m(k') for all e ∈ [1,d], and we remark that a member of bit(d,m,k) may also be called a *bitableau of type* (d,m,k).

Given any d ∈ N and T ∈ bit(d), we define

sid(T) = {A ∈ bid(d) : for all e ∈ [1,d] and k ∈ [1,2]
we have A[e](k) ⊂ sim(T)[e](k)}

and we remark that a member of sid(T) may be called a *subtabloid* of T.

Given any T ∈ bit(2) and k ∈ [1,2], first we define

disid(k,T) = {A ∈ sid(T) : for all e ∈ [1,2] we have
[sim(T)[e](k) \ A[e](k)] ∩ A[e'](k) = ∅}

and we remark that a member of disid(k,T) may be called a *disjointed subtabloid* of (k,T), and second, we define

hasid(k,T) = {A ∈ sid(T) : A[1](k') ∩ A[2](k') = ∅}

and we remark that a member of hasid(k,T) may be called a *half-sided subtabloid* of (k,T), and third, we define

osid(k,T) = {A ∈ sid(T) : A[1](k') = ∅ = A[2](k')}

and we remark that a member of osid(k,T) may be called a *one-sided subtabloid* of (k,T), and fourth, we define

eosid(k,T) = {A ∈ osid(k,T) : card(A[1](k)) = card(A[2](k))}

and we remark that a member of eosid(k,T) may be called an *equicardinal one-sided subtabloid* of (k,T).

Given any T ∈ bit(2), first for every A ∈ sid(T) and e ∈ [1,2] we introduce the *satellitic binumber* of (T,A) *at* e, which we denote by sab(T,A)[e] and which we define by saying that sab(T,A)[e] is the unique member of N(2) such that for all k ∈ [1,2] we have

sab(T,A)[e](k) = len(T[e]) - card(A[e](k)) + card(A[e'](k)), and second, for every $A \in \text{sid}(T)$ we introduce the *satelloid* of (T,A), which we denote by $\text{sad}(T,A)$ and which we define by saying that $\text{sad}(T,A)$ is the unique member of $\text{bid}(2)$ such that for all $e \in [1,2]$ and $k \in [1,2]$ we have

$$\text{sad}(T,A)[e](k) = [\text{sim}(T)[e](k) \setminus A[e](k)] \cup A[e'](k),$$

and third, we define

$$\begin{aligned}\text{nex}(T) = \{A \in \text{sid}(T) : \ & \text{for all } e \in [1,2] \text{ and } k \in [1,2] \text{ we have} \\ & \text{sab}(T,A)[e](1) = \text{sab}(T,A)[e](2) \\ & \qquad = \text{card}(\text{sad}(T,A)[e](k))\}\end{aligned}$$

and we remark that a member of $\text{nex}(T)$ may be called a *numerically exchangeable subtabloid* of T, and fourth, we define

$$\begin{aligned}\text{pex}(T) = \{A \in \text{nex}(T) : \ & \text{for some } e \in [1,2] \text{ and some } k \in [1,2] \\ & \text{we have } A[e](k) \neq \emptyset\}\end{aligned}$$

and we remark that a member of $\text{pex}(T)$ may be called a *properly exchangeable subtabloid* of T, and fifth, for every $B \in \text{sid}(T)$ we define

$$\begin{aligned}\text{nex}(T,B) = \{A \in \text{nex}(T) : \ & \text{for all } e \in [1,2] \text{ and } k \in [1,2] \\ & \text{we have } A[e](k) \subset B[e](k)\}\end{aligned}$$

and we remark that a member of $\text{nex}(T,B)$ may be called a *numerically exchangeable subtabloid* of (T,B), and sixth, for every $B \in \text{sid}(T)$ we define

$$\text{pex}(T,B) = \text{nex}(T,B) \cap \text{pex}(T)$$

and we remark that a member of $\text{pex}(T,B)$ may be called a *properly exchangeable subtabloid* of (T,B), and seventh, for every $A \in \text{nex}(T)$ we introduce the *satellite* of (T,A), which we denote by $\text{sat}(T,A)$ and which we define by saying that $\text{sat}(T,A)$ is the unique member of $\text{tid}[2,2]$ such that

$$\text{sim}(\text{sat}(T,A)) = \text{sad}(T,A)$$

and eighth, as a useful characterization of $\text{nex}(T)$ we observe that

$$\left[\begin{aligned}\text{nex}(T) = \{A \in \ & \text{disid}(1,T) \cap \text{disid}(2,T) : \\ & \text{for all } e \in [1,2] \text{ we have} \\ & \text{sab}(T,A)[e](1) = \text{sab}(T,A)[e](2)\}.\end{aligned}\right. \tag{11.1.4}$$

(11.2) Preview of Laplace In (12.3) we put in proper perspective the classical Laplace development of a determinant. What we prove in (12.3) can be *approximately* described as follows.

Let there be given any $y \in \mathrm{mat}(R,n)$ where R is a ring and where $n \in N(2)$ with $n(1) = n(2)$. Let us cut up y into row blocks $x[e]_{1\le e\le d}$ with $x[e] \in \mathrm{mat}(R,u[e])$ and call it the case of $k = 2$ and let us depict it by the following picture of y:

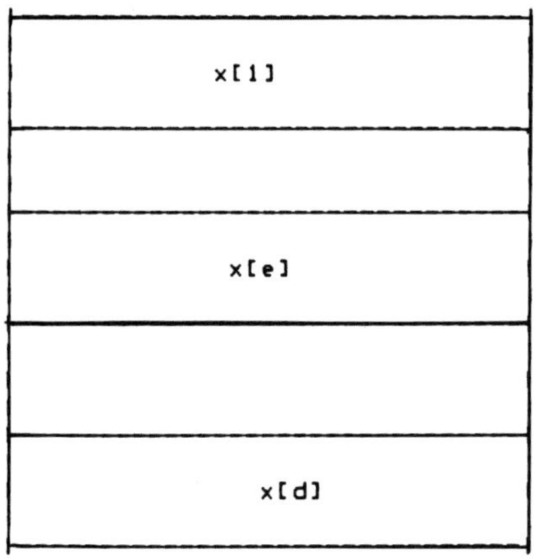

case of k = 2

or let us cut up y into column blocks $x[e]_{1\le e\le d}$ and call it the case of $k = 1$ and let us depict it by the following picture of y:

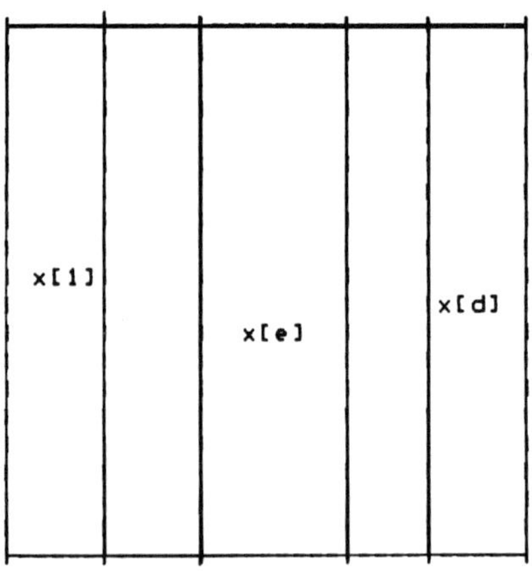

case of k = 1

For the sequence $u[e]_{1\le e\le d}$ with $u[e] \in N(2)$ we are clearly assuming that

$$\sum_{e\in[1,d]} u[e](k') = n(k') = u[\hat{e}](k) \qquad \text{for all } \hat{e} \in [1,d].$$

Let bid(d,u,k,n) denote the set of those disjoint partitions of [1,n(k)] which give rise to systems of mutually complementary unordered minors of $x[e]_{1\le e\le d}$; i.e., let bid(d,u,k,n) be the set of all A in bid(d) such that

$$A[e](k') = \emptyset \quad \text{and} \quad \text{card}(A[e](k)) = u[e](k') \qquad \text{for all } e \in [1,d]$$

and such that

$$[1,n(k)] = \coprod_{e\in[1,d]} A[e](k).$$

For every A in bid(d,u,k,n) let g(A) be a sequence of maps which gives rise to a corresponding system of ordered minors; i.e., for each $e \in [1,d]$ let there be given an injective map

$$g(A)[e]: [1,u[e](k')] \to [1,n(k)]$$

such that

$$\text{im}(g(A)[e]) = A[e](k).$$

Now (12.3) says that

$$\det y = \sum_{A\in \text{bid}(d,u,k,n)} S(A) \prod_{e\in[1,d]} \det \text{mas}(x[e],k,g(A)[e])$$

where S is a certain map

$$S: \text{bid}(d,u,k,n) \to \{1,-1\}.$$

(11.3) Preview of the Basic Case In (13.4.1), (13.4.3), (13.5.1), (13.5.2), and (13.5.3) we formalize (10.3.1) to (10.3.5), respectively. In the formalism,

$$S(T,A)[1] \quad \text{and} \quad S(T,A)[2]$$

and

$$(\det T[1])(\det T[2]) \quad \text{and} \quad (\det S(T,A)[1])(\det S(T,A)[2])$$

and

K(T,B) and L(T,B)

of (10.3) will, respectively, correspond to

sat(T,A)[1] and sat(T,A)[2]

and

mom(x,T) and mom(x,sat(T,A))

and

nex(T,B) and pex(T,B).

Moreover, in (13.1) we introduce the *basic residue* bar(x,m,k,B), which is the formalization of det y of (10.3). In (13.1) we also introduce the *basic signature* bag(m,k,A), which is the formalization of the signs ± of (10.3).

In the formal situation we will have m ∈ N(2) and k ∈ [1,2] with m(k) = 2m(k'), and T will be a certain member of bit(2,m,k) and B will be any member of osid(k,T). In the discussion of (10.3) we have taken k = 2 and m(k') = p. The passage from k = 2 to k = 1 can be achieved by interchanging rows and columns.

(11.4) Preview of the Maximal Size Case In (14.7.1), (14.7.3), (14.8.1), (14.8.2), and (14.8.3) we formalize (10.4.1) to (10.4.5), respectively. In the formalism,

S(T*,A*)[1] and S(T*,A*)[2]

and

(det T*[1])(det T*[2]) and (det S(T*,A*)[1])(det S(T*,A*)[2])

and

K(T*,B*) and L(T*,B*)

of (10.4) will, respectively, correspond to

sat(T*,A*)[1] and sat(T*,A*)[2]

and

mom(x*,T*) and mom(x*,sat(T*,A*))

and

nex(T*,B*) and pex(T*,B*).

Moreover, in (14.1) we introduce the *castic residue* car(x*,m*,k,B*), which is the formalization of det y of (10.4). In (14.1) we also introduce the *castic signature* cag(m*,k,T*,A*), which is the formalization of the signs ± of (10.4).

In the formal situation we will have m* ∈ N(2) and k ∈ [1,2], and T* will be any member of bit(2,m*,k) and B* will be any member of osid(k,T*). In the discussion of (10.4) we have taken k = 2 and m*(k') = p and m*(k) = q*. The passage from k = 2 to k = 1 can be achieved by interchanging rows and columns.

The conversion of the maximal size case to the basic case will be accomplished by the castic machinery introduced in (14.1).

(11.5) Preview of the Mixed Size Case In (15.7.1), (15.7.3), (15.8.1), (15.8.2), and (15.8.3) we formalize (10.5.1) to (10.5.5), respectively. In the formalism,

S(T,A)[1] and S(T,A)[2]

and

(det T[1])(det T[2]) and (det S(T,A)[1])(det S(T,A)[2])

and

K(T,B) and L(T,B)

of (10.5) will, respectively, correspond to

sat(T,A)[1] and sat(T,A)[2]

and

mom(x,T) and mom(x,sat(T,A))

and

nex(T,B) and pex(T,B).

Moreover, in (15.1) we introduce the *satellitic residue* sar(x,m,k,B), which is the formalization of det y of (10.5). In (15.1) we also introduce the *satellitic signature* sag(m,k,A), which is the formalization of the signs ± of (10.5). Let us note that, in the formalism, condition (1*) of (10.5) corresponds to the condition which says that B ∈ hasid(k,T). Let us also note that in (15.1) we introduce the *unification number* uni(m,k,T,A), which is the formalization of E(T,A) of (10.5), so the conditions (1*) and (*) of (10.5) will get formalized into the condition which we introduce in (15.1) and which in symbols will say that B ∈ amsid(m,k,T) and which in words will say that B is an *ample subtabloid* of (m,k,T).

In the formal situation we will have m ∈ N(2) and k ∈ [1,2], and T will be any member of bit(2,m) and as indicated above, B will be any member of hasid(k,T). In the discussion of (10.5) we have taken k = 2 and m(k') = p and m(k) = q. The passage from k = 2 to k = 1 can be achieved by interchanging rows and columns.

The conversion of the mixed size case to the maximal size case will be accomplished by the unification machinery which will be introduced in (15.1).

(11.6) Preview of the Full Depth Case In (16.7), (16.8), (16.2.3), and (16.11) we formalize (10.6.1) to (10.6.4), respectively.

In the formal situation we shall be given a bitableau T which is bounded by some m ∈ N(2); in the discussion of (10.6) we have taken m(1) = p and m(2) = q. In (16.1) we introduce the *depthwise obstruction number* don(T), which is the formalization of e* of (10.6). In (16.1) we also introduce the *valid truncation* vat(e*,T), which is the formalization of t of (10.6). In (16.1) we also introduce the *widthwise obstruction number* won(t); upon letting k = won(T), in (16.1) we introduce the *lengthwise obstruction number* lon(k,T), which is the formalization of v of (10.6); in (10.6) we have illustrated the case of k = 2. In (16.1) we introduce the *lengthwise obstruction subtabloid* lod(k,t), which is the formalization of B of (10.6). In (16.1) we also introduce the set fex(T), whose members will be called *fully exchangeable subtabloids* of T and which is the formalization of L(t,B) of (10.6). In (16.1) we introduce the *relevant integer* rin(m,T,S,n) and the *final integer* fin(m,T,S), which are the respective formalizations of g(T,S,n) and f(T,S,n) of (10.6). In (16.1) we also introduce the set sit(m,d), which is the formalization of M(d) of (10.6) and whose members are all those standard bitableaux which are bounded by m and whose depth is at most d. Finally, let us note that in the formalism, the sets Y(d) and Y*(d) and M*(d) of (10.6) correspond to tid[2,d] and qstid[2,d] and stid[2,d], respectively.

12. LAPLACE DEVELOPMENT OF A DETERMINANT

(12.1) Notation In this subsection we introduce some more notation which will be used in the rest of this chapter.

Given any partially ordered sets X and Y and given any map $f: X \to Y$, first we say that f is *increasing*, or *order preserving*, to mean that for all $r \in X$ and $s \in X$ we have that

$$r \leq s \Rightarrow f(r) \leq f(s),$$

and second, we say that f is *strictly increasing*, or *strictly order preserving*, to mean that for all $r \in X$ and $s \in X$ we have that

$$r < s \Rightarrow f(r) < f(s),$$

and third, we say that f is *bistrictly increasing*, or *bistrictly order preserving*, to mean that for all $r \in X$ and $s \in X$ we have that

$$r < s \Leftrightarrow f(r) < f(s),$$

and fourth, we say that f is *biincreasing*, or *biorder preserving*, to mean that for all $r \in X$ and $s \in X$ we have that

$$r \leq s \Leftrightarrow f(r) \leq f(s),$$

and fifth, we say that f is an *order isomorphism* to mean that f is bijective and biorder preserving, and sixth, we say that f is *decreasing*, or *order reversing*, to mean that for all $r \in X$ and $s \in X$ we have that

$$r \leq s \Rightarrow f(r) \geq f(s),$$

and seventh, we say that f is *strictly decreasing*, or *strictly order reversing*, to mean that for all $r \in X$ and $s \in X$ we have that

$$r < s \Rightarrow f(r) > f(s),$$

and eighth, we say that f is *bistrictly decreasing*, or *bistrictly order reversing*, to mean that for all $r \in X$ and $s \in X$ we have that

$$r < s \Leftrightarrow f(r) > f(s),$$

and ninth, we say that f is *bidecreasing*, or *biorder reversing*, to mean that for all $r \in X$ and $s \in X$ we have that

$$r \le s \Leftrightarrow f(r) \ge f(s),$$

and tenth, we say that f is an *antiorder isomorphism* to mean that f is bijective and biorder reversing, and we note that

f is an order isomorphism
$\Leftrightarrow$ f is surjective and biorder preserving
$\Rightarrow$ f is biorder preserving
$\Leftrightarrow$ f is injective and bistrictly order preserving
$\Rightarrow$ f is bistrictly order preserving
$\Rightarrow$ f is strictly order preserving
$\Rightarrow$ f is order preserving

and we note that

f is an antiorder isomorphism
$\Leftrightarrow$ f is surjective and biorder reversing
$\Rightarrow$ f is biorder reversing
$\Leftrightarrow$ f is injective and bistrictly order reversing
$\Rightarrow$ f is bistrictly order reversing
$\Rightarrow$ f is strictly order reversing
$\Rightarrow$ f is order reversing.

Given any partially ordered sets X and Y and given any mapping F: $X \to Y$ and given any one of the foregoing 10 properties L, we say that F is L to mean that the map $X \to Y$ *associated* with F is L; for example, given any sequence $F = (F_i)_{i \in N}$ with $F_i \in Z$, we say that F is strictly increasing to mean that $F_i < F_{i+1}$ for all $i \in N$.

We note that given any totally ordered set X and given any partially ordered set Y and any map f: $X \to Y$, we have that

f is injective and order preserving
$\Leftrightarrow$ f is biorder preserving

and we note that

f is injective and order reversing
$\Leftrightarrow$ f is biorder reversing.

Finally, we note that if X and Y are any sets and f: $X \to Y$ is any map such that

$$\text{card}(X) = \text{card}(\text{im}(f)) < \infty$$

then f is injective.

Given any $d \in N$, first we define

$$\text{bib}(d) = \text{the set of all sequences } u = u[e]_{1\le e\le d} \text{ where } u[e] \in N(2) \text{ for every } e \in [1,d]$$

and we remark that a member of bib(d) may be called a *biboard of depth* d, and second, for every $n \in N(2)$ and $k \in [1,2]$ we define

$$\text{bib}(d,n,k) = \{u \in \text{bib}(d) : u[e](k) = n(k) \text{ for all } e \in [1,d], \text{ and } \sum_{e\in[1,d]} u[e](k') = n(k')\}$$

and we remark that a member of bib(d,n,k) may be called a *biboard of type* (d,n,k).

Given any $d \in N$ and $u \in \text{bib}(d)$, we define

$$\text{bid}(d,u) = \{A \in \text{bid}(d) : A[e](k) \subset [1,u[e](k)] \text{ for all } e \in [1,d] \text{ and } k \in [1,2]\}$$

and we remark that a member of bid(d,u) may be called a *bitabloid of type* (d,u).

Given any $d \in N$ and $u \in \text{bib}(d)$ and $k \in [1,2]$, first we define

$$\text{bid}(d,u,k) = \{A \in \text{bid}(d,u) : A[e](k') = \emptyset \text{ for all } e \in [1,d]\}$$

and we remark that a member of bid(d,u,k) may be called a *bitabloid of type* (d,u,k), and second, we define

$$\text{bil}(d,u,k) = \text{the set of all sequences } h = h[e]_{1\le e\le d} \text{ where } h[e] : [1,u[e](k')] \to [1,u[e](k)] \text{ is a map for every } e \in [1,d]$$

and we remark that a member of bil(d,u,k) may be called a *bilabel of type* (d,u,k).

Given any n ∈ N(2) with n(1) = n(2) and given any d ∈ N and k ∈ [1,2] and u ∈ bib(d,n,k), first we define

$$\text{bid}(d,u,k,n) = \{A \in \text{bid}(d,u,k) : \text{card}(A[\hat{e}](k)) = u[\hat{e}](k') \text{ for all } \hat{e} \in [1,d], \text{ and } \coprod_{e\in[1,d]} A[e](k) = [1,n(k)]\}$$

and we remark that a member of bid(d,u,k,n) may be called a *bitabloid of type* (d,u,k,n), and second, we define

$$\text{bil}(d,u,k,n) = \{h \in \text{bil}(d,u,k) : \text{card}(\text{im}(h[\hat{e}])) = u[\hat{e}](k') \text{ for all } \hat{e} \in [1,d], \text{ and } \coprod_{e\in[1,d]} \text{im}(h[e]) = [1,n(k)]\}$$

and we remark that a member of bil(d,u,k,n) may be called a *bilabel of type* (d,u,k,n), and third, we introduce the *order preservator* of (d,u,k,n), which we denote by orp(d,u,k,n) and which we define by saying that

$$\text{orp}(d,u,k,n): \text{bid}(d,u,k,n) \to \text{bil}(d,u,k,n)$$

is the unique map such that for every A ∈ bid(d,u,k,n) and e ∈ [1,d] we have that

$$\text{orp}(d,u,k,n)(A)[e]: [1,u[e](k')] \to [1,u[e](k)]$$

is order preserving and

$$\text{im}(\text{orp}(d,u,k,n)(A)[e]) = A[e](k).$$

Given any n ∈ N(2) with n(1) = n(2) and given any d ∈ N and k ∈ [1,2] and u ∈ bib(d,n,k) and f ∈ bil(d,u,k,n) and g ∈ bil(d,u,k,n) first we introduce the *laplasic permutation* of (d,u,k,n,f,g), which we denote by lap(d,u,k,n,f,g) and which we define by saying that lap(d,u,k,n,f,g) is the unique member of W(n(k)) such that for all e ∈ [1,d] and i ∈ [1,u[e](k')] we have

$$\text{lap}(d,u,k,n,f,g)(f[e](i)) = g[e](i),$$

and second, for every sequence

$$v = v[e]_{1\le e\le d} \quad \text{with } v[e] \in W(u[e](k'))$$

we introduce the *extended laplasic permutation* of (d,u,k,n,f,g,v) which we denote by elap(d,u,k,n,f,g,v) and which we define by saying that elap(d,u,k,n,f,g,v) is the unique member of W(n(k)) such that for all $e \in [1,d]$ and $i \in [1,u[e](k')]$ we have

$$\text{elap}(d,u,k,n,f,g,v)(f[e](i)) = g[e](v[e](i)).$$

(12.2) Review of sgn *We shall now review the concept of sgn of a permutation.* First let us recall the concept of product of permutations; we use this concept of product only in this subsection. Given any p and d in N, and given any $w_1, w_2, \ldots, w_d$ in W(p), by the product of $w_d, w_{d-1}, \ldots, w_1$ we mean their composition, i.e., let $(\hat{w}_e)_{0\le e\le d}$ be the unique sequence in W(p) such that

$$\hat{w}_0(i) = i \qquad \text{for all } i \in [1,p]$$

and such that for each $e \in [1,d]$ we have

$$\hat{w}_e(i) = w_e(\hat{w}_{e-1}(i)) \qquad \text{for all } i \in [1,p],$$

and now by definition

$$\hat{w}_d = \text{the product of } w_d, w_{d-1}, \ldots, w_1.$$

Given any $p \in N$, first, for every injective map $b: [1,2] \to [1,p]$, let t(b) denote the corresponding *transposition*; i.e., let $t(b) \in W(p)$ be defined by putting

$$t(b)(i) = \begin{cases} b(2) & \text{if } i = b(1) \\ b(1) & \text{if } i = b(2) \\ i & \text{if } i \in [1,p] \setminus \{b(1),b(2)\}, \end{cases}$$

and second, for every $D \in N$, let

$$\left[\begin{array}{l} \hat{D}_p = \text{the set of all sequences } B = (B_E)_{1\le E\le D} \text{ of injective} \\ \qquad \text{maps } B_E\colon [1,2] \to [1,p] \end{array}\right.$$

and third, for every $D \in N$ and $B \in \hat{D}_p$ let $T(B) \in W(p)$ be defined by putting

$$T(B) = \text{the product of } t(B_D), t(B_{D-1}), \ldots, t(B_1),$$

and fourth, let

$$W_0(p) = \{w \in W(p) : w \text{ can be expressed as a product of an even number of transpositions}\}$$

and

$$W_1(p) = \{w \in W(p) : w \text{ can be expressed as a product of an odd number of transpositions}\}$$

i.e., for each $j \in [0,1]$, let

$$W_j(p) = \{w \in W(p) : w = T(B) \text{ for some } D \in N \text{ with } D \equiv j \bmod 2 \text{ and some } B \in \hat{D}_p\},$$

and fifth, let us review the concept of parity by recalling the well-known fact that

(1) $W(p) = W_0(p) \coprod W_1(p)$

and by saying that given any $w \in W(p)$,

(2) [w is said to be even or odd according as $w \in W_0(p)$ or $w \in W_1(p)$,

and sixth, let us review the concept of sgn by saying that given any $w \in W(p)$, by definition

(3) $\operatorname{sgn}(w) = 1$ or -1 according as w is even or odd.

By (2) and (3) it follows that

(4) [*the sgn of a permutation equals the sgn of its inverse*, i.e., for all $p \in N$ and $w \in W(p)$ we have $\operatorname{sgn}(w^{-1}) = \operatorname{sgn}(w)$.

By (2) and (3) it also follows that

(5) [*the sgn of a product is the product of the sgns*, i.e., if p and d are any nonnegative integers and if $w_1, w_2, \ldots, w_d, \hat{w}_d$ are any members of $W(p)$ such that $\hat{w}_d$ = the product of $w_d, w_{d-1}, \ldots, w_1$ then $\operatorname{sgn}(\hat{w}_d) = \prod_{e \in [1,d]} \operatorname{sgn}(w_e)$.

Note that (4) *is tacitly used in the proof of the fact that the two definitions* (11.1.1) *and* (11.1.2) *of det are equivalent.* Also note that the $d = 2$ case of (5) *is tacitly used in the proof of* (11.1.3).

As examples of sgn calculations, given any $p \in N$, first for every $s \in [1, p - 1]$ let $b_s: [1,2] \to [1,p]$ be the injective map obtained by taking $b_s(1) = s$ and $b_s(2) = s + 1$ and let $t_s = t(b_s)$ be the corresponding transposition, and second, for every $s \in [1,p]$ let c_s be the initial cycle of length s; i.e., let c_s be defined by putting

$$c_s(i) = \begin{cases} i + 1 & \text{if } i \in [1, s - 1] \\ 1 & \text{if } i = s \\ i & \text{if } i \in [1,p] \setminus [1,s] \end{cases}$$

and let us note that

$$(6) \quad c_s = \text{the product of } t_1, t_2, \ldots, t_{s-1}$$

and hence

$$(7) \quad \text{sgn}(c_s) = (-1)^{s-1},$$

and third, for every $s \in [1,p]$ let r_s be the initial reverse identity of length s; i.e., let $r_s \in W(p)$ be defined by putting

$$r_s(i) = \begin{cases} s + 1 - i & \text{if } i \in [1,s] \\ i & \text{if } i \in [1,p] \setminus [1,s] \end{cases}$$

and let us note that

$$(8) \quad \begin{bmatrix} \text{if } 1 \in [1,p], \text{ then } r_1 = c_1 \\ \text{whereas if } s \in [2,p], \text{ then} \\ r_s = \text{the product of } c_s \text{ and } r_{s-1} \end{bmatrix}$$

and hence, in view of (6), by induction on s it follows that for all $s \in [1,p]$ we have

$$(9) \quad \begin{bmatrix} r_s = \text{the product of} \\ \quad t_1, t_2, \ldots, t_{s-1}, t_1, t_2, \ldots, t_{s-2}, \ldots, t_1 \end{bmatrix}$$

and fourth, let us note that

$$(10)\quad \begin{cases} \text{[the number of transpositions on the RHS of (9)]} \\ = (s-1) + (s-2) + \cdots + 1 \\ = s(s-1)/2 \end{cases}$$

and hence for all $s \in [1,p]$ we have

$$(11)\quad \operatorname{sgn}(r_s) = (-1)^{s(s-1)/2},$$

and fifth, let us consider the p by p antiidentity matrix $I(p)$, i.e., let R be a ring and let $I(p) \in \operatorname{mat}(R,\operatorname{dup}(p))$ be defined by putting

$$(12)\quad I(p)_{ij} = \begin{cases} 1 & \text{if } i \in [1,p] \text{ and } j \in [1,p] \text{ and } i + j = p + 1 \\ 0 & \text{if } i \in [1,p] \text{ and } j \in [1,p] \text{ and } i + j \neq p + 1 \end{cases}$$

and let us note that

$$\prod_{i\in[1,p]} I(p)_{i,w(i)} = \begin{cases} 1 & \text{if } w = r_p \\ 0 & \text{if } w \in W(p) \setminus \{r_p\} \end{cases}$$

and hence by the definition of det we have

$$(13)\quad \det I(p) = \operatorname{sgn}(r_p)$$

and therefore by (11) we get

$$(14)\quad \det I(p) = (-1)^{p(p-1)/2};$$

thus

$$\begin{cases} \text{given any ring R and given any } p \in N, \\ \text{for the p by p antiidentity matrix} \\ I(p) \in \operatorname{mat}(R,\operatorname{dup}(p)) \text{ defined by (12),} \\ \text{we have } \det I(p) = (-1)^{p(p-1)/2}. \end{cases} \qquad (12.2.1)$$

To discuss another consequence of (2) and (3), given any nonnegative integers q and p and given any injective map $G: [1,q] \to [1,p]$, let $G^*: W(q) \to W(p)$ be the injective map *induced* by G; i.e., let $G^*: W(q) \to W(p)$ be the injective map which we obtain when for each $V \in W(q)$ we define $G^*(V) \in W(p)$ by putting

$$G^*(V)(G(i)) = G(V(i)) \quad \text{for all } i \in [1,q]$$

and

$$G^*(V)(i) = i \quad \text{for all } i \in [1,p] \setminus \mathrm{im}(G);$$

we claim that then

(15) for every $V \in W(q)$ we have $\mathrm{sgn}(G^*(V)) = \mathrm{sgn}(V)$;

to see this, it is sufficient to note that given any $D \in N$ and $B \in \hat{D}_q$, we clearly have

$$G^*(T(B)) = T(G^*(B))$$

where $G^*(B)$ is the member of $\hat{D}_p$ which we obtain when for each $E \in [1,D]$ we let

$$G^*(B)_E\colon \quad [1,2] \to [1,p]$$

to be the composition

$$[1,2] \xrightarrow{B_E} [1,q] \xrightarrow{G} [1,p].$$

Now

(21) $\left[\begin{array}{l}\text{given any } n \in N(2) \text{ with } n(1) = n(2) \\ \text{and given any } d \in N \text{ and } k \in [1,2] \text{ and} \\ \text{given any } u \in \mathrm{bib}(d,n,k) \text{ and } g \in \mathrm{bil}(d,u,k,n) \\ \text{and given any sequence} \\ v = v[e]_{1\le e\le d} \text{ with } v[e] \in W(u[e](k')),\end{array}\right.$

upon letting

$$g[e]^*\colon \quad W(u[e](k')) \to W(n(k))$$

be the injective map induced by $g[e]$ in the sense explained in the above paragraph, we clearly have

(22) $\left[\begin{array}{l}\mathrm{elap}(d,u,k,n,g,g,v) = \text{the product of} \\ g[d]^*(v[d]),\ g[d-1]^*(v[d-1]),\ \ldots,\ g[1]^*(v[1])\end{array}\right.$

and therefore by (5) we get

(23) $\mathrm{sgn}(\mathrm{elap}(d,u,k,n,g,g,v)) = \prod_{e\in[1,d]} \mathrm{sgn}(g[e]^*(v[e]))$

and by (15) we also get

(24) $\operatorname{sgn}(g[e]*(v[e])) = \operatorname{sgn}(v[e])$ for all $e \in [1,d]$

and by (23) and (24) we see that

(25) $$\operatorname{sgn}(\operatorname{elap}(d,u,k,n,g,g,v)) = \prod_{e\in[1,d]} \operatorname{sgn}(v[e])$$

and now given any

(26) $f \in \operatorname{bil}(d,u,k,n)$

we clearly have

(27') $$\left[\begin{array}{l}\operatorname{elap}(d,u,k,n,f,g,v)(i)\\ = \operatorname{elap}(d,u,k,n,g,g,v)(\operatorname{lap}(d,u,k,n,f,g)(i))\\ \text{for all } i \in [1,n(k)],\end{array}\right.$$

i.e., we have

(27) $$\left[\begin{array}{l}\operatorname{elap}(d,u,k,n,f,g,v)\\ = \text{the product of } \operatorname{elap}(d,u,k,n,g,g,v) \text{ and } \operatorname{lap}(d,u,k,n,f,g)\end{array}\right.$$

and by (5) and (27) we see that

(28) $$\left[\begin{array}{l}\operatorname{sgn}(\operatorname{elap}(d,u,k,n,f,g,v))\\ = \operatorname{sgn}(\operatorname{elap}(d,u,k,n,g,g,v)) \text{ times } \operatorname{sgn}(\operatorname{lap}(d,u,k,n,f,g))\end{array}\right.$$

and by (23) and (28) we see that

(29) $$\left[\begin{array}{l}\operatorname{sgn}(\operatorname{elap}(d,u,k,n,f,g,v))\\ = \operatorname{sgn}(\operatorname{lap}(d,u,k,n,f,g)) \prod_{e\in[1,d]} \operatorname{sgn}(v[e]).\end{array}\right.$$

Thus we conclude that

$$\left[\begin{array}{l}\text{if assumptions (21) and (26) are satisfied,}\\ \text{then conclusion (29) is valid.}\end{array}\right. \qquad (12.2.2)$$

(12.3) Laplace Development Let there be given any $n \in N(2)$ with $n(1) = n(2)$, and let there be given any $d \in N$ and $k \in [1,2]$ and $u \in \operatorname{bib}(d,n,k)$ and $f \in \operatorname{bil}(d,u,k,n)$. Also, let there be given any ring R, and let there be given any $y \in \operatorname{mat}(R,n)$, and for every $e \in [1,d]$ let $x[e]$ be the unique member of $\operatorname{mat}(R,u[e])$ such that for all $i \in [1,u[e](1)]$ and $j \in [1,u[e](2)]$ we have

$$(1)\quad x[e]_{ij} = \begin{cases} y_{f[e](i),j} & \text{if } k = 2 \\ y_{i,f[e](j)} & \text{if } k = 1. \end{cases}$$

Finally, let there be given any map

$$g\colon \operatorname{bid}(d,u,k,n) \to \operatorname{bil}(d,u,k,n)$$

such that

$$\left[\begin{array}{l} \operatorname{im}(g(A)[e]) = A[e](k) \\ \text{for all } A \in \operatorname{bid}(d,u,k,n) \text{ and } e \in [1,d]; \end{array}\right.$$

[e.g., we may take $g = \operatorname{orp}(d,u,k,n)$]. Then we have

$$(2)\quad \det y = \sum_{A\in \operatorname{bid}(d,u,k,n)} S(A) \prod_{e\in[1,d]} \det \operatorname{mas}(x[e],k,g(A)[e])$$

where

$$(3)\quad S(A) = \operatorname{sgn}(\operatorname{lap}(d,u,k,n,f,g(A))).$$

Proof: By the definition of det we see that for every

$$A \in \operatorname{bid}(d,u,k,n) \quad \text{and} \quad e \in [1,d]$$

we have

$$(10)\quad \det \operatorname{mas}(x[e],k,g(A)[e]) = \sum_{w\in W(u[e](k'))} \operatorname{sgn}(w)r(A,e,w)$$

where for all

$$\left[\begin{array}{l} A \in \operatorname{bid}(d,u,k,n) \text{ and } e \in [1,d] \\ \text{and } w \in W(u[e](k')) \end{array}\right.$$

we have put

$$(11)\quad r(A,e,w) = \begin{cases} \displaystyle\prod_{i\in[1,u[e](k')]} \operatorname{mas}(x[e],k,g(A)[e])_{i,w(i)} & \text{if } k = 2 \\ \displaystyle\prod_{j\in[1,u[e](k')]} \operatorname{mas}(x[e],k,g(A)[e])_{w(j),j} & \text{if } k = 1. \end{cases}$$

By (10) we get

$$(12)\quad \left[\begin{array}{l} [\text{RHS of } (2)] \\ \displaystyle = \sum_{A\in \operatorname{bid}(d,u,k,n)} S(A) \prod_{e\in[1,d]} \sum_{w\in W(u[e](k'))} \operatorname{sgn}(w)r(A,e,w). \end{array}\right.$$

Upon letting

$$V = \begin{bmatrix} \text{the set of all sequences } v = v[e]_{1\le e\le d} \\ \text{with } v[e] \in W(u[e](k')) \end{bmatrix}$$

and

$$L = \begin{bmatrix} \text{the set of all pairs } (A,v) \\ \text{with } A \in \text{bid}(d,u,k,n) \text{ and } v \in V, \end{bmatrix}$$

we clearly have

$$[\text{RHS of } (12)] = \sum_{(A,v)\in L} \left[S(A) \prod_{e\in[1,d]} \text{sgn}(v[e])\right]\left[\prod_{e\in[1,d]} r(A,e,v[e])\right]$$

and in view of (3), upon taking g(A) for g in (12.2.2) we see that for every $(A,v) \in L$ we have

$$\text{sgn}(\text{elap}(d,u,k,f,g(A),v)) = S(A) \prod_{e\in[1,d]} \text{sgn}(v[e])$$

and now by the two equations above we get

$$(13) \quad \begin{bmatrix} [\text{RHS of } (12)] \\ = \displaystyle\sum_{(A,v)\in L} \text{sgn}(\text{elap}(d,u,k,n,f,g(A),v)) \prod_{e\in[1,d]} r(A,e,v[e]). \end{bmatrix}$$

By the definition of mas we see that for all

$$\begin{bmatrix} A \in \text{bid}(d,u,k,n) \text{ and } e \in [1,d] \text{ and} \\ i \in [1,u[e](k')] \text{ and } j \in [1,u[e](k')] \end{bmatrix}$$

we have

$$\text{mas}(x[e],k,g(A)[e])_{ij} = \begin{cases} x[e]_{i,g(A)[e](j)} & \text{if } k = 2 \\ x[e]_{g(A)[e](i),j} & \text{if } k = 1 \end{cases}$$

and hence by (11) we see that for all

$$\begin{bmatrix} A \in \text{bid}(d,u,k,n) \text{ and } e \in [1,d] \\ \text{and } v \in V \end{bmatrix}$$

we have

$$r(A,e,v[e]) = \begin{cases} \displaystyle\prod_{i\in[1,u[e](k')]} x[e]_{i,g(A)[e](v[e](i))} & \text{if } k = 2 \\ \displaystyle\prod_{j\in[1,u[e](k')]} x[e]_{g(A)[e](v[e](j)),j} & \text{if } k = 1 \end{cases}$$

and hence by (1) we see that for all

$$\begin{bmatrix} A \in \mathrm{bid}(d,u,k,n) \text{ and } e \in [1,d] \\ \text{and } v \in V \end{bmatrix}$$

we have

$$r(A,e,v[e]) = \begin{cases} \prod_{i\in[1,u[e](k')]} y_{f[e](i),g(A)[e](v[e](i))} & \text{if } k = 2 \\ \prod_{j\in[1,u[e](k')]} y_{g(A)[e](v[e](j)),f[e](j)} & \text{if } k = 1 \end{cases}$$

and therefore we see that for all $(A,v) \in L$ we have

$$\prod_{e\in[1,d]} r(A,e,v[e]) = \begin{cases} \prod_{e\in[1,d]} \prod_{i\in[1,u[e](k')]} y_{f[e](i),g(A)[e](v[e](i))} & \text{if } k = 2 \\ \prod_{e\in[1,d]} \prod_{j\in[1,u[e](k')]} y_{g(A)[e](v[e](j)),f[e](j)} & \text{if } k = 1 \end{cases}$$

and consequently, by the definition of elap we see that for all $(A,v) \in L$ we have

$$\prod_{e\in[1,d]} r(A,e,v[e]) = \begin{cases} \prod_{i\in[1,n(k)]} y_{i,\mathrm{elap}(d,u,k,n,f,g(A),v)(i)} & \text{if } k = 2 \\ \prod_{j\in[1,n(k)]} y_{\mathrm{elap}(d,u,k,n,f,g(A),v)(j),j} & \text{if } k = 1 \end{cases}$$

and upon noting that

$$\begin{bmatrix} (A,v) \mapsto \mathrm{elap}(d,u,k,n,f,g(A),v) \\ \text{gives a bijection of } L \text{ onto } W(n(k)), \end{bmatrix}$$

by the equation above we see that

$$[\text{RHS of (13)}] = \begin{cases} \sum_{w\in W(n(k))} \mathrm{sgn}(w) \prod_{i\in[1,n(k)]} y_{i,w(i)} & \text{if } k = 2 \\ \sum_{w\in W(n(k))} \mathrm{sgn}(w) \prod_{j\in[1,n(k)]} y_{w(j),j} & \text{if } k = 1 \end{cases}$$

and now, upon noting that the RHS above is clearly equal to det y, we conclude that

(14) [RHS of (13)] = det y.

By (12), (13), and (14) we get (2), and this completes the proof.

13. THE BASIC CASE

(13.1) Notation In this subsection we introduce some further terminology to be used in the rest of this chapter.

Given any $m \in N(2)$ and $k \in [1,2]$ with $m(k) = 2m(k')$, first we introduce the *basic tableau* of (m,k), which we denote by bat(m,k) and which we define by saying that bat(m,k) is the unique member of bit(2,m,k) such that for all $e \in [1,2]$ and $i \in [1,m(k')]$ we have

$$\text{bat}(m,k)[e](k,i) = i + (e - 1)m(k'),$$

and second, for every $B \in \text{osid}(k,\text{bat}(m,k))$ and $x \in \text{mat}(R,m)$, where R is a ring, we introduce the *basic residual matrix* of (x,m,k,B), which we denote by barem(x,m,k,B) and which we define by saying that barem(x,m,k,B) is the unique member of mat(R,dup(2m(k'))) such that in the case $k = 2$, for all $e \in [1,2]$ and $i \in [1,m(k')]$ we have

$$\text{barem}(x,m,k,B)_{i+(e-1)m(k'),j} = \begin{cases} x_{ij} & \text{if } j \in [1 + (e - 1)m(k'), em(k')] \cup B[e'](k) \\ 0 & \text{if } j \in [1 + (e' - 1)m(k'), e'm(k')] \setminus B[e'](k) \end{cases}$$

whereas in the case $k = 1$, for all $e \in [1,2]$ and $j \in [1,m(k')]$ we have

$$\text{barem}(x,m,k,B)_{i,j+(e-1)m(k')} = \begin{cases} x_{ij} & \text{if } i \in [1 + (e - 1)m(k'), em(k')] \cup B[e'](k) \\ 0 & \text{if } i \in [1 + (e' - 1)m(k'), e'm(k')] \setminus B[e'](k), \end{cases}$$

and third, for every $B \in \text{osid}(k,\text{bat}(m,k))$ and $x \in \text{mat}(R,m)$, where R is a ring, we introduce the *basic residue* of (x,m,k,B), which we denote by bar(x,m,k,B) and which we define by putting

$$\text{bar}(x,m,k,B) = -\det \text{barem}(x,m,k,B),$$

and fourth, for every $A \in \text{nex}(\text{bat}(m,k)) \cap \text{osid}(k,\text{bat}(m,k))$ we introduce the *basic permutation* of (m,k,A), which we denote by bap(m,k,A) and which we define by saying that bap(m,k,A) is the unique member of W(2m(k')) such that for all $e \in [1,2]$ and $i \in [1,m(k')]$ we have

$$\mathrm{bap}(m,k,A)(i + (e-1)m(k')) = \mathrm{sat}(\mathrm{bat}(m,k),A)[e](k,i),$$

and fifth, for every $A \in \mathrm{nex}(\mathrm{bat}(m,k)) \cap \mathrm{osid}(k,\mathrm{bat}(m,k))$ we introduce the *basic signature* of (m,k,A), which we denote by bag(m,k,A) and which we define by putting

$$\mathrm{bag}(m,k,A) = -\mathrm{sgn}(\mathrm{bap}(m,k,A)),$$

and sixth, we note that the definitions of bap and bag above make sense because of the observation, where we have put

$$T = \mathrm{bat}(m,k),$$

and which says that

$$\left[\begin{array}{l} \text{we have } \mathrm{nex}(T) \cap \mathrm{osid}(k,T) = \mathrm{eosid}(k,T) \\ \text{and for every } A \in \mathrm{nex}(T) \cap \mathrm{osid}(k,T) \\ \text{we have } \mathrm{sat}(T,A) \in \mathrm{bit}(2,m,k) \text{ and} \\ [1,2m(k')] = \coprod_{e\in[1,2]} \mathrm{sad}(T,A)[e](k). \end{array}\right. \tag{13.1.1}$$

FIRST MAIN LEMMA (13.2) Given any ring R, and given any $m \in N(2)$ and $k \in [1,2]$ with $m(k) = 2m(k')$, and given any $B \in \mathrm{osid}(k,\mathrm{bat}(m,k))$ and $x \in \mathrm{mat}(R,m)$, upon letting $T = \mathrm{bat}(m,k)$, we have

$$\det \mathrm{barem}(x,m,k,B) = \sum_{A\in\mathrm{nex}(T,B)} \mathrm{sgn}(\mathrm{bap}(m,k,A))\mathrm{mom}(x,\mathrm{sat}(T,A)).$$

Proof: Let

$$n = \mathrm{dup}(2m(k')) \qquad \text{and} \qquad y = \mathrm{barem}(x,m,k,B),$$

and let u be the unique member of bib(2) such that

$$u[1] = m = u[2].$$

Now clearly,

$$u \in \mathrm{bib}(2,n,k) \qquad \text{and} \qquad y \in \mathrm{mat}(R,n).$$

Let f be the unique member of bil(2,u,k,n) such that

$$f[e](i) = i + (e-1)m(k') \quad \text{for all } e \in [1,2] \text{ and } i \in [1,m(k')].$$

For every $e \in [1,2]$ let $x[e]$ be the unique member of $\mathrm{mat}(R,u[e])$ such that for all $i \in [1,u[e](1)]$ and $j \in [1,u[e](2)]$ we have

$$x[e]_{ij} = \begin{cases} y_{f[e](i),j} & \text{if } k = 2 \\ y_{i,f[e](j)} & \text{if } k = 1. \end{cases}$$

Finally, let

$$g = \mathrm{orp}(2,u,k,n).$$

Now our assertion follows from the $d = 2$ case of Laplace (12.3) by noting that

(1) $\left[\begin{array}{l} A \mapsto \mathrm{sad}(T,A) \text{ gives an injective map} \\ \mathrm{nex}(T,B) \to \mathrm{bid}(2,u,k,n) \end{array}\right.$

and

(2) $\left[\begin{array}{l} \text{for every } A \in \mathrm{nex}(T,B) \text{ we have} \\ \mathrm{bap}(m,k,A) = \mathrm{lap}(2,u,k,n,f,g(\mathrm{sad}(T,A))) \end{array}\right.$

and

(3) $\left[\begin{array}{l} \text{for every } A \in \mathrm{nex}(T,B) \text{ and } e \in [1,2] \text{ we have} \\ \mathrm{sub}(x,\mathrm{sat}(T,A)[e]) = \mathrm{mas}(x[e],k,g(\mathrm{sad}(T,A))[e]) \end{array}\right.$

and

(4) $\left[\begin{array}{l} \text{for every } A \in \mathrm{bid}(2,u,k,n) \setminus \{\mathrm{sad}(T,A) : A \in \mathrm{nex}(T,B)\} \\ \text{we have } \prod_{e \in [1,2]} \det \mathrm{mas}(x[e],k,g(A)[e]) = 0. \end{array}\right.$

Out of the four claims above, the first three are easily verified by using observation (13.1.1); we now proceed to verify (4).

For every $e \in [1,2]$ let $\tilde{B}[e] \subset [1,u[e](k)]$ be defined by putting

$$\tilde{B}[e] = [1 + (e - 1)m(k'), em(k')] \cup B[e'](k).$$

Now if $A \in \mathrm{bid}(2,u,k,n)$ and $e \in [1,2]$ are such that

$$A[e](k) \setminus \tilde{B}[e] \neq \emptyset,$$

then in the case $k = 2$, for some $j \in [1,m(k')]$ we would have

$$g(A)[e](j) \in A[e](k) \setminus \tilde{B}[e]$$

and for any such j and for all $i \in [1,m(k')]$ we would get

$$\begin{aligned} \mathrm{mas}(x[e],k,g(A)[e])_{ij} &= x[e]_{i,g(A)[e](j)} \\ &= y_{f[e](i),g(A)[e](j)} \\ &= 0 \end{aligned}$$

and hence we would have that the jth column of mas(x[e],k,g(A)[e]) consists entirely of zeros and therefore we would have that

$$\det \mathrm{mas}(x[e],k,g(A)[e]) = 0,$$

and similarly, in the case $k = 1$, for some $i \in [1,m(k')]$ we would have

$$g(A)[e](i) \in A[e](k) \setminus \tilde{B}[e]$$

and for any such i and for all $j \in [1,m(k')]$ we would get

$$\begin{aligned} \mathrm{mas}(x[e],k,g(A)[e])_{ij} &= x[e]_{g(A)[e](i),j} \\ &= y_{g(A)[e](i),f[e](j)} \\ &= 0 \end{aligned}$$

and hence we would have that the ith row of mas(x[e],k,g(A)[e]) consists entirely of zeros and therefore we would have that

$$\det \mathrm{mas}(x[e],k,g(A)[e]) = 0;$$

thus we conclude that

(10) $\left[\begin{array}{l} \text{if } A \in \mathrm{bid}(2,u,k,n) \text{ and } e \in [1,2] \text{ are such that} \\ A[e](k) \setminus \tilde{B}[e] \neq \emptyset \\ \text{then } \det \mathrm{mas}(x[e],k,g(A)[e]) = 0. \end{array}\right.$

By (10) we see that

(11) $\left[\begin{array}{l} \text{if } A \in \mathrm{bid}(2,u,k,n) \text{ is such that} \\ \text{either } A[1](k) \setminus \tilde{B}[1] \neq \emptyset \text{ or } A[2](k) \setminus \tilde{B}[2] \neq \emptyset \\ \text{then } \prod_{e \in [1,2]} \det \mathrm{mas}(x[e],k,g(A)[e]) = 0. \end{array}\right.$

By remembering (13.1.1) we also see that

$\left[\begin{array}{l} \text{if } A \in \mathrm{bid}(2,u,k,n) \text{ is such that} \\ A[1](k) \setminus \tilde{B}[1] = \emptyset = A[2](k) \setminus \tilde{B}[2] \end{array}\right.$

(12) then, upon letting
$\tilde{A}[e](\hat{k}) = B[e](\hat{k}) \setminus A[e](\hat{k})$ for all $e \in [1,2]$ and $\hat{k} \in [1,2]$,
we would have $\tilde{A} \in \text{nex}(T,B)$ and $\text{sad}(T,\tilde{A}) = A$.

Now (4) follows from (11) and (12).

SECOND MAIN LEMMA (13.3) Given any ring R, and given any $m \in N(2)$ and $k \in [1,2]$ with $m(k) = 2m(k')$, and given any $B \in \text{osid}(k,\text{bat}(m,k))$ and $x \in \text{mat}(R,m)$, let us put $T = \text{bat}(m,k)$, and let us assume that there exists a finite set b^* together with injective maps

$$v[1]: b^* \to [1,m(k')] \quad \text{and} \quad v[2]: b^* \to [1 + m(k'), 2m(k')]$$

such that

$$B[1](k) \cap \text{im}(v[1]) = \emptyset = B[2](k) \cap \text{im}(v[2])$$

and such that

$$\text{card}(B[1](k)) + \text{card}(B[2](k)) + \text{card}(b^*) > m(k')$$

and such that

in the case $k = 2$ we have
$x_{i,v[1](j)} = x_{i,v[2](j)}$ for all $i \in [1,m(k')]$ and $j \in b^*$

whereas

in the case $k = 1$ we have
$x_{v[1](i),j} = x_{v[2](i),j}$ for all $j \in [1,m(k')]$ and $i \in b^*$.

Then

$$\det \text{barem}(x,m,k,B) = 0.$$

Proof: Let $n = \text{dup}(2m(k'))$. Let y be the unique member of $\text{mat}(R,n)$ such that in the case $k = 2$, for all $i \in [1,m(k')]$ we have

$$y_{i+(e-1)m(k'),j} = \begin{cases} x_{ij} & \text{if } e = 1 \text{ and } j \in [1,m(k')] \setminus B[1](k) \\ 0 & \text{if } e = 1 \text{ and } j \in B[1](k) \cup B[2](k) \\ -x_{ij} & \text{if } e = 1 \text{ and } j \in [1 + m(k'), 2m(k')] \setminus B[2](k) \\ x_{ij} & \text{if } e = 2 \text{ and } j \in B[1](k) \end{cases}$$

$$\begin{cases} 0 & \text{if } e = 2 \text{ and } j \in [1,m(k')] \setminus B[1](k) \\ x_{ij} & \text{if } e = 2 \text{ and } j \in [1 + m(k'), 2m(k')] \end{cases}$$

whereas in the case k = 1, for $j \in [1,m(k')]$ we have

$$y_{i,j+(e-1)m(k')}$$

$$= \begin{cases} x_{ij} & \text{if } e = 1 \text{ and } i \in [1,m(k')] \setminus B[1](k) \\ 0 & \text{if } e = 1 \text{ and } i \in B[1](k) \cup B[2](k) \\ -x_{ij} & \text{if } e = 1 \text{ and } i \in [1 + m(k'), 2m(k')] \setminus B[2](k) \\ x_{ij} & \text{if } e = 2 \text{ and } i \in B[1](k) \\ 0 & \text{if } e = 2 \text{ and } i \in [1,m(k')] \setminus B[1](k) \\ x_{ij} & \text{if } e = 2 \text{ and } i \in [1 + m(k'), 2m(k')]. \end{cases}$$

Now, in the case k = 2, when for each $i \in [1,m(k')]$ we subtract the $(i + m(k'))$th row of barem(x,m,k,B) from its ith row we get y, and similarly, in the case k = 1, when for each $j \in [1,m(k')]$ we subtract the $(j + m(k'))$th column of barem(x,m,k,B) from its jth column we get y; hence by the alternating property of det we see that

(1) det barem(x,m,k,B) = det y.

Upon letting u be the unique member of bib(2) such that

$$u[1] = m = u[2],$$

we clearly have $u \in \text{bib}(2,n,k)$. Let f be the unique member of bil(2,u,k,n) such that

$$f[e](i) = i + (e - 1)m(k') \quad \text{for all } e \in [1,2] \text{ and } i \in [1,m(k')].$$

For every $e \in [1,2]$ let x[e] be the unique member of mat(R,u[e]) such that for all $i \in [1,u[e](1)]$ and $j \in [1,u[e](2)]$ we have

$$x[e]_{ij} = \begin{cases} y_{f[e](i),j} & \text{if } k = 2 \\ y_{i,f[e](j)} & \text{if } k = 1. \end{cases}$$

Finally, let g = orp(2,u,k,n). Now by the d = 2 case of Laplace (12.3) we get

$$\text{(2)} \quad \det y = \sum_{A \in \text{bid}(2,u,k,n)} S(A) \prod_{e \in [1,2]} \det \text{mas}(x[e],k,g(A)[e])$$

where

(3) $S(A) = \mathrm{sgn}(\mathrm{lap}(2,u,k,n,f,g(A)))$.

By assumption v[1] and v[2] are injective and hence

(4) $\mathrm{card}(\mathrm{im}(v[1])) = \mathrm{card}(b^*) = \mathrm{card}(\mathrm{im}(v[2]))$.

Also, recall that by assumption

(5) $B[1](k) \cap \mathrm{im}(v[1]) = \emptyset = B[2](k) \cap \mathrm{im}(v[2])$

and

(6) $\mathrm{card}(B[1](k)) + \mathrm{card}(B[2](k)) + \mathrm{card}(b^*) > m(k')$.

Let

$$\tilde{B}[1] = [1,m(k')] \setminus [B[1](k) \cup \mathrm{im}(v[1])]$$

and

$$\tilde{B}[2] = [1 + m(k'), 2m(k')] \setminus [B[2](k) \cup \mathrm{im}(v[2])].$$

Now by (5) we see that

$$\begin{aligned}&[1,2m(k')]\\ &= B[1](k) \amalg B[2](k) \amalg \tilde{B}[1] \amalg \tilde{B}[2] \amalg \mathrm{im}(v[1]) \amalg \mathrm{im}(v[2])\end{aligned}$$

and hence in view of (4) we get

$$\begin{aligned}2m(k') = {} & \mathrm{card}(B[1](k)) + \mathrm{card}(B[2](k)) + \mathrm{card}(\tilde{B}[1])\\ & + \mathrm{card}(\tilde{B}[2]) + 2\mathrm{card}(b^*)\end{aligned}$$

and therefore in view of (6) we conclude that

(7) $\mathrm{card}(\tilde{B}[1]) + \mathrm{card}(\tilde{B}[2]) + \mathrm{card}(b^*) < m(k')$.

For every $H \subset [1,2m(k')]$ let

$$D(H) = \{q \in b^* : v[1](q) \in H \text{ and } v[2](q) \in H\}.$$

Given any $H \subset [1,2m(k')]$, it is obvious that

(8) $\begin{cases}\text{if } D(H) = \emptyset, \text{ then}\\ \mathrm{card}(H \cap [\mathrm{im}(v[1]) \cup \mathrm{im}(v[2])]) \le \mathrm{card}(b^*)\end{cases}$

and it is obvious that

(9) $\left[\begin{array}{l}\text{if } H \subset \tilde{B}[1] \cup \tilde{B}[2] \cup \text{im}(v[1]) \cup \text{im}(v[2]), \\ \text{then } \text{card}(H) \le \text{card}(\tilde{B}[1]) + \text{card}(\tilde{B}[2]) \\ \qquad\qquad \text{card}(A[1])(k)) = m(k');\end{array}\right.$

and it is obvious that

(10) $\left[\begin{array}{l}\text{if } H \cap [B[1](k) \cup B[2](k)] = \emptyset, \text{ then} \\ H \subset \tilde{B}[1] \cup \tilde{B}[2] \cup \text{im}(v[1]) \cup \text{im}(v[2])\end{array}\right.$

and now by (7) to (10) it follows that

(11) $\left[\begin{array}{l}\text{if } \text{card}(H) \ge m(k'), \text{ then either} \\ H \cap [B[1](k) \cup B[2](k)] \ne \emptyset \text{ or } D(H) \ne \emptyset.\end{array}\right.$

For every A ∈ bid(2,u,k,n), by definition we have

$$\text{card}(A[7])(k)) = m(k');$$

therefore, by (11) we conclude that

(12) $\left[\begin{array}{l}\text{given any } A \in \text{bid}(2,u,k,n), \text{ we have that} \\ \text{either } A[1](k) \cap [B[1](k) \cup B[2](k)] \ne \emptyset \\ \text{or } D(A[1](k)) \ne \emptyset.\end{array}\right.$

Clearly,

(13) $\left[\begin{array}{l}\text{if } A \in \text{bid}(2,u,k,n) \text{ is such that} \\ A[1](k) \cap [B[1](k) \cup B[2](k)] \ne \emptyset, \\ \text{then for some } t \in [1,m(k')] \text{ we have} \\ g(A)[1](t) \in [B[1](k) \cup B[2](k)].\end{array}\right.$

If A ∈ bid(2,u,k,n) is such that D(A[1](k)) ≠ ∅, then first we can find q in b* such that

$$v[1](q) \in A[1](k) \quad \text{and} \quad v[2](q) \in A[1](k),$$

and second, because of the two inclusion relations above, we can find p[1] and p[2] in [1,m(k')] such that

$$g(A)[1](p[1]) = v[1](q) \quad \text{and} \quad g(A)[1](p[2]) = v[2](q),$$

and third, it is obvious that v[1](q) ≠ v[2](q) and hence we must have p[1] ≠ p[2]; thus

$$(14)\quad \left[\begin{array}{l}\text{if } A \in \text{bid}(2,u,k,n) \text{ is such that } D(A[1](k)) \neq \emptyset, \text{ then} \\ \text{for some } q \in b^* \text{ and some } p[1] \neq p[2] \text{ in } [1,m(k')] \text{ we have} \\ g(A)[1](p[1]) = v[1](q) \text{ and } g(A)[1](p[2]) = v[2](q).\end{array}\right.$$

If $A \in \text{bid}(2,u,k,n)$ is such that for some $t \in [1,m(k')]$ we have

$$g(A)[1](t) \in [B[1](k) \cup B[2](k)],$$

then in the case $k = 2$, for all $i \in [1,m(k')]$ we would have

$$\begin{aligned} \text{mas}(x[1],k,g(A)[1])_{it} &= x[1]_{i,g(A)[1](t)} \\ &= y_{i,g(A)[1](t)} \\ &= 0 \end{aligned}$$

and hence we would have that the tth column of mas(x[1],k,g(A)[1]) consists entirely of zeros, and therefore we would have that

$$\det \text{mas}(x[1],k,g(A)[1]) = 0,$$

and similarly, in the case $k = 1$, for all $j \in [1,m(k')]$ we would have

$$\begin{aligned} \text{mas}(x[1],k,g(A)[1])_{tj} &= x[1]_{g(A)[1](t),j} \\ &= y_{g(A)[1](t),j} \\ &= 0 \end{aligned}$$

and hence we would have that the tth row of mas(x[1],k,g(A)[1]) consists entirely of zeros, and therefore we would have that

$$\det \text{mas}(x[1],k,g(A)[1]) = 0;$$

thus

$$(15)\quad \left[\begin{array}{l}\text{if } A \in \text{bid}(2,u,k,n) \text{ is such that for some } t \in [1,m(k')] \\ \text{we have } g(A)[1](t) \in [B[1](k) \cup B[2](k)], \\ \text{then } \det \text{mas}(x[1],k,g(A)[1]) = 0.\end{array}\right.$$

If $A \in \text{bid}(2,u,k,n)$ is such that for some q in b^* and some $p[1] \neq p[2]$ in $[1,m(k')]$ we have

$$g(A)[1](p[1]) = v[1](q) \quad \text{and} \quad g(A)[1](p[2]) = v[2](q),$$

then in the case k = 2, for all i $\in$ [1,m(k')] we would have

$$\begin{aligned}\text{mas}(x[1],k,g(A)[1])_{i,p[1]} &= x[1]_{i,g(A)[1](p[1])} \\ &= y_{i,v[1](q)} \\ &= x_{i,v[1](q)} \\ &= x_{i,v[2](q)}\end{aligned}$$

and

$$\begin{aligned}\text{mas}(x[1],k,g(A)[1])_{i,p[2]} &= x[1]_{i,g(A)[1](p[2])} \\ &= y_{i,v[2](q)} \\ &= -x_{i,v[2](q)}\end{aligned}$$

and hence we would have that the p[2]th column of mas(x[1],k,g(A)[1]) is the negative of its p[1]th column and therefore we would have that

$$\det \text{mas}(x[1],k,g(A)[1]) = 0,$$

and similarly, in the case k = 1, for all j $\in$ [1,m(k')] we would have

$$\begin{aligned}\text{mas}(x[1],k,g(A)[1])_{p[1],j} &= x[1]_{g(A)[1](p[1]),j} \\ &= y_{v[1](q),j} \\ &= x_{v[1](q),j} \\ &= x_{v[2](q),j}\end{aligned}$$

and

$$\begin{aligned}\text{mas}(x[1],k,g(A)[1])_{p[2],j} &= x[1]_{g(A)[1](p[2]),j} \\ &= y_{v[2](q),j} \\ &= -x_{v[2](q),j}\end{aligned}$$

and hence we would have that the p[2]th row of mas(x[1],k,g(A)[1]) is the negative of its p[1]th row and therefore we would have that

$$\det \text{mas}(x[1],k,g(A)[1]) = 0;$$

thus

$$(16)\quad \begin{cases} \text{if } A \in \mathrm{bid}(2,u,k,n) \text{ is such that for some } q \text{ in } b^* \\ \text{and some } p[1] \neq p[2] \text{ in } [1,m(k')] \text{ we have} \\ g(A)[1](p[1]) = v[1](q) \text{ and } g(A)[1](p[2]) = v[2](q), \\ \text{then } \det \mathrm{mas}(x[1],k,g(A)[1]) = 0. \end{cases}$$

By (12) to (16) we see that

$$(17)\quad \begin{cases} \text{for every } A \in \mathrm{bid}(2,u,k,n) \text{ we have} \\ \det \mathrm{mas}(x[1],k,g(A)[1]) = 0. \end{cases}$$

By (2) and (17) we get det y = 0 and therefore by (1) we get

$$\det \mathrm{barem}(x,m,k,B) = 0$$

and this completes the proof.

PROPOSITION (13.4) Let there be given any ring R, and let there be given any $m \in N(2)$ and $k \in [1,2]$ with $m(k) = 2m(k')$, and let there be given any $B \in \mathrm{osid}(k,\mathrm{bat}(m,k))$ and $x \in \mathrm{mat}(R,m)$, and let us put $T = \mathrm{bat}(m,k)$. Then:

(13.4.1) We have

$$\mathrm{bar}(x,m,k,B) = \sum_{A \in \mathrm{nex}(T,B)} \mathrm{bag}(m,k,A)\mathrm{mom}(x,\mathrm{sat}(T,A)).$$

(13.4.2) Upon letting E be the empty bitabloid of depth 2, i.e., upon letting E be the bitabloid of depth 2 obtained by putting

$$E[e](k') = \emptyset = E[e](k) \qquad \text{for all } e \in [1,2],$$

first we have

$$\mathrm{nex}(T,B) = \{E\} \coprod \mathrm{pex}(T,B),$$

and second, we have

$$\mathrm{sat}(T,E) = T,$$

and third, we have that bap(m,k,E) is the identity permutation of $[1,2m(k')]$, i.e.,

$$\mathrm{bap}(m,k,E)(i) = i \qquad \text{for all } i \in [1,2m(k')],$$

and fourth, we have

$$\mathrm{bag}(m,k,E) = -1.$$

(13.4.3) Finally, we have

$$\mathrm{bar}(x,m,k,B) + \mathrm{mom}(x,T) = \sum_{A \in \mathrm{pex}(T,B)} \mathrm{bag}(m,k,A)\mathrm{mom}(x,\mathrm{sat}(T,A)).$$

Proof: In view of the definitions of bar and bag, (13.4.1) follows from (13.2). The first three parts of (13.4.2) are obvious, and its fourth part follows from its third part. (13.4.3) follows from (13.4.1) and (13.4.2).

PROPOSITION (13.5) Let there be given any ring R, and let there be given any $m \in N(2)$ and $k \in [1,2]$ with $m(k) = 2m(k')$, and let there be given any $B \in \mathrm{osid}(k,\mathrm{bat}(m,k))$ and $x \in \mathrm{mat}(R,m)$, and let us put $T = \mathrm{bat}(m,k)$, and let us assume that there exists a finite set b^* together with injective maps

$$v[1]: b^* \to [1,m(k')] \quad \text{and} \quad v[2]: b^* \to [1 + m(k'), 2m(k')]$$

such that

$$B[1](k) \cap \mathrm{im}(v[1]) = \emptyset = B[2](k) \cap \mathrm{im}(v[2])$$

and such that

$$\mathrm{card}(B[1](k)) + \mathrm{card}(B[2](k)) + \mathrm{card}(b^*) > m(k')$$

and such that

$$\left[\begin{array}{l}\text{in the case } k = 2 \text{ we have}\\ x_{i,v[1](j)} = x_{i,v[2](j)} \text{ for all } i \in [1,m(k')] \text{ and } j \in b^*,\end{array}\right.$$

whereas

$$\left[\begin{array}{l}\text{in the case } k = 1 \text{ we have}\\ x_{v[1](i),j} = x_{v[2](i),j} \text{ for all } j \in [1,m(k')] \text{ and } i \in b^*.\end{array}\right.$$

Then:

(13.5.1) We have $\mathrm{bar}(x,m,k,B) = 0$

(13.5.2) Also, we have

$$\sum_{A \in \mathrm{nex}(T,B)} \mathrm{bag}(m,k,A)\mathrm{mom}(x,\mathrm{sat}(T,A)) = 0.$$

(13.5.3) Finally, we have

$$\text{mom}(x,T) = \sum_{A \in \text{pex}(T,B)} \text{bag}(m,k,A)\text{mom}(x,\text{sat}(T,A)).$$

Proof: In view of the definition of bar, (13.5.1) follows from (13.3). (13.5.2) follows from (13.4.1) and (13.5.1). (13.5.3) follows from (13.4.3) and (13.5.1).

14. CASTIFICATION AND THE MAXIMAL SIZE CASE

(14.1) Notation In this subsection we introduce some further terminology to be used in the rest of this chapter.

Given any $m^* \in N(2)$ and $k \in [1,2]$, first we introduce the *castic binumber* of (m^*,k), which we denote by cab(m^*,k) and which we define by saying that cab(m^*,k) is the unique member of $N(2)$ such that

$$\text{cab}(m^*,k)(k') = m^*(k') \qquad \text{and} \qquad \text{cab}(m^*,k)(k) = 2m^*(k'),$$

and second, we introduce the *cast* of (m^*,k), which we denote by cat(m^*,k) and which we define by saying that cat(m^*,k) is the unique member of bit$(2,\text{cab}(m^*,k),k)$ obtained by putting

$$\text{cat}(m^*,k) = \text{bat}(\text{cab}(m^*,k),k),$$

and third, for every $T^* \in \text{bit}(2,m^*,k)$ and $x^* \in \text{mat}(R,m^*)$, where R is a ring, we introduce the *castic matrix* of (x^*,m^*,k,T^*), which we denote by cam(x^*,m^*,k,T^*) and which we define by saying that cam(x^*,m^*,k,T^*) is the unique member of mat$(R,\text{cab}(m^*,k))$ such that for all $e \in [1,2]$ we have

$$\text{sub}(\text{cam}(x^*,m^*,k,T^*),\text{cat}(m^*,k)[e]) = \text{sub}(x^*,T^*[e]),$$

and fourth, for every $T^* \in \text{bit}(2,m^*,k)$ and $A^* \in \text{osid}(k,T^*)$ we introduce the *castoid* of (m^*,k,T^*,A^*), which we denote by cad(m^*,k,T^*,A^*) and which we define by saying that cad(m^*,k,T^*,A^*) is the unique member of osid$(k,\text{cat}(m^*,k))$ such that for all $e \in [1,2]$ we have

$$\{T^*[e](k, i - (e - 1)m^*(k')) : i \in \text{cad}(m^*,k,T^*,A^*)[e](k)\}$$
$$= A^*[e](k),$$

and fifth, for every $T^* \in \mathrm{bit}(2,m^*,k)$ and $B^* \in \mathrm{osid}(k,T^*)$ and $x^* \in \mathrm{mat}(R,m^*)$, where R is a ring, we introduce the *castic residual matrix* of (x^*,m^*,k,T^*,B^*), which we denote by $\mathrm{carem}(x^*,m^*,k,T^*,B^*)$ and which we define by saying that $\mathrm{carem}(x^*,m^*,k,T^*,B^*)$ is the unique member of $\mathrm{mat}(R,\mathrm{dup}(2m^*(k')))$ obtained by putting

$$\begin{aligned}&\mathrm{carem}(x^*,m^*,k,T^*,B^*)\\&= \mathrm{barem}(\mathrm{cam}(x^*,m^*,k,T^*),\mathrm{cab}(m^*,k),k,\mathrm{cad}(m^*,k,T^*,B^*)),\end{aligned}$$

and sixth, for every $T^* \in \mathrm{bit}(2,m^*,k)$ and $B^* \in \mathrm{osid}(k,T^*)$ and $x^* \in \mathrm{mat}(R,m^*)$, where R is a ring, we introduce the *castic residue* of (x^*,m^*,k,T^*,B^*), which we denote by $\mathrm{car}(x^*,m^*,k,T^*,B^*)$ and which we define by putting

$$\mathrm{car}(x^*,m^*,k,T^*,B^*) = -\det \mathrm{carem}(x^*,m^*,k,T^*,B^*),$$

and seventh, for every $T^* \in \mathrm{bit}(2,m^*,k)$ we introduce the *reverse cast* of (m^*,k,T^*), which we denote by $\mathrm{rat}(m^*,k,T^*)$ and which we define by saying that

$$\mathrm{rat}(m^*,k,T^*):\quad [1,2m^*(k')] \to [1,m^*(k')]$$

is the map such that for any given $i \in [1,2m^*(k')]$, we have

$$\mathrm{rat}(m^*,k,T^*)(i) = \begin{cases} T^*[1](k,i) & \text{if } i \le m^*(k') \\ T^*[2](k,\ i - m^*(k')) & \text{if } i > m^*(k'), \end{cases}$$

and eighth, for every $T^* \in \mathrm{bit}(2,m^*,k)$ we introduce the *enlarged reverse cast* of (m^*,k,T^*), which we denote by $\mathrm{erat}(m^*,k,T^*)$ and which we define by saying that

$$\mathrm{erat}(m^*,k,T^*):\quad \mathrm{osid}(k,\mathrm{cat}(m^*,k)) \to \mathrm{osid}(k,T^*)$$

is the map such that for all $A \in \mathrm{osid}(k,\mathrm{cat}(m^*,k))$ and $e \in [1,2]$ we have

$$\mathrm{erat}(m^*,k,T^*)(A)[e](k') = \emptyset$$

and

$$\mathrm{erat}(m^*,k,T^*)(A)[e](k) = \{\mathrm{rat}(m^*,k,T^*)(i) : i \in A[e](k)\},$$

and ninth, for every $T^* \in \mathrm{bit}(2,m^*,k)$ and $A^* \in \mathrm{nex}(T^*) \cap \mathrm{osid}(k,T^*)$ and $e \in [1,2]$ we introduce the *castic permutation* of (m^*,k,T^*,A^*)

at e, which we denote by cap(m*,k,T*,A*)[e] and which we define by saying that cap(m*,k,T*,A*)[e] is the unique member of W(m*(k')) such that for all i ∈ [1,m*(k')] we have

$$\text{sat}(T^*,A^*)[e](k,\text{cap}(m^*,k,T^*,A^*)[e](i)) = \text{rat}(m^*,k,T^*)(\text{sat}(\text{cat}(m^*,k),\text{cad}(m^*,k,T^*,A^*))[e](k,i)),$$

and tenth, for every T* ∈ bit(2,m*,k) and A* ∈ nex(T*) ∩ osid(k,T*) we introduce the *castic signature* of (m*,k,T*,A*), which we denote by cag(m*,k,T*,A*) and which we define by putting

$$\text{cag}(m^*,k,T^*,A^*) = \text{bag}(\text{cab}(m^*,k),k,\text{cad}(m^*,k,T^*,A^*)) \times \prod_{e\in[1,2]} \text{sgn}(\text{cap}(m^*,k,T^*,A^*)[e]),$$

and eleventh, for discussing the validity of the definitions of cap and cag above we make the following observations, where we are given any

$$T^* \in \text{bit}(2,m^*,k)$$

and where we have put

$$m = \text{cab}(m^*,k) \quad \text{and} \quad T = \text{cat}(m^*,k)$$

and

$$r = \text{rat}(m^*,k,T^*) \quad \text{and} \quad h = \text{erat}(m^*,k,T^*),$$

and which say that

$$\left[\begin{array}{l} \text{we have nex}(T^*) \cap \text{osid}(k,T^*) \\ = \text{disid}(k,T^*) \cap \text{eosid}(k,T^*) \text{ and} \\ \text{for every } A^* \in \text{nex}(T^*) \cap \text{osid}(k,T^*) \\ \text{we have sat}(T^*,A^*) \in \text{bit}(2,m^*,k) \end{array}\right. \qquad (14.1.1)$$

and

$$\left[\begin{array}{l} \text{for every } A^* \in \text{osid}(k,T^*) \text{ we have} \\ h(\text{cad}(m^*,k,T^*,A^*)) = A^* \\ \text{and for every } A \in \text{osid}(k,T) \text{ we have} \\ \text{cad}(m^*,k,T^*,h(A)) = A \end{array}\right. \qquad (14.1.2)$$

and

$$\left[\begin{array}{l}\text{for every } A \in \text{osid}(k,T) \text{ and } e \in [1,2]\\ \text{we have that } i \mapsto r(i) \text{ gives a bijection}\\ A[e](k) \to h(A)[e](k)\end{array}\right. \tag{14.1.3}$$

and

$$\left[\begin{array}{l}\text{for every } A^* \in \text{osid}(k,T^*) \text{ and } e \in [1,2] \text{ we have}\\ \text{card}(A^*[e](k)) = \text{card}(\text{cad}(m^*,k,T^*,A^*)[e](k))\end{array}\right. \tag{14.1.4}$$

and

$$\left[\begin{array}{l}\text{we have that } A^* \mapsto \text{cad}(m^*,k,T^*,A^*) \text{ gives a}\\ \text{bijection osid}(k,T^*) \to \text{osid}(k,T)\end{array}\right. \tag{14.1.5}$$

and

$$\left[\begin{array}{l}\text{we have that } A^* \mapsto \text{cad}(m^*,k,T^*,A^*) \text{ gives a}\\ \text{bijection eosid}(k,T^*) \to \text{eosid}(k,T)\end{array}\right. \tag{14.1.6}$$

and

$$\left[\begin{array}{l}\text{we have that } A^* \mapsto \text{cad}(m^*,k,T^*,A^*) \text{ gives an}\\ \text{injection nex}(T^*) \cap \text{osid}(k,T^*) \to \text{nex}(T) \cap \text{osid}(k,T)\end{array}\right. \tag{14.1.7}$$

and

$$\left[\begin{array}{l}\text{for any } A^* \in \text{nex}(T^*) \cap \text{osid}(k,T^*)\text{, upon letting}\\ A = \text{cad}(m^*,k,T^*,A^*)\text{, by (14.1.7) we have}\\ A \in \text{nex}(T) \cap \text{osid}(k,T)\\ \text{and hence sat}(T,A) \text{ and bag}(m,k,A) \text{ make sense}\\ \text{and now for all } e \in [1,2] \text{ we clearly have}\\ \text{sim}(\text{sat}(T^*,A^*))[e](k)\\ = \{r(\text{sat}(T,A)[e](k,i)) : i \in [1,m^*(k')]\}\end{array}\right. \tag{14.1.8}$$

and

$$\left[\begin{array}{l}\text{for any } A^* \in \text{nex}(T^*) \cap \text{osid}(k,T^*)\text{, upon letting}\\ A = \text{cad}(m^*,k,T^*,A^*)\text{, by (14.1.7) we have}\\ A \in \text{nex}(T) \cap \text{osid}(k,T)\\ \text{and hence sat}(T,A) \text{ and bag}(m,k,A) \text{ make sense}\\ \text{and by (14.1.8) we also see that}\\ \text{cap}(m^*,k,T^*,A^*) \text{ and cag}(m^*,k,T^*,A^*) \text{ make sense,}\end{array}\right. \tag{14.1.9}$$

and twelfth, we note that out of the nine observations above, the first three are obvious and the fourth follows from the second and the third, whereas the fifth follows from the second and, in turn, the sixth follows from the fourth and the fifth, and moreover, in view of (13.1.1), the seventh follows from the first and the sixth, and finally, the proofs of the eighth and the ninth are embedded in themselves, and thirteenth, as an application of the observations above we note that

$$\left[\begin{array}{l} \text{given any } T^* \in \text{bit}(2,m^*,k) \text{ and } B^* \in \text{osid}(k,T^*), \\ \text{for every } A^* \in \text{nex}(T^*,B^*) \text{ we obviously have} \\ A^* \in \text{nex}(T^*) \cap \text{osid}(k,T^*) \text{ and hence by (14.1.9)} \\ \text{we see that cag}(m^*,k,T^*,A^*) \text{ makes sense,} \end{array}\right. \qquad (14.1.10)$$

and fourteenth, we illustrate the castic definitions by the following figure, in which we have taken

k = 2

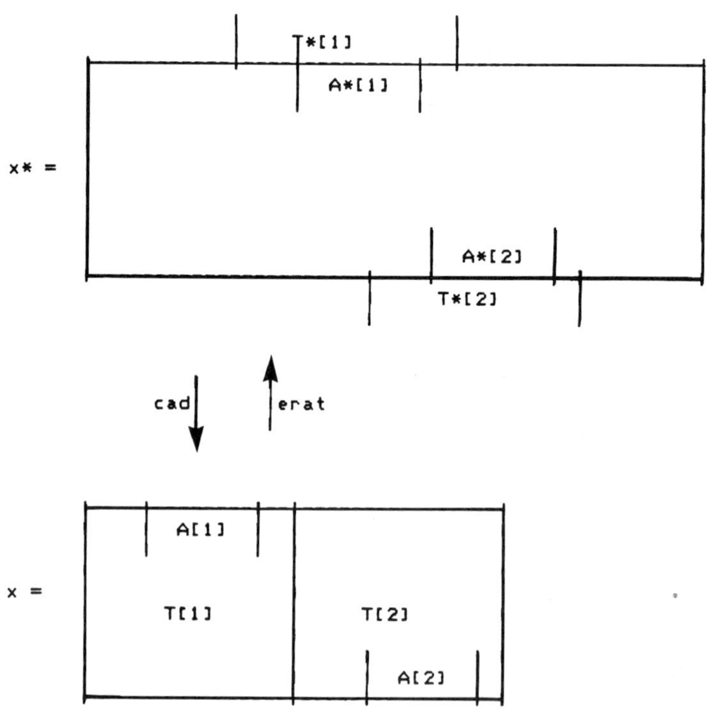

and in which we are given any

$$T^* \in \mathrm{bit}(2,m^*,k) \text{ and } x^* \in \mathrm{mat}(R,m^*), \text{ where } R \text{ is a ring,}$$

and in which we have put

$$T = \mathrm{cat}(m^*,k) \quad \text{and} \quad x = \mathrm{cam}(x^*,m^*,k,T^*)$$

and in which we are given any

$$A^* \in \mathrm{osid}(k,T^*) \quad \text{and} \quad A \in \mathrm{osid}(k,T)$$

such that

$$A = \mathrm{cad}(m^*,k,T^*,A^*) \quad \text{and} \quad A^* = \mathrm{erat}(m^*,k,T^*)(A).$$

LEMMA (14.2) Let there be given any ring R and let there be given any $m^* \in N(2)$ and $k \in [1,2]$ and $T^* \in \mathrm{bit}(2,m^*,k)$ and $A^* \in \mathrm{nex}(T^*) \cap \mathrm{osid}(k,T^*)$ and $x^* \in \mathrm{mat}(R,m^*)$ and $e \in [1,2]$, and let us put

$$s^* = \mathrm{sat}(T^*,A^*)[e]$$

and

$$s = \mathrm{sat}(\mathrm{cat}(m^*,k),\mathrm{cad}(m^*,k,T^*,A^*))[e]$$

and

$$c = \mathrm{cap}(m^*,k,T^*,A^*)[e] \quad \text{and} \quad x = \mathrm{cam}(x^*,m^*,k,T^*)$$

[where we refer to (14.1.9) for seeing that s and c make sense]. Then we have

$$\mathrm{mas}(\mathrm{sub}(x^*,s^*),k,c) = \mathrm{sub}(x,s) \tag{14.2.1}$$

and

$$\mathrm{sgn}(c)\mathrm{mor}(x^*,s^*) = \mathrm{mor}(x,s). \tag{14.2.2}$$

Proof: Upon letting $r = \mathrm{rat}(m^*,k,T^*)$, by the definition of cam we see that

$$(1) \quad \begin{cases} \text{in the case } k = 2 \text{ we have} \\ x_{ij} = x^*_{i,r(j)} \text{ for all } i \in [1,m^*(k')] \text{ and } j \in [1,2m^*(k')], \end{cases}$$

whereas

(2) $\left[\begin{array}{l}\text{in the case } k = 1 \text{ we have} \\ x_{ij} = r^*_{r(i),j} \text{ for all } j \in [1,m^*(k')] \text{ and } i \in [1,2m^*(k')].\end{array}\right.$

Now for all $i \in [1,m^*(k')]$ and $j \in [1,m^*(k')]$, in the case $k = 2$ we have

$$\begin{aligned}
&\text{mas}(\text{sub}(x^*,s^*),k,c)_{ij} \\
&= \text{sub}(x^*,s^*)_{i,c(j)} && \text{(by definition of mas)} \\
&= x^*_{i,s^*(k,c(j))} && \text{(by definition of sub)} \\
&= x^*_{i,r(s(k,j))} && \text{(by definition of cap)} \\
&= x_{i,s(k,j)} && \text{[by (1)]} \\
&= \text{sub}(x,s)_{ij} && \text{(by definition of sub)}
\end{aligned}$$

and similarly, in the case $k = 1$ we have

$$\begin{aligned}
&\text{mas}(\text{sub}(x^*,s^*),k,c)_{ij} \\
&= \text{sub}(x^*,s^*)_{c(i),j} && \text{(by definition of mas)} \\
&= x^*_{s^*(k,c(i)),j} && \text{(by definition of sub)} \\
&= x^*_{r(s(k,i)),j} && \text{(by definition of cap)} \\
&= x_{s(k,i),j} && \text{[by (2)]} \\
&= \text{sub}(x,s)_{ij} && \text{(by definition of sub)}
\end{aligned}$$

and this proves (14.2.1). In view of (11.1.3), upon taking det on both sides of (14.2.1) we get (14.2.2).

LEMMA (14.3) Let there be given any ring R and let there be given any $m^* \in N(2)$ and $k \in [1,2]$ and $T^* \in \text{bit}(2,m^*,k)$ and $A^* \in \text{nex}(T^*) \cap \text{osid}(k,T^*)$ and $x^* \in \text{mat}(R,m^*)$, and let us put

$$m = \text{cab}(m^*,k) \quad \text{and} \quad T = \text{cat}(m^*,k)$$

and

$$A = \text{cad}(m^*,k,T^*,A^*) \quad \text{and} \quad x = \text{cam}(x,m^*,k,T^*).$$

Then we have

$$\text{cag}(m^*,k,T^*,A^*)\text{mom}(x^*,\text{sat}(T^*,A^*)) = \text{bag}(m,k,A)\text{mom}(x,\text{sat}(T,A))$$

[where we refer (14.1.9) for seeing that cag(m*,k,T*,A*) and bag(m,k,A) and sat(T,A) make sense].

Proof: In view of the definitions of cag and mom, our assertion follows from (14.2.2) by producting both sides of (14.2.2) as e varies over [1,2] and then multiplying both sides by bag(m,k,A).

LEMMA (14.4) Let there be given any ring R and let there be given any $m^* \in N(2)$ and $k \in [1,2]$ and $T^* \in \mathrm{bit}(2,m^*,k)$ and $x^* \in \mathrm{mat}(R,m^*)$, and let us put

$$T = \mathrm{cat}(m^*,k) \qquad \text{and} \qquad x = \mathrm{cam}(m^*,k,T^*),$$

and let there be given any $A \in \mathrm{nex}(T) \cap \mathrm{osid}(k,T)$ such that

$$A \notin \{\mathrm{cad}(m^*,k,T^*,A^*) \;:\; A^* \in \mathrm{disid}(k,T^*) \cap \mathrm{osid}(k,T^*)\}.$$

Then $\mathrm{mom}(x,\mathrm{sat}(T,A)) = 0$.

Proof: Upon letting $r = \mathrm{rat}(m^*,k,T^*)$, by the definition of cam we see that

$$(1)\quad \begin{cases} \text{in the case } k = 2 \text{ we have} \\ x_{ij} = x^*_{i,r(j)} \text{ for all } i \in [1,m^*(k')] \text{ and } j \in [1,2m^*(k')], \end{cases}$$

whereas

$$(2)\quad \begin{cases} \text{in the case } k = 1 \text{ we have} \\ x_{ij} = x^*_{r(i),j} \text{ for all } j \in [1,m^*(k')] \text{ and } i \in [1,2m^*(k')]. \end{cases}$$

We obtain $A^* \in \mathrm{osid}(k,T^*)$ by putting

$$A^*[e](k') = \emptyset \text{ and } A^*[e](k) = r(A[e](k)) \qquad \text{for all } e \in [1,2]$$

and now clearly we have $\mathrm{cad}(m^*,k,T^*,A^*) = A$ and hence by assumption we must have $A^* \notin \mathrm{disid}(k,T^*)$, and therefore for some $e \in [1,2]$ there exists

$$q \in [\mathrm{sim}(T^*)[e](k) \setminus A^*[e](k)] \cap A^*[e'](k).$$

Henceforth let us fix e and q as above, and let us put

$$s = \mathrm{sat}(T,A)[e].$$

Now there exist unique elements

$p[e] \in \text{sim}(T)[e](k) \setminus A[e](k) \quad \text{and} \quad p[e'] \in A[e'](k)$

such that

(3) $r(p[e]) = q = r(p[e'])$.

Clearly, $p[e] \neq p[e']$ are elements of $\text{sim}(\text{sat}(T,A))[e](k)$ and therefore there exist unique elements $w[e] \neq w[e']$ in $[1,m^*(k')]$ such that

(4) $s(k,w[e]) = p[e] \quad \text{and} \quad s(k,w[e']) = p[e']$.

Now, in the case $k = 2$, for all $i \in [1,m^*(k')]$ we have

$$\begin{aligned}
&\text{sub}(x,s)_{i,w[e]} \\
&= x_{i,s(k,w[e])} && \text{(by definition of sub)} \\
&= x^*_{i,r(s(k,w[e]))} && \text{[by (1)]} \\
&= x^*_{i,q} && \text{[by (3) and (4)]} \\
&= x^*_{i,r(s(k,w[e']))} && \text{[by (3) and (4)]} \\
&= x_{i,s(k,w[e'])} && \text{[by (1)]} \\
&= \text{sub}(x,s)_{i,w[e']} && \text{(by definition of sub)}
\end{aligned}$$

and thus $w[e] \neq w[e']$ and the $w[e]$th column of $\text{sub}(x,s)$ is equal to its $w[e']$th column and hence

$$\det \text{sub}(x,s) = 0$$

and similarly, in the case $k = 1$, for all $j \in [1,m^*(k')]$ we have

$$\begin{aligned}
&\text{sub}(x,s)_{w[e],j} \\
&= x_{s(k,w[e]),j} && \text{(by definition of sub)} \\
&= x^*_{r(s(k,w[e])),j} && \text{[by (2)]} \\
&= x^*_{q,j} && \text{[by (3) and (4)]} \\
&= x^*_{r(s(k,w[e'])),j} && \text{[by (3) and (4)]} \\
&= x_{s(k,w[e']),j} && \text{[by (2)]} \\
&= \text{sub}(x,s)_{w[e'],j} && \text{(by definition of sub)}
\end{aligned}$$

and thus $w[e] \neq w[e']$ and the $w[e]$th row of $sub(x,s)$ is equal to its $w[e']$th row and hence

$$\det sub(x,s) = 0.$$

Thus we always have $\det sub(x,s) = 0$ and therefore by the definition of mom we get

$$mom(x,sat(T,A)) = 0$$

and this completes the proof.

LEMMA (14.5) Let there be given any ring R and let there be given any $m^* \in N(2)$ and $k \in [1,2]$ and $T^* \in bit(2,m^*,k)$ and $B^* \in osid(k,T^*)$ and $x^* \in mat(R,m^*)$, and let

$$m = cab(m^*,k) \quad \text{and} \quad T = cat(m^*,k)$$

and

$$B = cad(m^*,k,T^*,B^*) \quad \text{and} \quad x = cam(x^*,m^*,k,T^*)$$

[and let us note that obviously

(1) $nex(T^*,B^*) \subset nex(T^*) \cap osid(k,T^*)$

and hence $cad(m^*,k,T^*,A^*)$ is defined for every $A^* \in nex(T^*,B^*)$]. Then:

(14.5.1) $A^* \mapsto cad(m^*,k,T^*,A^*)$ gives an injection

$nex(T^*,B^*) \to nex(T,B)$.

(14.5.2) For every $A^* \in nex(T^*,B^*)$, upon letting

$A = cad(m^*,k,T^*,A^*)$,

we have

$$cag(m^*,k,T^*,A^*)mom(x^*,sat(T^*,A^*)) = bag(m,k,A)mom(x,sat(T,A))$$

[where we refer to (1) and (14.1.9) for seeing that $cag(m^*,k,T^*,A^*)$ and $bag(m,k,A)$ and $sat(T,A)$ make sense].

(14.5.3) For every

$A \in nex(T,B) \setminus \{cad(m^*,k,T^*,A^*) : A^* \in nex(T^*,B^*)\}$

we have $mom(x,sat(T,A)) = 0$.

Proof: Now obviously,

$$(2)\quad \begin{cases} \text{if } A^* \in \text{sid}(T^*) \text{ is such that } A^*[e](\hat{k}) \subset B^*[e](\hat{k}) \text{ for all} \\ e \in [1,2] \text{ and } \hat{k} \in [1,2], \text{ then } A^* \in \text{osid}(k,T^*) \text{ and} \\ \text{upon letting } A = \text{cad}(m^*,k,T^*,A^*) \text{ we have} \\ A[e](\hat{k}) \subset B[e](\hat{k}) \text{ for all } e \in [1,2] \text{ and } \hat{k} \in [1,2]. \end{cases}$$

By (1) and (2) and (14.1.7) we get (14.5.1). By (1) and (14.3) we get (14.5.2). Given any A as in (14.5.3), by (14.1.5) and (14.1.7) we see that

$$A \in \text{nex}(T) \cap \text{osid}(k,T)$$

and

$$A \notin \{\text{cad}(m^*,k,T^*,A^*) : A^* \in \text{disid}(k,T^*) \cap \text{osid}(k,T^*)\}$$

and therefore (14.5.3) follows from (14.4).

LEMMA (14.6) Given any ring R and given any $m^* \in N(2)$ and $k \in [1,2]$ and $T^* \in \text{bit}(2,m^*,k)$ and $B^* \in \text{osid}(k,T^*)$ and $x^* \in \text{mat}(R,m^*)$, let

$$m = \text{cab}(m^*,k) \quad \text{and} \quad B = \text{cad}(m^*,k,T^*,B^*) \quad \text{and}$$
$$x = \text{cam}(x^*,m^*,k,T^*)$$

and

$$b^* = \bigcap_{e\in[1,2]} [\text{sim}(T^*)[e](k) \setminus B^*[e](k)]$$

we assume that

$$\text{card}(B^*[1](k)) + \text{card}(B^*[2](k)) + \text{card}(b^*) > m^*(k')$$

and let

$$v[1]: b^* \to [1,m(k')] \quad \text{and} \quad v[2]: b^* \to [1 + m(k'), 2m(k')]$$

be the unique injective maps such that for all $q \in b^*$ we have

$$T^*[1](k,v[1](q)) = q = T^*[2](k, v[2](q) - m(k')).$$

Then

$$\text{card}(B[1](k)) + \text{card}(B[2](k)) + \text{card}(b^*) > m(k')$$

and

$$\begin{cases}\text{in the case } k = 2 \text{ we have}\\ x_{i,v[1](j)} = x_{i,v[2](j)} \text{ for all } i \in [1,m(k')] \text{ and } j \in b^*,\end{cases}$$

whereas

$$\begin{cases}\text{in the case } k = 1 \text{ we have}\\ x_{v[1](i),j} = x_{v[2](i),j} \text{ for all } j \in [1,m(k')] \text{ and } i \in b^*.\end{cases}$$

Proof: By (14.1.4) we have

$$\text{card}(B^*[e](k)) = \text{card}(B[e](k)) \quad \text{for all } e \in [1,2]$$

and hence we see that

$$\text{card}(B[1](k)) + \text{card}(B[2](k)) + \text{card}(b^*) > m(k').$$

Upon letting $r = \text{rat}(m^*,k,T^*)$, by the definition of cam we see that

$$(1)\quad \begin{cases}\text{in the case } k = 2 \text{ we have}\\ x_{ij} = x^*_{i,r(j)} \text{ for all } i \in [1,m(k')] \text{ and } j \in [1,2m(k')],\end{cases}$$

whereas

$$(2)\quad \begin{cases}\text{in the case } k = 1 \text{ we have}\\ x_{ij} = x^*_{r(i),j} \text{ for all } j \in [1,m(k')] \text{ and } i \in [1,2m(k')].\end{cases}$$

and by the definitions of $v[1]$ and $v[2]$ we have

$$(3)\quad r(v[1](q)) = q = r(v[2](q)) \quad \text{for all } q \in b^*.$$

Now, in the case $k = 2$, for all $i \in [1,m(k')]$ and $j \in b^*$ we have

$$\begin{aligned} x_{i,v[1](j)} &= x^*_{i,r(v[1](j))} && \text{[by (1)]}\\ &= x^*_{i,r(v[2](j))} && \text{[by (3)]}\\ &= x_{i,v[2](j)} && \text{[by (1)]}\end{aligned}$$

and similarly, in the case $k = 1$, for all $j \in [1,m(k')]$ and $i \in b^*$ we have

$$\begin{aligned} x_{v[1](i),j} &= x^*_{r(v[1](i)),j} && \text{[by (2)]}\\ &= x^*_{r(v[2](i)),j} && \text{[by (3)]}\\ &= x_{v[2](i),j} && \text{[by (2)]}\end{aligned}$$

PROPOSITION (14.7) Let there be given any ring R and let there be given any $m^* \in N(2)$ and $k \in [1,2]$ and $T^* \in \mathrm{bit}(2,m^*,k)$ and $B^* \in \mathrm{osid}(k,T^*)$ and $x^* \in \mathrm{mat}(R,m^*)$ [and let us refer to (14.1.10)]. Then:

(14.7.1) We have

$$\mathrm{car}(x^*,m^*,k,T^*,B^*) = \sum_{A^*\in \mathrm{nex}(T^*,B^*)} \mathrm{cag}(m^*,k,T^*,A^*)\mathrm{mom}(x^*,\mathrm{sat}(T^*,A^*)).$$

(14.7.2) Upon letting E be the empty bitabloid of depth 2, i.e., upon letting E be the bitabloid of depth 2 obtained by putting

$$E[e](k') = \emptyset = E[e](k) \quad \text{for all } e \in [1,2],$$

first we have

$$\mathrm{nex}(T^*,B^*) = \{E\} \coprod \mathrm{pex}(T^*,B^*),$$

and second, we have

$$\mathrm{sat}(T^*,E) = T^*,$$

and third, we have

$$\mathrm{cad}(m^*,k,T^*,E) = E,$$

and fourth, for every $e \in [1,2]$ we have that $\mathrm{cap}(m^*,k,T^*,E)$ is the identity permutation of $[1,m^*(k')]$, i.e.,

$$\mathrm{cap}(m^*,k,T^*,E)[e](i) = i \quad \text{for all } i \in [1,m^*(k')]$$

and fifth, we have

$$\mathrm{cag}(m^*,k,T^*,E) = -1.$$

(14.7.3) Finally, we have

$$\mathrm{car}(x^*,m^*,k,T^*,B^*) + \mathrm{mom}(x^*,T^*) = \sum_{A^*\in \mathrm{pex}(T^*,B^*)} \mathrm{cag}(m^*,k,T^*,A^*)\mathrm{mom}(x^*,\mathrm{sat}(T^*,A^*)).$$

Proof: Upon letting

$$m = \mathrm{cab}(m^*,k) \quad \text{and} \quad T = \mathrm{cat}(m^*,k)$$

and

$$B = \mathrm{cad}(m^*,k,T^*,B^*) \quad \text{and} \quad x = \mathrm{cam}(x^*,m^*,k,T^*),$$

by the definitions of bar and car we see that

$$[\text{LHS of } (14.7.1)] = [\text{LHS of } (13.4.1)]$$

and by (14.1.7), (14.1.8), (14.4), and (14.5) we see that

$$[\text{RHS of } (14.7.1)] = [\text{RHS of } (13.4.1)]$$

and therefore by (13.4.1) we get (14.7.1). The first four parts of (14.7.2) are obvious and its fifth part follows from its fourth part together with (13.4.2). (14.7.3) follows from (14.7.1) and (14.7.2).

PROPOSITION (14.8) Let there be given any ring R and let there be given any $m^* \in N(2)$ and $k \in [1,2]$ and $T^* \in \mathrm{bit}(2,m^*,k)$ and $B^* \in \mathrm{osid}(k,T^*)$ and $x^* \in \mathrm{mat}(R,m^*)$, and let us assume that

$$\mathrm{card}(B^*[1](k)) + \mathrm{card}(B^*[2](k)) + \mathrm{card}(b^*) > m^*(k')$$

where we have put

$$b^* = \bigcap_{e\in[1,2]} [\mathrm{sim}(T^*)[e](k) \setminus B^*[e](k)]$$

[and let us refer to (14.1.10)]. Then:

(14.8.1) We have $\mathrm{car}(x^*,m^*,k,T^*,B^*) = 0$.

(14.8.2) Also, we have

$$\sum_{A^*\in \mathrm{nex}(T^*,B^*)} \mathrm{cag}(m^*,k,T^*,A^*)\mathrm{mom}(x^*,\mathrm{sat}(T^*,A^*)) = 0.$$

(14.8.3) Finally, we have

$$\mathrm{mom}(x^*,T^*) = \sum_{A^*\in \mathrm{pex}(T^*,B^*)} \mathrm{cag}(m^*,k,T^*,A^*)\mathrm{mom}(x^*,\mathrm{sat}(T^*,A^*)).$$

Proof: In view of the definitions of bar and car, (14.8.1) follows from (13.5.1) and (14.6). (14.8.2) follows from (14.7.1) and (14.8.1). (14.8.3) follows from (14.7.3) and (14.8.1).

15. UNIFICATION AND THE MIXED SIZE CASE

(15.1) Notation In this subsection we introduce some more terminology to be used in the rest of this chapter.

Given any $m \in N(2)$ and $k \in [1,2]$, first we introduce the *binumbric unification* of (m,k), which we denote by $ubi(m,k)$ and which we define by saying that $ubi(m,k)$ is the unique member of $N(2)$ such that

$$ubi(m,k)(k') = m(k')$$

and

$$ubi(m,k)(k) = m(k) + m(k'),$$

and second, for every ring R and every $x \in mat(R,m)$ we introduce the *matrical unification* of (x,m,k), which we denote by $umi(x,m,k)$ and which we define by saying that $umi(x,m,k)$ is the unique member of $mat(R,ubi(m,k))$ such that in the case $k = 2$ for all $i \in [1,m(k')]$ we have

$$umi(x,m,k)_{ij} = \begin{cases} x_{ij} & \text{if } j \in [1,m(k)] \\ 1 & \text{if } j \in [m(k)+1,\ m(k)+m(k')] \text{ and } j = m(k)+m(k')+1-i \\ 0 & \text{if } j \in [m(k)+1,\ m(k)+m(k')] \text{ and } j \neq m(k)+m(k')+1-i, \end{cases}$$

whereas in the case $k = 1$, for all $j \in [1,m(k')]$ we have

$$umi(x,m,k)_{ij} = \begin{cases} x_{ij} & \text{if } i \in [1,m(k)] \\ 1 & \text{if } i \in [m(k)+1,\ m(k)+m(k')] \text{ and } i = m(k)+m(k')+1-j \\ 0 & \text{if } i \in [m(k)+1,\ m(k)+m(k')] \text{ and } i \neq m(k)+m(k')+1-j \end{cases}$$

and third, for every $a \in bic(m)$ we introduce the *vectorial complementation* of $[m,k,a]$, which we denote by $com[m,k,a]$ and which we define by saying that

$$com[m,k,a]: \quad [1 + len(a),\ m(k')] \to [1,m(k')]$$

is the unique strictly decreasing (or strictly order reversing) map such that

$$im(com[m,k,a]) = [1,m(k')] \setminus sim[a](k'),$$

and fourth, for every $d \in N$ and $T \in bit(d,m)$ and $e \in [1,d]$ we introduce the *tabular complementation* of (m,k,T) *at* e which we denote by com(m,k,T)[e] and which we define by putting

$$\mathrm{com}(m,k,T)[e] = \mathrm{com}[m,k,T[e]],$$

and fifth, for every a ∈ bic(m) we introduce the *vectorial unification* of [m,k,a], which we denote by uti[m,k,a] and which we define by saying that uti[m,k,a] is the unique member of bic(ubi(m,k),k) such that

$$\mathrm{uti}[m,k,a](k,i) = \begin{cases} a(k,i) & \text{if } i \in [1,\mathrm{len}(a)] \\ m(k) + m(k') + 1 - \mathrm{com}[m,k,a](i) & \text{if } i \in [1+\mathrm{len}(a),\ m(k')], \end{cases}$$

and sixth, for every d ∈ N and T ∈ bit(d,m) we introduce the *tabular unification* of (m,k,T), which we denote by uti(m,k,T) and which we define by saying that uti(m,k,T) is the unique member of bit(d, ubi(m,k),k) such that for all e ∈ [1,d] we have

$$\mathrm{uti}(m,k,T)[e] = \mathrm{uti}[m,k,T[e]],$$

and seventh, for every a ∈ bic(m) we introduce the *permutationic vectorial unification* of [m,k,a], which we denote by upi[m,k,a] and which we define by saying that upi[m,k,a] is the unique member of W(m(k')) such that, on the one hand, for all i ∈ [1,len(a)] we have

$$\mathrm{upi}[m,k,a](a(k',i)) = i,$$

whereas on the other hand, for all i ∈ [1 + len(a), m(k')] we have

$$\mathrm{upi}[m,k,a](\mathrm{com}[m,k,a](m(k') + 1 + \mathrm{len}(a) - i)) = i,$$

and eighth, for every a ∈ bic(m) we introduce the *signatoric vectorial unification* of [m,k,a], which we denote by ugi[m,k,a] and which we define by putting

$$\mathrm{ugi}[m,k,a] = (-1)^{[m(k')-\mathrm{len}(a)][m(k')-\mathrm{len}(a)-1]/2}\mathrm{sgn}(\mathrm{upi}[m,k,a])$$

and we note that then

$$\mathrm{ugi}[m,k,a] \in \{1,-1\},$$

and ninth, for every d ∈ N and T ∈ bit(d,m) we introduce the *signatoric tabular unification* of (m,k,T), which we denote by ugi(m,k,T) and which we define by putting

$$\text{ugi}(m,k,T) = \prod_{e\in[1,d]} \text{ugi}[m,k,T[e]]$$

and we note that then

$$\text{ugi}(m,k,T) \in \{1,-1\},$$

and tenth, for every $T \in \text{bit}(2,m)$ and $A \in \text{sid}(T)$ and $e \in [1,2]$ we introduce the *flipping unification* of (m,k,T,A) *at* e, which we denote by ufi(m,k,T,A)[e] and which we define by putting

$$\text{ufi}(m,k,T,A)[e] = A[e'](k') \cap \text{im}(\text{com}(m,k,T)[e])$$

and we note that then

$$\text{ufi}(m,k,T,A)[e] \subset [1,m(k')],$$

and eleventh, for every $T \in \text{bit}(2,m)$ and $A \in \text{sid}(T)$ we introduce the *tabloidal unification* of (m,k,T,A), which we denote by udi(m,k,T,A) and which we define by saying that udi(m,k,T,A) is the unique member of osid(k,uti(m,k,T)) such that for all $e \in [1,2]$ we have

$$\begin{aligned}&\text{udi}(m,k,T,A)[e](k)\\ &= A[e](k) \cup \{m(k) + m(k') + 1 - i : i \in \text{ufi}(m,k,T,A)[e]\},\end{aligned}$$

and twelfth, for every $T \in \text{bit}(2,m)$ and $B \in \text{sid}(T)$ and $x \in \text{mat}(R,m)$, where R is a ring, we introduce the *satellitic residual matrix* of (x,m,k,T,B), which we denote by sarem(x,m,k,T,B) and which we define by saying that sarem(x,m,k,T,B) is the unique member of mat(R,2m(k')) obtained by putting

$$\begin{aligned}&\text{sarem}(x,m,k,T,B)\\ &= \text{carem}(\text{umi}(x,m,k),\text{ubi}(m,k),k,\text{uti}(m,k,T),\text{udi}(m,k,T,B)),\end{aligned}$$

and thirteenth, for every $T \in \text{bit}(2,m)$ and $B \in \text{sid}(T)$ and $x \in \text{mat}(R,m)$, where R is a ring, we introduce the *satellitic residue* of (x,m,k,T,B), which we denote by sar(x,m,k,T,B) and which we define by putting

$$\text{sar}(x,m,k,T,B) = \frac{-\det \text{sarem}(x,m,k,T,B)}{\text{ugi}(m,k,T)},$$

and fourteenth, for every $T \in \text{bit}(2,m)$ and $B \in \text{sid}(T)$ we introduce the *numerically flipped unification* of (m,k,T,B), which we denote by uni(m,k,T,B) and which we define by putting

$$\begin{aligned}&\text{uni}(m,k,T,B)\\&= \text{card}(B[1](k)) + \text{card}(B[2](k))\\&\quad + \text{card}\left(\bigcap_{e\in[1,2]} [\text{sim}(T)[e](k) \setminus B[e](k)]\right)\\&\quad + \text{card}(\text{ufi}(m,k,T,B)[1]) + \text{card}(\text{ufi}(m,k,T,B)[2])\\&\quad + \text{card}\left(\bigcap_{e\in[1,2]} [\text{im}(\text{com}(m,k,T)[e]) \setminus \text{ufi}(m,k,T,B)[e]]\right),\end{aligned}$$

and fifteenth, for every $T \in \text{bit}(2,m)$ we define

$$\text{amsid}(m,k,T) = \{B \in \text{hasid}(k,T) : \text{uni}(m,k,T,B) > m(\kappa')\}$$

and we remark that a member of amsid(m,k,T) may be called an *ample subtabloid* of (m,k,T), and sixteenth, for every $T \in \text{bit}(2,m)$ we define

$$\text{amosid}(m,k,T) = \{B \in \text{osid}(k,T) : \text{uni}(m,k,T,B) > m(k')\}$$

and we remark that a member of amosid(m,k,T) may be called an *ample one-sided subtabloid* of (m,k,T), and seventeenth, for every $T \in \text{bit}(2,m)$ and $A \in \text{nex}(T) \cap \text{hasid}(k,T)$ we introduce the *satellitic signature* of (m,k,T,A), which we denote by sag(m,k,T,A), and which we define by putting

$$\begin{aligned}&\text{sag}(m,k,T,A)\\&= \frac{\text{ugi}(m,k,\text{sat}(T,A))\,\text{cag}(\text{ubi}(m,k),k,\text{uti}(m,k,T),\text{udi}(m,k,T,A))}{\text{ugi}(m,k,T)}\end{aligned}$$

and we note that then

$$\text{sag}(m,k,T,A) \in \{1,-1\},$$

and eighteenth, concerning the definition of amosid above we observe that

$$\left[\begin{aligned}&\text{for every } T \in \text{bit}(2,m) \text{ and } B \in \text{osid}(k,t) \text{ we have}\\&\text{uni}(m,k,T,B) = \text{card}(B[1](k) + \text{card}(B[2](k)\\&\qquad\qquad + \text{card}\left(\bigcap_{e\in[1,2]} [\text{sim}(T)[e](k) \setminus B[e](k)]\right),\end{aligned}\right. \tag{15.1.1}$$

and ninteenth, we note that the definition of sag above makes sense because of the observation which says that

$$\left[\begin{array}{l}\text{given any } T \in \text{bit}(2,m) \text{ and } A \in \text{nex}(T) \cap \text{hasid}(k,T), \\ \text{upon letting } T^* = \text{uti}(m,k,T) \text{ and } A^* = \text{udi}(m,k,T,A), \\ \text{we have } A^* \in \text{nex}(T^*) \cap \text{osid}(k,T^*)\end{array}\right. \qquad (15.1.2)$$

and which will be proved in Proposition (15.3.1), and twentieth, we note that some other observations concerning uti and udi will be proved in Propositions (15.3.2) to (15.3.4), and twenty-first, we note that Propositions (15.3.1) to (15.3.4) will actually be proved in *the long technical* Lemma (15.2), *which the reader may like to skip in a first reading*, and twenty-second, as an observation which may be used *after we have proved* (15.3.1), we note that

$$\left[\begin{array}{l}\text{given any } T \in \text{bit}(2,m) \text{ and } B \in \text{hasid}(k,T), \\ \text{for every } A \in \text{nex}(T,B) \text{ we obviously have} \\ A \in \text{nex}(T) \cap \text{osid}(k,T) \\ \text{and hence by (15.3.1) we see that} \\ \text{sag}(m,k,T,A) \text{ makes sense,}\end{array}\right. \qquad (15.1.3)$$

and twenty-third, to approximately illustrate some of the unificational definitions, let us take

$$k = 2$$

and let there be given any ring R and let there be given any

$$x \in \text{mat}(R,m) \text{ and } T \in \text{bit}(2,m) \text{ and } A \in \text{sid}(T)$$

and let us put

$$\begin{aligned} x^* &= \text{umi}(x,m,k) \\ &= x \text{ augmented by the } m(k') \text{ by } m(k') \text{ anti-identity matrix} \end{aligned}$$

and

$$T^* = \text{uti}(m,k,T) \quad \text{and} \quad A^* = \text{udi}(m,k,T,A)$$

so that

$$\begin{aligned} A^*[1](k) &= \text{udi}(m,k,T,A)[1](k) \\ &= A[1](k) \cup Y \end{aligned}$$

where

$$Y = \{m(k') + m(k) + 1 - i : i \in y\}$$

with

$$y = \text{ufi}(m,k,T,A)[1]$$
$$= A[2](k') \setminus \text{sim}(T)[1](k')$$

and let us depict the situation by the picture

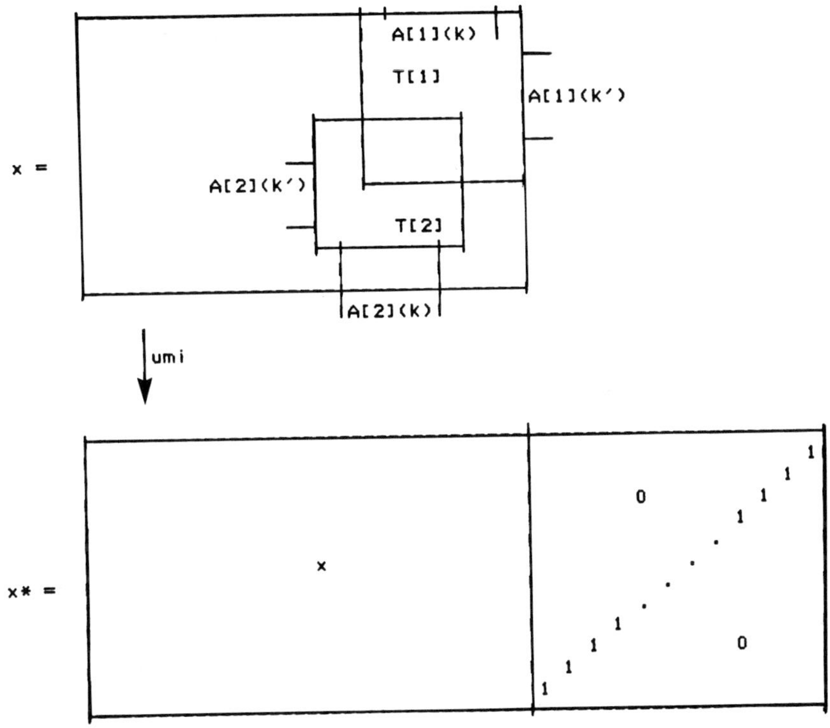

and the pictures

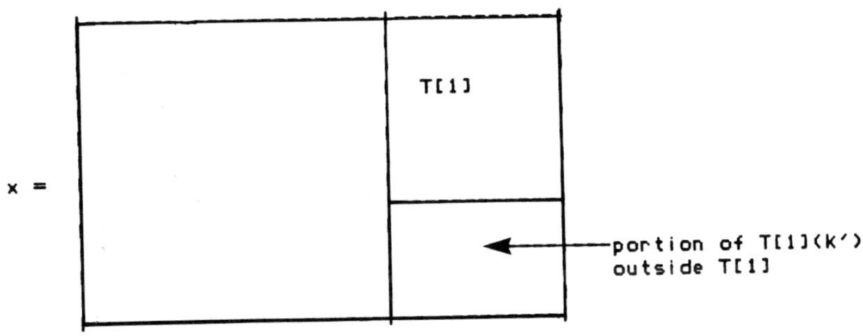

and

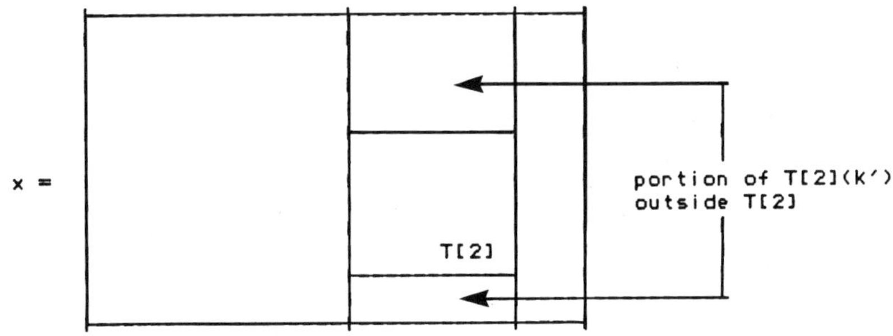

and

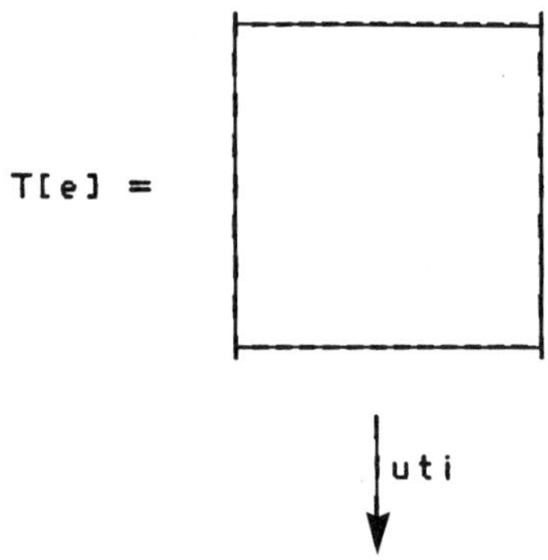

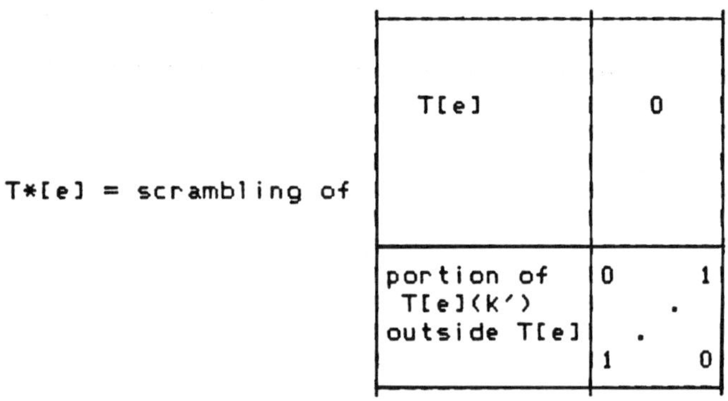

and the picture

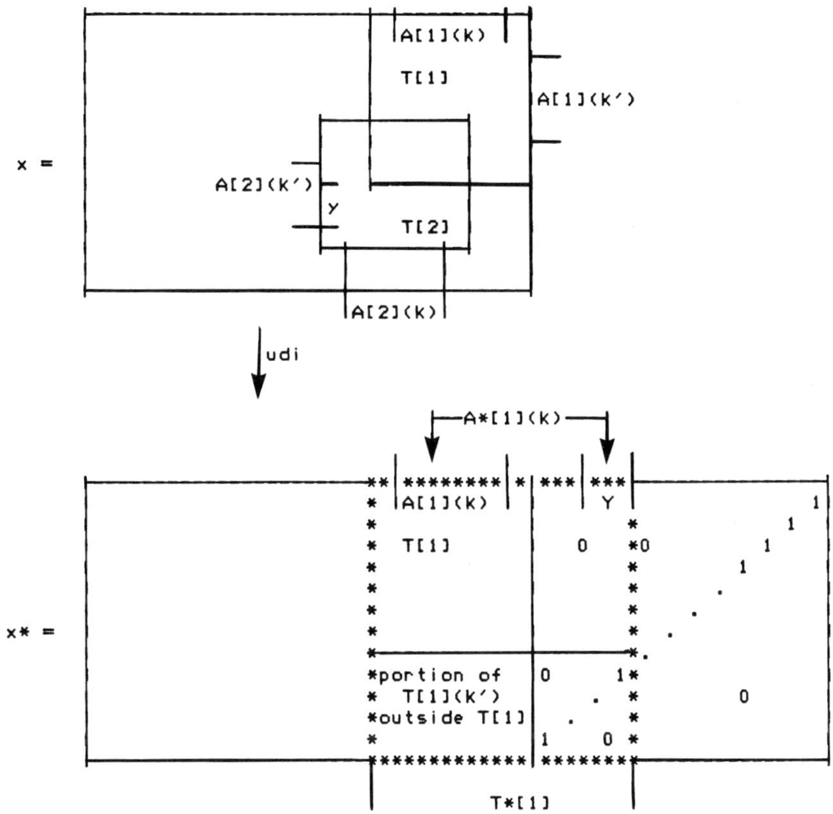

LEMMA (15.2) Given any m ∈ N(2) and k ∈ [1,2], first let

$$p = m(k') \quad \text{and} \quad q = m(k)$$

and second, let

$$D = \{A \in \text{bid}(2) : \text{for all } e \in [1,2] \text{ we have } A[e](k') \subset [1,p] \text{ and } A[e](k) \subset [1,q]\}$$

and

$$D^* = \{A^* \in \text{bid}(2) : \text{for all } e \in [1,2] \text{ we have } A^*[e](k') = \emptyset \text{ and } A^*[e](k) \subset [1, q + p]\},$$

and third, let

$$d = \{t \in D : \text{for all } e \in [1,2] \text{ we have } \operatorname{card}(t[e](k')) = \operatorname{card}(t[e](k))\},$$

and fourth, for every $t \in D$ let

$$D(t) = \{A \in D : \text{for all } e \in [1,2] \text{ we have } A[e](k') \subset t[e](k') \text{ and } A[e](k) \subset t[e](k)\}$$

and for every $t^* \in D^*$ let

$$D^*(t^*) = \{A^* \in D^* : \text{for all } e \in [1,2] \text{ we have } A^*[e](k) \subset t^*[e](k)\},$$

and fifth, for every $t \in D$ and $\hat{k} \in [1,2]$ let

$$D(\hat{k},t) = \{A \in D(t) : \text{for all } e \in [1,2] \text{ we have } [t[e](\hat{k}) \setminus A[e](\hat{k})] \cap A[e'](\hat{k}) = \emptyset\}$$

and for every $t^* \in D^*$ let

$$D^{**}(t^*) = \{A^* \in D^*(t^*) : \text{for all } e \in [1,2] \text{ we have } [t^*[e](k) \setminus A^*[e](k)] \cap A^*[e'](k) = \emptyset\},$$

and sixth, for every $t \in D$ let

$$X(t) = \{A \in D(k',t) \cap D(k,t) : \text{for all } e \in [1,2] \text{ we have } \operatorname{card}(t[e](k')) - \operatorname{card}(A[e](k')) + \operatorname{card}(A[e'](k')) = \operatorname{card}(t[e](k)) - \operatorname{card}(A[e](k)) + \operatorname{card}(A[e'](k))\}$$

and for every $t^* \in D^*$ let

$$X^*(t^*) = \{A^* \in D^{**}(t^*) : \operatorname{card}(A^*[1](k)) = \operatorname{card}(A^*[2](k))\},$$

and seventh, for every $t \in D$ and $B \in D$ let

$$X(t,B) = X(t) \cap D(B)$$

and for every $t^* \in D^*$ and $B^* \in D^*$ let

$$X^*(t^*,B^*) = X^*(t^*) \cap D^*(B^*),$$

and eighth,

$$\left[\begin{array}{l}\text{for every } Y \in D^* \text{ and } e \in [1,2] \\ \text{let } Y_e(k') = Y[e](k) \cap [q+1, q+p] \\ \text{and } Y_e(k) = Y[e](k) \cap [1,q],\end{array}\right.$$

and ninth, for every $t^* \in D^*$ and $\hat{k} \in [1,2]$ let

$$D^*(\hat{k},t^*) = \{A^* \in D^*(t^*) : \text{for all } e \in [1,2] \text{ we have } [t^*_e(\hat{k}) \setminus A^*_e(\hat{k})] \cap A^*_{e'}(\hat{k}) = \emptyset\},$$

and tenth, let

$$H = \{A \in D : A[1](k') \cap A[2](k') = \emptyset\}$$

and

$$H^* = \{A^* \in D^* : A^*_1(k') \cap A^*_2(k') = \emptyset\},$$

and eleventh,

$$\begin{bmatrix} \text{for every } t \in D \text{ and } A \in D \\ \text{let } S(t,A) \text{ be the unique member of } D \text{ such that} \\ \text{for all } e \in [1,2] \text{ and } \hat{k} \in [1,2] \text{ we have} \\ S(t,A)[e](\hat{k}) = [t[e](\hat{k}) \setminus A[e](\hat{k})] \cup A[e'](\hat{k}), \end{bmatrix}$$

whereas

$$\begin{bmatrix} \text{for every } t^* \in D^* \text{ and } A^* \in D^* \\ \text{let } S^*(t^*,A^*) \text{ be the unique member of } D^* \\ \text{such that for all } e \in [1,2] \text{ we have} \\ S^*(t^*,A^*)[e](k) = [t^*[e](k) \setminus A^*[e](k)] \cup A^*[e'](k), \end{bmatrix}$$

and twelfth, let f and f* be the maps

$$[1,p] \underset{f^*}{\overset{f}{\rightleftarrows}} [q+1, q+p]$$

obtained by putting

$$f(i) = q + p + 1 - i \qquad \text{for all } i \in [1,P]$$

and

$$f^*(i) = q + p + 1 - i \qquad \text{for all } i \in [q+1, q+p],$$

and thirteenth, let g and g* be the maps

$$P([1,p]) \underset{g^*}{\overset{g}{\rightleftarrows}} P([q+1, q+p])$$

obtained by putting

$$g(a) = \{f(i) : i \in a\} \quad \text{for all } a \in P([1,p])$$

and

$$g^*(a^*) = \{f^*(i) : i \in a^*\} \quad \text{for all } a^* \in P([q+1, q+p]),$$

and fourteenth, let v and v* be the maps

$$P([1,p]) \xrightarrow{v} P([1,p])$$

and

$$P([q+1, q+p]) \xrightarrow{v^*} P([q+1, q+p])$$

obtained by putting

$$v(a) = [1,p] \setminus a \quad \text{for all } a \in P([1,p])$$

and

$$v^*(a^*) = [q+1, q+p] \setminus a^* \quad \text{for all } a^* \in P([q+1, q+p]),$$

and fifteenth,

$$\left[\begin{array}{l}\text{for every } t \in D \text{ and } A \in D \text{ let } U(t,A) \text{ be the} \\ \text{unique member of } D^* \text{ such that for all } e \in [1,2] \text{ we have} \\ U(t,A)_e(k') = g(v(t[e](k')) \cap A[e'](k')) \\ \text{and } U(t,A)_e(k) = A[e](k)\end{array}\right.$$

and

$$\left[\begin{array}{l}\text{for every } t \in D \text{ and } A^* \in D^* \text{ let } U^*(t,A^*) \text{ be the} \\ \text{unique member of } D \text{ such that for all } e \in [1,2] \text{ we have} \\ U^*(t,A^*)[e](k') = t[e](k') \cap g^*(A^*_{e'}(k')) \\ \text{and } U^*(t,A^*)[e](k) = A^*_e(k),\end{array}\right.$$

and sixteenth, let G and G* be the unique maps

$$D \underset{G^*}{\overset{G}{\rightleftarrows}} D^*$$

such that

$$\left[\begin{array}{l}\text{for all } A \in D \text{ and } e \in [1,2] \text{ we have} \\ G(A)_e(k') = g(A[e](k')) \text{ and } G(A)_e(k) = A[e](k)\end{array}\right.$$

and

$$\left[\begin{array}{l}\text{for all } A^* \in D^* \text{ and } e \in [1,2] \text{ we have}\\ G^*(A^*)[e](k') = g^*(A^*_e(k')) \text{ and } G^*(A^*)[e](k) = A^*_e(k),\end{array}\right.$$

and seventeenth, let V and I be the unique maps

$$D \xrightarrow{V} D \qquad \text{and} \qquad D \xrightarrow{I} D$$

such that

$$\left[\begin{array}{l}\text{for all } A \in D \text{ and } e \in [1,2] \text{ we have}\\ V(A)[e](k') = v(A[e](k')) \text{ and } V(A)[e](k) = A[e](k)\end{array}\right.$$

and

$$\left[\begin{array}{l}\text{for all } A \in D \text{ and } e \in [1,2] \text{ we have}\\ I(A)[e](k') = A[e'](k') \text{ and } I(A)[e](k) = A[e](k),\end{array}\right.$$

and eighteenth, for every $t \in D$ and $B \in D$ let

$$\begin{aligned} E(t,B) = {} & \mathrm{card}(B[1](k)) + \mathrm{card}(B[2](k)) \\ & + \mathrm{card}\left\{\bigcap_{e\in[1,2]} [t[e](k) \setminus B[e](k)]\right\} \\ & + \mathrm{card}(v(t[1](k')) \cap B[2](k')) \\ & + \mathrm{card}(v(t[2](k')) \cap B[1](k')) \\ & + \mathrm{card}\left\{\bigcap_{e\in[1,2]} [v(t[e](k')) \setminus (v(t[e](k')) \cap B[e'](k'))]\right\} \end{aligned}$$

and for every $t^* \in D^*$ and $B^* \in D^*$ let

$$\begin{aligned} E^*(t^*,B^*) = {} & \mathrm{card}(B^*[1](k)) + \mathrm{card}(B^*[2](k)) \\ & + \mathrm{card}\left\{\bigcap_{e\in[1,2]} [t^*[e](k) \setminus B^*[e](k)]\right\}. \end{aligned}$$

Now we have the following.

(1) We have

$$v(v(a)) = a \qquad \text{for all } a \in P([1,p]).$$

(1*) We have

$$v^*(v^*(a^*)) = a^* \qquad \text{for all } a^* \in P([q+1, q+p]).$$

(2) We have that v is bijective.

(2*) We have that v^* is bijective.

(3) We have

$$V(V(A)) = A \quad \text{for all } A \in D.$$

(4) We have that V is bijective.

(5) We have

$$I(I(A)) = A \quad \text{for all } A \in D.$$

(6) We have that I is bijective.

(7) We have

$$f^*(f(i)) = i \quad \text{for all } i \in [1,p].$$

(7*) We have

$$f(f^*(i)) = i \quad \text{for all } i \in [q+1, q+p].$$

(8) We have that f is bijective.

(8*) We have that f* is bijective.

(9) We have

$$[1, q+p] = [1,q] \coprod [q+1, q+p].$$

(10) We have

$$g^*(g(a)) = a \quad \text{for all } a \in P([1,p]).$$

(10*) We have

$$g(g^*(a^*)) = a^* \quad \text{for all } a^* \in P([q+1, q+p]).$$

(11) We have that g is bijective.

(11*) We have that g* is bijective.

(12) We have

$$g(v(a)) = v^*(g(a)) \quad \text{for all } a \in P([1,p]).$$

(12*) We have

$$g^*(v^*(a^*)) = v(g^*(a^*)) \quad \text{for all } a^* \in P([q+1, q+p]).$$

(13) Given any $t \in D$, upon letting $t^* = G(V(t))$, for every $e \in [1,2]$ we have

$$t^*_e(k') = g(v(t[e](k'))).$$

(14) Given any $t \in D$, upon letting $t^* = G(V(t))$, for every $e \in [1,2]$ we have

$$g^*(v^*(t^*_e(k'))) = t[e](k').$$

(15) We have

$$G^*(G(A)) = A \quad \text{for all } A \in D.$$

(15*) We have

$$G(G^*(A^*)) = A^* \quad \text{for all } A^* \in D^*.$$

(16) We have that G is bijective.

(16*) We have that G* is bijective.

(17) We have

$$G(S(t,A)) = S^*(G(t),G(A)) \quad \text{for all } t \in D \text{ and } A \in D.$$

(18) We have

$$D^{**}(t^*) = D^*(k',t^*) \cap D^*(k,t^*) \quad \text{for all } t^* \in D^*.$$

(19) Given any $t \in D$, upon letting $t^* = G(V(t))$, we have

$$U(t,A) \in D^*(k,t^*) \quad \text{for all } A \in D(k,t).$$

(19*) Given any $t \in D$, upon letting $t^* = G(V(t))$, we have

$$U^*(t,A^*) \in D(k,t) \quad \text{for all } A^* \in D^*(k,t^*).$$

(20) Given any $t \in d$ and $A \in D(k',t) \cap D(k,t)$, we have that

$$A \in X(t) \Leftrightarrow \left[\begin{array}{l} \text{card}(A[1](k)) + \text{card}(A[2](k')) \\ = \text{card}(A[2](k)) + \text{card}(A[1](k')). \end{array}\right.$$

(21) For every $t \in D$ and $A \in H$ and $e \in [1,2]$, we have

$$S(t,A)[e](k') = [t[e](k') \cup A[e'](k')] \setminus A[e](k').$$

(22) For every $t \in D$ and $A \in H$ we have

$$V(S(t,A)) = S(V(t),I(A)).$$

(23) For every $t \in D$ and $A \in H$ we have

$$G(V(S(t,A))) = S^*(G(V(t)),G(I(A))).$$

(24) For every $t \in D$ and $A \in D(t)$ we have

$$U(t,A) \in H^* \cap D^*(k',G(V(t))).$$

(24*) For every $t \in D$ and $A^* \in D^*(G(V(t)))$ we have

$$U^*(t,A^*) \in H \cap D(k',t).$$

(25) For every $t \in D$ and $A \in H \cap D(k',t)$ and $e \in [1,2]$, we have

$$A[e'](k) \subset v(t[e](k')).$$

(25*) For every $t \in D$ and $A^* \in H^* \cap D^*(k',G(V(t)))$ and $e \in [1,2]$, we have

$$g^*(A^*_{e'}(k')) \subset t[e](k').$$

(26) For every $t \in D$ and $A \in H \cap D(k',t)$ we have

$$U(t,A) = G(I(A)).$$

(26*) For every $t \in D$ and $A^* \in H^* \cap D^*(k',G(V(t)))$ we have

$$U^*(t,A^*) = I(G^*(A^*)).$$

(27) For every $t \in D$ and $A \in H \cap D(k',t)$ we have

$$G(V(S(t,A))) = S^*(G(V(t)),U(t,A)).$$

(28) For every $t \in D$ and $A \in H \cap D(k',t)$ we have

$$U^*(t,U(t,A)) = A.$$

(28*) For every $t \in D$ and $A^* \in H^* \cap D^*(k',t)$ we have

$$U(t,U^*(t,A^*)) = A^*.$$

(29) For every $t \in D$ and $A \in H \cap D(k',t)$ and $e \in [1,2]$, we have

$$\mathrm{card}(U(t,A)[e](k)) = \mathrm{card}(A[e](k)) + \mathrm{card}(A[e'](k')).$$

(29*) For every $t \in D$ and $A^* \in H^* \cap D^*(k',G(V(t))$ and $e \in [1,2]$, we have

$$\mathrm{card}(A^*[e](k)) = \mathrm{card}(U^*(t,A^*)[e](k)) + \mathrm{card}(U^*(t,A^*)[e'](k')).$$

(30) Given any $t \in D$, we have that $A \mapsto U(t,A)$ gives a bijection

$$H \cap D(k',t) \to H^* \cap D^*(k',G(V(t))).$$

(30*) Given any $t \in D$, we have that $A^* \mapsto U^*(t,A^*)$ gives a bijection

$$H^* \cap D^*(k',G(V(t))) \to H \cap D(k',t)$$

whose inverse is the bijection mentioned in (30).

(31) For every $t \in D$ and $B \in D$ we have

$$E^*(G(V(t)),U(t,B)) = E(t,B).$$

(32) For every $t \in D$ and $B \in D$ and $A \in D(B)$, we have

$$U(t,A) \in D^*(U(t,B)).$$

(32*) For every $t \in D$ and $B \in D$ and $A^* \in D^*(U(t,B))$, we have

$$U^*(t,A^*) \in D(B).$$

(33) For every $t \in d$ and $A \in H \cap X(t)$ we have

$$U(t,A) \in H^* \cap X^*(G(V(t))).$$

(33*) For every $t \in d$ and $A^* \in H^* \cap X^*(G(V(t)))$ we have

$$U^*(t,A^*) \in H \cap X(t).$$

(34) For every $t \in d$ and $A \in H \cap X(t)$ we have

$$G(V(S(t,A))) = S^*(G(V(t)),U(t,A)).$$

(35) For every $t \in d$ and $A \in H \cap X(t)$ we have

$$U^*(t,U(t,A)) = A.$$

(35*) For every $t \in d$ and $A^* \in H^* \cap X^*(G(V(t)))$ we have

$$U(t,U^*(t,A^*)) = A^*.$$

(36) Given any $t \in d$, we have that $A \mapsto U(t,A)$ gives a bijection

$$H \cap X(t) \to H^* \cap X^*(G(V(t))).$$

(36*) Given any $t \in d$, we have that $A^* \mapsto U^*(t,A^*)$ gives a bijection

$$H^* \cap X^*(G(V(t))) \to H \cap X(t)$$

whose inverse is the bijection mentioned in (36).

(37) For every $t \in d$ and $B \in H$ and $A \in X(t,B)$, we have

$$U(t,A) \in X^*(G(V(t)),U(t,B)).$$

(37*) For every $t \in d$ and $B \in H$ and $A^* \in X^*(G(V(t)),U(t,B))$, we have

$$U^*(t,A^*) \in X(t,B).$$

(38) For every $t \in d$ and $B \in H$ and $A \in X(t,B)$, we have

$$G(V(S(t,A))) = S^*(G(V(t)),U(t,A)).$$

(39) For every $t \in d$ and $B \in H$ and $A \in X(t,B)$, we have

$$U^*(t,U(t,A)) = A.$$

(39*) For every $t \in d$ and $B \in H$ and $A^* \in X^*(G(V(t)),U(t,B))$, we have

$$U(t,U^*(t,A^*)) = A^*.$$

(40) Given any $t \in d$ and $B \in H$, we have that $A \mapsto U(t,A)$ gives a bijection

$$X(t,B) \to X^*(G(V(t)),U(t,B)).$$

(40*) Given any $t \in d$ and $B \in H$, we have that $A^* \mapsto U^*(t,A^*)$ gives a bijection

$$X^*(G(V(t)),U(t,B)) \to X(t,B)$$

whose inverse is the bijection mentioned in (40).

(41) Given any $T \in \text{bit}(2,m)$ and $A \in \text{nex}(T) \cap \text{hasid}(k,T)$, upon letting

$$T^* = \text{uti}(m,k,T) \quad \text{and} \quad A^* = \text{udi}(m,k,T,A),$$

we have

$$A^* \in \text{nex}(T^*) \cap \text{osid}(k,T^*).$$

(42) Given any $T \in \text{bit}(2,m)$ and $B \in \text{hasid}(k,T)$, upon letting

$$T^* = \text{uti}(m,k,T) \quad \text{and} \quad B^* = \text{udi}(m,k,T,B),$$

we have that $A \mapsto \text{udi}(m,k,T,A)$ gives a bijection

$$\text{nex}(T,B) \to \text{nex}(T^*,B^*)$$

and moreover, for every $A \in \text{nex}(T,B)$, upon letting

$$A^* = \text{udi}(m,k,T,A),$$

we have

$$uti(m,k,sat(T,A)) = sat(T^*,A^*).$$

(43) Given any $T \in bit(2,m)$ and $B \in sid(T)$, upon letting

$$m^* = ubi(m,k) \quad \text{and} \quad T^* = uti(m,k,T) \quad \text{and}$$
$$B^* = udi(m,k,T,B),$$

we have

$$uni(m^*,k,T^*,B^*) = uni(m,k,T,B).$$

(44) Given any $T \in bit(2,m)$ and $B \in amsid(T)$, upon letting

$$m^* = ubi(m,k) \quad \text{and} \quad T^* = uti(m,k,T) \quad \text{and}$$
$$B^* = udi(m,k,T,B),$$

we have

$$B^* \in amosid(m^*,k,T^*).$$

Proof: (1) and (1*) are obvious and they yield (2) and (2*), respectively. (3) is obvious and it yields (4). (5) is obvious and it yields (6). (7) and (7*) are obvious and they collectively yield (8) and (8*). (9) is obvious. (7) and (7*) yield (10) and (10*), respectively. (10) and (10*) collectively yield (11) and (11*). (12) and (12*) follow from (8) and (8*), respectively. (13) follows from the definitions of G and V. Given any $t \in D$, upon letting $t^* = G(V(t))$, for every $e \in [1,2]$ we have

$$\begin{aligned} g^*(v^*(t^*_e(k'))) &= g^*(v^*(g(v(t[e](k'))))) && \text{[by (13)]} \\ &= g^*(v^*(v^*(g(t[e](k'))))) && \text{[by (12)]} \\ &= g^*(g(t[e](k'))) && \text{[by (1*)]} \\ &= t[e](k') && \text{[by (10)]} \end{aligned}$$

and this proves (14). In view of (9), we see that (7) and (7*) yield (15) and (15*), respectively. (15) and (15*) collectively yield (16) and (16*). In view of the definitions of G and S and S*, we see that (17) follows from (8) and (9). (18) follows from (9). (19) and (19*) are obvious. (20) is obvious.

Now let us recall the well-known laws of the boolean algebra of sets whose proofs are quite obvious and according to which first, for all a and b in P([1,p]) we have

(51) $v(a \cup b) = v(a) \cap v(b)$

and

(52) $v(a \cap b) = v(a) \cup v(b)$

and

(53) $a \setminus b = a \cap v(b)$

and

(54) $a \cap b = \emptyset \Leftrightarrow a \subset v(b) \Leftrightarrow v(b) \cup a = v(b)$,

and second, for all a, b, c in P([1,p]) we have

(55) $(a \cup b) \cap c = (a \cap c) \cup (b \cap c)$

and

(56) $(a \cap b) \cup c = (a \cup c) \cap (b \cup c)$.

In view of (54), by the definition of H we see that

(57) $\begin{cases} \text{for all } A \in H \text{ and } e \in [1,2] \text{ we have} \\ v(A[e](k')) \cup A[e'](k') = v(A[e](k')). \end{cases}$

Now given any $t \in D$ and $A \in H$ and $e \in [1,2]$, we get

$$\begin{aligned} &S(t,A)[e](k') \\ &= [t[e](k') \setminus A[e](k')] \cup A[e'](k') && \text{(by definition of S)} \\ &= [t[e](k') \cap v(A[e](k'))] \cup A[e'](k') && \text{[by (53)]} \\ &= [t[e](k') \cup A[e'](k')] \cap [v(A[e](k')) \cup A[e'](k')] \\ & && \text{[by (52)]} \\ &= [t[e](k') \cup A[e'](k')] \cap v(A[e](k')) && \text{[by (57)]} \\ &= [t[e](k') \cup A[e'](k')] \setminus A[e](k') && \text{[by (53)]} \end{aligned}$$

and this proves (21). Again, given any $t \in D$ and $A \in H$, for every $e \in [1,2]$ we get

$$\begin{aligned} &V(S(t,A))[e](k') \\ &= v([t[e](k') \cup A[e'](k')] \setminus A[e](k')) && \text{[by (21) and definition of V]} \\ &= v([t[e](k') \cup A[e'](k')] \cap v(A[e](k')) && \text{[by (53)]} \end{aligned}$$

$$= v([t[e](k') \cup A[e'](k')] \cup v(v(A[e](k'))) \qquad \text{[by (52)]}$$
$$= v([t[e](k') \cup A[e'](k')]) \cup A[e](k') \qquad \text{[by (1)]}$$
$$= [v(t[e](k')) \cap v(A[e'](k'))] \cup A[e](k') \qquad \text{[by (51)]}$$
$$= [v(t[e](k')) \setminus A[e'](k')] \cup A[e](k') \qquad \text{[by (53)]}$$
$$= S(V(t),I(A))[e](k') \qquad \text{(by definitions of S, V, I)}$$

and we also get

$$V(S(t,A))[e](k) = S(V(t),I(A))[e](k) \qquad \text{(by definitions of S, V, I)}$$

and therefore we see that

$$V(S(t,A)) = S(V(t),I(A))$$

and this proves (22).

By (17) and (22) we get (23).

Given any $t \in D$ and $A \in D(t)$ and $e \in [1,2]$, upon letting

$$\begin{bmatrix} \tilde{t}[e] = t[e](k') \text{ and } \tilde{t}[e'] = t[e'](k') \\ \text{and } \tilde{A}[e] = A[e](k') \text{ and } \tilde{A}[e'] = A[e'](k'), \end{bmatrix}$$

first, by the definition of U we have

$$U(t,A)_e(k') = g(v(\tilde{t}[e]) \cap \tilde{A}[e'])$$

and

$$U(t,A)_{e'}(k') = g(v(\tilde{t}[e']) \cap \tilde{A}[e])$$

and by (13) we also have

$$G(V(t))_e(k') = g(v(\tilde{t}[e]))$$

and therefore we get

$$\begin{bmatrix} U(t,A)_e(k') \cap U(t,A)_{e'}(k') \\ = g(v(\tilde{t}[e]) \cap \tilde{A}[e']) \cap g(v(\tilde{t}[e']) \cap \tilde{A}[e]) \end{bmatrix}$$

and

$$\begin{bmatrix} [G(V(t))_e(k') \setminus U(t,A)_e(k')] \cup U(t,A)_{e'}(k') \\ = [f(v(\tilde{t}[e])) \setminus g(v(\tilde{t}[e]) \cap \tilde{A}[e'])] \cap g(v(\tilde{t}[e']) \cap \tilde{A}[e]) \end{bmatrix}$$

and in view of (8), by the two equations above we get, respectively,

$$(61)\quad \begin{bmatrix} U(t,A)_e(k') \cap U(t,A)_{e'}(k') \\ = g(v(\tilde{t}[e]) \cap \tilde{A}[e'] \cap v(\tilde{t}[e']) \cap \tilde{A}[e]) \end{bmatrix}$$

and

$$(62)\quad \begin{bmatrix} [G(V(t))_e(k') \setminus U(t,A)_e(k')] \cap U(t,A)_{e'}(k') \\ = g([v(\tilde{t}[e]) \setminus (v(\tilde{t}[e]) \cap \tilde{A}[e'])] \cap (v(\tilde{t}[e']) \cap \tilde{A}[e])), \end{bmatrix}$$

and second, it is obvious that

$$(63)\quad v(\tilde{t}[e]) \cap \tilde{A}[e'] \subset v(\tilde{t}[e]),$$

and third, it is obvious that

$$(64)\quad v(\tilde{t}[e]) \setminus (v(\tilde{t}[e]) \cap \tilde{A}[e']) \subset v(\tilde{t}[e]),$$

and fourth, we have

$$\begin{aligned} v(\tilde{t}[e']) \cap \tilde{A}[e] &\subset \tilde{A}[e] && \text{(obviously)} \\ &\subset \tilde{t}[e] && [\text{because } A \in D(t)] \end{aligned}$$

and hence we get

$$(65)\quad v(\tilde{t}[e']) \cap \tilde{A}[e] \subset \tilde{t}[e],$$

and fifth, we have

$$\begin{aligned} & v(\tilde{t}[e]) \cap \tilde{A}[e'] \cap v(\tilde{t}[e']) \cap \tilde{A}[e] \\ & \subset v(\tilde{t}[e]) \cap \tilde{t}[e] && [\text{by (63) and (65)}] \\ & = \emptyset && \text{(obviously)} \end{aligned}$$

and hence we get

$$(66)\quad v(\tilde{t}[e]) \cap \tilde{A}[e'] \cap v(\tilde{t}[e']) \cap \tilde{A}[e] = \emptyset,$$

and sixth, we have

$$\begin{aligned} & [v(\tilde{t}[e]) \setminus (v(\tilde{t}[e]) \cap \tilde{A}[e'])] \cap (v(\tilde{t}[e']) \\ & \cap \tilde{A}[e]) \subset v(\tilde{t}[e]) \cap \tilde{t}[e] && [\text{by (64) and (65)}] \\ & = \emptyset && \text{(obviously)} \end{aligned}$$

and hence we get

$$(67)\quad [v(\tilde{t}[e]) \setminus (v(\tilde{t}[e]) \cap \tilde{A}[e'])] \cap (v(\tilde{t}[e']) \cap \tilde{A}[e]) = \emptyset,$$

and seventh, upon applying g to (66) and (67) we see, respectively, that

$$[\text{RHS of } (61)] = \emptyset \quad \text{and} \quad [\text{RHS of } (62)] = \emptyset$$

and therefore by (61) and (62) we get, respectively, that

(68) $[\text{LHS of } (61)] = \emptyset$

and

(69) $[\text{LHS of } (62)] = \emptyset$,

and finally, in view of (68) and (69) we conclude that (24) has been proved.

Given any $t \in D$ and $A^* \in D^*(G(V(t)))$ and $e \in [1,2]$, upon letting

$$\begin{cases} t^* = G(V(t)) \text{ and } \tilde{t}[e] = t^*_e(k') \text{ and } \tilde{t}[e'] = t^*_{e'}(k') \\ \text{and } \tilde{A}[e] = A^*_e(k') \text{ and } \tilde{A}[e'] = A^*_{e'}(k'), \end{cases}$$

first, by (14) and the definition of U^* we see that

$$U^*(t,A^*)[e](k') = g^*(v^*(\tilde{t}[e])) \cap g^*(\tilde{A}[e'])$$

and

$$U^*(t,A^*)[e'](k') = g^*(v^*(\tilde{t}[e'])) \cap g^*(\tilde{A}[e])$$

and in view of (8*), by the two equations above we get, respectively,

$$U^*(t,A^*)[e](k') = g^*(v^*(\tilde{t}[e]) \cap \tilde{A}[e'])$$

and

$$U^*(t,A^*)[e'](k') = g^*(v^*(\tilde{t}[e']) \cap \tilde{A}[e])$$

and by (14) we also have

$$t[e](k') = g^*(v^*(\tilde{t}[e]))$$

and therefore we get

$$\begin{cases} U^*(t,A^*)[e](k') \cap U^*(t,A)_{e'}(k') \\ = g^*(v^*(\tilde{t}[e]) \cap \tilde{A}[e']) \cap g^*(v^*(\tilde{t}[e'] \cap \tilde{A}[e]) \end{cases}$$

and

$$\begin{bmatrix} [t[e](k') \setminus U^*(t,A^*)[e](k')] \cap U^*(t,A)_{e'}(k') \\ = [g^*(v^*(\tilde{t}[e])) \setminus g^*(v^*(\tilde{t}[e]) \cap \tilde{A}[e'])] \cap g^*(v^*(\tilde{t}[e']) \cap \tilde{A}[e]) \end{bmatrix}$$

and now, again in view of (8*), by the two equations above we get, respectively,

$$\text{(61*)} \quad \begin{bmatrix} U^*(t,A^*)[e](k') \cap U^*(t,A)_{e'}(k') \\ = g^*(v^*(\tilde{t}[e]) \cap \tilde{A}[e'] \cap v^*(\tilde{t}[e']) \cap \tilde{A}[e]) \end{bmatrix}$$

and

$$\text{(62*)} \quad \begin{bmatrix} [t[e](k') \setminus U^*(t,A^*)[e](k')] \cap U^*(t,A)_{e'}(k') \\ = g^*([v^*(\tilde{t}[e]) \setminus (v^*(\tilde{t}[e]) \cap \tilde{A}[e'])] \cap (v^*(\tilde{t}[e']) \cap \tilde{A}[e])), \end{bmatrix}$$

and second, it is obvious that

$$\text{(63*)} \quad v^*(\tilde{t}[e]) \cap \tilde{A}[e'] \subset v^*(\tilde{t}[e]),$$

and third, it is obvious that

$$\text{(64*)} \quad v^*(\tilde{t}[e]) \setminus (v^*(\tilde{t}[e]) \cap \tilde{A}[e']) \subset v^*(\tilde{t}[e]),$$

and fourth, we have

$$\begin{aligned} v^*(\tilde{t}[e']) \cap \tilde{A}[e] &\subset \tilde{A}[e] && \text{(obviously)} \\ &\subset \tilde{t}[e] && [\text{because } A^* \in D^*(t^*)] \end{aligned}$$

and hence we get

$$\text{(65*)} \quad v^*(\tilde{t}[e']) \cap \tilde{A}[e] \subset \tilde{t}[e],$$

and fifth, we have

$$\begin{aligned} & v^*(\tilde{t}[e]) \cap \tilde{A}[e'] \cap v^*(\tilde{t}[e']) \cap \tilde{A}[e] \\ & \subset v^*(\tilde{t}[e]) \cap \tilde{t}[e] && [\text{by (63*) and (65*)}] \\ & = \emptyset && \text{(obviously)} \end{aligned}$$

and hence we get

$$\text{(66*)} \quad v^*(\tilde{t}[e]) \cap \tilde{A}[e'] \cap v^*(\tilde{t}[e']) \cap \tilde{A}[e] = \emptyset,$$

and sixth, we have

$$\begin{aligned} & [v^*(\tilde{t}[e]) \setminus (v^*(\tilde{t}[e]) \cap \tilde{A}[e'])] \cap (v^*(\tilde{t}[e']) \cap \tilde{A}[e]) \\ & \subset v^*(\tilde{t}[e]) \cap \tilde{t}[e] && [\text{by (64*) and (65*)}] \\ & = \emptyset && \text{(obviously)} \end{aligned}$$

and hence we get

(67*) $[v^*(\tilde{t}[e]) \setminus (v^*(\tilde{t}[e]) \cap \tilde{A}[e'])] \cap (v^*(\tilde{t}[e']) \cap \tilde{A}[e]) = \emptyset$,

and seventh, by applying g^* to (66*) and (67*) we get, respectively,

$$[\text{RHS of } (61^*)] = \emptyset \quad \text{and} \quad [\text{RHS of } (62^*)] = \emptyset$$

and therefore by (61*) and (62*) we see, respectively, that

(68*) $[\text{LHS of } (61^*)] = \emptyset$

and

(69*) $[\text{LHS of } (62^*)] = \emptyset$,

and finally, in view of (68*) and (69*) we conclude that (24*) has been proved.

Given any $t \in D$ and $A \in H \cap D(k',t)$ and $e \in [1,2]$, first by the definition of $D(k',t)$ we have

$$A \in D(k',t) \subset D(t)$$

and hence by the definition of $D(t)$ we have

$$A[e](k') \subset t[e](k')$$

and therefore we get

(71) $t[e](k') = A[e](k') \cup [t[e](k') \setminus A[e](k')]$,

and second, we have

$$\begin{aligned} A[e'](k') &\subset [1,p] && \text{(because } A \in H \subset D) \\ &= t[e](k') \cup v(t[e](k')) && \text{(because } t \in D) \\ &= A[e](k') \cup [t[e](k') \setminus A[e](k')] \cup v(t[e](k')) \\ & && [\text{by } (71)] \end{aligned}$$

and hence we get

(72) $A[e'](k') \subset A[e](k') \cup [t[e](k') \setminus A[e](k')] \cup v(t[e](k'))$,

and third, because $A \in H$, we have

(73) $A[e](k') \cap A[e'](k') = \emptyset$,

and fourth, because $A \in D(k',t)$, we have

(74) $[t[e](k') \setminus A[e](k')] \cap A[e'](k') = \emptyset$,

and fifth, by (72) to (74) we get

$$A[e'](k') \subset v(t[e](k'))$$

and this proves (25).

Given any $t \in D$ and $A^* \in H^* \cap D^*(k',G(V(t)))$ and $e \in [1,2]$, upon letting

$$t^* = G(V(t)),$$

first, by the definition of $D^*(k',t^*)$ we have

$$A^* \in D^*(k',t^*) \subset D^*(t^*)$$

and hence by the definition of $D^*(t^*)$ we have

$$A^*_e(k') \subset t^*_e(k')$$

and therefore we get

$$(71^*) \quad t^*_e(k') = A^*_e(k') \cup [t^*_e(k') \setminus A^*_e(k')],$$

and second, we have

$$\begin{aligned} A^*_{e'}(k') &\subset [q + 1, q + p] && \text{(because } A^* \in H^* \subset D^*) \\ &= t^*_e(k') \cup v^*(t^*_e(k')) && \text{(because } t^* \in D^*) \\ &= A^*_e(k') \cup [t^*_e(k') \setminus A^*_e(k')] \cup v^*(t^*_e(k')) \\ & && [\text{by } (71^*)] \end{aligned}$$

and hence we get

$$(72^*) \quad A^*_{e'}(k') \subset A^*_e(k') \cup [t^*_e(k') \setminus A^*_e(k')] \cup v^*(t^*_e(k')),$$

and third, because $A^* \in H^*$, we have

$$(73^*) \quad A^*_e(k') \cap A^*_{e'}(k') = \emptyset,$$

and fourth, because $A^* \in D^*(k',t^*)$, we have

$$(74^*) \quad [t^*_e(k') \setminus A^*_e(k')] \cap A^*_{e'}(k') = \emptyset,$$

and fifth, by (72*) and (74*) we get

$$A^*_{e'}(k') \subset v^*(t^*_e(k'))$$

and upon applying f^* to both sides of the inclusion above we see that

$$g^*(A^*_{e'}(k')) \subset g^*(v^*(t^*_e(k')))$$

and therefore by (14) we conclude that

$$g^*(A^*_{e'}(k')) \subset t[e](k')$$

and this proves (25*).

By the definitions of G and I and U we see that

$$(75)\quad \begin{cases} \text{if } t \in D \text{ and } A \in D \text{ are such that} \\ A[1](k') \subset v(t[2](k')) \text{ and } A[2](k') \subset v(t[1](k')) \\ \text{then } U(t,A) = G(I(A)), \end{cases}$$

whereas by the definitions of G* and I and U* we see that

$$(75^*)\quad \begin{cases} \text{if } t \in D \text{ and } A^* \in D^* \text{ are such that} \\ g^*(A^*_1(k')) \subset t[2](k') \text{ and } g^*(A^*_2(k')) \subset t[1](k') \\ \text{then } U^*(t,A^*) = I(G^*(A^*)). \end{cases}$$

By (75) and (25) we get (26). By (75*) and (25*) we get (26*). By (23) and (26) we get (27).

Given any $t \in D$ and $A \in H \cap D(k',t)$ we have

$$\begin{aligned} U^*(t,U(t,A)) &= U^*(t,G(I(A))) && \text{[by (26)]} \\ &= I(G^*(G(I(A))) && \text{[by (24) and (26*)]} \\ &= I(I(A)) && \text{[by (15)]} \\ &= A && \text{[by (5)]} \end{aligned}$$

and this proves (28).

Given any $t \in D$ and $A^* \in H^* \cap D^*(k',G(V(t)))$ we have

$$\begin{aligned} U(t,U^*(t,A^*)) &= U(t,I(G^*(A^*))) && \text{[by (26*)]} \\ &= G(I(I(G^*(A^*))) && \text{[by (24*) and (26)]} \\ &= G(G^*(A^*)) && \text{[by (5)]} \\ &= A^* && \text{[by (15*)]} \end{aligned}$$

and this proves (28*).

Given any $t \in D$ and $A \in D$ and $e \in [1,2]$, by (9) and the definition of U we see that

$$U(t,A)[e](k) = A[e](k) \perp\!\!\perp g(v(t[e](k')) \cap A[e'](k'))$$

and hence

$$\text{card}(U(t,A)[e](k)) = \text{card}(A[e](k)) + \text{card}(g(v(t[e](k') \cap A[e'](k')))$$

and by (8) we also get

$$\text{card}(g(v(t[e](k') \cap A[e'](k'))) = \text{card}(v(t[e](k')) \cap A[e'](k'));$$

thus

(76) $\left[\begin{array}{l}\text{for every } t \in D \text{ and } A \in D \text{ and } e \in [1,2], \text{ we have} \\ \text{card}(U(t,A)[e](k)) = \text{card } A[e](k) \\ \qquad + \text{card}(v(t[e](k')) \cap A[e'](k')).\end{array}\right.$

Given any $t \in D$ and $A \in H \cap D(k',t)$ and $e \in [1,2]$, by (24) we get

$$v(t[e](k')) \cap A[e'](k') = A[e'](k')$$

and hence by (76) we get

$$\text{card}(U(t,A)[e](k)) = \text{card}(A[e](k)) + \text{card}(A[e'](k'))$$

and this proves (29).

Given any $t \in D$ and $A^* \in H^* \cap D^*(k',G(V(t)))$, upon letting $A = U^*(t,A^*)$, by (24*) and (27*) we see that

$$A \in H \cap D(k',t) \quad \text{and} \quad U(t,A) = A^*$$

and therefore by (29) we see that for every $e \in [1,2]$ we have

$$\text{card}(A^*[e](k)) = \text{card}(U^*(t,A^*)[e](k)) + \text{card}(U^*(t,A^*)[e'](k'))$$

and this proves (29*).

By (24), (24*), (28) and (28*) we get (30) and (30*).

Given any $t \in D$ and $B \in D$, upon letting

$$t^* = G(V(t)) \quad \text{and} \quad B^* = U(t,B),$$

first, by (76) we see that

(81) $\text{card}(B^*[1](k)) = \text{card}(B[1](k)) + \text{card}(v(t[1](k')) \cap B[2](k'))$

and

(82) $\text{card}(B^*[2](k)) = \text{card}(B[2](k)) + \text{card}(v(t[2](k')) \cap B[1](k')),$

and second, by (9) we see that

(83) $$\left[\begin{array}{l} \bigcap_{e\in[1,2]} [t^*[e](k) \setminus B^*[e](k)] \\ = \left(\bigcap_{e\in[1,2]} [t^*_e(k) \setminus B^*_e(k)]\right) \coprod \left(\bigcap_{e\in[1,2]} [t^*_e(k') \setminus B^*_e(k')]\right) \end{array}\right.$$

and by the definitions of G and V and U we see that

(84) $$\left[\begin{array}{l} \text{for all } e \in [1,2] \text{ we have} \\ t^*_e(k) = t[e](k) \text{ and } B^*[e](k) = B[e](k) \end{array}\right.$$

and

(85) $$\left[\begin{array}{l} \text{for all } e \in [1,2] \text{ we have} \\ t^*_e(k') = g(v(t[e](k'))) \text{ and} \\ B^*_e(k') = g(v(t[e](k')) \cap B[e'](k')), \end{array}\right.$$

and fourth, by (8) and (85) we see that

(86) $$\left[\begin{array}{l} \operatorname{card}\left(\bigcap_{e\in[1,2]} [t^*_e(k') \setminus B^*_e(k')]\right) \\ = \operatorname{card}\left(\bigcap_{e\in[1,2]} [v(t[e](k')) \setminus (v(t[e](k') \cap B[e'](k'))]\right) \end{array}\right.$$

and fifth, by (83), (84), and (86) we see that

(87) $$\left[\begin{array}{l} \operatorname{card}\left(\bigcap_{e\in[1,2]} [t^*[e](k) \setminus B^*[e](k)]\right) \\ = \operatorname{card}\left(\bigcap_{e\in[1,2]} [t[e](k) \setminus B[e](k)]\right) \\ \quad + \operatorname{card}\left(\bigcap_{e\in[1,2]} [v(t[e](k')) \setminus (v(t[e](k')) \cap B[e'](k'))]\right), \end{array}\right.$$

and sixth, by the definitions of E* and E we see that

$$E^*(t^*,B^*) = [\text{LHS of } (81)] + [\text{LHS of } (82)] + [\text{LHS of } (87)]$$

and

$$E(t,B) = [\text{RHS of } (81)] + [\text{RHS of } (82)] + [\text{RHS of } (87)]$$

and therefore by (81), (82), and (87) we conclude that

$$E^*(t^*,B^*) = E(t,B)$$

and this proves (31).

By the definition of U we get (32).

Given any $t \in D$ and $B \in D$ and $A^* \in D^*(U(t,B))$, upon letting $B^* = U(t,B)$, for every $e \in [1,2]$ we have

$$\begin{aligned}
&U^*(t,A^*)[e](k') \\
&= t[e](k') \cap g^*(A^*_{e'}(k')) && \text{(by definition of } U^*) \\
&\subset g^*(A^*_{e'}(k')) && \text{(obviously)} \\
&\subset g^*(B^*_{e'}(k')) && [\text{because } A^* \in D^*(B^*)] \\
&= g^*(g(v(t[e](k')) \cap B[e](k'))) && \text{(by definition of } U) \\
&= v(t[e](k')) \cap B[e](k') && [\text{by } (10)] \\
&\subset B[e](k') && \text{(obviously)}
\end{aligned}$$

and

$$\begin{aligned}
U^*(t,A^*)[e](k) &= A^*_e(k) && \text{(by definition of } U^*) \\
&\subset B^*_e(k) && [\text{because } A^* \in D^*(B^*)] \\
&= B[e](k) && \text{(by definition of } U)
\end{aligned}$$

and therefore we have

$$U^*(t,A^*) \subset D(B)$$

and this proves (32*).

By (18), (19), (20), (24), and (29) we get (33). By (18*), (19*), (20), (24*), and (29*) we get (33*). By (27) we get (34). (35) and (35*) follow from (28) and (28*), respectively. By (33), (33*), (35), and (35*) we get (36) and (36*). By (32) and (33) we get (37). By (32*) and (33*) we get (37*). (38) follows from (34). (39) and (39*) follow from (35) and (35*), respectively. By (37), (37*), (39), and (39*) we get (40) and (40*).

Given any $T \in \text{bit}(2,m)$, upon taking

$$t = \text{sim}(T)$$

and upon letting

$$m^* = \text{ubi}(m,k) \quad \text{and} \quad T^* = \text{uti}(m,k,T) \quad \text{and} \quad t^* = G(V(t)),$$

in view of the various relevant definitions, and in view of observations (11.1.4), (13.1.1), (14.1.1), and (15.1.1), we see that

$$\left[\begin{array}{l} \text{t} \in \text{d and sid(T) = D(t) and hasid(k,T) = H} \cap \text{D(t)} \\ \text{and} \\ \text{disid(1,T) = D(1,t) and disid(2,T) = D(2,t)} \\ \text{and} \\ \text{nex(T) = X(t)} \subset \text{D(t)} \end{array}\right.$$

and

$$\left[\begin{array}{l} \text{sim(T*)[1](k) = t*[1](k) and sim(T*)[2](k) = t*[2](k)} \\ \text{and} \\ \text{osid(k,T*) = D*(t*) and disid(k,T*)} \cap \text{osid(k,T*) = D**(t*)} \\ \text{and} \\ \text{nex(T*)} \cap \text{osid(k,T*) = X*(t*)} \subset \text{D*(t*)} \end{array}\right.$$

and

$$\left[\begin{array}{l} \text{for all A} \in \text{D(t) we have} \\ \text{udi(m,k,T,A) = U(t,A) and sad(m,k,T,A) = S(t,A)} \end{array}\right.$$

and

$$\left[\begin{array}{l} \text{for all A*} \in \text{D*(t*) and e} \in \text{[1,2] we have} \\ \text{sad(m*,k,T*,A*)[e](k) = S*(t*,A*)[e](k)} \end{array}\right.$$

and

$$\left[\begin{array}{l} \text{for all A} \in \text{X(t) and e} \in \text{[1,2] we have} \\ \text{sim(uti(m,k,sat(T,A)))[e](k) = G(V(S(t,A)))[e](k)} \end{array}\right.$$

and

$$\left[\begin{array}{l} \text{for all B} \in \text{D(t) we have} \\ \text{nex(T,B) = X(t,B)} \subset \text{X(t) and uni(m,k,T,B) = E(t,B)} \end{array}\right.$$

and

$$\left[\begin{array}{l} \text{for all B*} \in \text{D*(t*) we have} \\ \text{nex(T*,B*) = X*(t*,B*)} \subset \text{X*(t*)} \\ \text{and uni(m*,k,T*,B*) = E*(t*,B*)} \end{array}\right.$$

and therefore, (41) follows from (33), whereas (42) follows from (38) and (40), and finally (43) and (44) follow from (31).

PROPOSITION (15.3) Let there be given any $m \in N(2)$ and $k \in [1,2]$ and $T \in \mathrm{bit}(2,m)$ and let us put

$$m^* = \mathrm{ubi}(m,k) \quad \text{and} \quad T^* = \mathrm{uti}(m,k,t).$$

Then we have the following.

(15.3.1) Given any $A \in \mathrm{nex}(T) \cap \mathrm{hasid}(k,T)$, upon letting

$$A^* = \mathrm{udi}(m,k,T,A),$$

we have

$$A^* \in \mathrm{nex}(T^*) \cap \mathrm{osid}(k,T^*).$$

(15.3.2) Given any $B \in \mathrm{hasid}(k,T)$, upon letting

$$B^* = \mathrm{udi}(m,k,T,B),$$

we have that $A \mapsto \mathrm{udi}(m,k,T,A)$ gives a bijection

$$\mathrm{nex}(T,B) \to \mathrm{nex}(T^*,B^*)$$

and moreover, for every $A \in \mathrm{nex}(T,B)$, upon letting

$$A^* = \mathrm{udi}(m,k,T,A),$$

we have

$$\mathrm{uti}(m,k,\mathrm{sat}(T,A)) = \mathrm{sat}(T^*,A^*).$$

(15.3.3) Given any $B \in \mathrm{sid}(T)$, upon letting

$$B^* = \mathrm{udi}(m,k,T,B),$$

we have

$$\mathrm{uni}(m^*,k,T^*,B^*) = \mathrm{uni}(m,k,T,B).$$

(15.3.4) Given any $B \in \mathrm{amsid}(T)$, upon letting

$$B^* = \mathrm{udi}(m,k,T,B),$$

we have

$$B^* \in \mathrm{amosid}(m^*,k,T^*).$$

Proof: (15.3.1) to (15.3.4) have been proved in parts (41) to (44), respectively, of Lemma (15.2).

LEMMA (15.4) Given any ring R and given any m ∈ N(2) and k ∈ [1,2] and a ∈ bic(m) and x ∈ mat(R,m), we have

$$\text{ugi}[m,k,a]\text{mor}(x,a) = \text{mor}(\text{uni}(x,m,k),\text{uti}[m,k,a]).$$

Proof: Let

$$n = \text{dup}(m(k')).$$

Now we obtain

$$u \in \text{bib}(2,n,k) \quad \text{and} \quad f \in \text{bil}(2,u,k,n)$$

by putting

$$u[e](\hat{k}) = \begin{cases} \text{len}(a) & \text{if } e = 1 \text{ and } \hat{k} = k' \\ m(k') & \text{if } e = 1 \text{ and } \hat{k} = k \\ m(k') - \text{len}(a) & \text{if } e = 2 \text{ and } \hat{k} = k' \\ m(k') & \text{if } e = 2 \text{ and } \hat{k} = k \end{cases}$$

and

$$f[e](i) = \begin{cases} a(k',i) & \text{if } e = 1 \text{ and } i \in [1,u[1](k')] \\ \text{com}[m,k,a](m(k') + 1 - i) & \text{if } e = 2 \text{ and } i \in [1,u[2](k')]. \end{cases}$$

Upon letting

$$y = \text{sub}(\text{umi}(x,m,k),\text{uti}[m,k,a])$$

we clearly have

$$y \in \text{mat}(R,n)$$

and by the definition of mor we get

(1) $\det y = \text{mor}(\text{umi}(x,m,k),\text{uti}[m,k,a]).$

For every e ∈ [1,2] let

$$x[e] \in \text{mat}(R,u[e])$$

be defined by saying that for all i ∈ [1,u[e](1)] and j ∈ [1,u[e](2)] we have

$$x[e]_{ij} = \begin{cases} y_{f[e](i),j} & \text{if } k = 2 \\ y_{i,f[e](j)} & \text{if } k = 1. \end{cases}$$

Let

$$g = \mathrm{orp}(d,u,k,n).$$

Also, let

$$A^* \in \mathrm{bid}(d,u,k,n)$$

be obtained by putting

$$A^*[e](\hat{k}) = \begin{cases} \emptyset & \text{if } e = 1 \text{ and } \hat{k} = k' \\ [1,\mathrm{len}(a)] & \text{if } e = 1 \text{ and } \hat{k} = k \\ \emptyset & \text{if } e = 2 \text{ and } \hat{k} = k' \\ [1 + \mathrm{len}(a),\ m(k')] & \text{if } e = 2 \text{ and } \hat{k} = k. \end{cases}$$

Finally, let I(u[2](k')) be the u[2](k') by u[2](k') anti-identity matrix; i.e., let

$$I(u[2](k')) \in \mathrm{mat}(R,\mathrm{dup}(u[2](k')))$$

be defined by saying that for any $i \in [1,u[2](k')]$ and $j \in [1,u[2](k')]$ we have

$$I(u[2](k'))_{ij} = \begin{cases} 1 & \text{if } i + j = u[2](k') + 1 \\ 0 & \text{if } i + j \neq u[2](k') + 1. \end{cases}$$

Now clearly,

$$\mathrm{mas}(x[1],k,g(A^*)[e]) = \begin{cases} \mathrm{sub}(x,a) & \text{if } e = 1 \\ I(u[2](k')) & \text{if } e = 2 \end{cases}$$

and

$$\mathrm{lap}(d,u,k,n,f,g(A^*)) = \mathrm{upi}[m,k,a]$$

and therefore by (12.2.1) we get

$$(2)\quad \left[\begin{array}{l} \mathrm{sgn}(\mathrm{lap}(d,u,k,n,f,g(A^*)) \displaystyle\prod_{e\in[1,2]} \det \mathrm{mas}(x[e],k,g(A^*)[e]) \\ = \mathrm{ugi}[m,k,a]\mathrm{mor}(x,a). \end{array}\right.$$

Given any

$$A \in \mathrm{bid}(d,u,k,n) \setminus \{A^*\}$$

clearly there exists $q \in [1,\mathrm{len}(a)]$ such that

$$g(A)[1](q) \in [1 + \mathrm{len}(a),\ m(k')]$$

and now

$$\left[\begin{array}{l}\text{in the case } k = 2 \text{ we get}\\ \text{mas}(x[1],k,g(A)[1])_{qj} = 0 \text{ for all } j \in [1,\text{len}(a)]\\ \text{and hence we have det mas}(x[1],k,g(A)[1]) = 0,\end{array}\right.$$

whereas

$$\left[\begin{array}{l}\text{in the case } k = 1 \text{ we get}\\ \text{mas}(x[1],k,g(A)[1])_{iq} = 0 \text{ for all } i \in [1,\text{len}(a)]\\ \text{and hence we have det mas}(x[1],k,g(A)[1]) = 0;\end{array}\right.$$

thus

$$(3)\quad \left[\begin{array}{l}\text{for all } A \in \text{bid}(d,u,k,n) \setminus \{A^*\}\\ \text{we have det mas}(x[1],k,g(A)[1]) = 0.\end{array}\right.$$

In view of (2) and (3), by Laplace (12.3) we conclude that

$$\det y = \text{ugi}[m,k,a]\text{mor}(x,a)$$

and in view of (1), this completes the proof.

LEMMA (15.5) Given any ring R and given any $m \in N(2)$ and $k \in [1,2]$ and $d \in N$ and $T \in \text{bit}(d,m)$ and $x \in \text{mat}(R,m)$, we have

$$\text{ugi}(m,k,T)\text{mom}(x,T) = \text{mom}(\text{umi}(x,m,k),\text{uti}(m,k,T)).$$

Proof: In (15.4) take $a = T[e]$ and then product both sides of (15.4) as e ranges over [1,d].

LEMMA (15.6) Given any ring R and given any $m \in N(2)$ and $k \in [1,2]$ and $T \in \text{bit}(2,m)$ and $B \in \text{hasid}(k,T)$ and $A \in \text{nex}(T,B)$ and $x \in \text{mat}(R,m)$, upon letting

$$m^* = \text{ubi}(m,k) \quad\text{and}\quad T^* = \text{uti}(m,k,T)$$

and

$$A^* = \text{udi}(m,k,T,A) \quad\text{and}\quad x^* = \text{umi}(x,m,k),$$

we have

$$\text{sag}(m,k,T,A)\text{mom}(x,\text{sat}(T,A))$$
$$= \frac{\text{cag}(m^*,k,T^*,A^*)\text{mom}(x^*,\text{uti}(m,k,\text{sat}(T,A)))}{\text{ugi}(m,k,T)}$$

[where we note that now obviously $A \in \text{nex}(T) \cap \text{hasid}(k,T)$ and hence by (15.1.2) we see that sag(m,k,T,A) and cag(m*,k,T*,A*) make sense].

Proof: Upon taking sat(T,A) for T in (15.5) we get

$$\text{ugi}(m,k,\text{sat}(T,A))\text{mom}(x,\text{sat}(T,A)) = \text{mom}(x^*,\text{uti}(m,k,\text{sat}(T,A)))$$

and now, in view of the definition of sag, our assertion follows by multiplying both sides of the equation above by

$$\frac{\text{cag}(m^*,k,T^*,A^*)}{\text{ugi}(m,k,T)}.$$

THEOREM (15.7) Let there be given any ring R and let there be given any $m \in N(2)$ and $k \in [1,2]$ and $T \in \text{bit}(2,m)$ and $B \in \text{hasid}(k,T)$ and $x \in \text{mat}(R,m)$ [and let us refer to (15.1.3)]. Then:

(15.7.1) we have

$$\text{sar}(x,m,k,B) = \sum_{A\in\text{nex}(T,B)} \text{sag}(m,k,T,A)\text{mom}(x,\text{sat}(T,A)).$$

(15.7.2) Upon letting E be the empty bitabloid of depth 2, i.e., upon letting E be the bitabloid of depth 2 obtained by putting

$$E[e](k') = \emptyset = E[e](k) \quad \text{for all } e \in [1,2],$$

first we have

$$\text{nex}(T,B) = \{E\} \coprod \text{pex}(T,B),$$

and second, we have

$$\text{sat}(T,E) = T,$$

and third, we have

$$\text{udi}(m,k,T,E) = E,$$

and fourth, we have

$$\text{sag}(m,k,T,E) = -1.$$

(15.7.3) Finally, we have

$$\text{sar}(x,m,k,B) + \text{mom}(x,T) = \sum_{A\in\text{pex}(T,B)} \text{sag}(m,k,T,A)\text{mom}(x,\text{sat}(T,A)).$$

Proof: Upon letting

$$m^* = \mathrm{ubi}(m,k) \quad \text{and} \quad T^* = \mathrm{uti}(m,k,T)$$

and

$$B^* = \mathrm{udi}(m,k,T,B) \quad \text{and} \quad x^* = \mathrm{umi}(x,m,k),$$

by the definitions of car and sar we see that

$$[\text{LHS of } (15.7.1)] = \frac{[\text{LHS of } (14.7.1)]}{\mathrm{ugi}(m,k,T)}$$

and upon letting

$$A^* = \mathrm{udi}(m,k,T,A) \quad \text{for all } A \in \mathrm{nex}(T,B)$$

by (15.3.2) and (15.6) we see that

$$[\text{RHS of } (15.7.1)] = \frac{[\text{RHS of } (14.7.1)]}{\mathrm{ugi}(m,k,T)}$$

and therefore by (14.7.1) we get (15.7.1). The first three parts of (15.7.2) are obvious and they together with (14.7.2) imply the fourth part (15.7.2). (15.7.3) follows from (15.7.1) and (15.7.2).

THEOREM (15.8) Let there be given any ring R and let there be given any $m \in N(2)$ and $k \in [1,2]$ and $T \in \mathrm{bit}(2,m)$ and $B \in \mathrm{amsid}(m,k,T)$ and $x \in \mathrm{mat}(R,m)$ [and let us refer to (15.1.3)]. Then:

(15.8.1) We have $\mathrm{sar}(x,m,k,T,B) = 0$.

(15.8.2) Also, we have

$$\sum_{A\in \mathrm{nex}(T,B)} \mathrm{sag}(m,k,T,A)\mathrm{mom}(x,\mathrm{sat}(T,A)) = 0.$$

(15.8.3) Finally, we have

$$\mathrm{mom}(x,T) = \sum_{A\in \mathrm{pex}(T,B)} \mathrm{sag}(m,k,T,A)\mathrm{mom}(x,\mathrm{sat}(T,A)).$$

Proof: In view of (15.1.1) and (15.3.4) and in view of the definitions of amosid and car and sar, (15.8.1) follows from (14.8.1) by taking

$$m^* = \mathrm{ubi}(m,k) \quad \text{and} \quad T^* = \mathrm{uti}(m,k,T)$$

and

$$B^* = \mathrm{udi}(m,k,T,B) \quad \text{and} \quad x^* = \mathrm{umi}(x,m,k).$$

(15.8.2) follows from (15.7.1) and (15.8.1). (15.8.3) follows from (15.7.3) and (15.8.1).

16. THE FULL DEPTH CASE

(16.1) Notation In this subsection we introduce some more terminology which will be used in the rest of this chapter.

Given any $q \in N^*$ and $T \in \text{tid}[q,2]$, first, for every $k \in [1,q]$ we introduce the *lengthwise obstruction set* of (k,T), which we denote by $\text{los}(k,T)$ and which we define by putting

$$\text{los}(k,T) = \{i \in [1,\text{len}(T[2])] : \text{either } i \le \text{len}(T[1]) \text{ and } T[1](k,i) \nsubseteq T[2](k,i), \text{ or } i > \text{len}(T[1])\},$$

and second, for every $k \in [1,q]$ we introduce the *lengthwise obstruction number* of (k,T), which we denote by $\text{lon}(k,T)$ and which we define by putting

$$\text{lon}(k,T) = \begin{cases} \text{the smallest member of } \text{los}(k,T) & \text{if } \text{los}(k,T) \neq \emptyset \\ \infty & \text{if } \text{los}(k,T) = \emptyset, \end{cases}$$

and third, we introduce the *widthwise obstruction set* of T which we denote by $\text{wos}(T)$ and which we define by putting

$$\text{wos}(T) = \{k \in [1,q] : \text{lon}(k,T) \neq \infty\}$$

and fourth, we introduce the *widthwise obstruction number* of T which we denote by $\text{won}(T)$ and which we define by putting

$$\text{won}(T) = \begin{cases} \text{the smallest member of } \text{wos}(T) & \text{if } \text{wos}(T) \neq \emptyset \\ \infty & \text{if } \text{wos}(T) = \emptyset, \end{cases}$$

and we note that then obviously

$$\text{won}(T) = \infty \Leftrightarrow T[1] \le T[2]. \tag{16.1.1}$$

Given any $T \in \text{bit}(2)$ and $k \in [1,2]$, for which $\text{lon}(k,T) \neq \infty$, we introduce the *lengthwise obstruction subtabloid* of (k,T), which we denote by $\text{lod}(k,T)$ and which we define by saying that $\text{lod}(k,T)$ is the unique member of $\text{hasid}(k,T)$ such that

$$\text{lod}(k,T)[1](k') = \emptyset$$

and

$$\text{lod}(k,T)[2](k') = \text{sim}(T)[2](k') \setminus \text{sim}(T)[1](k')$$

and

$$\text{lod}(k,T)[1](k) = \{T[1](k,i) : i \in [\text{lon}(k,T),\text{len}(T[1])]\}$$

and

$$\text{lod}(k,T)[2](k) = \{T[2](k,i) : i \in [1,\text{lon}(k,T)]\}.$$

Given any $q \in N^*$ and $T \in \text{tab}(q)$, first, for every $e \in [1, \text{dep}(T) - 1]$ we introduce the *valid truncation* of (e,T), which we denote by vat(e,T) and which we define by saying that vat(e,T) is the unique member of tid[q,2] such that

$$\text{vat}(e,T)[1] = T[e] \quad \text{and} \quad \text{vat}(e,T)[2] = T[e + 1],$$

and second, we introduce the *depthwise obstruction set* of T, which we denote by dos(T) and which we define by putting

$$\text{dos}(T) = \{e \in [1, \text{dep}(T) - 1] : \text{won}(\text{vat}(e,T)) \neq \infty\},$$

and third, we introduce the *depthwise obstruction number* of T, which we denote by don(T) and which we define by putting

$$\text{don}(T) = \begin{cases} \text{the smallest member of dos}(T) & \text{if dos}(T) \neq \emptyset \\ \infty & \text{if dos}(T) = \emptyset \end{cases}$$

and we note that then obviously

$$\text{don}(T) = \infty \Leftrightarrow T \text{ is quasistandard.} \tag{16.1.2}$$

Given any $T \in \text{tab}(2) \setminus \text{qstab}(2)$, first we define

$$\text{fex}(T) = \text{pex}(\text{vat}(e^*,T),\text{lod}(k,\text{vat}(e^*,T)))$$

where we have *temporarily* put

$$e^* = \text{don}(T) \quad \text{and} \quad k = \text{won}(\text{vat}(e^*,T)),$$

and we remark that a member of fex(T) may be called a *fully exchangeable subtabloid* of T, and second, for every $A \in \text{fex}(T)$ we introduce the *full satellite* of (T,A), which we denote by fat(T,A) and which we define by saying that fat(T,A) is the unique member of bit(dep(T)) such that

$$\text{fat}(T,A)[e] = \begin{cases} T[e] & \text{if } e \in [1,\text{dep}(T)] \setminus [e^*, e^* + 1] \\ \text{sat}(\text{vat}(e^*,T),A)[e + 1 - e^*] & \\ & \text{if } e \in [e^*, e^* + 1] \end{cases}$$

where we have *temporarily* put

$$e^* = \text{don}(T).$$

Given any $m \in N(2)$ and $T \in \text{tab}(2) \setminus \text{qstab}(2)$ and $A \in \text{fex}(T)$ we introduce the *full signature* of (m,T,A), which we denote by $\text{fag}(m,T,A)$ and which we define by putting

$$\text{fag}(m,T,A) = \text{sag}(m,k,\text{vat}(e^*,T),A)$$

where we have *temporarily* let

$$e^* = \text{don}(T) \quad \text{and} \quad k = \text{won}(\text{vat}(e^*,T))$$

and we note that then

$$\text{fag}(m,T,A) \in \{1,-1\}.$$

Let us note that

$$\left[\begin{array}{l} \text{given any } T \in \text{bit}(2) \text{ and} \\ B \in \text{hasid}(1,T) \cap \text{hasid}(2,T) \text{ we have} \\ \text{that } A \longmapsto \text{sat}(T,A) \text{ gives an injection} \\ \text{nex}(T,B) \to \text{bit}(2). \end{array}\right. \tag{16.1.3}$$

Let us also note that

$$\left[\begin{array}{l} \text{if } T \in \text{bit}(2) \text{ and } k \in [1,2] \\ \text{are such that } \text{lon}(k,T) \neq \infty \\ \text{then } \text{lod}(k,T) \in \text{hasid}(1,T) \cap \text{hasid}(2,T). \end{array}\right. \tag{16.1.4}$$

Finally, let us note that by (16.1.3) and (16.1.4) we see that

$$\left[\begin{array}{l} \text{given any } m \in N(2) \text{ and } T \in \text{tab}(2,m) \setminus \text{qstab}(2,m) \\ \text{we have that } A \longmapsto \text{fat}(T,A) \text{ gives} \\ \text{an injection } \text{fex}(T) \to \text{tid}(2,m,\text{dep}(T)). \end{array}\right. \tag{16.1.5}$$

Given any $m \in N(2)$, we define

$$\left[\begin{array}{l} \text{rin}(m,T,S,n) \in Z \text{ for all } n \in N \text{ and } T \in \text{tab}(2,m) \\ \text{and } S \in \text{tid}(2,m,\text{dep}(T)) \end{array}\right.$$

by the recursive relations which say that for all

$$T \in \mathrm{tab}(2,m) \quad \text{and} \quad S \in \mathrm{tid}(2,m,\mathrm{dep}(T))$$

we have

$$\mathrm{rin}(m,T,S,0) = \begin{cases} 1 & \text{if } S = T \\ 0 & \text{if } S \neq T \end{cases}$$

and for all

$$T \in \mathrm{tab}(2,m) \setminus \mathrm{qstab}(2,m) \quad \text{and} \quad A \in \mathrm{fex}(T)$$

we have

$$\mathrm{rin}(m,T,\mathrm{fat}(T,A),1) = \mathrm{fag}(m,T,A)$$

and for all

$$\begin{bmatrix} T \in \mathrm{tab}(2,m) \setminus \mathrm{qstab}(2,m) \\ \text{and} \\ S \in \mathrm{tid}(2,m,\mathrm{dep}(T)) \setminus \{\mathrm{fat}(T,A) : A \in \mathrm{fex}(T)\} \end{bmatrix}$$

we have

$$\mathrm{rin}(m,T,S,1) = 0$$

and for all

$$T \in \mathrm{qstab}(2,m)$$

we have

$$\mathrm{rin}(m,T,T,1) = 1$$

and for all

$$T \in \mathrm{qstab}(2,m) \quad \text{and} \quad S \in \mathrm{tid}(2,m,\mathrm{dep}(T)) \setminus \{T\}$$

we have

$$\mathrm{rin}(m,T,S,1) = 0$$

and for all

$$n \in N^* \setminus \{1\} \text{ and } T \in \mathrm{tid}(2,m) \text{ and } S \in \mathrm{tid}(2,m,\mathrm{dep}(T))$$

we have

$$\mathrm{rin}(m,T,S,n) = \sum_{U \in \mathrm{tid}(2,m,\mathrm{dep}(T))} \mathrm{rin}(m,T,U,n-1)\mathrm{rin}(m,U,S,1)$$

and we remark that rin(m,T,S,n) may be called the *relevant integer* of (m,T,S,n) and we note that the definition above makes sense because of (16.1.5).

Given any $q \in N^*$, for every S and T in pab(q) we define

$$S \le T$$

to mean that

$$\mathrm{dep}(S) \le \mathrm{dep}(T)$$

and either for all $e \in [1,\mathrm{dep}(S)]$ we have

$$S[e] = T[e]$$

or for some $e \in [1,\mathrm{dep}(S)]$ we have

$$S[e] < T[e] \text{ and } S[\hat{e}] = T[\hat{e}] \qquad \text{for each } \hat{e} \in [1, e - 1]$$

and we note that this gives a partial order on pab(q) which may be called the *lexicographic extension* to pab(q) of the partial order on pre(q) which was defined in Subsection (2.3) of Chapter 1.

Given any $m \in N(1)$, a univector bounded by m may be thought of as a strictly increasing sequence whose length is at most m(1) and whose entries belong to the set [1,m(1)] of cardinality m(1) and hence

$$\mathrm{card}(\mathrm{vec}(1,m)) \le 2^{m(1)};$$

thus

$$\text{for all } m \in N(1) \text{ we have } \mathrm{card}(\mathrm{vec}(1,m)) \le 2^{m(1)}. \qquad (16.1.6)$$

Given any $q \in N^*$ and $m \in N(q)$, for every $k \in [1,q]$ let us *temporarily* define $m^*[k] \in N(1)$ by putting $m^*[k](1) = m(k)$, and for every $a \in \mathrm{vec}(q,m)$ and $k \in [1,q]$ let us *temporarily* define $a^*[k] \in \mathrm{vec}(1,m^*[k])$ by putting $\mathrm{len}(a^*[k]) = \mathrm{len}(a)$ and $a^*[k](1,i) = a(k,i)$ for all $i \in [1,\mathrm{len}(a)]$; now vec(q,m) becomes a subset of the set of all sequences of length q whose successive entries belong to the sets vec(1,m*[1]), vec(1,m*[2]), ..., vec(1,m*[q]); therefore, by (16.1.6) we conclude that

$$\left[\begin{array}{l}\text{for all } q \in N^* \text{ and } m \in N(q) \text{ we have} \\ \text{card}(\text{vec}(q,m)) \le \prod_{k\in[1,q]} 2^{m(k)}.\end{array}\right. \tag{16.1.7}$$

Given any $q \in N^*$ and $m \in N(q)$ and $d \in N$, a member T of tid(q,m,d) is simply a sequence

$$T[e]_{1\le e\le d} \text{ with } T[e] \in \text{vec}(q,m);$$

therefore, by (16.1.7) we see that

$$\left[\begin{array}{l}\text{for all } q \in N^* \text{ and } m \in N(q) \text{ and } d \in N \text{ we have} \\ \text{card}(\text{tid}(q,m,d)) \le \left[\prod_{k\in[1,q]} 2^{m(k)}\right]^d \\ \text{and hence tid}(q,m,d) \text{ is a finite set.}\end{array}\right. \tag{16.1.8}$$

Given any finite partially ordered set Y, for every $T \in Y$ we introduce the *height of* T *in* Y which we denote by hit(T,Y) and which we define by saying that hit(T,Y) is the unique member of N such that there exists a sequence

$$S_0 < S_1 < \cdots < S_{\text{hit}(T,Y)} = T$$

in Y but there does not exist any sequence

$$S_0 < S_1 < \cdots < S_{1+\text{hit}(T,Y)} = T$$

in Y.

Given any $m \in N(2)$, first, for every $d \in N$ we define

$$\text{sit}(m,d) = \{S \in \text{stab}(2,m) : \text{dep}(S) \le d\}$$

and we note that a member of sit(m,d) is a *standard bitableau bounded by* m *and of depth at most* d, and second, for every $d \in N$ and $S \in$ sit(m,d) we introduce the *valid enlargement* of (m,d,S), which we denote by val(m,d,S) and which we define by saying that val(m,d,S) is the unique member of qstid(2,m,d) such that

$$\text{val}(m,d,S)[e] = S[e] \qquad \text{for all } e \in [1,\text{dep}(S)]$$

and

$$\text{len}(\text{val}(m,d,S)[e]) = 0 \qquad \text{for all } e \in [1 + \text{dep}(S), d],$$

and third, for every $T \in \text{tab}(2,m)$ and $S \in \text{sit}(m,\text{dep}(T))$ we introduce the *final integer* of (m,T,S), which we denote by $\text{fin}(m,T,S)$ and which we define by saying that $\text{fin}(m,T,S)$ is the unique member of Z obtained by putting

$$\text{fin}(m,T,S) = \text{rin}(m,T,\text{val}(m,\text{dep}(T),S),\text{hit}(T,\text{tid}(2,m,\text{dep}(T)))).$$

LEMMA (16.2) Given any $T \in \text{bit}(2)$ and $A \in \text{pex}(T)$ and $k \in [1,2]$, we have the following.

(16.2.1) If $A[1](k') = \emptyset$ and there exists an injective map $h\colon A[1](k) \to A[2](k)$ such that $h(i) < i$ for all $i \in A[1](k)$, then $\text{sat}(T,A)[1] < T[1]$.

(16.2.2) If $A \in \text{pex}(T,B)$ where $B \in \text{sid}(T)$ is such that $B[1](k') = \emptyset$ and $i < j$ for all $i \in B[2](k)$ and $j \in B[1](k)$, then $\text{sat}(T,A)[1] < T[1]$.

(16.2.3) If $\text{lon}(k,T) \neq \infty$ and $A \in \text{pex}(T,\text{lod}(k,T))$, then $\text{sat}(T,A)[1] < T[1]$.

Proof: First, for a moment suppose that

$$A[1](k') = \emptyset = A[1](k) \quad \text{and} \quad \text{card}(A[2](k)) = 1;$$

then by (11.1.4) we see that for every $\hat{k} \in [1,2]$ we have

$$A[2](\hat{k}) = \{v(\hat{k})\} \quad \text{with } v(\hat{k}) \in \text{sim}(T)[2](\hat{k}) \setminus \text{sim}(T)[1](\hat{k})$$

and clearly there exists $r(\hat{k}) \in [1, \text{len}(T[1]) + 1]$ such that

$$\begin{bmatrix} T[1](\hat{k},i) < v(\hat{k}) < T[1](\hat{k},j) \\ \text{for all } i \in [1, r(\hat{k}) - 1] \text{ and } j \in [r(\hat{k}),\text{len}(T[1])] \end{bmatrix}$$

and now obviously we have

$$\text{len}(\text{sat}(T,A)[1]) = \text{len}(T[1]) + 1$$

and

$$\text{sat}(T,A)[1](\hat{k},i) = \begin{cases} T[1](\hat{k},i) & \text{if } i \in [1, r(\hat{k}) - 1] \\ v(\hat{k}) & \text{if } i = r(\hat{k}) \\ T[1](\hat{k}, i - 1) & \text{if } i \in [r(\hat{k}) + 1, \text{len}(T[1]) + 1] \end{cases}$$

and hence

$$\text{sat}(T,A)[1](\hat{k},i) \le T[1](\hat{k},i) \qquad \text{for all } i \in [1,\text{len}(T[1])]$$

and therefore

$$\text{sat}(T,A)[1] < T[1];$$

thus

$$(1) \quad \left[\begin{array}{l}\text{if } A[1](k') = \emptyset = A[1](k) \text{ and card}(A[2](k)) = 1 \\ \text{then sat}(T,A)[1] < T[1].\end{array}\right.$$

Second, for a moment suppose that

$$A[1](k') = \emptyset \quad \text{and} \quad A[1](k) = \{w[1]\} \quad \text{and} \quad A[2](k) = \{w[2]\}$$

where

$$w[1] \in \text{sim}(T)[1](k) \qquad \text{and} \qquad w[2] \in \text{sim}(T)[2](k)$$

with

$$w[2] < w[1];$$

then by (11.1.4) we see that

$$A[2](k') = \emptyset \qquad \text{and} \qquad w[2] \notin \text{sim}(T)[1](k)$$

and clearly, there exists a unique $s[1] \in [1,\text{len}(T[1])]$ such that

$$T[1](k,s[1]) = w[1]$$

and there exists a unique $s[2] \in [1,s[1]]$ such that

$$\left[\begin{array}{l}T[1](k,i) < w[2] < T[1](k,j) \\ \text{for all } i \in [1, s[2] - 1] \text{ and } j \in [s[2],s[1]]\end{array}\right.$$

and obviously, we have

$$\text{len}(\text{sat}(T,A)[1]) = \text{len}(T[1])$$

and

$$\text{sat}(T,A)[1](k,i) = \begin{cases} T[1](k,i) & \text{if } i \in [1, s[2] - 1] \\ w[2] & \text{if } i = s[2] \\ T[1](k, i - 1) & \text{if } i \in [s[2] + 1, s[1]] \\ T[1](k,i) & \text{if } i \in [s[1] + 1, \text{len}(T[1])] \end{cases}$$

and

$$\mathrm{sat}(T,A)[1](k',i) = T[1](k',i) \quad \text{for all } i \in [1,\mathrm{len}(T[1])]$$

and

$$\mathrm{sat}(T,A)[1](k,i) = T[1](k,i) \quad \text{for all } i \in [1, s[2] - 1]$$

and

$$\mathrm{sat}(T,A)[1](k,i) < T[1](k,i) \quad \text{for all } i \in [s[2],s[1]]$$

and

$$\mathrm{sat}(T,A)[1](k,i) = T[1](k,i) \quad \text{for all } i \in [s[1] + 1, \mathrm{len}(T[1])]$$

and therefore

$$\mathrm{sat}(T,A)[1] < T[1];$$

thus

$$(2)\quad \left[\begin{array}{l} \text{if } A[1](k') = \emptyset \\ \text{and } A[1](k) = \{w[1]\} \text{ and } A[2](k) = \{w[2]\} \\ \text{where } w[1] \in \mathrm{sim}(T)[1](k) \text{ and } w[2] \in \mathrm{sim}(T)[2](k) \\ \text{with } w[2] < w[1], \\ \text{then } \mathrm{sat}(T,A)[1] < T[1]. \end{array}\right.$$

Third, for a moment suppose that $A[1](k') = \emptyset$ and there exists an injective map $h: A[1](k) \to A[2](k)$ such that $h(i) < i$ for all $i \in A[1](k)$. We claim that then $\mathrm{sat}(T,A)[1] < T[1]$. Let us note that by (11.1.4) we have

$$\mathrm{card}(A[1](k)) + \mathrm{card}(A[2](k')) = \mathrm{card}(A[2](k)) > 0$$

and let us proceed to prove the claim by induction on $\mathrm{card}(A[2](k))$. Now in case $\mathrm{card}(A[2](k)) = 1$, the claim follows from (1) or (2) according as $A[1](k) = \emptyset$ or $A[1](k) \neq \emptyset$. So let $\mathrm{card}(A[2](k)) > 1$ and assume that the claim is true for all values of $\mathrm{card}(A[2](k))$ smaller than the given one. Now on the one hand, in case

$$A[1](k) = \emptyset,$$

let us fix any

$$v(k') \in A[2](k') \quad \text{and} \quad v(k) \in A[2](k) \setminus \mathrm{im}(h)$$

and let us define $a \in \mathrm{bid}(2)$ by putting

$$a[1](k') = \emptyset \quad \text{and} \quad a[2](k') = \{v(k')\}$$

and

$$a[1](k) = \emptyset \quad \text{and} \quad a[2](k) = \{v(k)\}$$

and on the other hand, in case

$$A[1](k) \neq \emptyset,$$

let us fix any

$$w[1] \in A[1](k)$$

and let us take

$$w[2] = h(w[1])$$

and let us define $a \in \mathrm{bid}(2)$ by putting

$$a[1](k') = \emptyset = a[2](k')$$

and

$$a[1](k) = \{w[1]\} \quad \text{and} \quad a[2](k) = \{w[2]\}.$$

In both the cases let us define $A^* \in \mathrm{bid}(2)$ by putting

$$A^*[e](\hat{k}) = A[e](\hat{k}) \setminus a[e](\hat{k}) \quad \text{for all } e \in [1,2] \text{ and } \hat{k} \in [1,2]$$

and let $h^*: A^*[1](k) \to A^*[2](k)$ be the map obtained by putting $h^*(i) = h(i)$ for all $i \in A^*[1](k)$. Now clearly,

$$\mathrm{card}(A^*[2](k)) = \mathrm{card}(A[2](k)) - 1$$

and $A^* \in \mathrm{pex}(T)$ and $A^*[1](k') = \emptyset$ and $h(i) < i$ for all $i \in A^*[1](k)$ and therefore upon letting

$$t = \mathrm{sat}(T,A^*),$$

by the induction hypothesis we get

$$t[1] < T[1].$$

Also, clearly, $a \in \mathrm{pex}(t)$ and hence by (1) or (2), according as $a[1](k) = \emptyset$ or $a[1](k) \neq \emptyset$, we see that

$$\mathrm{sat}(t,a)[1] < t[1].$$

By the two inequalities above we get

$$sat(t,a)[1] < T[1].$$

Also, obviously, $sat(t,a) = sat(T,A)$ and hence

$$sat(T,A)[1] < T[1].$$

This completes the induction. Thus the claim has been established and hence (16.2.1) has been proved.

Fourth, let us note that if $A \in pex(T,B)$ where $B \in sid(T)$ is such that $B[1](k') = \emptyset$ and $i < j$ for all $i \in B[2](k)$ and $j \in B[1](k)$, then obviously $A[1](k') = \emptyset$ and by (11.1.4) we have $card(A[1](k)) < card(A[2](k))$ and hence there exists an injective map $h: A[1](k) \to A[2](k)$ such that $h(i) < i$ for all $i \in A[1](k)$; therefore, (16.2.2) follows from (16.2.1).

Finally, let us note that if $lon(k,T) \neq \infty$, then upon letting $B = lod(k,T)$ we obviously have $B[1](k') = \emptyset$ and $i < j$ for all $i \in B[2](k)$ and $j \in B[1](k)$; therefore, (16.2.3) follows from (16.2.2).

LEMMA (16.3) For every $T \in tab(2) \setminus qstab(2)$ and $A \in fex(T)$ we have $fat(T,A) < T$.

Proof: Upon letting

$$e^* = don(T) \quad \text{and} \quad k = won(vat(e^*,T))$$

by the definitions of lon, won, lod, vat, don, and fex we see that

$$vat(e^*,T) \in bit(2) \quad \text{and} \quad k \in [1,2]$$

and

$$lon(k,vat(e^*,T)) \neq \infty$$

and

$$A \in pex(vat(e^*,T),lod(k,vat(e^*,T)))$$

and hence upon taking $vat(e^*,T)$ for T in (16.2.3) we get

$$sat(vat(e^*,T),A)[1] < vat(e^*,T)[1]$$

and by the definitions of vat and don and fat we also have

$$fat(T,A) \in tid[2,dep(T)] \quad \text{and} \quad e^* \in [1,dep(T)]$$

and

$$\text{fat}(T,A)[e] = T[e] \qquad \text{for all } e \in [1, e^* - 1]$$

and

$$\text{fat}(T,A)[e^*] = \text{sat}(\text{vat}(e^*,T),A)[1]$$

and

$$T[e^*] = \text{vat}(e^*,T)[1]$$

and therefore by the definition of lexicographic extension of partial order we conclude that

$$\text{fat}(T,A) < T.$$

LEMMA (16.4) Given any $m \in N(2)$ and $T \in \text{tab}(2,m) \setminus \text{qstab}(2,m)$ we have the following.

(16.4.1) If $S \in \text{tid}(2,m,\text{dep}(T))$ is such that $\text{rin}(m,T,S,1) \neq 0$, then $S < T$.

(16.4.2) If $S \in \text{tid}(2,m,\text{dep}(T))$ is such that $\text{rin}(m,T,S,1) \neq 0$, then $S < T$. Moreover, there exists $S \in \text{tid}(2,m,\text{dep}(T))$ such that $\text{rin}(m,T,S,1) \neq 0$.

Proof: In view of (16.3) and the definition of $\text{rin}(m,T,S,1)$ we get (16.4.1). In view of (16.3) and the definition of $\text{rin}(m,T,S,1)$ we also get (16.4.2) by noting that since $T \in \text{tab}(2,m) \setminus \text{qstab}(2,m)$, we clearly have $\text{fex}(T) \neq \emptyset$.

LEMMA (16.5) Let Y be a finite partially ordered set. Let $Y^* \subset Y$. For every $T \in Y$ and $S \in Y$ and $n \in N$ let there be given any $g(T,S,n) \in Z$. Assume that

$$(1)\quad g(T,S,0) = \begin{cases} 1 & \text{if } S = T \in Y \\ 0 & \text{if } S \in Y \text{ and } S \neq T \in Y \end{cases}$$

and

$$(2)\quad g(T,S,1) = \begin{cases} 1 & \text{if } S = T \in Y^* \\ 0 & \text{if } S \in Y \text{ and } S \neq T \in Y^* \\ 0 & \text{if } S \in Y \text{ and } S \nleq T \in Y \setminus Y^*. \end{cases}$$

Also, assume that for all $T \in Y$ and $S \in Y$ and $n \in N^*\setminus\{1\}$ we have

$$(3)\quad g(T,S,n) = \sum_{U \in Y} g(T, U, n-1)g(U,S,1).$$

Let $X \subset Y$ be such that

$$(4)\quad g(T,S,1) = 0 \qquad \text{for all } T \in X \text{ and } S \in Y \setminus X.$$

Then we have the following.

(16.5.1) For every $T \in X$ and $n \in N$ with $n \geq 1 + \mathrm{hit}(T,X)$ we have

$$g(T,S,n) = \begin{cases} g(T, S, 1 + \mathrm{hit}(T,X)) & \text{if } S \in X \cap Y^* \\ 0 & \text{if } S \in Y \setminus (X \cap Y^*). \end{cases}$$

(16.5.2) If for every $T^* \in X \setminus Y^*$ we have

$$\{S^* \in Y : g(T^*,S^*,1) \neq 0\} \neq \emptyset$$

then for every $T \in X$ and $n \in N$ with $n \geq \mathrm{hit}(T,X)$ we have

$$g(T,S,n) = \begin{cases} g(T,S,\mathrm{hit}(T,X)) & \text{if } S \in X \cap Y^* \\ 0 & \text{if } S \in Y \setminus (X \cap Y^*). \end{cases}$$

Proof: Upon taking U for T in (2) we see that

$$(5)\quad g(U,S,1) = \begin{cases} 1 & \text{if } S = U \in X \cap Y^* \\ 0 & \text{if } S \in Y \text{ and } S \neq U \in X \cap Y^*. \end{cases}$$

By (2) and (4) we see that if $T \in X$ and $S \in Y \setminus (X \cap Y^*)$ are such that $g(T,S,1) \neq 0$, then $S \in X$ and $S < T$ and hence $\mathrm{hit}(S,X) \leq \mathrm{hit}(T,X) - 1$; thus

$$(6)\quad \left[\begin{array}{l} \text{if } T \in X \text{ and } S \in Y \setminus (X \cap Y^*) \text{ are such that } g(T,S,1) \neq 0, \\ \text{then } S \in X \text{ and } \mathrm{hit}(S,X) \leq \mathrm{hit}(T,X) - 1. \end{array}\right.$$

Upon taking U for T in (6) we see that

$$(7)\quad \left[\begin{array}{l} \text{if } U \in X \text{ and } S \in Y \setminus (X \cap Y^*) \text{ are such that } g(U,S,1) \neq 0, \\ \text{then } S \in X \text{ and } \mathrm{hit}(S,X) \leq \mathrm{hit}(U,X) - 1. \end{array}\right.$$

Henceforth let there be fixed any $T \in X$ and let us put $n^* = \mathrm{hit}(T,X)$.

By induction on n we shall show that given any $n \in N$, we have that

$$(8)\quad \left[\begin{array}{l}\text{if } S \in Y \setminus (X \cap Y^*) \text{ is such that } g(T,S,n) \neq 0, \\ \text{then } S \in X \text{ and } \mathrm{hit}(S,X) \leq n^* - n.\end{array}\right.$$

In the case of $n = 0$ our assertion follows from (1), whereas in the case of $n = 1$ we are reduced to (6). So let $n > 1$ and assume that the assertion is true for all values of n smaller than the given one. Now, upon letting

$$V = (X \cap Y^*) \cup \{U \in X : \mathrm{hit}(U,X) \leq n^* - n + 1\},$$

by the induction hypothesis we have

$$g(T, U, n - 1) = 0 \qquad \text{for all } U \in Y \setminus V$$

and hence by (3) we see that for every $S \in Y$ we have

$$g(T,S,n) = \sum_{U \in V} g(T, U, n - 1)g(U,S,1)$$

and therefore we see that

$$\left[\begin{array}{l}\text{if } S \in Y \text{ is such that } g(T,S,n) \neq 0, \\ \text{then } g(U,S,1) \neq 0 \text{ for some } U \in V\end{array}\right.$$

and consequently, by (5) and (7) we conclude that

$$\left[\begin{array}{l}\text{if } S \in Y \setminus (X \cap Y^*) \text{ is such that } g(T,S,n) \neq 0, \\ \text{then } \mathrm{hit}(S,X) \leq n^* - n.\end{array}\right.$$

This completes the induction and thus (8) has been established.

By (2) and (4) we see that

$$(9)\quad \left[\begin{array}{l}\text{if for every } T^* \in X \setminus Y^* \text{ we have} \\ \{S^* \in Y : g(T^*,S^*,1) \neq 0\} \neq \emptyset, \\ \text{then for every } T^* \in X \setminus Y^* \text{ we have } \mathrm{hit}(T^*,X) \neq 0.\end{array}\right.$$

Since $\mathrm{hit}(S,X) \geq 0$ for all $S \in X$, by (8) we see that

$$(10)\quad \left[\begin{array}{l}\text{for every } S \in Y \setminus (X \cap Y^*) \text{ and } n \in N \text{ with } n \geq 1 + n^* \\ \text{we have } g(T,S,n) = 0\end{array}\right.$$

and by (8) and (9) we see that

$$(11)\quad \left[\begin{array}{l}\text{if for every } T^* \in X \setminus Y^* \text{ we have } \{S^* \in Y : g(T^*,S^*,1) \neq 0\} \neq \emptyset, \\ \text{then for every } S \in Y \setminus (X \cap Y^*) \text{ and } n \in N \text{ with } n \geq n^* \\ \text{we have } g(T,S,n) = 0.\end{array}\right.$$

By induction on $n - n^* - 1$ we show that given any $n \in N$ with $n \geq 1 + n^*$, we have

(12) $g(T,S,n) = g(T, S, 1 + n^*)$ for all $S \in X \cap Y^*$

and in view of (10), this will complete the proof of (16.5.1). The case of $n - n^* - 1 = 0$ being obvious, let $n - n^* - 1 > 0$ and assume that the assertion is true for all values of $n - n^* - 1$ smaller than the given one. Now by the induction hypothesis we have

$$g(T, U, n - 1) = g(T, U, 1 + n^*) \quad \text{for all } U \in X \cap Y^*$$

and by (10) we have

$$g(T, U, n - 1) = 0 \quad \text{for all } U \in Y \setminus (X \cap Y^*)$$

and hence by (3) we see that for every $S \in Y$ we have

$$g(T,S,n) = \sum_{U \in X \cap Y^*} g(T, U, 1 + n^*)g(U,S,1)$$

and therefore by (5) we see that for every $S \in X \cap Y^*$ we have

$$g(T,S,n) = g(T, S, 1 + n^*).$$

This completes the induction and thus (12) has been established and hence (16.5.1) has been proved.

Henceforth assume that for every $T^* \in X \setminus Y^*$ we have $\text{hit}(T^*,X) \neq 0$. By induction on $n - n^*$ we show that given any $n \in N$ with $n \geq n^*$, we have

(13) $g(T,S,n) = g(T,S,n^*)$ for all $S \in X \cap Y^*$

and in view of (10), this will complete the proof of (16.5.2). The case of $n - n^* = 0$ being obvious, let $n - n^* > 0$, and assume that the assertion is true for all values of $n - n^*$ smaller than the given one. Now by the induction hypothesis we have

$$g(T, U, n - 1) = g(T,U,n^*) \quad \text{for all } U \in X \cap Y^*$$

and by (11) we have

$$g(T, U, n - 1) = 0 \quad \text{for all } U \in Y \setminus (X \cap Y^*)$$

and hence by (3) we see that for every $S \in Y$ we have

$$g(T,S,n) = \sum_{U \in X \cap Y^*} g(T,U,n^*)g(U,S,1)$$

and therefore by (5) we see that for every $S \in X \cap Y^*$ we have

$$g(T,S,n) = g(T,S,n^*).$$

This completes the induction and thus (13) has been established and hence (16.5.2) has been proved.

LEMMA (16.6) Given any $m \in N(2)$ and $T \in tab(2,m)$, upon letting

$$d = dep(T) \quad \text{and} \quad n^* = hit(T,tid(2,m,d)),$$

we have the following.

(16.6.1) For every $n \in N$ with $n \geq 1 + n^*$ we have

$$rin(m,T,S,n) = \begin{cases} rin(m, T, S, 1 + n^*) & \text{if } S \in qstid(2,m,d) \\ 0 & \text{if } S \in tid(2,m,d) \setminus qstid(2,m,d). \end{cases}$$

(16.6.2) For every $n \in N$ with $n \geq n^*$ we have

$$rin(m,T,S,n) = \begin{cases} rin(m,T,S,n^*) & \text{if } S \in qstid(2,m,d) \\ 0 & \text{if } S \in tid(2,m,d) \setminus qstid(2,m,d). \end{cases}$$

Proof: Given any $L \in [1,2]$, in view of (16.4.L) we see that (16.6.L) follows from (16.5.L) by taking

$$X = Y = tid(2,m,d) \quad \text{and} \quad Y^* = qstid(2,m,d)$$

and by taking

$$g(T,S,n) = rin(m,T,S,n) \quad \text{for all } T \in Y \text{ and } S \in Y \text{ and } n \in N.$$

LEMMA (16.7) Let there be given any $m \in N(2)$ and $T \in bit(2,m)$ and $k \in [1,2]$ such that $lon(k,T) \neq \infty$, and let us put $B = lod(k,T)$. Then we have $B \in amsid(m,k,T)$.

Proof: By the definition of lod we clearly have

(1) $card(B[1](k)) + card(B[2](k)) = 1 + len(T[1])$

and

(2) $\bigcap_{e\in[1,2]} [\mathrm{sim}(T)[e](k) \setminus B[e](k)] = \emptyset.$

By the definition of lod we have $B[1](k') = \emptyset$ and hence by the definition of ufi we get

(3) $\mathrm{ufi}(m,k,T,B)[2] = \emptyset.$

By the definition of lod we have

$$B[2](k') = \mathrm{sim}(T)[2](k') \setminus \mathrm{sim}(T)[1](k')$$

and hence by the definition of ufi we get

$$\mathrm{ufi}(m,k,T,B)[1] = \mathrm{sim}(T)[2](k') \setminus \mathrm{sim}(T)[1](k'),$$

and therefore in view of (3) we see that

$$\mathrm{im}(\mathrm{com}(m,k,T)[1]) = \mathrm{ufi}(m,k,T,B)[1] \coprod H$$

where

$$H = \bigcap_{e\in[1,2]} [\mathrm{im}(\mathrm{com}(m,k,T)[e] \setminus \mathrm{ufi}(m,k,T,B)[e])],$$

and so we conclude that

(4) $$\begin{cases} \mathrm{card}(\mathrm{im}(\mathrm{com}(m,k,T))) \\ = \mathrm{card}(\mathrm{ufi}(m,k,T,B)[1]) \\ \quad + \mathrm{card}\left(\bigcap_{e\in[1,2]} [\mathrm{im}(\mathrm{com}(m,k,T)[e]) \setminus \mathrm{ufi}(m,k,T,B)[e]]\right). \end{cases}$$

Also, clearly,

(5) $\mathrm{card}(\mathrm{im}(\mathrm{com}(m,k,T))) = m(k') - \mathrm{len}(T[1]).$

In view of the definition of uni, by (1) to (5) we get

(6) $\mathrm{uni}(m,k,T,B) = 1 + m(k').$

By the definition of lod we have $B[1](k') = \emptyset$ and hence we get

(7) $B \in \mathrm{hasid}(k,T).$

By (6) and (7) we see that $B \in \mathrm{amsid}(m,k,T)$.

THEOREM (16.8) Let there be given any $m \in N(2)$ and $T \in \mathrm{bit}(2,m)$ and $k \in [1,2]$ such that $\mathrm{lon}(k,T) \neq \infty$, and let us put $B = \mathrm{lod}(k,T)$. Also let there be given any ring R and any $x \in \mathrm{mat}(R,m)$. Then we have

$$\mathrm{mom}(x,T) = \sum_{A \in \mathrm{nex}(T,B)} \mathrm{sag}(m,k,T,A)\mathrm{mom}(x,\mathrm{sat}(T,A)).$$

Proof: Follows from (15.8.3) and (16.7).

THEOREM (16.9) Given any ring R and given any $m \in N(2)$ and $x \in \mathrm{mat}(R,m)$ and $T \in \mathrm{tab}(2,m) \setminus \mathrm{qstab}(2,m)$, we have

$$\mathrm{mom}(x,T) = \sum_{A \in \mathrm{fex}(T)} \mathrm{fag}(m,T,A)\mathrm{mom}(x,\mathrm{fat}(T,A)).$$

Proof: Follows from (16.8).

THEOREM (16.10) Given any ring R and given any $m \in N(2)$ and $x \in \mathrm{mat}(R,m)$ and $T \in \mathrm{tab}(2,m)$ and $n \in N$ we have

$$\mathrm{mom}(x,T) = \sum_{S \in \mathrm{tid}(2,m,\mathrm{dep}(T))} \mathrm{rin}(m,T,S,n)\mathrm{mom}(x,S).$$

Proof: In view of the definition of rin, our assertion follows from (16.9).

THEOREM (16.11). Given any ring R and given any $m \in N(2)$ and $x \in \mathrm{mat}(R,m)$ and $T \in \mathrm{tab}(2,m)$, we have

$$\mathrm{mom}(x,T) = \sum_{S \in \mathrm{sit}(m,\mathrm{dep}(T))} \mathrm{fin}(m,T,S)\mathrm{mom}(x,S).$$

Proof: Follows from (16.6) and (16.10).

REMARK (16.12) If we replace (16.4.2) and (16.5.2) and (16.6.2) by their simpler versions (16.4.1) and (16.5.1) and (16.6.1), respectively, and if for every $m \in N(2)$ and $T \in \mathrm{tab}(2,m)$ and $S \in \mathrm{sit}(m, \mathrm{dep}(T))$, we define

$$\mathrm{fim}(m,T,S) = \mathrm{rin}(m,T,\mathrm{val}(m,\mathrm{dep}(T),S),\ 1 + \mathrm{hit}(T,\mathrm{tid}(2,m,\mathrm{dep}(T)))),$$

then we get the apparently weaker version of (16.11) obtained by changing fin to fim. By (16.6.2) we of course see that the apparently weaker version of (16.11) is actually equivalent to its original version.

17. PARTIAL ORDER AND THE FULL DEPTH CASE

In Theorem (17.2) we provide some information concerning the partial order of the terms with nonzero coefficients in the expansion given in Theorem (16.11). Then in Theorem (17.3) we state the resulting version of Theorem (16.11).

LEMMA (17.1) Given any $m \in N(2)$ and $T \in tab(2,m)$, upon letting $d = dep(T)$, we have the following.

(17.1.0) If $T \in qstab(2,m)$, then for some $S \in tid(2,m,d)$ we have $rin(m,T,S,1) \neq 0$.

(17.1.1) If $S \in tid(2,m,d)$ is such that $rin(m,T,S,n) \neq 0$ for some $n \in N$, then $S \leq T$.

(17.1.2) If $T \notin qstab(2,m)$ and $S \in tid(2,m,d)$ is such that $rin(m,T,S,n) \neq 0$ for some $n \in N^*$, then $S < T$.

(17.1.3) If $T \in qstab(2,m)$ and $S \in tid(2,m,d)$ is such that $rin(m,T,S,n) \neq 0$ for some $n \in N$, then $S = T$.

(17.1.4) If $T \in qstab(2,m)$, then for every $n \in N$ we have $rin(m,T,T,n) = 1$.

Proof: In view of the definition of rin, our assertions follow from (16.4.2) by induction on n.

THEOREM (17.2) Given any $m \in N(2)$ and $T \in tab(2,m)$, upon letting $d = dep(T)$, we have the following.

(17.2.1) If $S \in sit(m,d)$ is such that $fin(m,T,S) \neq 0$, then $S \leq T$.

(17.2.2) If $T \notin stab(2,m)$ and $S \in sit(m,d)$ is such that $fin(m,T,S) \neq 0$, then $S < T$.

(17.2.3) If $T \in stab(2,m)$ and $S \in sit(m,d)$ is such that $fin(m,T,S) \neq 0$, then $S = T$.

(17.2.4) If $T \in stab(2,m)$, then we have $fin(m,T,T) = 1$.

Proof: Follows from (17.1).

THEOREM (17.3) Given any $m \in N(2)$ and $T \in tab(2,m)$, there exists a subset $S^*(m,T)$ of $sit(m,dep(T))$ together with a nonzero integer $f^*(m,T,S)$ for every S in $S^*(m,T)$ such that for every S in $S^*(m,T)$

we have $S \le T$ and such that for every ring R and every x in mat(R,m) we have

$$\text{mom}(x,T) = \sum_{S \in S^*(m,T)} f^*(m,T,S)\text{mom}(x,S).$$

Proof: Follows from (16.11) and (17.2.1) by taking

$$S^*(m,T) = \{S \in \text{sit}(m,\text{dep}(T)) : \text{fin}(m,T,S) \neq 0\}$$

and

$$f^*(m,T,S) = \text{fin}(m,T,S).$$

18. CONTENT AND THE FULL DEPTH CASE

In Theorem (18.4) we provide some information concerning the content and area of the terms with nonzero coefficients in the expansion given in Theorem (16.11), and then in Theorem (18.5) we state the resulting version of Theorem (16.11). In Theorem (18.6) we comment on the number of steps required to achieve straightening, and then in Remark (18.7) we pose some related problems.

LEMMA (18.1) For every $T \in \text{bit}(2)$ and $A \in \text{nex}(T)$ we have the following.

(18.1.1) $\text{con}(\text{sat}(T,A)) = \text{con}(T)$.

(18.1.2) $\text{are}(\text{sat}(T,A)) = \text{are}(T)$.

Proof: Follows from the definitions of con and sat given in (11.1).

LEMMA (18.2) For every $T \in \text{tab}(2) \setminus \text{qstab}(2)$ and $A \in \text{fex}(T)$ we have the following.

(18.2.1) $\text{con}(\text{fat}(T,A)) = \text{con}(T)$.

(18.2.2) $\text{are}(\text{fat}(T,A)) = \text{are}(T)$.

Proof: In view of the definitions of fex and fat given in (16.1), our assertion follows from (18.1).

LEMMA (18.3) Given any $m \in N(2)$ and $T \in \text{tab}(2,m)$ and $S \in \text{tid}(2,m,\text{dep}(T))$, we have the following.

(18.3.1) If $\text{rin}(m,T,S,n) \neq 0$ for some $n \in N$, then $\text{con}(S) = \text{con}(T)$.

(18.3.2) If rin(m,T,S,n) $\neq$ 0 for some n $\in$ N, then are(S) = are(T).

Proof: In view of the definition of rin given in (16.1), our assertion follows from (18.2) by induction on n.

THEOREM (18.4) Given any m $\in$ N(2) and T $\in$ tab(2,m) and S $\in$ sit(m, dep(T)), we have the following.

(18.4.1) If fin(m,T,S) $\neq$ 0, then con(S) = con(T).

(18.4.2) If fin(m,T,S) $\neq$ 0, then are(S) = are(T).

Proof: In view of the definitions of sit and fin given in (16.1), our assertion follows from (18.3).

THEOREM (18.5) Given any m $\in$ N(2) and T $\in$ tab(2,m), there exists a subset S*(m,T) of sit(m,dep(T)) together with a nonzero integer f*(m,T,S) for every S in S*(m,T) such that for every S in S*(m,T) we have S $\leq$ T and con(S) = con(T) and are(S) = are(T) and such that for every ring R and every x in mat(R,m) we have

$$\mathrm{mom}(x,T) = \sum_{S\in S^*(m,T)} f^*(m,T,S)\,\mathrm{mom}(x,S).$$

Proof: Follows from (16.11) and (17.2.1) and (18.4) by taking

$$S^*(m,T) = \{S \in \mathrm{sit}(m,\mathrm{dep}(T)) : \mathrm{fin}(m,T,S) \neq 0\}$$

and

$$f^*(m,T,S) = \mathrm{fin}(m,T,S).$$

THEOREM (18.6) Given any m $\in$ N(2) and T $\in$ tab(2,m), upon letting d = dep(T), let us define step(m,T) to be the smallest member of N such that for every n $\in$ N with n $\geq$ step(m,T) and for every S $\in$ tid(2,m,d) we have

$$\mathrm{rin}(m,T,S,n) = \mathrm{rin}(m,T,S,\mathrm{step}(m,T))$$

[by (16.6.2) we see that this definition makes sense]. Then for every n $\in$ N with n $\geq$ step(m,T) and for every S $\in$ sit(m,d) we have

$$\mathrm{fin}(m,T,S) = \mathrm{rin}(m,T,\mathrm{val}(m,d,S),n).$$

Moreover, upon letting

$$\hat{T} = \{S \in \text{tid}(2,m,d) : \text{con}(S) = \text{con}(T)\},$$

we have

$$\text{step}(m,T) \le \text{hit}(T,\hat{T}) \le \text{hit}(T,\text{tid}(2,m,d)).$$

Proof: In view of (16.6.2) and (18.3.1), our assertion follows from (16.5.2) by taking Y = tid(2,m,d) and Y* = qstid(2,m,d) and $X = \hat{T}$ and g(T,S,n) = rin(m,T,S,n).

REMARK (18.7) Given any m ∈ N(2) and T ∈ tab(2,m), upon letting step(m,T) be as in (18.6), we may pose the problem of evaluating step(m,T); note that step(m,T) may be called the number of steps required to straighten T, i.e., the number of times we have to apply (16.8) to get (16.11). More significantly, we may pose the problem of determining the coefficients fin(m,T,S) occurring in the expansion given in (16.11). In particular, we may pose the problem of finding those S for which fin(m,T,S) ≠ 0.

19. AN ALTERNATIVE TREATMENT OF THE FULL DEPTH CASE

Note that Theorem (17.3) was mostly derived from (16.8), and in turn (16.8) was derived from (14.8) with the intervention of (15.8). Now, in view of (15.4), we can deduce (17.3) directly from (14.8), without the intervention of (15.8), by means of the following Proposition (19.1), which says that uti gives order isomorphisms. The details of the deduction may be left to the reader. The material of this section will not be used anywhere else in this monograph.

PROPOSITION (19.1) Let there be given any m ∈ N(2) and k ∈ [1,2], and let us put p = m(k') and q = m(k), and let us also put m* = ubi(m,k), i.e., let m* ∈ N(2) be defined by putting m*(k') = p and m*(k) = q + p. Then we have the following.

(19.1.1). $a \mapsto$ uti[m,k,a] gives an order isomorphism bic(m) → bic(m,k).

(19.1.2) $T \mapsto$ uti(m,k,T) gives a bijection tab(2,m) → tip(2,m,p).

(19.1.3) For every d ∈ N we have that $T \mapsto$ uti(m,k,T) gives a bijection bit(d,m) → bit(d,m,k).

(19.1.4) $T \mapsto \mathrm{uti}(m,k,T)$ gives a bijection qstab(2,m) $\to$ qstip(2,m,p).

(19.1.5) For every $d \in N$ we have that $T \mapsto \mathrm{uti}(m,k,T)$ gives a bijection qstid(2,m,d) $\to$ qstap[2,m,p,d].

Proof: Upon letting f and f* be the maps

$$[1,p] \underset{f^*}{\overset{f}{\rightleftarrows}} [q + 1, q + p]$$

obtained by putting

$$f(i) = q + p + 1 - i \quad \text{for all } i \in [1,p]$$

and

$$f^*(i) = q + p + 1 - i \quad \text{for all } i \in [q + 1, q + p]$$

we clearly have

(1) $f^*(f(i)) = i \quad$ for all $i \in [1,p]$

and

(2) $f(f^*(i)) = i \quad$ for all $i \in [q + 1, q + p]$.

For every $a \in \mathrm{bic}(m)$ let

$$c(a)\colon [1 + \mathrm{len}(a),p] \to [1,p]$$

be the unique strictly decreasing map such that

$$\mathrm{im}(c[a]) = [1,p] \setminus \mathrm{sim}[a](k').$$

For every $a^* \in \mathrm{bic}(m^*,k)$ let

$$L(a^*) = \mathrm{card}([1,q] \cap \mathrm{sim}[a^*](k))$$

and let

$$c^*[a^*]\colon [1,L(a^*)] \to [q + 1, q + p]$$

be the unique strictly decreasing map such that

$$\mathrm{im}(c^*[a^*]) = [q + 1, q + p] \setminus \mathrm{sim}[a^*](k).$$

Let u and u* be the unique maps

$$\mathrm{bic}(m) \underset{u^*}{\overset{u}{\rightleftarrows}} \mathrm{bic}(m^*,k)$$

such that for every $a \in \text{bic}(m)$ we have

$$u(a)(k,i) = \begin{cases} a(k,i) & \text{if } i \in [1,\text{len}(a)] \\ f(c[a](i)) & \text{if } i \in [1+\text{len}(a),p] \end{cases}$$

and such that for all $a^* \in \text{bic}(m^*,k)$ we have

$$\text{len}(u^*(a^*)) = L(a^*)$$

and

$$u^*(a^*)(\hat{k},i) = \begin{cases} a^*(k,i) & \text{if } \hat{k} = k \text{ and } i \in [1,L(a^*)] \\ f^*(c^*[a^*](i)) & \text{if } \hat{k} = k' \text{ and } i \in [1,L(a^*)]. \end{cases}$$

Because of (1) and (2) we see that

(3) $u^*(u(a)) = a \quad \text{for all } a \in \text{bic}(m)$

and

(4) $u(u^*(a^*)) = a^* \quad \text{for all } a^* \in \text{bic}(m^*,k).$

By the definition of uti we also see that

(5) $\text{uti}[m,k,a] = u(a) \quad \text{for all } a \in \text{bic}(m).$

By (3) to (5) we conclude that

(6) $a \mapsto \text{uti}[m,k,a]$ gives a bijection $\text{bic}(m) \to \text{bic}(m^*,k).$

As preparation for showing that u is an order isomorphism, first we note that if $a \in \text{bic}(m)$ is such that $\text{len}(a) = 0$, then

$$[1,p] \setminus \text{sim}[a](k') = \{1 < 2 < \cdots < p\}$$

and hence

$$c[a](i) = p + 1 - i \quad \text{for all } i \in [1,p]$$

and therefore

$$u(a)(k,i) = q + i \quad \text{for all } i \in [1,p];$$

thus

(7) $\left[\begin{array}{l} \text{if } a \in \text{bic}(m) \text{ is such that } \text{len}(a) = 0, \text{ then} \\ u(a)(k,i) = q + i \text{ for all } i \in [1,p]. \end{array}\right.$

Second, we note that for every $a^* \in \text{bic}(m^*,k)$ we have that $i \mapsto a^*(k,i)$ gives a strictly increasing map $[1,p] \to [1, q + p]$ and hence

$$a^*(k,i) \le q + i \quad \text{for all } i \in [1,p];$$

thus

(8) $\left[\begin{array}{l}\text{for every } a^* \in \text{bic}(m^*,k) \text{ we have} \\ a^*(k,i) \le q + i \text{ for all } i \in [1,p].\end{array}\right.$

Third, we note that if a and b in bic(m) are such that $\text{len}(b) = 0$ then clearly $a \le b$ and for all $i \in [1,p]$ we have

$$\begin{aligned} u(a)(k,i) &\le q + i && [\text{by taking } a^* = u(a) \text{ in (8)}] \\ &= u(b)(k,i) && [\text{by taking } b \text{ for } a \text{ in (7)}] \end{aligned}$$

and hence $u(a) \le u(b)$; thus

(9) $\left[\begin{array}{l}\text{if a and b in bic(m) are such that } \text{len}(b) = 0, \\ \text{then } a \le b \text{ and } u(a) \le u(b).\end{array}\right.$

Fourth, we note that if a and b in bic(m) are such that $\text{len}(a) = 0 \ne \text{len}(b)$, then clearly $a \nleq b$ and also

$$\begin{aligned} u(a)(k,1) &= q + 1 && [\text{by (7)}] \\ &> b(k,1) && (\text{obviously}) \\ &= u(b)(k,1) && (\text{obviously}) \end{aligned}$$

and hence $u(a) \nleq u(b)$; thus

(10) $\left[\begin{array}{l}\text{if a and b in bic(m) are such that } \text{len}(a) = 0 \ne \text{len}(b), \\ \text{then } a \nleq b \text{ and } u(a) \nleq u(b).\end{array}\right.$

Fifth, we note that if a and b in bic(m) are such that $\text{len}(a) \ne 0 \ne \text{len}(b)$ and $a(k,1) \nleq b(k,1)$, then clearly $a \nleq b$ and also

$$u(a)(k,1) = a(k,1) \nleq b(k,1) = u(b)(k,1)$$

and hence $u(a) \nleq u(b)$; thus

(11) $\left[\begin{array}{l}\text{if a and b in bic(m) are such that} \\ \text{len}(a) \ne 0 \ne \text{len}(b) \text{ and } a(k,1) \nleq b(k,1) \\ \text{then } a \nleq b \text{ and } u(a) \nleq u(b).\end{array}\right.$

Sixth, for a moment let there be given any a and b in bic(m) such that $\text{len}(a) \ne 0 \ne \text{len}(b)$, and to simplify notation, let us put

$$r = a(k',1) \quad \text{and} \quad s = b(k',1)$$

and let $\tilde{a}$ and $\tilde{b}$ in bic(m) be defined by saying that

(12) $\mathrm{len}(\tilde{a}) = \mathrm{len}(a) - 1 \quad \text{and} \quad \mathrm{len}(\tilde{b}) = \mathrm{len}(b) - 1$

and

(13) $\tilde{a}(\hat{k},i) = a(\hat{k}, i + 1) \quad \text{for all } \hat{k} \in [1,2] \text{ and } i \in [1,\mathrm{len}(\tilde{a})]$

and

(14) $\tilde{b}(\hat{k},i) = b(\hat{k}, i + 1) \quad \text{for all } \hat{k} \in [1,2] \text{ and } i \in [1,\mathrm{len}(\tilde{b})]$.

With this notation we clearly have

(15) $[[1,p] \setminus \mathrm{sim}[a](k')] \cap [1,r] = [1, r - 1]$

and hence

$$c[a](i) = p + 1 - i \quad \text{for all } i \in [p - r + 2, p]$$

and

$$c[a](i) > p + 1 - i \quad \text{for } i = p - r + 1$$

and therefore,

(16) $u(a)(k,i) = q + i \quad \text{for all } i \in [p - r + 2, p]$

and

(17) $u(a)(k,i) < q + i \quad \text{for } i = p - r + 1$

and similarly,

(18) $[[1,p] \setminus \mathrm{sim}[b](k')] \cap [1,s] = [1, s - 1]$

and hence

$$c[b](i) = p + 1 - i \quad \text{for all } i \in [p - s + 2, p]$$

and

$$c[b](i) > p + 1 - i \quad \text{for } i = p - s + 1$$

and therefore,

(19) $u(b)(k,i) = q + i \quad \text{for all } i \in [p - s + 2, p]$

and

(20) $u(b)(k,i) < q + i \quad \text{for } i = p - s + 1$

Now if $r \nleq s$, then clearly $a \nleq b$ and also

$$u(a)(k, p+1-s) = q + p + 1 - s \qquad \text{[by (16)]}$$
$$> u(b)(k, p+1-s) \qquad \text{[by (20)]}$$

and hence $u(a) \nleq u(b)$; thus

$$\left[\begin{array}{l}\text{if } a \text{ and } b \text{ in } \mathrm{bic}(m) \text{ are such that} \\ \mathrm{len}(a) \neq 0 \neq \mathrm{len}(b) \text{ and } a(k',1) \nleq b(k',1), \\ \text{then } a \nleq b \text{ and } u(a) \nleq u(b).\end{array}\right.$$

With the notation above we also have

$$[1,p] \setminus \mathrm{sim}[\tilde{a}](k') = \{r\} \coprod [[1,p] \setminus \mathrm{sim}[a](k')]$$

and hence by (15) we see that

$$c[\tilde{a}](i) = \begin{cases} c[a](i+1) & \text{for all } i \in [\mathrm{len}(a), p - r] \\ c[a](i) & \text{for all } i \in [p - r + 2, p] \\ p + 1 - i & \text{for all } i \in [p - r + 1, p] \end{cases}$$

and therefore,

$$\text{(22)} \quad u(\tilde{a})(k,i) = \begin{cases} u(a)(k, i+1) & \text{if } i \in [1, p - r] \\ u(a)(k,i) & \text{if } i \in [p - r + 2, p] \\ q + i & \text{if } i \in [p - r + 1, p] \end{cases}$$

and similarly,

$$[1,p] \setminus \mathrm{sim}[\tilde{b}](k') = \{s\} \coprod [[1,p] \setminus \mathrm{sim}[b](k')]$$

and hence by (18) we see that

$$c[\tilde{b}](i) = \begin{cases} c[b](i+1) & \text{for all } i \in [\mathrm{len}(b), p - s] \\ c[b](i) & \text{for all } i \in [p - s + 2, p] \\ p + 1 - i & \text{for all } i \in [p - s + 1, p] \end{cases}$$

and therefore,

$$\text{(23)} \quad u(\tilde{b})(k,i) = \begin{cases} u(b)(k, i+1) & \text{if } i \in [1, p - s] \\ u(b)(k,i) & \text{if } i \in [p - s + 2, p] \\ q + i & \text{if } i \in [p - s + 1, p]. \end{cases}$$

If $r \leq s$, then for all $i \in [p - s + 2, p]$ we have

$$\begin{aligned} u(a)(k,i) &\le q + i && \text{[by taking } a^* = u(a) \text{ in (8)]} \\ &= u(b)(k,i) && \text{[by (19)];} \end{aligned}$$

thus

$$(24)\quad \begin{bmatrix} \text{if } r \le s, \text{ then} \\ u(a)(k,i) \le u(b)(k,i) \text{ for all } i \in [p - s + 2, p]. \end{bmatrix}$$

If $r \le s$, then for all $i \in [p - s + 1, p]$ we have

$$\begin{aligned} u(\tilde{a})(k,i) &\le q + i && \text{[by taking } a^* = u(\tilde{a}) \text{ in (8)]} \\ &= u(\tilde{b})(k,i) && \text{[by (23)];} \end{aligned}$$

thus

$$(25)\quad \begin{bmatrix} \text{if } r \le s, \text{ then} \\ u(\tilde{a})(k,i) \le u(\tilde{b})(k,i) \text{ for all } i \in [p - s + 1, p]. \end{bmatrix}$$

If $a(k,1) \le b(k,1)$ and $r \le s$ and $u(\tilde{a}) \le u(\tilde{b})$, then for all $i \in [1, p - s]$ we have

$$\begin{aligned} u(a)(k, i + 1) &= u(\tilde{a})(k,i) && \text{[by (22)]} \\ &\le u(\tilde{b})(k,i) && \text{[because } u(\tilde{a}) \le u(\tilde{b})\text{]} \\ &= u(b)(k, i + 1) && \text{[by (23)]} \end{aligned}$$

and clearly, we have

$$u(a)(k,1) = a(k,1) \le b(k,1) = u(b)(k,i);$$

thus

$$(26)\quad \begin{bmatrix} \text{if } a(k,1) \le b(k,1) \text{ and } r \le s \text{ and } u(\tilde{a}) \le u(\tilde{b}), \text{ then} \\ u(a)(k,i) \le u(b)(k,i) \text{ for all } i \in [1, p - s + 1]. \end{bmatrix}$$

If $r \le s$ and $u(a) \le u(b)$, then for all $i \in [1, p - s]$ we have

$$\begin{aligned} u(\tilde{a})(k,i) &= u(a)(k, i + 1) && \text{[by (22)]} \\ &\le u(b)(k, i + 1) && \text{[because } u(a) \le u(b)\text{]} \\ &= u(\tilde{b})(k,i) && \text{[by (23)];} \end{aligned}$$

thus

$$(27)\quad \begin{bmatrix} \text{if } r \le s \text{ and } u(a) \le u(b), \text{ then} \\ u(\tilde{a})(k,i) \le u(\tilde{b})(k,i) \text{ for all } i \in [1, p - s]. \end{bmatrix}$$

By (24) and (26) we see that

(28) $\left[\begin{array}{l}\text{if a and b in bic(m) are such that len(a)} \neq 0 \neq \text{len(b)} \\ \text{and a(k,1)} \le \text{b(k,1) and a(k',1)} \le \text{b(k',1) and} \\ \text{u}(\tilde{a}) \le \text{u}(\tilde{b})\text{, where } \tilde{a} \text{ and } \tilde{b} \text{ are the members of bic(m)} \\ \text{defined by (12) to (14), then we have u(a)} \le \text{u(b).}\end{array}\right.$

By (25) and (27) we see that

(29) $\left[\begin{array}{l}\text{if a and b in bic(m) are such that len(a)} \neq 0 \neq \text{len(b)} \\ \text{and a(k,1)} \le \text{b(k,1) and a(k',1)} \le \text{b(k',1) and} \\ \text{u}(\tilde{a}) \not\le \text{u}(\tilde{b})\text{, where } \tilde{a} \text{ and } \tilde{b} \text{ are the members of bic(m)} \\ \text{defined by (12) to (14), then we have u(a)} \not\le \text{u(b).}\end{array}\right.$

By induction on len(b) we show that given any a and b in bic(m) we have that

(30) $\left[\begin{array}{l}\text{if a} \le \text{b, then u(a)} \le \text{u(b),} \\ \text{whereas} \\ \text{if a} \not\le \text{b, then u(a)} \not\le \text{u(b).}\end{array}\right.$

For len(b) = 0, the assertion follows from (9). So let len(b) > 0 and assume that the assertion is true for all values of len(b) smaller than the given one. Now first if len(a) = 0, then the assertion follows from (10). Second, if len(a) > 0 and a(k,1) $\not\le$ b(k,1), then the assertion follows from (11). Third, if len(a) > 0 and a(k',1) $\not\le$ b(k',1), then the assertion follows from (21). Finally, suppose that len(a) > 0 and a(k,1) $\le$ b(k,1) and a(k',1) $\le$ b(k',1) and let $\tilde{a}$ and $\tilde{b}$ be the members of bic(m) defined by (12) and (14). Now obviously,

$\left[\begin{array}{l}\text{if a} \le \text{b, then } \tilde{a} \le \tilde{b}, \\ \text{whereas} \\ \text{if a} \not\le \text{b, then } \tilde{a} \not\le \tilde{b}\end{array}\right.$

and clearly len($\tilde{b}$) < len(b) and hence, in view of the induction hypothesis, the assertion follows from (28) and (29). This completes the induction.

By (3), (4), (5), and (30) we see that

(31) $\left[\begin{array}{l}\text{a} \mapsto \text{uti[m,k,a] gives an order isomorphism} \\ \text{bic(m)} \to \text{bic(m*,k)}\end{array}\right.$

and this completes the proof of (19.1.1). (19.1.2) and (19.1.3) follow from (6). (19.1.4) and (19.1.5) follow from (31).

4
Applications to Ideal Theory

20. HILBERTIAN IDEALS AND GENERALIZED SECOND FUNDAMENTAL THEOREM OF INVARIANT THEORY

(20.1) Notation and Summary In this section we follow the terminology introduced in Section 2 of Chapter 1 and Subsection (11.1) of Chapter 3.

Now we introduce some notation to be used in the rest of this section. So let there be fixed any $m \in N^*(2)$ and let there be fixed any field K. Also, let there be fixed any $X \in mat(R,m)$, where R is an overfield of K, such that the m(1)m(2) elements X_{ij}, with (i,j) ranging over rec(m(1),m(2)), are independent indeterminates over K.

Let K[X] be the ring of polynomials in the m(1)m(2) indeterminates X_{ij} with coefficients in K, and let K(X) be the corresponding field of rational functions; i.e., let

K[X] = the subring of R generated over K by the m(1)m(2) elements X_{ij} with (i,j) ranging over rec(m(1),m(2))

and

K(X) = the quotient field of K[X] in R.

For every $V \in N$ we put

$K[X]_V$ = the set of all homogeneous polynomials of degree V in the m(1)m(2) indeterminates X_{ij}, together with the zero polynomial

and we note that in the usual sense which we shall recall toward the end of Subsection (20.2), K[X] is a homogeneous ring whose Vth homogeneous component is $K[X]_V$, so in particular,

$$K[X] = \sum_{V \in N} K[X]_V$$

where the sum is direct, and we also note that

$$K[X]_0 = K.$$

Before introducing any more notation, let us summarize the results to be used and the results to be proved. First, let us recall that in Section 16 of Chapter 3, for every $T \in \text{tab}(2,m)$ and $S \in \text{sit}(m,\text{dep}(T))$, where $\text{sit}(m,\text{dep}(T))$ is the set of all members of $\text{stab}(2,m)$ whose depth is less equal the depth of T, we have defined an integer $\text{fin}(m,T,S)$ and in Theorem (16.11) of Chapter 3 we have proved the existence part of the Straightening Law, which says that

$$\left[\begin{array}{l}\text{for every } T \in \text{tab}(2,m) \text{ we have} \\ \text{mom}(X,T) = \displaystyle\sum_{S \in \text{sit}(m,\text{dep}(T))} \text{fin}(m,T,S)\text{mom}(X,S)\end{array}\right. \tag{20.1.1}$$

and in Theorem (18.4.2) of Chapter 3 we have noted that

$$\left[\begin{array}{l}\text{given any } T \in \text{tab}(2,m) \text{ and } S \in \text{sit}(m,\text{dep}(T)) \text{ we} \\ \text{have that if } \text{fin}(m,T,S) \neq 0 \text{ then } \text{are}(S) = \text{are}(T)\end{array}\right. \tag{20.1.2}$$

and in Theorem (18.4.1) of Chapter 3 we have noted that

$$\left[\begin{array}{l}\text{given any } T \in \text{tab}(2,m) \text{ and } S \in \text{sit}(m,\text{dep}(T)) \text{ we} \\ \text{have that if } \text{fin}(m,T,S) \neq 0 \text{ then } \text{con}(S) = \text{con}(T)\end{array}\right. \tag{20.1.3}$$

and in Theorem (17.2.1) of Chapter 3 we have noted that

$$\left[\begin{array}{l}\text{given any } T \in \text{tab}(2,m) \text{ and } S \in \text{sit}(m,\text{dep}(T)) \\ \text{we have that if } \text{fin}(m,T,S) \neq 0 \text{ then } S \le T\end{array}\right. \tag{20.1.4}$$

where the meaning of the partial order $S \le T$ for bitableaux is as given in Section 16; concerning this partial order, the only thing that we need to know here is that if $S \le T$ and $\text{dep}(S) \neq 0 \neq \text{dep}(T)$, then $S[1] \le T[1]$ in the sense of the partial order for bivectors as defined in Subsection (2.2) of Chapter 1; consequently, by (20.1.2) and (20.1.4) we see that

$$\left[\begin{array}{l}\text{given any } T \in \text{tab}(2,m) \text{ and } S \in \text{sit}(m,\text{dep}(T)) \\ \text{we have that if are}(T) \neq 0 \text{ and fin}(m,T,S) \neq 0 \\ \text{then dep}(S) \neq 0 \neq \text{dep}(T) \text{ and } S[1] \leq T[1].\end{array}\right. \tag{20.1.5}$$

Let us recall that in Theorem (9.11) of Chapter 2 we have proved that

$$\text{for every } V \in N \text{ we have Stib}(2,m,V) = \text{Mon}[[2,m,V]]. \tag{20.1.6}$$

Let us also recall that in Theorem (9.9) of Chapter 2 we have proved that

$$\left[\begin{array}{l}\text{for every } p \in N \text{ and } a \in \text{vec}(2,m,p) \text{ and } V \in N \\ \text{we have Stab}(2,m,p,a,V) = \text{Mon}(2,m,p,a,V)\end{array}\right. \tag{20.1.7}$$

and let us note that although in Theorem (9.9) we have proved the statement above only for $p \neq 0$, in the remaining case of $p = 0$ we obviously have

$$\text{Stab}(2,m,0,a,0) = 1 = \text{Mon}(2,m,0,a,0)$$

and

$$\text{Stab}(2,m,0,a,V) = 0 = \text{Mon}(2,m,0,a,V) \qquad \text{for all } V \in N^*.$$

Finally, let us recall that in Theorem (9.7) of Chapter 2 we have proved that

$$\left[\begin{array}{l}\text{for every } p \in N^* \text{ and } a \in \text{vec}(2,m,p) \text{ and } V \in N \\ \text{and } L \in \{1,1^*,2,2^*,3,3^*\} \text{ and } k \in [1,2] \text{ we have} \\ \text{Stab}(2,m,p,a,V) = F^{(Lk)}(m,p,a,V)\end{array}\right. \tag{20.1.8}$$

where, in view of (2.10), (2.11.11), (2.11.12), and (2.11.13) of Chapter 1 and (9.13) and (9.17) of Chapter 2, we have

$$F^{(Lk)}(m,p,a,V) = \sum_{D\in[0,C]} (-1)^D F_D^{(Lk)}(m,p,a) \begin{bmatrix} V \\ C - D \end{bmatrix} \tag{20.1.9}$$

with

$$\left[\begin{array}{l}C = C(m,p,a) \\ \quad = p - 1 + \sum_{k\in[1,2]} \sum_{i\in[1,p]} [m(k) - a(k,i)]\end{array}\right. \tag{20.1.10}$$

and

$$\begin{bmatrix} V \\ C - D \end{bmatrix} = \begin{pmatrix} V + C - D \\ C - D \end{pmatrix} = \frac{(V + C - D)!}{V!\ (C - D)!} \tag{20.1.11}$$

and

$$F_D^{(Lk)}(m,p,a) \in N \qquad \text{for all } D \in [0,C] \tag{20.1.12}$$

and

$$F_0^{(Lk)}(m,p,a) \in N^*. \tag{20.1.13}$$

In Theorem (20.3), as a consequence of (20.1.6), we prove the uniqueness part of the Straightening, i.e., we prove the uniqueness of the expansion given in (20.1.1), and then in Theorem (20.4) we show that the resulting K-basis of K[X] partitions well for various ideals in K[X], and then, as a consequence of (20.1.7), in Theorem (20.10) we establish certain modifications of the said basis which are parametrized by mon(2,m).

Finally, for any $p \in N^*$ and $a \in vec(2,m,p)$, we consider a certain ideal I(p,a) in K[X] generated by a certain family of different size minors of X and, as a consequence of (20.1.8), in Theorem (20.14) we show that I(p,a) is hilbertian because its Hilbert function is $F^{(Lk)}(m,p,a,V)$, where the terms "hilbertian" and "Hilbert function" are defined at the end of Subsection (20.2). In Theorem (20.14) we also show that the ideal I(p,a) is prime and the quotient field of the residue class ring K[X]/I(p,a) is a pure transcendental extension of K. In Theorem (20.16) we show that certain more general versions of the ideal are also prime. In Theorem (20.15) we state a special case of this which says that given any nonnegative integer p, the ideal I[p + 1] generated by all the p + 1 by p + 1 minors of X is a prime ideal in K[X]; we also give a direct proof of this special case which does not use (20.1.7) and which is independent of items (20.6) to (20.14); the reader may like to read this direct proof before reading items (20.6) to (20.14). This special case was proved by Eduardo Pascal in the nineteenth century as part of the Second Fundamental Theorem of Invariant Theory; the relevant paper of Pascal appeared in Volume 4a of Series V of *Mem. del R.*

Acc. dei Lincei for the year 1888. So Theorems (20.14) to (20.17) may be called Generalizations of the Second Fundamental Theorem. In (20.18) and (20.19) we make some relevant remarks, and in (20.20) we pose a related problem. It may be noted that various proofs of Pascal's Theorem, i.e., our Theorem (20.15), have recently been given by Mount in Volume 42 of the *Journal of the London Mathematical Society* for the year 1967, and by Eagon-Hochster in Volume 93 of the *American Journal of Mathematics* for the year 1971, and by DeConcini-Eisenbud-Procesi in Volume 56 of *Invent. Math.* for the year 1980, and by Abhyankar in Volume 42 of *Rendiconti del Seminario Matematico di Universita et Politecnico di Torino* for the year 1984.

(20.2) More Notation In this subsection we introduce some more notation to be used in the rest of this section.

Given any subset Y of rec(m(1),m(2)), let K[Y] be the ring of polynomials in the indeterminates X_{ij} with (i,j) ranging over Y and with coefficients in K, and let K(Y) be the corresponding field of rational functions; i.e., let

K[Y] = the subring of R generated over K by the elements X_{ij} as (i,j) ranges over Y

and

K(Y) = the quotient field of K[Y] in R.

Note that then

$$K[X] = K[\mathrm{rec}(m(1),m(2))] \quad \text{and} \quad K(X) = K(\mathrm{rec}(m(1),m(2))).$$

Given any subset Y of rec(m(1),m(2)), for every $V \in N$ we put

$$K[Y]_V = K[Y] \cap K[X]_V$$

and we note that K[Y] is a homogeneous subring of K[X] and in particular we have

$$K[Y] = \sum_{V \in N} K[Y]_V$$

where the sum is direct, and we also note that

$$K[Y]_0 = K.$$

By a *saturated subset* of rec(m(1),m(2)) we mean a subset Y of rec(m(1),m(2)) having the property which says that if b ∈ vec(2,m) is such that

$$(b(1,i),b(2,i)) \in Y \quad \text{for all } i \text{ in } [1,\text{len}(b)],$$

then

$$(b(1,i),b(2,j)) \in Y \quad \text{for all } i \text{ and } j \text{ in } [1,\text{len}(b)].$$

Given any p ∈ [0,min(m(1),m(2))], by E[p] we denote the *minimal element* of vec(2,m,p); i.e., we let E[p] be the unique member of vec(2,m,p) such that for all U ∈ [1,p] we have

$$E[p](1,U) = U = E[p](2,U).$$

Given any p ∈ [0,min(m(1),m(2))], first we let

$$G[p + 1] = \text{vec}(2, m, p + 1)$$

and

$$\hat{G}[p + 1] = \{\text{mor}(X,b) : b \in G[p + 1]\}$$

and we let I[p + 1] be the ideal in K[X] generated by $\hat{G}[p + 1]$; i.e., we let

$$I[p + 1] = \hat{G}[p + 1]K[X]$$

and we let

$$I[p + 1]_V = I[p + 1] \cap K[X]_V \quad \text{for all } V \in N$$

and we note that in the sense which we shall recall toward the end of this subsection, I[p + 1] is a homogeneous ideal in K[X] and its Vth homogeneous component is $I[p + 1]_V$, and second, for every subset Y of rec(m(1),m(2)) we let

$$G^*[p + 1, Y] = \{b \in G[p + 1] : (b(1,i),b(2,j)) \in Y \text{ for all } i \text{ and } j \text{ in } [1,\text{len}(b)]\}$$

and

$$\hat{G}^*[p + 1, Y] = \{\text{mor}(X,b) : b \in G^*[p + 1, Y]\}$$

and

$$I^*[p + 1, Y] = \hat{G}^*[p + 1, Y]K[Y]$$

and

$$I^{**}[p + 1, Y, X] = \hat{G}^*[p + 1, Y]K[X]$$

and

$$I^*[p + 1, Y]_V = I^*[p + 1, Y] \cap K[Y]_V \quad \text{for all } V \in N$$

and

$$I^{**}[p + 1, Y, X]_V = I^{**}[p + 1, Y, X] \cap K[X]_V \quad \text{for all } V \in N$$

and we note that $I^*[p + 1, Y]$ and $I^{**}[p + 1, Y, X]$ are homogeneous ideals in $K[Y]$ and $K[X]$, respectively, and their Vth homogeneous components are $I^*[p + 1, Y]_V$ and $I^{**}[p + 1, Y, X]_V$, respectively, and third, we let

$$y[p + 1] = \{(i,j) \in \text{rec}(m(1),m(2)) : \text{either } i \in [1,p] \text{ or } j \in [1,p]\},$$

and fourth, for every

$$(i,j) \in \text{rec}(m(1),m(2)) \setminus y[p + 1]$$

we define $A[p + 1, i, j]$ to be the unique member of $\text{vec}(2, m, p + 1)$ such that for all $u \in [1,p]$ we have

$$A[p + 1, i, j](1,u) = u = A[p + 1, i, j](2,u)$$

and such that for $u = p + 1$ we have

$$A[p + 1, i, j](1,u) = i \quad \text{and} \quad A[p + 1, i, j](2,u) = j$$

and we define $\tilde{A}[X, p + 1, i, j]$ to be the unique member of $K[y[p + 1]]$ such that (upon expanding $\text{mor}(X,A[p + 1, i, j])$ by the last column we get)

$$\text{mor}(X,A[p + 1, i, j]) = \text{mor}(X,E[p])X_{ij} - \tilde{A}[X, p + 1, i, j],$$

and fifth, we define

$$f[p + 1]: K[X] \to K(y[p + 1])$$

to be unique K-homomorphism of rings such that for all

$$(i,j) \in y[p + 1]$$

we have

$$f[p + 1](X_{ij}) = X_{ij}$$

and for all

$$(i,j) \in \operatorname{rec}(m(1),m(2)) \setminus y[p + 1]$$

we have

$$f[p + 1](X_{ij}) = \frac{\tilde{A}[X,\ p + 1,\ i,\ j]}{\operatorname{mor}(X,E[p])}$$

where we note that the denominator is nonzero because it is the determinant of a u by u matrix whose u^2 entries are independent indeterminates.

As usual, by ker(f) we shall denote the kernel of a map f: H → H* where H and H* are additive abelian groups and f is a group homomorphism.

In this section we want to show that given any saturated subset Y of rec(m(1),m(2)) we have K[Y] ∩ ker(f[p + 1]) = I*[p + 1, Y] and so in particular ker(f[p + 1]) = I[p + 1]. Actually, we want to prove a generalization of this. In order that the arguments be applicable to the generalization, we first generalize the notation above.

Given any p ∈ N and a ∈ vec(2,m,p), first for every k ∈ [1,2] and u ∈ [1,p] we let

$$G(p,a,k,u) = \{b \in \operatorname{vec}(2,m,u) : b(k,u) < a(k,u)\}$$

and

$$\hat{G}(p,a,k,u) = \{\operatorname{mor}(X,b) : b \in G(p,a,k,u)\}$$

and

$$I(p,a,k,u) = \hat{G}(p,a,k,u)K[X]$$

and

$$I(p,a,k,u)_V = I(p,a,k,u) \cap K[X]_V \qquad \text{for all } V \in N$$

and we note that I(p,a,k,u) is a homogeneous ideal in K[X] and its Vth homogeneous component is $I(p,a,k,u)_V$, and second, we let

$$G(p,a) = G[p + 1] \cup \left[\bigcup_{k\in[1,2]} \bigcup_{u\in[1,p]} G(p,a,k,u) \right]$$

and

$$\hat{G}(p,a) = \{mor(X,b) : b \in G(p,a)\}$$

and

$$I(p,a) = \hat{G}(p,a)K[X]$$

and

$$I(p,a)_V = I(p,a) \cap K[X]_V \qquad \text{for all } V \in N$$

and we note that I(p,a) is a homogeneous ideal in K[X] and its Vth homogeneous component is $I(p,a)_V$, and third, for every $k \in [1,2]$ and $u \in [1,p]$ and for every subset Y of rec(m(1),m(2)) we let

$$G^*(p,a,k,u,Y) = \{b \in G(p,a,k,u) : (b(1,i),b(2,j)) \in Y \text{ for all i and j in } [1,len(b)]\}$$

and

$$\hat{G}^*(p,a,k,u,Y) = \{mor(X,b) : b \in G^*(p,a,k,u,Y)\}$$

and

$$I^*(p,a,k,u,Y) = \hat{G}^*(p,a,k,u,Y)K[Y]$$

and

$$I^{**}(p,a,k,u,Y,X) = \hat{G}^*(p,a,k,u,Y)K[X]$$

and

$$I^*(p,a,k,u,Y)_V = I^*(p,a,k,u,Y) \cap K[Y]_V \qquad \text{for all } V \in N$$

and

$$I^{**}(p,a,k,u,Y,X)_V = I^{**}(p,a,k,u,Y,X) \cap K[X]_V \qquad \text{for all } V \in N$$

and we note that I*(p,a,k,u,Y) and I**(p,a,k,u,Y,X) are homogeneous ideals in K[Y] and K[X], respectively, and their Vth homogeneous components are $I^*(p,a,k,u,Y)_V$ and $I^{**}(p,a,k,u,Y,X)_V$, respectively, and fourth, for every subset Y of rec(m(1),m(2)) we let

$$G^*(p,a,Y) = \{b \in G(p,a) : (b(1,i),b(2,j)) \in Y \text{ for all } i \text{ and } j \text{ in } [1,\text{len}(b)]\}$$

and

$$\hat{G}^*(p,a,Y) = \{\text{mor}(X,b) : b \in G^*(p,a,Y)\}$$

and

$$I^*(p,a,Y) = \hat{G}^*(p,a,Y)K[Y]$$

and

$$I^{**}(p,a,Y,X) = \hat{G}^*(p,a,Y)K[X]$$

and

$$I^*(p,a,Y)_V = I^*(p,a,Y) \cap K[Y]_V \quad \text{for all } V \in N$$

and

$$I^{**}(p,a,Y,X)_V = I^{**}(p,a,Y,X) \cap K[X]_V \quad \text{for all } V \in N$$

and we note that $I^*(p,a,Y)$ and $I^{**}(p,a,Y,X)$ are homogeneous ideals in K[Y] and K[X], respectively, and their Vth homogeneous components are $I^*(p,a,Y)_V$ and $I^{**}(p,a,Y,X)_V$, respectively.

Given any $p \in N$ and $a \in \text{vec}(2,m,p)$, first for every $u \in [0,p]$, we define E(p,a,u) to be the unique member of vec(2,m,u) such that for all $U \in [1,u]$ we have

$$E(p,a,u)(1,U) = a(1,U) \quad \text{and} \quad E(p,a,u)(2,U) = a(2,U)$$

and second, for every $u \in [1,p]$ we define

$$z(p,a,u) = \{(i,j) \in \text{rec}(m(1),m(2)) : i \in [a(1,u),m(1)] \text{ and } j \in [a(2,u),m(2)]\}$$

and

$$z_0(p,a,u) = \{(i,j) \in \text{rec}(m(1),m(2)) : i \in [a(1,u) + 1, m(1)] \text{ and } j \in [a(2,u) + 1, m(2)]\}$$

and third, we define

$$z(p, a, p + 1) = \emptyset,$$

and fourth, we note that

$$\left[\begin{array}{l}\text{for all } u \in [1,p] \text{ we have}\\ \text{rec}(m(1),m(2)) \supset z(p,a,u) \supset z_0(p,a,u) \supset z(p, a, u + 1),\end{array}\right.$$

and fifth, for every $u \in [1,p]$ we define

$$y(p,a,u) = \bigcup_{U\in[1,u]} [z(p,a,U) \setminus z_0(p,a,U)],$$

and sixth, for every

$$(i,j) \in z(p,a,1) \setminus y(p,a,p)$$

we define $u^*(p,a,i,j)$ to be the unique member of $[1,p]$ such that

$$(i,j) \in z_0(p,a,u^*(p,a,i,j)) \setminus z(p,a,u^*(p,a,i,j) + 1)$$

and we define $A(p,a,i,j)$ to be the unique member of $\text{vec}(2,m,u^*(p,a,i,j) + 1)$ such that for all $u \in [1,u^*(p,a,i,j)]$ we have

$$A(p,a,i,j)(1,u) = a(1,u) \quad \text{and} \quad A(p,a,i,j)(2,u) = a(2,u)$$

and such that for $u = u^*(p,a,i,j) + 1$ we have

$$A(p,a,i,j)(1,u) = i \quad \text{and} \quad A(p,a,i,j)(2,u) = j$$

and we define $\tilde{A}(X,p,a,i,j)$ to be the unique member of $K[y(p,a,u^*(p,a,i,j))]$ such that [upon expanding $\text{mor}(X,A(p,a,i,j))$ by the last column we get]

$$\text{mor}(X,A(p,a,i,j)) = \text{mor}(X,E(p,a,u^*(p,a,i,j)))X_{ij} - \tilde{A}(X,p,a,i,j)$$

and finally we define

$$f(p,a)\colon K[X] \to K(y(p,a,p))$$

to be the unique K-homomorphism of rings such that for all

$$(i,j) \in \text{rec}(m(1),m(2)) \setminus z(p,a,1)$$

we have

$$f(p,a)(X_{ij}) = 0$$

and for all

$$(i,j) \in y(p,a,p)$$

we have

$$f(p,a)(X_{ij}) = X_{ij}$$

and for all

$$(i,j) \in z(p,a,1) \setminus y(p,a,p)$$

we have

$$f(p,a)(X_{ij}) = \frac{\tilde{A}(X,p,a,i,j)}{mor(X,E(p,a,u^*(p,a,i,j)))}$$

where we note that the denominator is nonzero because it is the determinant of a $u^*(p,a,i,j)$ by $u^*(p,a,i,j)$ matrix whose $u^*(p,a,i,j)^2$ entries are independent indeterminates.

Let us note that given any $p \in [0,\min(m(1),m(2))]$, we have

$$I[p+1] = I(p,E[p]) \quad \text{and} \quad f[p+1] = f(p,E[p])$$

and

$$y[p+1] = y(p,E[p],p)$$

and for every subset Y of rec(m(1),m(2)) we have

$$I^*[p+1, Y] = I^*(p,E[p],Y)$$

and

$$I^{**}[p+1, Y, X] = I^{**}(p,E[p],Y,X).$$

Given any $t \in mon(2,m)$ we let X^t be the monomial in the m(1)m(2) indeterminates X_{ij} obtained by putting

$$X^t = \prod_{r \in cub(2,m)} X_{r(1),r(2)}^{t(r)}.$$

Given any V in N, and given any t and t* in mon[[2,m,V]] we define $t < t^*$ to mean that there exists $r \in cub(2,m)$ such that $t(r) > t^*(r)$, and such that $t(\tilde{r}) = t^*(\tilde{r})$ for every $\tilde{r} \in cub(2,m)$ for which $\tilde{r}(1) < r(1)$, and such that $t(\hat{r}) = t^*(\hat{r})$ for every $\hat{r} \in cub(2,m)$ for which $\hat{r}(1) = r(1)$ and $\hat{r}(2) < r(2)$; note that this gives a total order on mon[[2,m,V]], so in particular we may speak of the smallest element of any nonempty subset of mon[[2,m,V]].

Given any $b \in \text{vec}(2,m)$ we introduce the *associated protomonomial* asom(b) by saying that asom(b) is the unique member of mon(2,m) such that

$$\text{asom}(b)(r) = \begin{cases} 1 & \text{if } r = b_i \text{ for some } i \in [1,\text{len}(b)] \\ 0 & \text{if } r \in \text{cub}(2,m) \setminus \{b_i : i \in [1,\text{len}(b)]\} \end{cases}$$

where we have *temporarily* let b_i be the unique member of cub(2,m) such that $b_i(1) = b(1,i)$ and $b_i(2) = b(2,i)$.

Given any $t \in \text{mon}(2,m)$ and $b \in \text{vec}(2,m)$ such that $\text{supp}(\text{asom}(b)) \subset \text{supp}(t)$, first we introduce the *associated protomonomial* asom(b,t) by saying that asom(b,t) is the unique member of mon(2,m) such that for all $r \in \text{cub}(2,m)$ we have

$$t(r) = \text{asom}(b)(r) + \text{asom}(b,t)(r),$$

and second, we introduce the *modified minor* momor(X,b,t) by putting

$$\text{momor}(X,b,t) = \text{mor}(X,b)X^{d(b,t)}$$

where we have *temporarily* put $d(b,t) = \text{asom}(b,t)$.

Given any $p \in N$ and $a \in \text{vec}(2,m,p)$ and $t \in \text{mon}(2,m)$, we define

$$\hat{k}(p,a,t) \in [0,3] \quad \text{and} \quad \hat{u}(p,a,t) \in [0, p+1]$$

by saying that, first, if

$$\text{ind}(\text{supp}(t) \cap \text{truc}(2,m,p,a,1,i)) \geq i \quad \text{for some } i \in [1,p]$$

then $\hat{k}(p,a,t) = 1$ and $\hat{u}(p,a,t)$ is the smallest such i, and second, if

$$\text{ind}(\text{supp}(t) \cap \text{truc}(2,m,p,a,1,i)) < i \quad \text{for all } i \in [1,p]$$

and

$$\text{ind}(\text{supp}(t) \cap \text{truc}(2,m,p,a,2,j)) \geq j \quad \text{for some } j \in [1,p]$$

then $\hat{k}(p,a,t) = 2$ and $\hat{u}(p,a,t)$ is the smallest such j, and third, if

$$\text{ind}(\text{supp}(t) \cap \text{truc}(2,m,p,a,1,i)) < i \quad \text{for all } i \in [1,p]$$

and

$$\text{ind}(\text{supp}(t) \cap \text{truc}(2,m,p,a,2,j)) < j \quad \text{for all } j \in [1,p]$$

and

$$\text{ind}(\text{supp}(t)) > p$$

then $\hat{k}(p,a,t) = 3$ and $\hat{u}(p,a,t) = p + 1$, and finally, if

$$\text{ind}(\text{supp}(t) \cap \text{truc}(2,m,p,a,1,i)) < i \quad \text{for all } i \in [1,p]$$

and

$$\text{ind}(\text{supp}(t) \cap \text{truc}(2,m,p,a,2,j)) < j \quad \text{for all } j \in [1,p]$$

and

$$\text{ind}(\text{supp}(t)) \leq p,$$

then $\hat{k}(p,a,t) = 0$ and $\hat{u}(p,a,t) = 0$.

Given any $p \in N$ and $a \in \text{vec}(2,m,p)$ and $t \in \text{mon}(2,m)$, we introduce the *associated bivector* asob(p,a,t) by saying that asob(p,a,t) is a member of vec(2,m) for which we have

$$\text{supp}(\text{asom}(\text{asob}(p,a,t))) \subset \text{supp}(t)$$

and which is defined thus. First, if $\hat{k}(p,a,t) \in [1,2]$, then clearly,

$$\{\text{asom}(b) : b \in \text{vec}(2,m,\hat{u}(p,a,t)) \text{ such that } \text{supp}(\text{asom}(b)) \subset \text{supp}(t) \text{ and } b(\hat{k}(p,a,t),\hat{u}(p,a,t)) < a(\hat{k}(p,a,t),\hat{u}(p,a,t))\}$$

is a nonempty subset of $\text{mon}[[2,m,\hat{u}(p,a,t)]]$, and we define asob(p,a, t) to be the unique member of $\text{vec}(2,m,\hat{u}(p,a,t))$ such that asom(asob(p, a,t)) is the smallest element of the above-mentioned nonempty subset of $\text{mon}[[2,m,\hat{u}(p,a,t)]]$. Second, if $\hat{k}(p,a,t) = 3$, then clearly,

$$\{\text{asom}(b) : b \in \text{vec}(2, m, p + 1) \text{ such that } \text{supp}(\text{asom}(b)) \subset \text{supp}(t)\}$$

is a nonempty subset of mon[[2, m, p + 1]], and we define asob(p,a,t) to be the unique member of vec(2, m, p + 1) such that asom(asob(p,a, t)) is the smallest member of the above-mentioned nonempty subset of mon[[2, m, p + 1]]. Finally, if $\hat{k}(p,a,t) = 0$, then we define asob(p,a,t) to be the only existing member of vec(2,m,0).

We note that given any $p \in N$ and $a \in \text{vec}(2,m,p)$ and $t \in \text{mon}(2,m)$, we have

$$t \in \text{mon}(2,m,p,a) \Leftrightarrow \hat{k}(p,a,t) = 0$$

and hence in particular we have

$$t \in \text{mon}(2,m,p,a) \Rightarrow \text{momor}(X,\text{asob}(p,a,t),t) = X^t.$$

Given any map $w: M \to H$, where M and H are any sets, and given any subset M^* of M, by $w|M^*$ we shall denote the restriction of w to M^*; i.e., by $w|M^*$ we shall denote the map $M^* \to H$ induced by w.

Given any set M, by a map $v: M \to K$ with *finite support* we mean a map $v: M \to K$ such that

$$\text{card}(\{s \in M : v(s) \neq 0\}) < \infty.$$

Given any map $w: M \to H$, where M is a set and H is a K-vector-space, we say that w is K-*independent* to mean that w has the property which says that if $v: M \to K$ is any map with finite support such that

$$\sum_{s \in M} v(s)w(s) = 0$$

then $v(s) = 0$ for all $s \in M$. Given any map $w: M \to H$, where M is a set and H is a K-vector-space, and given any K-vector-subspace H^* of H, we say that w is a K-*generator* of H^* to mean that $w(s) \in H^*$ for all $s \in M$, and for every $x \in H^*$ there exists a map $x^*: M \to K$ with finite support such that

$$x = \sum_{s \in M} x^*(s)w(s).$$

Given any map $w: M \to H$, where M is a set and H is a K-vector-space, and given any K-vector-subspace H^* of H, we say that w is K-*basis* of H^* to mean that w is K-independent and w is a K-generator of H^*. Given any K-vector-space H, by [H:K] we denote the cardinality of any set M for which there exists a map $w: M \to H$ such that w is a K-basis of H; (it is well known that the said cardinality is independent of M). Given any map $w: M \to H$, where M is a set and H is a K-vector-space, and given any K-vector-subspace H^* of H, and given any

K-vector-subspace H' of H*, we say that w is K-*basis* of H* *modulo* H' to mean that the composition of w followed by the canonical map $H \to H/H'$ is a K-basis of the K-vector-space which is the image of H^*/H' under the natural injective map of H^*/H' into H/H'.

Let us now recall the terminology concerning homogeneous rings. By a *homogeneous ring* we mean a ring H together with a family $(H_V)_{V \in N}$ of additive subgroups of H such that the underlying additive group of H is the direct sum of the said family, and such that for all V and V* in N and for all $x \in H_V$ and $x^* \in H_{V^*}$ we have $xx^* \in H_{V+V^*}$, and such that H_0 is a field, and such that $H = H_0[H_1]$, and finally, such that $0 < [H_1:H_0] < \infty$. Here H_V is called the Vth *homogeneous component* of H. By a *homogeneous subring* of H we mean a homogeneous ring $\tilde{H}$ such that $\tilde{H}$ is a subring of H, and $\tilde{H}_0 = H_0$, and for every $V \in N$ we have $\tilde{H}_V = \tilde{H} \cap H_V$. By a *homogeneous ideal* in H we mean an ideal $\hat{H}$ in H such that $\hat{H}$ is generated by

$$\bigcup_{V \in N} (\hat{H} \cap H_V);$$

we put $\hat{H}_V = \hat{H} \cap H_V$ and we call $\hat{H}_V$ the Vth *homogeneous component* of $\hat{H}$.

Finally, recall that given any homogeneous ideal $\hat{H}$ in the polynomial ring H in a finite number of indeterminates over K, the map $N \to N$ which sends V to $[H_V/\hat{H}_V:K]$ is called the *Hilbert function* of $\hat{H}$ in H, and by a theorem of Hilbert there exists a unique polynomial h(V) in an indeterminate V with rational coefficients such that for all large enough nonnegative integers V we have $h(V) = [H_V/\hat{H}_V:K]$; note that h(V) is called the *Hilbert polynomial* of $\hat{H}$ in H; this motivates the definition according to which we say that $\hat{H}$ is *hilbertian* to mean that *for all nonnegative integers* V we have $h(V) = [H_V/\hat{H}_V:K]$.

THEOREM (20.3) Let $q\colon \mathrm{tab}(2,m) \to K[X]$ be the map defined by putting $q(T) = \mathrm{mom}(X,T)$ for all $T \in \mathrm{tab}(2,m)$. Then we have the following.

(20.3.1) For every $V \in N$ we have that $q|\mathrm{stib}(2,m,V)$ is a K-basis of $K[X]_V$.

(20.3.2) We have that $q|\mathrm{stab}(2,m)$ is a K-basis of $K[X]$.

(20.3.3) Given any $T \in \mathrm{tab}(2,m)$, upon letting $\hat{T}: \mathrm{stab}(2,m) \to K$ be the unique map with finite support such that

$$q(T) = \sum_{S \in \mathrm{stab}(2,m)} \hat{T}(S) q(S)$$

[where the existence and uniqueness of $\hat{T}$ follows from (20.3.2)], we have that

$$\hat{T}(S) = \begin{cases} \mathrm{fin}(m,T,S) & \text{if } S \in \mathrm{stab}(2,m) \text{ and } \mathrm{dep}(S) \le \mathrm{dep}(T) \\ 0 & \text{if } S \in \mathrm{stab}(2,m) \text{ and } \mathrm{dep}(S) > \mathrm{dep}(T). \end{cases}$$

(20.3.4) Given any $T \in \mathrm{tab}(2,m)$ and $T^* \in \mathrm{tab}(2,m)$, upon letting $\hat{T}: \mathrm{stab}(2,m) \to K$ and $\hat{T}^*: \mathrm{stab}(2,m) \to K$ be the unique maps with finite support such that

$$q(T) = \sum_{S \in \mathrm{stab}(2,m)} \hat{T}(S) q(S)$$

and

$$q(T^*) = \sum_{S \in \mathrm{stab}(2,m)} \hat{T}^*(S) q(S)$$

[where the existence and uniqueness of $\hat{T}$ and $\hat{T}^*$ follows from (20.3.2)], we have that if $\mathrm{dep}(T) = \mathrm{dep}(T^*)$ and there exists a permutation $s: [1,\mathrm{dep}(T)] \to [1,\mathrm{dep}(T^*)]$ such that $T[e] = T^*[s(e)]$ for all $e \in [1, \mathrm{dep}(T)]$, then $\hat{T} = \hat{T}^*$.

(20.3.5) Given any $T \in \mathrm{tab}(2,m)$ with $\mathrm{are}(T) \neq 0$, upon letting $\hat{T}: \mathrm{stab}(2,m) \to K$ be the unique map with finite support such that

$$q(T) = \sum_{S \in \mathrm{stab}(2,m)} \hat{T}(S) q(S)$$

[where the existence and uniqueness of $\hat{T}$ follows from (20.3.2)], and given any $S \in \mathrm{stab}(2,m)$ with $\hat{T}(S) \neq 0$, we have that $\mathrm{dep}(S) \neq 0$ and $S[1] \le T[e]$ for all $e \in [1,\mathrm{dep}(T)]$.

Proof: Consider the map $q^*: \mathrm{mon}(2,m) \to K[X]$ defined by putting $q^*(t) = X^t$ for all $t \in \mathrm{mon}(2,m)$. Let there be given any $V \in N$. Now obviously,

(1) $q^*|\mathrm{mon}[[2,m,V]]$ is a K-generator of $K[X]_V$

and more precisely q*|mon[[2,m,V]] is a K-basis of $K[X]_V$ and hence

(2) $[K[X]_V:K] = \text{Mon}[[2,m,V]]$.

Also, clearly,

$$\begin{aligned}\{q^*(t) : t \in \text{mon}[[2,m,V]]\} &= \{q(T) : T \in \text{tap}[2,m,1,V]\}\\ &\subset \{q(T) : T \in \text{tib}(2,m,V)\}\\ &\subset K[X]_V\end{aligned}$$

and hence by (1) we see that

(3) q|tib(2,m,V) is a K-generator of $K[X]_V$.

In view of (3), by (20.1.1) and (20.1.2) we see that

(4) q|stib(2,m,V) is a K-generator of $K[X]_V$.

Finally, in view of (2) and (4), by (20.1.6) we conclude that

(5) q|stib(2,m,V) is a K-basis of $K[X]_V$,

which proves (20.3.1), and this being so for all V ∈ N, we also get (20.3.2). Now (20.3.3) follows from (20.1.1) and (20.3.2). Also, (20.3.4) follows from (20.3.2). Finally, (20.3.5) follows from (20.1.5), (20.3.3), and (20.3.4).

THEOREM (20.4) Let q: tab(2,m) → K[X] be the map defined by putting q(T) = mom(X,T) for all T ∈ tab(2,m). Also, let there be given any p ∈ N and a ∈ vec(2,m,p). Then we have the following.

(20.4.1) We have that q|[stab(2,m) \ stab(2,m,p,a)] is a K-basis of I(p,a), whereas q|stab(2,m,p,a) is a K-basis of K[X] modulo I(p,a).

(20.4.2) For every V ∈ N we have that q|[stib(2,m,V) \ stab(2,m,p,a,V)] is a K-basis of $I(p,a)_V$, whereas q|stab(2,m,p,a,V) is a K-basis of $K[X]_V$ modulo $I(p,a)_V$.

(20.4.3) For every V ∈ N we have

$$[K[X]_V/I(p,a)_V:K] = \text{Stab}(2,m,p,a,V).$$

Proof: Given any $r \in I(p,a)$, in view of the definition of $I(p,a)$ we can write

$$r = \sum_{i \in M} \mathrm{mor}(X,b_i)\mathrm{mom}(X,U_i)r_i$$

where M is a finite set and where for every $i \in M$ we have

$$b_i \in G(p,a) \quad \text{and} \quad U_i \in \mathrm{tap}(2,m,1) \quad \text{and} \quad r_i \in K.$$

For each $i \in M$, upon letting T_i be the unique member of $\mathrm{tab}(2,m)$ such that $\mathrm{dep}(T_i) = 1 + \mathrm{dep}(U_i)$ and

$$T_i[e] = \begin{cases} b_i & \text{if } e = 1 \\ U_i[e-1] & \text{if } e \in [2, 1 + \mathrm{dep}(U_i)], \end{cases}$$

by (20.3.5) we get

$$q(T_i) = \sum_{S \in \mathrm{stab}(2,m)} \hat{T}_i(S)q(S)$$

where $\hat{T}_i$: $\mathrm{stab}(2,m) \to K$ is a map with finite support such that

$$\begin{bmatrix} \text{for every } S \in \mathrm{stab}(2,m) \text{ with } \hat{T}_i(S) \neq 0 \\ \text{we have } \mathrm{dep}(S) \neq 0 \text{ and } S[1] \le T_i[1] \end{bmatrix}$$

and now, because $b_i \in G(p,a)$, by the definition of $G(p,a)$ we have $a \nleq b_i$ and so, because $b_i = T_i[1]$, we get

$$a \nleq T_i[1]$$

and hence we see that

$$\begin{bmatrix} \text{for every } S \in \mathrm{stab}(2,m) \text{ with } \hat{T}_i(S) \neq 0 \\ \text{we have } \mathrm{dep}(S) \neq 0 \text{ and } a \nleq S[1] \end{bmatrix}$$

and therefore by the definition of $\mathrm{stab}(2,m,p,a)$ we get that

$$\begin{bmatrix} \text{for every } S \in \mathrm{stab}(2,m) \text{ with } \hat{T}_i(S) \neq 0 \\ \text{we have } S \in [\mathrm{stab}(2,m) \setminus \mathrm{stab}(2,m,p,a)] \end{bmatrix}$$

and now upon letting $\hat{r}$: $\mathrm{stab}(2,m) \to K$ be the map with finite support obtained by putting

$$\hat{r}(S) = \sum_{i \in M} \hat{T}_i(S) r_i \qquad \text{for all } S \in \text{stab}(2,m)$$

we conclude that

$$r = \sum_{S \in \text{stab}(2,m)} \hat{r}(S) q(S)$$

and

$$\left[\begin{array}{l}\text{for every } S \in \text{stab}(2,m) \text{ with } \hat{r}(S) \neq 0 \\ \text{we have } S \in [\text{stab}(2,m) \setminus \text{stab}(2,m,p,a)].\end{array}\right.$$

Thus we have shown that

$$(1) \quad \left[\begin{array}{l}\text{given any } r \in I(p,a)\text{, there exists a map} \\ \hat{r}\colon \text{stab}(2,m) \to K \text{ with finite support such that} \\ r = \sum_{S \in \text{stab}(2,m)} \hat{r}(S) q(S) \text{ and such that} \\ \text{for every } S \in \text{stab}(2,m) \text{ with } \hat{r}(S) \neq 0 \text{ we have} \\ S \in [\text{stab}(2,m) \setminus \text{stab}(2,m,p,a)].\end{array}\right.$$

Given any $S \in [\text{stab}(2,m) \setminus \text{stab}(2,m,p,a)]$, we see that $\text{dep}(S) \neq 0$ and either, as the first case, for some $u \in [1,p]$ we have

$$\left[\begin{array}{l}p \geq \text{len}(S[1]) \geq u \\ \text{and } S[1](1,i) \geq a(1,i) \text{ for all } i \in [1, u-1] \\ \text{and } S[1](1,u) < a(1,u)\end{array}\right.$$

or, as the second case, for some $u \in [1,p]$ we have

$$\left[\begin{array}{l}p \geq \text{len}(S[1]) \geq u \\ \text{and } S[1](1,i) \geq a(1,i) \text{ for all } i \in [1,\text{len}(S[1])] \\ \text{and } S[1](2,j) \geq a(2,j) \text{ for all } j \in [1, u-1] \\ \text{and } S[1](2,u) < a(2,u)\end{array}\right.$$

or, as the third case, we have $\text{len}(S[1]) > p$; in the first case, upon expanding $\text{mor}(X,S[1])$ by its first u rows we see that

$$\text{mor}(X,S[1]) = \sum_{b \in G(p,a,1,u)} \text{mor}(X,b) S^*(b) \quad \text{with } S^*(b) \in K[X]$$

and in the second case, upon expanding $\text{mor}(X,S[1])$ by its first u columns we see that

$$\operatorname{mor}(X,S[1]) = \sum_{b\in G(p,a,2,u)} \operatorname{mor}(X,b)S^*(b) \quad \text{with } S^*(b) \in K[X]$$

and in the third case, upon expanding mor(X,S[1]) by its first p + 1 rows we see that

$$\operatorname{mor}(X,S[1]) = \sum_{b\in G[p+1]} \operatorname{mor}(X,b)S^*(b) \quad \text{with } S^*(b) \in K[X];$$

in this manner, in all the three cases we get

$$\operatorname{mor}(X,S[1]) \in I(p,a)$$

and hence we get

$$\operatorname{mom}(X,S) \in I(p,a).$$

Thus we have shown that

(2) $\begin{bmatrix}\text{for every } S \in [\operatorname{stab}(2,m) \setminus \operatorname{stab}(2,m,p,a)] \\ \text{we have } q(S) \in I(p,a).\end{bmatrix}$

In view of (1) and (2), by (20.3.2) we get (20.4.1). Moreover, by (20.3.1), (20.3.2), and (20.4.1) we get (20.4.2). Finally, by (20.4.2) we get (20.4.3).

LEMMA (20.5) Let there be given any $p \in N$ and $a \in \operatorname{vec}(2,m,p)$. Now, given any $B \in K[X]$, we have that if $\operatorname{mor}(X,a)B \in I(p,a)$, then $B \in I(p,a)$.

Proof: If p = 0, then mor(X,a) = 1 and hence in this case our assertion is obvious. So henceforth we may suppose that $p \neq 0$. Now given any $B \in K[X]$, by (20.3.2) there exists a map $\hat{B}: \operatorname{stab}(2,m) \to K$ with finite support such that

$$B = \sum_{S\in \operatorname{stab}(2,m)} \hat{B}(S)q(S).$$

Upon letting

$$(10) \quad B_0 = \sum_{S\in[\operatorname{stab}(2,m)\setminus \operatorname{stab}(2,m,p,a)]} \hat{B}(S)q(S)$$

and

$$(11) \quad B_1 = \sum_{S\in \operatorname{stab}(2,m,p,a)} \hat{B}(S)q(S)$$

we get

(12) $B = B_0 + B_1$

and

(13) $\mathrm{mor}(X,a)B = \mathrm{mor}(X,a)B_0 + \mathrm{mor}(X,a)B_1$.

In view of (10), by (20.4.1) we see that

(14) $B_0 \in I(p,a)$

and hence

$$\mathrm{mor}(X,a)B_0 \in I(p,a)$$

and therefore by (13) we see that

(15) if $\mathrm{mor}(X,a)B \in I(p,a)$, then $\mathrm{mor}(X,a)B_1 \in I(p,a)$.

Upon letting

$$\mathrm{stab}^*(2,m,p,a) = \{S \in \mathrm{stab}(2,m,p,a) : \mathrm{dep}(S) \neq 0 \text{ and } S[1] = a\}$$

first we get a bijective map

(16) $g\colon \mathrm{stab}(2,m,p,a) \to \mathrm{stab}^*(2,m,p,a)$

by saying that for every $S \in \mathrm{stab}(2,m,p,a)$ we have that $g(S)$ is the unique member of $\mathrm{stab}(2,m,p,a)$ such that $\mathrm{dep}(g(S)) = 1 + \mathrm{dep}(S)$ and

(17) $g(S)[e] = \begin{cases} a & \text{if } e = 1 \\ S[e-1] & \text{if } e \in [2, 1 + \mathrm{dep}(S)] \end{cases}$

and second, we get a map

$$B^*\colon \mathrm{stab}^*(2,m,p,a) \to K$$

with finite support by saying that

(18) $B^*(g(S)) = \hat{B}(S) \quad$ for all $S \in \mathrm{stab}(2,m,p,a)$

and now

$$\mathrm{mor}(X,a)B_1 = \sum_{S \in \mathrm{stab}(2,m,p,a)} \hat{B}(S)[\mathrm{mor}(X,a)q(S)] \qquad [\text{by } (11)]$$

$$\mathrm{mor}(X,a)B_1 = \sum_{S\in \mathrm{stab}(2,m,p,a)} B^*(g(S))q(g(S)) \qquad \text{[by (17) and (18)]}$$

$$= \sum_{S^*\in \mathrm{stab}^*(2,m,p,a)} B^*(S^*)q(S^*) \qquad \text{[by (16)]}$$

and thus

(19) $$\mathrm{mor}(X,a)B_1 = \sum_{S^*\in \mathrm{stab}^*(2,m,p,a)} B^*(S^*)q(S^*).$$

Now stab*(2,m,p,a) $\subset$ stab(2,m,p,a) and hence, in view of (20.4.2), by (19) we conclude that

(20) $\left[\begin{array}{l}\text{if } \mathrm{mor}(X,a)B_1 \in I(p,a)\text{, then}\\ B^*(S^*) = 0 \text{ for all } S^* \in \mathrm{stab}^*(2,m,p,a).\end{array}\right.$

By (16) and (18) we see that

(21) $\left[\begin{array}{l}\text{if } B^*(S^*) = 0 \text{ for all } S^* \in \mathrm{stab}^*(2,m,p,a),\\ \text{then } \hat{B}(S) = 0 \text{ for all } S \in \mathrm{stab}(2,m,p,a).\end{array}\right.$

By (11) we see that

(22) $\left[\begin{array}{l}\text{if } \hat{B}(S) = 0 \text{ for all } S \in \mathrm{stab}(2,m,p,a),\\ \text{then } B_1 = 0.\end{array}\right.$

By (12) and (14) we see that

(23) if $B_1 = 0$, then $B \in I(p,a)$.

Finally, by (15) and (20) to (23) we conclude that if mor(X,a)B $\in$ I(p,a), then B $\in$ I(p,a), and this completes the proof.

LEMMA (20.6) Given any b $\in$ vec(2,m) we have

$$\mathrm{mor}(X,b) = X^d + \sum_{u\in U} b^*(u)X^u$$

where d = asom(b) and where U is a subset of mon[[2,m,len(b)]] and $b^*: U \to K$ is a map such that for every u $\in$ U we have asom(b) $<$ u and $b^*(u) \in \{1,-1\}$.

Proof: Obvious.

LEMMA (20.7) Let there be given any $V \in N$ and $M^* \subset \text{mon}[[2,m,V]]$. Let $g: M^* \to \text{vec}(2,m)$ be any map such that for every $t \in M^*$ we have $\text{supp}(\text{asom}(g(t))) \subset \text{supp}(t)$. Let $w^*: M^* \to K[X]$ be the map obtained by taking $w^*(t) = \text{momor}(X,g(t),t)$ for every $t \in M^*$. Then w^* is K-independent.

Proof: For every $t \in \text{mon}(2,m)$ let

$$U(t) = \{s \in \text{mon}[[2,m,\text{abs}(t)]] : t < s\}.$$

Given any $t \in M^*$ let

$$d(t) = \text{asom}(g(t)) \quad \text{and} \quad e(t) = \text{asom}(g(t),t)$$

and let

$$n: U(d(t)) \to U(t)$$

be the (injective) map obtained by saying that for every $u \in U(d(t))$ we have that $n(u)$ is the unique member of $U(t)$ such that

$$n(u)(r) = u(r) + e(t)(r) \quad \text{for all } r \in \text{cub}(2,m).$$

Given any $t \in M^*$, by (20.6) there exists a map $t^*: U(d(t)) \to K$ such that

$$\text{mor}(X,g(t)) = X^{d(t)} + \sum_{u \in U(d(t))} t^*(u)X^u$$

and now upon multiplying the equation above by $X^{e(t)}$ we get

$$\text{momor}(X,g(t),t) = X^t + \sum_{s \in U(t)} \hat{t}(s)X^s$$

where $\hat{t}: U(t) \to K$ is the unique map such that

$$\hat{t}(n(u)) = t^*(u) \quad \text{for all } u \in U(d(t))$$

and

$$\hat{t}(s) = 0 \quad \text{for all } s \in U(t) \setminus \text{im}(n).$$

Given any nonempty subset M^{**} of M^* and any map $v: M^{**} \to K$ such that for every $t \in M^{**}$ we have $v(t) \neq 0$, we shall prove that

$$\sum_{t \in M^{**}} v(t)w^*(t) \neq 0$$

and this will show that w^* is K-independent. So let c be the smallest element of M^{**}, and for every $t \in M^{**} \setminus \{c\}$ let $\tilde{t}: U(c) \to K$ be the map obtained by putting

$$\tilde{t}(s) = \begin{cases} \hat{t}(s) & \text{if } s \in U(t) \\ 1 & \text{if } s = t \\ 0 & \text{if } t \neq s \in U(c) \setminus U(t). \end{cases}$$

Now upon defining $c^{**}: U(c) \to K$ to be the map obtained by putting

$$c^{**}(s) = v(c)\hat{c}(s) + \sum_{t \in M^{**}\setminus\{c\}} v(t)\tilde{t}(s) \qquad \text{for all } s \in U(c)$$

we see that

$$\sum_{t \in M^{**}} v(t)w^*(t) = v(c)X^c + \sum_{s \in U(c)} c^{**}(s)X^s$$

and obviously the RHS of the equation above is nonzero and hence so is the LHS.

LEMMA (20.8) Let H be an overring of K, let $\tilde{H}$ be a subring of H with $K \subset \tilde{H}$, let $\hat{H}$ be an ideal in H, and let $\tilde{G}$ be a subset of $\tilde{H} \cap \hat{H}$. Let M be a set, and let $\tilde{M}$ and $\hat{M}$ be subsets of M. Let $w: M \to H$ be a K-basis of H such that $w|\tilde{M}$ is a K-basis of $\tilde{H}$, and $w|\hat{M}$ is a K-basis of $\hat{H}$. Assume that for every $t \in \tilde{M} \cap \hat{M}$ we have $w(t) \in \tilde{G}\tilde{H}$. Then

$$\tilde{G}\tilde{H} = \tilde{H} \cap \hat{H}.$$

Proof: Since $\tilde{G} \subset \tilde{H} \cap \hat{H}$, we obviously have $\tilde{G}\tilde{H} \subset \tilde{H} \cap \hat{H}$. To prove the other inclusion, let there be given any $x \in \tilde{H} \cap \hat{H}$. Since w is a K-basis of H, there exists a map $x^*: M \to K$ with finite support such that

$$x = \sum_{t \in M} x^*(t)w(t).$$

Let

$$M^* = \{t \in M : x^*(t) \neq 0\}.$$

Since $x \in \tilde{H}$, we must have $M^* \subset \tilde{M}$; similarly, since $x \in \hat{H}$, we must have $M^* \subset \hat{M}$. Thus $M^* \subset \tilde{M} \cap \hat{M}$. Consequently, for every $t \in M^*$ we must have $w(t) \in \tilde{G}\tilde{H}$, and hence $x \in \tilde{G}\tilde{H}$. Therefore, $\tilde{G}\tilde{H} = \tilde{H} \cap \hat{H}$.

LEMMA (20.9) Let H be a homogeneous ring with $H_0 = K$. Let $\tilde{H}$ be a homogeneous subring of H, and let $\hat{H}$ be a homogeneous ideal in H. Let $\tilde{G} \subset \tilde{H} \cap \hat{H}$. Let M be a set and let $M_V \subset M$ for every $V \in N$ be such that

$$M = \coprod_{V \in N} M_V.$$

For every $V \in N$, let $\tilde{M}_V$ and $\hat{M}_V$ be subsets of M_V and let $w_V: M_V \to H$ be a K-basis of H_V such that $w_V|\tilde{M}_V$ is a K-basis of $\tilde{H}_V$ whereas $w_V|\hat{M}_V$ is a K-basis of $\hat{H}_V$. Assume that for every $V \in N$ and $t \in \tilde{M}_V \cap \hat{M}_V$ we have $w_V(t) \in \tilde{G}\tilde{H}$. Then

$$\tilde{G}\tilde{H} = \tilde{H} \cap \hat{H}.$$

Proof: Follows from (20.8) by taking

$$\tilde{M} = \bigcup_{V \in N} \tilde{M}_V \quad \text{and} \quad \hat{M} = \bigcup_{V \in N} \hat{M}_V$$

and by letting $w: M \to H$ be the unique map such that for every $V \in N$ we have $w|M_V = w_V$.

THEOREM (20.10) Given any $p \in N$ and $a \in \text{vec}(2,m,p)$ and $V \in N$, let

$$M_V = \text{mon}[[2,m,V]] \quad \text{and} \quad \hat{M}_V = M_V \setminus \text{mon}(2,m,p,a,V)$$

and let $w_V: M_V \to K[X]$ be the map defined by putting

$$w_V(t) = \text{momor}(X,\text{asob}(p,a,t),t) \quad \text{for all } t \in M_V.$$

Then

(1) w_V is a K-basis of $K[X]_V$

and

(2) $w_V|\hat{M}_V$ is a K-basis of $I(p,a)_V$

whereas

(3) $w_V|\text{mon}(2,m,p,a,V)$ is a K-basis of $K[X]_V$ modulo $I(p,a)_V$.

Moreover, given any subset Y of rec(m(1),m(2)), we have that

(4) $\hat{G}^*(p,a,Y) \subset K[Y] \cap I(p,a)$ and $\hat{G}^*(p,a,Y)K[Y] = I^*(p,a,Y)$

and upon letting

$$\tilde{M}_V = \{t \in M_V : \mathrm{supp}(t) \subset Y\}$$

we have that

(5) $\left[\begin{array}{l}\text{if } Y \text{ is a saturated subset of } \mathrm{rec}(m(1),m(2)),\\ \text{then } w_V|\tilde{M}_V \text{ is a K-basis of } K[Y]_V\\ \text{and for every } t \in \tilde{M}_V \cap \hat{M}_V \text{ we have } w_V(t) \in I^*(p,a,Y).\end{array}\right.$

Proof: Let $q^*: \mathrm{mon}(2,m) \to K[X]$ be the map defined by putting $q^*(t) = X^t$ for all $t \in \mathrm{mon}(2,m)$. Now, clearly, $q^*|M_V$ is a K-basis of $K[X]_V$ and hence

(11) $[K[X]_V:K] = \mathrm{card}(M_V)$.

By taking $M^* = M_V$ and

$$g(t) = \mathrm{asob}(p,a,t) \quad \text{for all } t \in M_V$$

in (20.7) we see that

(12) w_V is K-independent

and hence by (11) we get (1). By the definitions of asob and asom and momor we see that for every $t \in \hat{M}_V$ we have

$$w_V(t) = \mathrm{mor}(X,b)X^s$$

where

$$b = \mathrm{asob}(p,a,t) \in G(p,a) \quad \text{and} \quad s = \mathrm{asom}(b,t) \in \mathrm{mon}(2,m)$$

and

$$\mathrm{len}(b) + \mathrm{abs}(s) = V$$

and hence $w_V(t) \in I(p,a)_V$; thus

(13) for every $t \in \hat{M}_V$ we have $w_V(t) \in I(p,a)_V$.

By (20.1.6), (20.1.7), and (20.4.2) we also have

(14) $[I(p,a)_V:K] = \operatorname{card}(\hat{M}_V)$.

By (12), (13), and (14) we get (2). (3) follows from (1) and (2).

Given any subset Y of rec(m(1),m(2)), by the definitions of $\hat{G}^*$ and I* we get (4), and upon letting

$$\tilde{M}_V = \{t \in M_V : \operatorname{supp}(t) \subset Y\}$$

we obviously see that $q^*|\tilde{M}_V$ is a K-basis of $K[Y]_V$ and hence

(15) $[K[Y]_V:K] = \operatorname{card}(\tilde{M}_V)$.

By the definitions of asob and asom and momor we see that if Y is a saturated subset of rec(m(1),m(2)) then for every $t \in \tilde{M}_V$ we have

$$w_V(t) = \operatorname{mor}(X,b)X^s$$

where

$$b = \operatorname{asob}(p,a,t) \in \operatorname{vec}(2,m) \quad \text{with } \operatorname{mor}(X,b) \in K[Y]$$

and

$$s = \operatorname{asom}(b,t) \in \operatorname{mon}(2,m) \quad \text{with } X^s \in K[Y]$$

and

$$\operatorname{len}(b) + \operatorname{abs}(s) = V$$

and hence $w_V(t) \in K[Y]_V$; thus

(16) $\left[\begin{array}{l}\text{if } Y \text{ is a saturated subset of } \operatorname{rec}(m(1),m(2)),\\ \text{then for every } t \in \tilde{M}_V \text{ we have } w_V(t) \in K[Y]_V.\end{array}\right.$

Again by the definitions of asob and asom and momor we see that if Y is a saturated subset of rec(m(1),m(2)), then for every $t \in \tilde{M}_V \cap \hat{M}_V$ we have

$$w_V(t) = \operatorname{mor}(X,b)X^s$$

where

$$b = \operatorname{asob}(p,a,t) \in G^*(p,a,Y)$$

and

$$s = \mathrm{asom}(b,t) \in \mathrm{mon}(2,m) \quad \text{with } X^s \in K[Y]$$

and

$$\mathrm{len}(b) + \mathrm{abs}(s) = V$$

and hence $w_V(t) \in I^*(p,a,Y)_V$; thus

(17) $\left[\begin{array}{l}\text{if } Y \text{ is a saturated subset of } \mathrm{rec}(m(1),m(2)),\\ \text{then for every } t \in \tilde{M}_V \cap \hat{M}_V \text{ we have}\\ w_V(t) \in I^*(p,a,Y)_V.\end{array}\right.$

Now (5) follows from (15), (16), and (17).

THEOREM (20.11) Given any $p \in N$ and $a \in \mathrm{vec}(2,m,p)$, for every saturated subset Y of $\mathrm{rec}(m(1),m(2))$ we have $I^*(p,a,Y) = K[Y] \cap I(p,a)$.
Proof: In view of (20.10), our assertion follows from (20.9) by taking $H = K[X]$ and $\tilde{H} = K[Y]$ and $\hat{H} = I(p,a)$ and $\tilde{G} = \hat{G}^*(p,a,Y)$ and $M = \mathrm{mon}(2,m)$.

LEMMA (20.12) Given any $p \in N$ and $a \in \mathrm{vec}(2,m,p)$, we have $I(p,a) \subset \ker(f(p,a))$.
Proof: Let

$$x \in \mathrm{mat}(K(y(p,a,p)),m)$$

be defined by putting

$$x_{ij} = f(p,a)(X_{ij}) \quad \text{for all } (i,j) \in \mathrm{rec}(m(1),m(2)).$$

First, recall that, by the definition of $f(p,a)$,

$\left[\begin{array}{l}\text{for every } (i,j) \in \mathrm{rec}(m(1),m(2)) \setminus z(p,a,1)\\ \text{we have } f(p,a)(X_{ij}) = 0\end{array}\right.$

and hence

(1) $\left[\begin{array}{l}\text{for every } (i,j) \in \mathrm{rec}(m(1),m(2)) \setminus z(p,a,1)\\ \text{we have } x_{ij} = 0.\end{array}\right.$

Second, recall that, by the definition of $f(p,a)$,

(2) $\begin{bmatrix} \text{for every } (i,j) \in y(p,a,p) \\ \text{we have } f(p,a)(X_{ij}) = X_{ij}. \end{bmatrix}$

Third, recall that, by the definition of f(p,a),

(3) $\begin{bmatrix} \text{for every } (i,j) \in z(p,a,1) \setminus y(p,a,p) \\ \text{we have } f(p,a)(X_{ij}) = \dfrac{\tilde{A}(X,p,a,i,j)}{\text{mor}(X,E(p,a,u^*(p,a,i,j)))} \end{bmatrix}$

where, by the definition of u*,

(4) $\begin{bmatrix} \text{for every } (i,j) \in z(p,a,1) \setminus y(p,a,p) \\ \text{we have } u^*(p,a,i,j) \in [1,p] \end{bmatrix}$

and by the definition of $\tilde{A}$,

(5) $\begin{bmatrix} \text{for every } (i,j) \in z(p,a,1) \setminus y(p,a,p) \\ \text{we have } \tilde{A}(X,p,a,i,j) \in K[y(p,a,u^*(p,a,i,j))] \end{bmatrix}$

and by the definition of E,

(6) $\begin{bmatrix} \text{for every } u \in [0,p] \\ \text{we have } \text{mor}(X,E(p,a,u)) \neq 0 \end{bmatrix}$

because mor(X,E(p,a,u)) is the determinant of a u by u matrix whose u^2 entries are independent indeterminates. Upon cross multiplying and then subtracting RHS from LHS, by (3) we see that

(7) $\begin{bmatrix} \text{for every } (i,j) \in z(p,a,1) \setminus y(p,a,p) \\ \text{we have } \text{mor}(X,E(p,a,u^*(p,a,i,j)))x_{ij} - \tilde{A}(X,p,a,i,j) = 0. \end{bmatrix}$

For every $u \in [1,p]$ we clearly have $y(p,a,u) \subset y(p,a,p)$, and hence by (2), (4), and (5) we see that

(8) $\begin{bmatrix} \text{for every } (i,j) \in z(p,a,1) \setminus y(p,a,p) \\ \text{we have } f(p,a)(\tilde{A}(X,p,a,i,j)) = \tilde{A}(X,p,a,i,j). \end{bmatrix}$

By the definition of E, for every $u \in [0,p]$ we have $E(p,a,u) \in \text{vec}(2, m,u)$ with

$$(E(p,a,u)(1,i),E(p,a,u)(2,j)) \in y(p,a,p)$$

for all $(i,j) \in \text{rec}(u,u)$, and hence by (2) we see that

(9) $\begin{bmatrix} \text{for every } u \in [0,p] \\ \text{we have } \text{mor}(x,E(p,a,u)) = \text{mor}(X,E(p,a,u)). \end{bmatrix}$

Recalling the definition of $\tilde{A}$, for every

$$(i,j) \in z(p,a,1) \setminus y(p,a,p)$$

we have

$$mor(X,A(p,a,i,j)) = mor(X,E(p,a,u^*(p,a,i,j)))X_{ij} - \tilde{A}(X,p,a,i,j)$$

and upon applying f(p,a) to the equation above we get

$$mor(x,A(p,a,i,j)) = mor(x,E(p,a,u^*(p,a,i,j)))x_{ij} - f(p,a)(\tilde{A}(X,p,a,i,j))$$

and hence by (7), (8), and (9) we conclude that

$$mor(x,A(p,a,i,j)) = 0;$$

thus

(10) $\left[\begin{array}{l}\text{for every } (i,j) \in z(p,a,1) \setminus y(p,a,p) \\ \text{we have } mor(x,A(p,a,i,j)) = 0.\end{array}\right.$

In view of the definition of A, by (10) we see that

(11) $\left[\begin{array}{l}\text{for all } i \in [a(1,p) + 1, m(1)] \text{ and } j \in [a(2,p) + 1, m(2)] \text{ we} \\ \text{have } A(p,a,i,j) \in vec(2, m, p + 1) \text{ and } mor(x,A(p,a,i,j)) = 0\end{array}\right.$

and

(12) $\left[\begin{array}{l}\text{for all } u \text{ in } [1, p - 1] \text{ and for all} \\ (i,j) \in z_0(p,a,u) \setminus z(p, a, u + 1) \text{ we have} \\ A(p,a,i,j) \in vec(2, m, u + 1) \text{ and } mor(x,A(p,a,i,j)) = 0.\end{array}\right.$

By (6) and (9) we see that

(13) $\left[\begin{array}{l}\text{for every } u \in [0,p] \\ \text{we have } mor(x,E(p,a,u)) \neq 0\end{array}\right.$

and hence in particular

(14) $mor(x,E(p,a,p)) \neq 0$

and

(15) $\left[\begin{array}{l}\text{for every } u \in [1, p - 1] \\ \text{we have } mor(x,E(p,a,u)) \neq 0.\end{array}\right.$

Given any $u \in [0,p]$, clearly, mor(x,E(p,a,u)) is the determinant of the u by u matrix formed by the intersection of the u rows of x with row numbers a(1,1), a(1,2), ..., a(1,u), and the u columns of x with column numbers a(2,1), a(2,2), ..., a(2,u), and hence by (13) we see that the said rows as well as columns of x are (linearly) independent; [here and in the rest of this proof we are tacitly using basic facts from the theory of finite-dimensional vector spaces over a field, where in our case the field is K(y(p,a,p))]; so in particular

(16) [the rows of x, with row numbers a(1,1), a(1,2), ..., a(1, p - 1), are independent

and

(16*) [the columns of x, with column numbers a(2,1), a(2,2), ..., a(2, p - 1), are independent

and

(17) [given any $u \in [1, p - 1]$, the rows of x, with row numbers a(1,1), a(1,2), ..., a(1, u - 1), are independent

and

(17*) [given any $u \in [1, p - 1]$, the columns of x, with column numbers a(2,1), a(2,2), ..., a(2, u - 1), are independent.

By induction on u we shall show that given any $u \in [0, p - 1]$, for every

$$b \in G(p, a, 1, u + 1) \cup G(p, a, 2, u + 1)$$

we have mor(x,b) = 0.

The case of u = 0 follows from (1) by noting that for every

$$b \in G(p,a,1,1) \cup G(p,a,2,1)$$

we clearly have

$$\text{mor}(x,b) = x_{b(1,1),b(2,1)}$$

and

$$(b(1,1),b(2,1)) \in \text{rec}(m(1),m(2)) \setminus z(p,a,1).$$

So now let $u \in [1, p - 1]$ and assume that the assertion is true for all values of u smaller than the given one. By the induction hypothesis, for every $b \in G(p,a,1,u)$ we have $mor(x,b) = 0$, so the matrix formed by the first $a(1,u) - 1$ rows of x has rank $< u$, and hence by (17) we see that for every $i \in [1, a(1,u) - 1]$ we have that the ith row of x (linearly) depends on the rows of x whose row numbers are $a(1,1), a(1,2), \ldots, a(1, u - 1)$, and therefore

(21) $\left[\begin{array}{l}\text{for every } i \in [1,a(1,u)] \text{ we have that the ith row} \\ \text{of x depends on the rows of x whose row numbers are} \\ a(1,1), a(1,2), \ldots, a(1,u).\end{array}\right.$

Again, by the induction hypothesis, for every $b \in G(p,a,2,u)$ we have $mor(x,b) = 0$, so the matrix formed by the first $a(2,u) - 1$ columns of x has rank $< u$, and hence by (17*) we see that for every $j \in [1, a(2,u) - 1]$ we have that the jth column of x (linearly) depends on the columns of x whose column numbers are $a(2,1), a(2,2), \ldots, a(2, u - 1)$, and therefore

(21*) $\left[\begin{array}{l}\text{for every } j \in [1,a(2,u)] \text{ we have that the jth column} \\ \text{of x depends on the columns of x whose column numbers} \\ \text{are } a(2,1), a(2,2), \ldots, a(2,u).\end{array}\right.$

Given any $i \in [a(1,u) + 1, a(1, u + 1) - 1]$, let us consider the matrix M formed by the rows of x whose row numbers are $a(1,1)$, $a(1,2), \ldots, a(1,u), i$. Now $mor(x,E(p,a,u))$ is the determinant of the u by u matrix formed by the intersection of the first u rows of M and those columns of M whose column numbers are $a(2,1), a(2,2)$, $\ldots, a(2,u)$, and hence by (15) we see that

(22) the first u rows of M are independent.

The columns of M are simply truncations of the corresponding columns of x and hence by (21*) we see that

(23) $\left[\begin{array}{l}\text{for every } j \in [1,a(2,u)] \text{ we have that the jth column of M} \\ \text{depends on the columns of M whose column numbers are} \\ a(2,1), a(2,2), \ldots, a(2,u).\end{array}\right.$

For a moment let us fix any $j \in [a(2,u) + 1, m(2)]$ and let us consider the $u + 1$ by $u + 1$ matrix B formed by the columns of M whose column numbers are $a(2,1), a(2,2), \ldots, a(2,u), j$. Now $\text{mor}(x,E(p,a,u))$ is the determinant of the u by u matrix formed by the intersection of the first u rows and the first u columns of B, and hence by (15) we see that

(24) the first u columns of B are independent.

Clearly, $\det B = \text{mor}(x,A(p,a,i,j))$ and hence by (12) and (24) we see that the $(u + 1)$th column of B depends on its first u columns; now the 1st, 2nd, ..., uth, $(u + 1)$th columns of B are simply the columns of M whose column numbers, respectively, are $a(2,1), a(2,2), \ldots, a(2,u), j$, and therefore we conclude that

(25) for every $j \in [a(2,u) + 1, m(2)]$ we have that the jth column of M depends on the columns of M whose column numbers are $a(2,1), a(2,2), \ldots, a(2,u)$.

By (23) and (25) we see that every column of M depends on its columns with column numbers $a(2,1), a(2,2), \ldots, a(2,u)$, and therefore M has rank $\le u$, and hence by (22) we conclude that the $(u + 1)$th row of M depends on its first u rows; now the 1st, 2nd, ..., uth, $(u + 1)$th rows of M are simply the rows of x whose row numbers, respectively, are $a(1,1), a(1,2), \ldots, a(1,u), i$, and thus we see that

(26) for every $i \in [a(1,u) + 1, a(1, u + 1) - 1]$ we have that the ith row of x depends on the rows of x whose row numbers are $a(1,1), a(1,2), \ldots, a(1,u)$.

By (21) and (26) we see that for every $i \in [1, a(1, u + 1) - 1]$ we have that the ith row of x depends on its rows whose row numbers are $a(1,1), a(1,2), \ldots, a(1,u)$, and hence the matrix formed by the first $a(1, u + 1) - 1$ rows of x has rank $\le u$ and therefore

(27) for every $b \in G(p, a, 1, u + 1)$ we have $\text{mor}(x,b) = 0$.

Given any $j \in [a(2,u) + 1, a(2, u + 1) - 1]$, let us consider the matrix M* formed by the columns of x whose column numbers are $a(2,1), a(2,2), \ldots, a(2,u), j$. Now $\text{mor}(x,E(p,a,u))$ is the determinant of the u by u matrix formed by the intersection of the first u columns of M* and those rows of M* whose row numbers are a(1,1), a(1,2), ..., a(1,u), and hence by (15) we see that

(22*) the first u columns of M* are independent.

The rows of M* are simply truncations of the corresponding rows of x, and hence by (21) we see that

(23*) $\left[\begin{array}{l}\text{for every i} \in \text{[1,a(1,u)] we have that the ith row of M*} \\ \text{depends on the rows of M* whose row numbers are} \\ \text{a(1,1), a(1,2), ..., a(1,u).}\end{array}\right.$

For a moment let us fix any i ∈ [a(1,u) + 1, m(1)] and let us consider the u + 1 by u + 1 matrix B* formed by the rows of M* whose row numbers are a(1,1), a(1,2), ..., a(1,u), i. Now mor(x,E(p,a,u)) is the determinant of the u by u matrix formed by the intersection of the first u columns and the first u rows of B* and hence by (15) we see that

(24*) the first u rows of B* are independent.

Clearly, det B* = mor(x,A(p,a,i,j)) and hence by (12) and (24*) we see that the (u + 1)th row of B* depends on its first u rows; now the 1st, 2nd, ..., uth, (u + 1)th rows of B* are simply the rows of M* whose row numbers, respectively, are a(1,1), a(1,2), ..., a(1,u), i, and therefore we conclude that

(25*) $\left[\begin{array}{l}\text{for every i} \in \text{[a(1,u) + 1, m(1)] we have that the ith row} \\ \text{of M* depends on the rows of M* whose row numbers are} \\ \text{a(1,1), a(1,2), ..., a(1,u).}\end{array}\right.$

By (23*) and (25*) we see that every row of M* depends on its rows with row numbers a(1,1), a(1,2), ..., a(1,u), and therefore M* has rank ≤ u, and hence by (22*) we conclude that the (u + 1)th column of M* depends on its first u columns; now the 1st, 2nd, ..., uth, (u + 1)th columns of M* are simply the columns of x whose column numbers, respectively, are a(2,1), a(2,2), ..., a(2,u), j, and thus we see that

(26*) [for every $j \in [a(2,u) + 1, a(2, u + 1) - 1]$ we have that the jth column of x depends on the columns of x whose column numbers are $a(2,1), a(2,2), \ldots, a(2,u)$.

By (21*) and (26*) we see that for every $j \in [1, a(2, u + 1) - 1]$ we have that the jth column of x depends on its columns whose column numbers are $a(2,1), a(2,2), \ldots, a(2,u)$, and hence the matrix formed by the first $a(2, u + 1) - 1$ columns of x has rank $\leq u$ and therefore

(27*) for every $b \in G(p, a, 2, u + 1)$ we have $\operatorname{mor}(x,b) = 0$.

By (27) and (27*) we see that the induction on u is complete and hence

(30) [given any $u \in [0, p - 1]$, for every $b \in G(p, a, 1, u + 1) \cup G(p, a, 2, u + 1)$ we have $\operatorname{mor}(x,b) = 0$.

By (30), for every $b \in G(p,a,1,p)$ we have $\operatorname{mor}(x,b) = 0$, so the matrix formed by the first $a(1,p) - 1$ rows of x has rank $< p$, and hence by (16) we see that for every $i \in [1, a(1,p) - 1]$ we have that the ith row of x (linearly) depends on the rows of x whose row numbers are $a(1,1), a(1,2), \ldots, a(1, p - 1)$, and therefore

(31) [for every $i \in [1,a(1,p)]$ we have that the ith row of x depends on the rows of x whose row numbers are $a(1,1), a(1,2), \ldots, a(1,p)$.

Again by (30), for every $b \in G(p,a,2,p)$ we have $\operatorname{mor}(x,b) = 0$, so the matrix formed by the first $a(2,p) - 1$ columns of x has rank $< p$, and hence by (16*) we see that for every $j \in [1, a(2,p) - 1]$ we have that the jth column of x (linearly) depends on the columns of x whose column numbers are $a(2,1), a(2,2), \ldots, a(2, p - 1)$, and therefore

(31*) [for every $j \in [1,a(2,p)]$ we have that the jth column of x depends on the columns of x whose column numbers are $a(2,1), a(2,2), \ldots, a(2,p)$.

Given any $i \in [a(1,p) + 1, m(1)]$, let us consider the matrix $\hat{M}$ formed by the rows of x whose row numbers are $a(1,1), a(1,2), \ldots, a(1,p), i$. Now $\operatorname{mor}(x,E(p,a,p))$ is the determinant of the p by p matrix formed by the intersection of the first p rows of $\hat{M}$ and those columns of $\hat{M}$ whose column numbers are a(2,1), a(2,2), ..., a(2,p), and hence by (14) we see that

(32) the first p rows of $\hat{M}$ are independent.

The columns of $\hat{M}$ are simply truncations of the corresponding columns of x and hence by (31*) we see that

(33) [for every j $\in$ [1,a(2,p)] we have that the jth column of $\hat{M}$ depends on the columns of $\hat{M}$ whose column numbers are a(2,1), a(2,2), ..., a(2,p).

For a moment let us fix any j $\in$ [a(2,p) + 1, m(2)] and let us consider the p + 1 by p + 1 matrix $\hat{B}$ formed by the columns of $\hat{M}$ whose column numbers are a(2,1), a(2,2), ..., a(2,p), j. Now mor(x,E(p, a,p)) is the determinant of the p by p matrix formed by the intersection of the first p rows and the first p columns of $\hat{B}$, and hence by (14) we see that

(34) the first p columns of $\hat{B}$ are independent.

Clearly, det $\hat{B}$ = mor(x,A(p,a,i,j)) and hence by (11) and (34) we see that the (p + 1)th column of $\hat{B}$ depends on its first p columns; now the 1st, 2nd, ..., pth, (p + 1)th columns of $\hat{B}$ are simply the columns of $\hat{M}$ whose column numbers, respectively, are a(2,1), a(2,2), ..., a(2,p), j, and therefore we conclude that

(35) [for every j $\in$ [a(2,p) + 1, m(2)] we have that the jth column of $\hat{M}$ depends on the columns of $\hat{M}$ whose column numbers are a(2,1), a(2,2), ..., a(2,p).

By (33) and (35) we see that every column of $\hat{M}$ depends on its columns with column numbers a(2,1), a(2,2), ..., a(2,p), and therefore $\hat{M}$ has rank $\le$ p, and hence by (32) we conclude that the (p + 1)th row of $\hat{M}$ depends on its first p rows; now the 1st, 2nd, ..., pth, (p + 1)th rows of $\hat{M}$ are simply the rows of x whose row numbers, respectively, are a(1,1), a(1,2), ..., a(1,p), i, and thus we see that

(36) [for every i $\in$ [a(1,p) + 1, m(1)] we have that the ith row of x depends on the rows of x whose row numbers

⌊are a(1,1), a(1,2), ..., a(1,p).

By (31) and (36) we see that every row of x depends on its rows whose row numbers are a(1,1), a(1,2), ..., a(1,p), and hence x has rank $\le$ p and therefore

(37) for every $b \in G[p + 1]$ we have $\text{mor}(x,b) = 0$.

By (30) and (37) we see that for every $b \in G(p,a)$ we have $\text{mor}(x,b) = 0$, and therefore $I(p,a) \subset \ker(f(p,a))$.

THEOREM (20.13) Given any $p \in N$ and $a \in \text{vec}(2,m,p)$, we have $I(p,a) = \ker(f(p,a))$.

Proof: We shall make induction on p. If p = 0, then $z(p,a,1) = \emptyset$ and hence

$$f(p,a)(X_{ij}) = 0 \qquad \text{for all } (i,j) \in \text{rec}(m(1),m(2))$$

and also $G(p,a) = G(p + 1) = G(1) = \text{vec}(2,m,1)$ and hence

$$\hat{G}(p,a) = \{X_{ij} : (i,j) \in \text{rec}(m(1),m(2))\}$$

and therefore

$$I(p,a) = \{X_{ij} : (i,j) \in \text{rec}(m(1),m(2))\}K[X] = \ker(f(p,a)).$$

So let p > 0 and assume that the assertion is true for all values of p smaller than the given one. Now $E(p, a, p - 1) \in \text{vec}(2, m, p - 1)$ and hence by the induction hypothesis

(41) $I(p - 1, E(p, a, p - 1)) = \ker(f(p - 1, E(p, a, p - 1)))$.

Upon letting

$$Y = \text{rec}(m(1),m(2)) \setminus z(p,a,p)$$

we clearly have that Y is a saturated subset of rec(m(1),m(2)) and hence upon taking p - 1 and E(p, a, p - 1) for p and a in (20.11), we get

$$I^*(p - 1, E(p, a, p - 1), Y) = K[Y] \cap I(p - 1, E(p, a, p - 1))$$

and therefore by (41) we see that

$$(42) \quad K[Y] \cap \ker(f(p - 1, E(p, a, p - 1))) = I^*(p - 1, E(p, a, p - 1), Y).$$

By the definition of f we see that

$$f(p,a)(X_{ij}) = f(p - 1, E(p, a, p - 1))(X_{ij}) \quad \text{for all } (i,j) \in Y$$

and hence

$$f(p,a)|K[Y] = f(p - 1, E(p, a, p - 1))|K[Y]$$

and therefore

$$(43) \quad \ker(f(p,a)|K[Y]) = \ker(f(p - 1, E(p, a, p - 1))|K[Y]).$$

Obviously,

$$K[Y] \cap \ker(f(p,a)) = \ker(f(p,a)|K[Y])$$

and

$$K[Y] \cap \ker(f(p - 1, E(p, a, p - 1))) = \ker(f(p - 1, E(p, a, p - 1))|K[Y])$$

and hence by (42) and (43) we get

$$(44) \quad K[Y] \cap \ker(f(p,a)) = I^*(p - 1, E(p, a, p - 1), Y).$$

Upon letting

$$\hat{Y} = \text{rec}(m(1),m(2)) \setminus z_o(p,a,p)$$

we obviously have

$$\hat{Y} \setminus Y \subset y(p,a,p)$$

and by the definition of the K-homomorphism

$$f(p,a)\colon \ K[X] \to K(y(p,a,p))$$

we clearly have

$$f(p,a)(X_{ij}) = X_{ij} \qquad \text{for all } (i,j) \in \hat{Y} \setminus Y$$

and

$$\text{im}(f(p,a)|K[Y]) \subset K(Y \cap y(p,a,p))$$

and hence we conclude that

$$K[\hat{Y}] \cap \ker(f(p,a)) = (K[Y] \cap \ker(f(p,a)))K[\hat{Y}]$$

and therefore by (44) we conclude that

$$(45) \quad K[\hat{Y}] \cap \ker(f(p,a)) = I^*(p - 1, E(p, a, p - 1), Y)K[\hat{Y}].$$

By the definition of I^* we have

$$\begin{aligned} &I^*(p - 1, E(p, a, p - 1), Y) \\ &= \{\mathrm{mor}(X,b) : b \in G^*(p - 1, E(p, a, p - 1), Y)\}K[Y] \end{aligned}$$

and by the definition of G^* we see that

$$G^*(p - 1, E(p, a, p - 1), Y) = \bigcup_{k\in[1,2]} \bigcup_{u\in[1,p]} G(p,a,k,u)$$

and therefore by (45) we get

$$(46) \quad \begin{bmatrix} K[\hat{Y}] \cap \ker(f(p,a)) \\ \subset \left\{\mathrm{mor}(X,b) : b \in \bigcup_{k\in[1,2]} \bigcup_{u\in[1,p]} G(p,a,k,u)\right\}K[\hat{Y}]. \end{bmatrix}$$

By the definition of G we have

$$(47) \quad G(p,a) = G[p + 1] \cup \left[\bigcup_{k\in[1,2]} \bigcup_{u\in[1,p]} G(p,a,k,u)\right]$$

and by the definition of I we have

$$(48) \quad I(p,a) = \{\mathrm{mor}(X,b) : b \in G(p,a)\}K[X]$$

and by the definition of A we have

$$(49) \quad A(p,a,i,j) \in G[p + 1] \qquad \text{for all } (i,j) \in z_0(p,a,p).$$

In view of (20.12), by (47), (48), and (49) we see that

$$(50) \quad \mathrm{mor}(X,A(p,a,i,j)) \in \ker(f(p,a)) \qquad \text{for all } (i,j) \in z_0(p,a,p).$$

Clearly, $E(p,a,p) = a$ and hence by the definition of $\tilde{A}$, for every $(i,j) \in z_0(p,a,p)$ we have

$$\mathrm{mor}(X,A(p,a,i,j)) = \mathrm{mor}(X,a)X_{ij} - \tilde{A}(X,p,a,i,j)$$

and

$$\tilde{A}(X,p,a,i,j) \in K[\hat{Y}]$$

and therefore, given any $B \in K[X]$, there exists

$$e \in N \text{ and } B^*(i,j) \in K[X] \quad \text{for all } (i,j) \in z_0(p,a,p)$$

and

(51) $B^{**} \in K[\hat{Y}]$

such that

(52) $\operatorname{mor}(X,a)^e B = B^{**} + \sum_{(i,j) \in z_0(p,a,p)} \operatorname{mor}(X,A(p,a,i,j))B^*(i,j)$

and now by (50) and (52) we see that

(53) if $B \in \ker(f(p,a))$, then $B^{**} \in \ker(f(p,a))$

and by (51) and (53) we see that

(54) if $B \in \ker(f(p,a))$, then $B^{**} \in K[\hat{Y}] \cap \ker(f(p,a))$

and by (46), (47), (48), (49), (52), and (54) we see that

(55) if $B \in \ker(f(p,a))$, then $\operatorname{mor}(X,a)^e B \in I(p,a)$

and, in view of (20.5), by (55) we see that

(56) if $B \in \ker(f(p,a))$, then $B \in I(p,a)$.

Thus we have shown that $\ker(f(p,a)) \subset I(p,a)$ and hence by (20.12) we see that $I(p,a) = \ker(f(p,a))$. This completes the induction on p.

THEOREM (20.14) Given any $p \in N$ and $a \in \operatorname{vec}(2,m,p)$, we have the following.

(20.14.1) $I(p,a) = \ker(f(p,a))$, and $K(y(p,a,p))$ is the quotient field of $\operatorname{im}(f(p,a))$, and $K(y(p,a,p))$ is a pure transcendental extension of K whose transcendence degree equals $\operatorname{card}(y(p,a,p))$, and we have

$$\operatorname{card}(y(p,a,p)) - 1 = \begin{cases} C(m,p,a) & \text{if } p \neq 0 \\ -1 & \text{if } p = 0 \end{cases}$$

where $C(m,p,a)$ is as in (20.1.10).

(20.14.2) $I(p,a)$ is the sum of the $2p + 1$ ideals

$I(p,a,1,1)$, $I(p,a,2,1)$, ..., $I(p,a,1,p)$, $I(p,a,2,p)$, $I[p + 1]$

each of which is a homogeneous prime ideal in $K[X]$.

(20.14.3) $I(p,a)$ is a homogeneous prime ideal in $K[X]$, and the quotient field of the residue class ring $K[X]/I(p,a)$ is a pure transcendental extension of (the image of) K whose transcendence degree equals $\mathrm{card}(y(p,a,p))$, and we have

$$\mathrm{card}(y(p,a,p)) - 1 = \begin{cases} C(m,p,a) & \text{if } p \neq 0 \\ -1 & \text{if } p = 0 \end{cases}$$

where $C(m,p,a)$ is as in (20.1.10).

(20.14.4) If $p \neq 0$, then for all $V \in N$ and $L \in \{1,1^*,2,2^*,3,3^*\}$ and $k \in [1,2]$ we have

$$[K[X]_V/I(p,a)_V : K] = F^{(Lk)}(m,p,a,V)$$

with $F^{(Lk)}(m,p,a,V)$ as in (20.1.9) to (20.1.13), and hence $I(p,a)$ is hilbertian and its Hilbert function, as well as its Hilbert polynomial, in $K[X]$ is $F^{(Lk)}(m,p,a,V)$.

Proof: In (20.13) we have shown that $I(p,a) = \ker(f(p,a))$, and the rest of (20.14.1) is obvious. Since $I(p,a) = \ker(f(p,a))$ and $I(p,a)$ is a homogeneous ideal in $K[X]$ and $f(p,a)$ is a ring homomorphism of $K[X]$ into the integral domain $K(y(p,a,p))$, we see that

(11) $I(p,a)$ is a homogeneous prime ideal in $K[X]$.

Clearly, $I[p + 1] = I(p,E[p])$ and hence by taking $E[p]$ for a in (11) we conclude that

(12) $I[p + 1]$ is a homogeneous prime ideal in $K[X]$.

Given any $k \in [1,2]$ and $u \in [1,p]$ with $a(k,u) > u$, let n be the unique member of $N^*(2)$ such that $n(k) = a(k,u) - 1$ and $n(k') = m(k')$ where $k' = 2$ or 1 according as $k = 1$ or 2, and let $Y^* = \mathrm{rec}(n(1), n(2))$, and let B be the unique member of $\mathrm{mat}(R,n)$ such that for all $(i,j) \in Y^*$ we have $B_{ij} = X_{ij}$; now if we change m and X and p to n and B and $u - 1$, respectively, then $I[p + 1]$ becomes $I^*(p,a,k,u,Y^*)$

and hence by (12) we see that I*(p,a,k,u,Y*) is a prime ideal in K[Y*], and by the definitions of I and I* we see that I(p,a,k,u) is the extension of I*(p,a,k,u,Y*) to K[X]; now K[X] is a polynomial ring in a certain number of indeterminates over K[Y*] and hence the extension of any prime ideal in K[Y*] to K[X] is a prime ideal in K[X]; therefore, I(p,a,k,u) is a prime ideal in K[X]; also, clearly, I(p,a,k,u) is a homogeneous ideal in K[X], so we conclude that

(13) $\begin{cases}\text{if } k \in [1,2] \text{ and } u \in [1,p] \text{ are such that } a(k,u) > u, \text{ then} \\ I(p,a,k,u) \text{ is a homogeneous prime ideal in } K[X].\end{cases}$

If k ∈ [1,2] and u ∈ [1,p] are such that $a(k,u) \ngtr u$, then clearly I(p,a,k,u) is the zero ideal in K[X] and hence it is a homogeneous prime ideal in K[X]; therefore, in view of (13) we see that

(14) $\begin{cases}\text{for all } k \in [1,2] \text{ and } u \in [1,p] \text{ we have that} \\ I(p,a,k,u) \text{ is a homogeneous prime ideal in } K[X].\end{cases}$

Obviously, I(p,a) is the sum of the 2p + 1 ideals

I(p,a,1,1), I(p,a,2,1), ..., I(p,a,1,p), I(p,a,2,p), I[p + 1]

and therefore by (12) and (14) we get (20.14.2). In view of (11), by (20.14.1) we get (20.14.3). Finally, in view of (20.1.8) to (20.1.13), by (20.4.3) we get (20.14.4).

THEOREM (20.15) Given any p ∈ [0,min(m(1),m(2))], we have the following.

(20.15.1) I[p + 1] = I(p,E[p]) and f[p + 1] = f(p,E[p]) and y[p + 1] = y(p,E[p],p).

(20.15.2) I[p + 1] = ker(f[p + 1]), and K(y[p + 1]) is the quotient field of im(f[p + 1]), and K(y[p + 1]) is a pure transcendental extension of K whose transcendence degree equals card(y[p + 1]), and we have

$$\operatorname{card}(y(p,E[p],p)) - 1 = \begin{cases} C(m,p,E[p]) = p(m(1) + m(2) - p) - 1 & \text{if } p \neq 0 \\ -1 & \text{if } p = 0 \end{cases}$$

where C(m,p,E[p]) is as in (20.1.10).

(20.15.3) $I[p+1]$ is a homogeneous prime ideal in $K[X]$, and the quotient field of the residue class ring $K[X]/I[p+1]$ is a pure transcendental extension of (the image of) K whose transcendence degree is $\mathrm{card}(y[p+1])$, and we have

$$\mathrm{card}(y(p,E[p],p)) - 1 = \begin{cases} C(m,p,E[p]) = p(m(1) + m(2) - p) - 1 & \text{if } p \neq 0 \\ -1 & \text{if } p = 0 \end{cases}$$

where $C(m,p,E[p])$ is as in (20.1.10).

(20.15.4) If $p \neq 0$, then for all $V \in N$ and $L \in \{1,1^*,2,2^*,3,3^*\}$ and $k \in [1,2]$ we have

$$[K[X]_V/I[p+1]_V:K] = F^{(Lk)}(m,p,E[p],V)$$

with $F^{(Lk)}(m,p,E[p],V)$ as in (20.1.9) to (20.1.13), and hence $I[p+1]$ is hilbertian and its Hilbert function, as well as its Hilbert polynomial, in $K[X]$ is $F^{(Lk)}(m,p,E[p],V)$.

Proof: (20.15.1) is obvious and in view of it the rest follows from (20.14). Alternatively, *without using* (20.6) *to* (20.14), we can give the following

Direct Proof: First, from the proof of (20.12), let us extract a proof of the inclusion $I[p+1] \subset \ker(f[p+1])$. So let

$$x \in \mathrm{mat}(K(y[p+1]),m)$$

be defined by putting

$$x_{ij} = f[p+1](X_{ij}) \qquad \text{for all } (i,j) \in \mathrm{rec}(m(1),m(2)).$$

Recall that by the definition of $f[p+1]$,

$$(2)\qquad \left[\begin{array}{l}\text{for every } (i,j) \in y[p+1] \\ \text{we have } f[p+1](X_{ij}) = X_{ij}\end{array}\right.$$

and

$$(3)\qquad \left[\begin{array}{l}\text{for every } (i,j) \in \mathrm{rec}(m(1),m(2)) \setminus y[p+1] \\ \text{we have } f[p+1](X_{ij}) = \dfrac{\hat{A}[X,\, p+1,\, i,\, j]}{\mathrm{mor}(X,E[p])}\end{array}\right.$$

where by the definition of $\tilde{A}$

$$(5)\quad \left[\begin{array}{l}\text{for every } (i,j) \in \mathrm{rec}(m(1),m(2)) \setminus y[p+1] \\ \text{we have } \tilde{A}[X, p-1, i, j] \in K[y[p+1]]\end{array}\right.$$

and by the definition of E we have

$$(6)\quad \mathrm{mor}(X,E[p]) \neq 0$$

because mor(X,E[p]) is the determinant of a p by p matrix whose p^2 entries are independent indeterminates. Upon cross multiplying and then subtracting RHS from LHS, by (3) we see that

$$(7)\quad \left[\begin{array}{l}\text{for every } (i,j) \in \mathrm{rec}(m(1),m(2)) \setminus y[p+1] \\ \text{we have } \mathrm{mor}(X,E[p])x_{ij} - \tilde{A}[X, p+1, i, j] = 0.\end{array}\right.$$

By (2) and (5) we see that

$$(8)\quad \left[\begin{array}{l}\text{for every } (i,j) \in \mathrm{rec}(m(1),m(2)) \setminus y[p+1] \\ \text{we have } f[p+1](\tilde{A}[X, p+1, i, j]) = \tilde{A}[X, p+1, i, j].\end{array}\right.$$

By the definition of E we have

$$(E[p](1,i),E[p](2,j)) \in y[p+1] \quad \text{for all } (i,j) \in \mathrm{rec}(p,p)$$

and hence by (2) we see that

$$(9)\quad \mathrm{mor}(x,E[p]) = \mathrm{mor}(X,E[p]).$$

Recalling the definition of $\tilde{A}$, for every

$$(i,j) \in \mathrm{rec}(m(1),m(2)) \setminus y[p+1]$$

we have

$$\mathrm{mor}(X,A[p+1, i, j]) = \mathrm{mor}(X,E[p])X_{ij} - \tilde{A}[X, p+1, i, j]$$

and upon applying f[p + 1] to the equation above we get

$$\mathrm{mor}(x,A[p+1, i, j]) = \mathrm{mor}(x,E[p])x_{ij} - f[p+1](\tilde{A}[X, p+1, i, j])$$

and hence by (7), (8), and (9) we get

$$\mathrm{mor}(x,A[p+1, i, j]) = 0;$$

thus

(10) $\left[\begin{array}{l}\text{for every } (i,j) \in \text{rec}(m(1),m(2)) \setminus y[p+1] \\ \text{we have mor}(x,A[p+1,\ i,\ j]) = 0.\end{array}\right.$

In view of the definition of A, by (10) we see that

(11) $\left[\begin{array}{l}\text{for all } i \in [p+1,\ m(1)] \text{ and } j \in [p+1,\ m(2)] \text{ we have} \\ A[p+1,\ i,\ j] \in \text{vec}(2,\ m,\ p+1) \text{ and mor}(x,A[p+1,\ i,\ j]) \\ = 0.\end{array}\right.$

By (6) and (9) we see that

(14) $\text{mor}(x,E[p]) \neq 0.$

Given any $i \in [p+1,\ m(1)]$, let us consider the matrix $\hat{M}$ formed by the rows of x whose row numbers are 1, 2, ..., p, i. Now mor(x,E[p]) is the determinant of the p by p matrix formed by the intersection of the first p rows and the first p columns of $\hat{M}$, and hence by (14) we see that

(32) the first p rows of $\hat{M}$ are independent.

For a moment let us fix any $j \in [p+1,\ m(2)]$ and let us consider the p + 1 by p + 1 matrix $\hat{B}$ formed by the columns of $\hat{M}$ whose column numbers are 1, 2, ..., p, j. Now mor(x,E[p]) is the determinant of the p by p matrix formed by the intersection of the first p rows and the first p columns of $\hat{B}$ and hence by (14) we see that

(34) the first p columns of $\hat{B}$ are independent.

Clearly, det $\hat{B}$ = mor(x,A[p + 1, i, j]) and hence by (11) and (34) we see that the (p + 1)th column of $\hat{B}$ depends on its first p columns; now the 1st, 2nd, ..., pth, (p + 1)th columns of $\hat{B}$ are simply the columns of $\hat{M}$ whose column numbers, respectively, are 1, 2, ..., p, j, and therefore we conclude that

(35) $\left[\begin{array}{l}\text{for every } j \in [p+1,\ m(2)] \text{ we have that the jth column} \\ \text{of } \hat{M} \text{ depends on its first p columns.}\end{array}\right.$

By (35) we see that every column of $\hat{M}$ depends on its first p columns, and therefore $\hat{M}$ has rank $\le p$, and hence by (32) we conclude that the (p + 1)th row of $\hat{M}$ depends on its first p rows; now the 1st, 2nd, ..., pth, (p + 1)th rows of $\hat{M}$ are simply the rows of x whose row numbers, respectively, are 1, 2, ..., p, i, and thus we see that

(36) $\begin{bmatrix} \text{for every } i \in [p+1, m(1)] \text{ we have that the ith row} \\ \text{of } x \text{ depends on its first } p \text{ rows.} \end{bmatrix}$

By (36) we see that every row of x depends on its first p rows, and hence x has rank $\le p$, and therefore

(37) for every $b \in G[p + 1]$ we have $\text{mor}(x,b) = 0$.

By (37) we see that

(40) $I[p + 1] \subset \ker(f[p + 1])$.

Second, from the proof of (20.13), let us extract a proof of the inclusion $\ker(f[p + 1]) \subset I[p + 1]$. By (2) we see that

(46) $K[y[p + 1]] \cap \ker(f[p + 1]) = \{0\}$.

By the definition of A we see that

(49) $\begin{bmatrix} \text{for every } (i,j) \in \text{rec}(m(1),m(2)) \setminus y[p + 1] \\ \text{we have } \text{mor}(X,A[p + 1, i, j]) \in I[p + 1] \end{bmatrix}$

and hence by (40) we see that

(50) $\begin{bmatrix} \text{for every } (i,j) \in \text{rec}(m(1),m(2)) \setminus y[p + 1] \\ \text{we have } \text{mor}(X,A[p + 1, i, j]) \in \ker(f[p + 1]). \end{bmatrix}$

By the definition of $\tilde{A}$, for every

$(i,j) \in \text{rec}(m(1),m(2)) \setminus y[p + 1]$

we have

$$\text{mor}(X,A[p + 1, i, j]) = \text{mor}(X,E[p])X_{ij} - \tilde{A}[X, p + 1, i, j]$$

and

$$\tilde{A}[X, p + 1, i, j] \in K[y[p + 1]]$$

and therefore, given any $B \in K[X]$, there exists

$e \in N$ and $B^*(i,j) \in K[X]$
for all $(i,j) \in \text{rec}(m(1),m(2)) \setminus y[p + 1]$

and

(51) $B^{**} \in K[y[p + 1]]$

such that

$$(52)\quad \begin{cases} \mathrm{mor}(X,E[p])^{e}B \\ = B^{**} + \sum_{(i,j)\in \mathrm{rec}(m(1),m(2))\ y[p+1]} \mathrm{mor}(X,A[p+1,\ i,j])B^{*}(i,j) \end{cases}$$

and now by (50) and (52) we see that

(54) if $B \in \ker(f[p + 1])$, then $B^{**} \in K[y[p + 1]] \cap \ker(f[p + 1])$

and by (46), (49), (52), and (54) we see that

(55) if $B \in \ker(f[p + 1])$, then $\mathrm{mor}(X,E[p])^{e}B \in I[p + 1]$

and, in view of (20.5) with a = E[p], by (55) we see that

(56) if $B \in \ker(f[p + 1])$, then $B \in I[p + 1]$.

Thus we have shown that $\ker(f[p + 1]) \subset I[p + 1]$ and hence by (40) we see that $I[p + 1] = \ker(f[p + 1])$. The remaining parts of (20.15.1) to (20.15.3) are now obvious. Finally, in view of (20.1.8) to (20.1.13) with a = E[p], by (20.4.3) we get (20.15.4).

THEOREM (20.16) Given any $p \in N$ and $a \in \mathrm{vec}(2,m,p)$ and given any saturated subset Y of rec(m(1),m(2)), we have the following.

(20.16.1) $I^*(p,a,Y) = \ker(f(p,a)|K[Y])$.

(20.16.2) I*(p,a,Y) is the sum of the 2p + 1 ideals

I*(p,a,1,1,Y), I*(p,a,2,1,Y), ..., I*(p,a,1,p,Y),
I*(p,a,2,p,Y), I*[p + 1, Y]

each of which is a homogeneous prime ideal in K[Y].

(20.16.3) I**(p,a,Y,X) is the sum of the 2p + 1 ideals

I**(p,a,1,1,Y,X), I**(p,a,2,1,Y,X), ..., I**(p,a,1,p,Y.X),
I**(p,a,2,p,Y,X), I**[p + 1, Y, X]

each of which is a homogeneous prime ideal in K[X].

(20.16.4) I*(p,a,Y) and I**(p,a,Y,X) are homogeneous prime ideals in K[Y] and K[X], respectively.

Proof: (20.16.1) follows from (20.11) and (20.13). Now I*(p,a,Y) is a homogeneous ideal in K[Y] and f(p,a)|K[Y] is a ring homomorphism of K[Y] into the integral domain K(y(p,a,p)), and hence by (20.16.1) we see that

(11) I*(p,a,Y) is a homogeneous prime ideal in K[Y].

Clearly, I*[p + 1, Y] = I*(p,E[p],Y) and hence by taking E[p] for a in (11) we conclude that

(12) I*[p + 1, Y] is a homogeneous prime ideal in K[Y].

Given any k ∈ [1,2] and u ∈ [1,p] with a(k,u) > u, let n be the unique member of N*(2) such that n(k) = a(k,u) - 1 and n(k') = m(k') where k' = 2 or 1 according as k = 1 or 2, and let Y* = rec(n(1),n(2)), and let B be the unique member of mat(R,n) such that for all (i,j) ∈ Y* we have $B_{ij} = X_{ij}$; now if we change m and X and p and Y to n and B and u - 1 and Y ∩ Y*, respectively, then I*[p + 1, Y] becomes I*(p,a,k,u,Y ∩ Y*) and hence by (12) we see that I*(p,a, k,u,Y ∩ Y*) is a prime ideal in K[Y ∩ Y*] and by the definition of I* we see that I*(p,a,k,u,Y) is the extension of I*(p,a,k,u,Y ∩ Y*) to K[Y]; now K[Y] is a polynomial ring in a certain number of indeterminates over K[Y ∩ Y*] and hence the extension of any prime ideal in K[Y ∩ Y*] to K[Y] is a prime ideal in K[Y]; therefore, I*(p,a,k, u,Y) is a prime ideal in K[Y]; also, clearly, I*(p,a,k,u,Y) is a homogeneous ideal in K[Y] and so we conclude that

(13) [if k ∈ [1,2] and u ∈ [1,p] are such that a(k,u) > u, then I*(p,a,k,u,Y) is a homogeneous prime ideal in K[Y].

If k ∈ [1,2] and u ∈ [1,p] are such that a(k,u) ≯ u, then clearly, I*(p,a,k,u,Y) is the zero ideal in K[Y] and hence it is a homogeneous prime ideal in K[Y]; therefore, in view of (13) we see that

(14) [for all k ∈ [1,2] and u ∈ [1,p] we have that I*(p,a,k,u,Y) is a homogeneous prime ideal in K[Y].

Obviously, I*(p,a,Y) is the sum of the 2p + 1 ideals

I*(p,a,1,1,Y), I*(p,a,2,1,Y), ..., I*(p,a,1,p,Y),
I*(p,a,2,p,Y), I*[p + 1, Y]

and therefore by (12) and (14) we get (20.16.2).

By the definitions of I* and I** we see that I**(p,a,Y,X) and I**[p + 1, Y, X] and I**(p,a,k,u,Y,X) are the respective extensions of I*(p,a,Y) and I*[p + 1, Y] and I*(p,a,k,u,Y) to K[X]; now K[X] is a polynomial ring in a certain number of indeterminates over K[Y] and hence the extension of any prime ideal in K[Y] to K[X] is a prime ideal in K[X]; also clearly, I**(p,a,Y,X) and I**[p + 1, Y, X] and I**(p,a,k,u,Y,X) are homogeneous ideals in K[X]; therefore, by (11) we see that

(15) I**(p,a,Y,X) is a homogeneous prime ideal in K[X]

and by (12) we see that

(16) I**[p + 1, Y, X] is a homogeneous prime ideal in K[X]

and by (14) we see that

(17) $\begin{cases} \text{for all } k \in [1,2] \text{ and } u \in [1,p] \text{ we have that} \\ \text{I**(p,a,k,u,Y,X) is a homogeneous prime ideal in K[X].} \end{cases}$

Obviously, I**(p,a,Y,X) is the sum of the 2p + 1 ideals

I**(p,a,1,1,Y,X), I**(p,a,2,1,Y,X), ..., I**(p,a,1,p,Y,X), I**(p,a,2,p,Y,X), I**[p + 1, Y, X]

and therefore, by (16) and (17) we get (20.16.3).

Finally, by (11) and (15) we get (20.16.4).

THEOREM (20.17) Given any p ∈ [0,min(m(1),m(2))] and given any saturated subset Y of rec(m(1),m(2)), we have the following.

(20.17.1) I*[p + 1, Y] = I*(p,E[p],Y) and I**[p + 1, Y, X] = I**(p,E[p],Y,X) and f[p + 1] = f(p,E[p]).

(20.17.2) I*[p + 1, Y] = ker(f[p + 1]|K[Y]).

(20.17.3) I*[p + 1, Y] and I**[p + 1, Y, X] are homogeneous prime ideals in K[Y] and K[X], respectively.

Proof: (20.17.1) is obvious and in view of it the rest follows from (20.16). Alternatively, (20.17.2) and (20.17.3) also follow from (20.11) and (20.15) where in (20.11) we take a = E[p].

REMARK (20.18) Given any p ∈ N and a ∈ vec(2,m,p) and L ∈ {1, 1*,2,2*,3,3*} and k ∈ [1,2], in view of (20.1.9) to (20.1.13) we know that $F^{(Lk)}(m,p,a,V)$ is a polynomial of degree C = C(m,p,a) in V with rational coefficients, and C! times the coefficient of V^C in

$F^{(Lk)}(m,p,a,V)$ equals the positive integer $F_0^{(Lk)}(m,p,a)$; now (20.14.3) and (20.14.4) give a reconfirmation of these assertions because of the theorem of Hilbert which says that if $\hat{H}$ is a homogeneous prime ideal in a polynomial ring in a finite number, say e + 1, indeterminates over K, and if d + 1 is the transcendence degree of the quotient field of $H/\hat{H}$ over (the image of) K, and if h(V) is the Hilbert polynomial of $\hat{H}$ in H, then d is the degree of h(V) in V and d! times the coefficient of V^d in h(V) is a positive integer which equals the order of the variety defined by $\hat{H}$ in the e-dimensional projective space over K, where we recall that the said order equals the number of points in which the said variety is met by a complementary dimensional [which means (e - d)-dimensional] linear space in the said projective space. Note that in our case of $\hat{H} = I(p,a)$, the quotient field of $H/\hat{H}$ is a pure transcendental extension of K, and so the said variety is rational; i.e., it is birationally equivalent to the d-dimensional projective space over K.

REMARK (20.19) In view of (20.15), Remark (20.18) applies to the special case of a = E[p].

REMARK (20.20) Given any p ∈ N and a ∈ vec(2,m,p) and given any saturated subset Y of rec(m(1),m(2)), in view of (20.14) and (20.16) we may ask whether I*(p,a,Y) and I**(p,a,Y,X) are hilbertian.

Notation

Index